Reproduction in Context

Reproduction in Context

Social and Environmental Influences on Reproductive Physiology and Behavior

edited by Kim Wallen and Jill E. Schneider

A Bradford Book
The MIT Press
Cambridge, Massachusetts
London, England

© 2000 Massachusetts Institute of Technology

All rights reserved. No part of this book may be reproduced in any form by any electronic or mechanical means (including photocopying, recording, or information storage and retrieval) without permission in writing from the publisher.

This book was set in Times New Roman by Asco Typesetters, Hong Kong and was printed and bound in the United States of America.

Library of Congress Cataloging-in-Publication Data

Reproduction in context: social and environmental influences on reproductive physiology and behavior / edited by Kim Wallen and Jill E. Schneider.
  p. cm.
"A Bradford Book."
Includes bibliographical references and index.
ISBN 0-262-23204-9 (hc.: alk. paper)
 1. Sex (Biology) 2. Reproduction. 3. Animal behavior. 4. Sex (Biology)—Social aspects. 5. Neuroendocrinology. I. Wallen, Kim. II. Schneider, Jill E.
QP251.R44438 1999
571.8′1—dc21   98-51874
                                                                CIP

The editors dedicate this book to their families—their parents, siblings, children, and especially their spouses—who have tolerated and supported them throughout this enterprise.

Contents

	Preface	ix
	Acknowledgments	xi
1	**Introduction to the Study of Reproduction in Context** Jill E. Schneider	1
IA	**ENVIRONMENTAL CONTEXT: ENERGY BALANCE AND REPRODUCTION**	13
2	**Puberty and Energy Reserves: A Walk on the Wild Side** Franklin H. Bronson	15
3	**Inhibition of Reproduction in Service of Energy Balance** Jill E. Schneider and George N. Wade	35
IB	**ENVIRONMENTAL CONTEXT: INTEGRATION OF ENVIRONMENTAL FACTORS IN BIRDS**	83
4	**Toward an Ecological Basis of Hormone-Behavior Interactions in Reproduction of Birds** John C. Wingfield, Jerry D. Jacobs, Anthony D. Tramontin, Nicole Perfito, Simone Meddle, Donna L. Maney, and Kiran Soma	85
5	**Neuroendocrine Mechanisms Mediating the Photoperiodic and Social Regulation of Seasonal Reproduction in Birds** Gregory F. Ball and George E. Bentley	129
6	**Sex and Brain Secretions** Rae Silver and Ann-Judith Silverman	159
IC	**ENVIRONMENTAL CONTEXT: INTEGRATION OF ENVIRONMENTAL FACTORS IN MAMMALS**	187
7	**Timing of Reproduction by the Integration of Photoperiod with Other Seasonal Signals** Theresa M. Lee and Michael R. Gorman	191
8	**Environmental and Social Influences on Seasonal Breeding and Immune Function** Randy J. Nelson and Sabra L. Klein	219

9	Feisty Females and Meek Males: Reproductive Strategies in the Spotted Hyena	257
	Kay E. Holekamp and Laura Smale	
IIA	**SOCIAL CONTEXT AND REPRODUCTIVE BEHAVIOR**	287
10	Risky Business: Social Context and Hormonal Modulation of Primate Sexual Desire	289
	Kim Wallen	
11	Contextual Influences on Sociosexual Behavior in Monogamous Primates	325
	Jeffrey A. French and Colleen M. Schaffner	
12	Human Pheromones: Primers, Releasers, Signalers, or Modulators?	355
	Martha K. McClintock	
IIB	**SOCIAL CONTEXT AND MECHANISMS UNDERLYING BEHAVIOR**	421
13	Neuronal Integration of Chemosensory and Hormonal Signals in the Control of Male Sexual Behavior	423
	Ruth I. Wood and Jennifer M. Swann	
14	Behavioral Activation of the Female Neuroendocrine Axis	445
	Patricia A. Schiml, Scott R. Wersinger, and Emilie F. Rissman	
15	Sexuality: The Environmental Organization of Phenotypic Plasticity	473
	David Crews	
	Contributors	501
	Author Index	503
	Subject Index	505

Preface

The idea for the theme of this book came from Drs. Antonio Nunez and Lynwood Clemens at Michigan State University in East Lansing, Michigan. Michigan State was the site of the first and the twenty-fifth Conference on Reproductive Behavior (CRB). The original conference was the first gathering of physiological psychologists, endocrinologists, behavioral biologists, and psychiatrists whose main scientific interest was the physiological basis of sexual behavior. The annual conference provided a meeting where scientists could share their progress in understanding the detailed hormonal and neural mechanisms that underlie measurable aspects of sex and parental behavior in animals, including human beings. Some of these scientists were driven by an interest in understanding the basic biology of animals, while others sought clues to the causes and cures of human sexual dysfunction, infertility, and sexually transmitted disease. Thus, the conference provided a means for essential scientific interactions in the public and private interest; clinical and social applications cannot be made without a thorough knowledge of basic biological principles. The next 25 years brought increasing technological sophistication to the study of reproductive biology.

However, not all investigators followed a reductionist path leading from the physiological to the cellular to the molecular level. In planning the twenty-fifth annual CRB, Lyn and Tony recognized another trend. Many investigators were taking an integrative approach to the study of reproduction. They were studying the effects of the environment (e.g., nutrition, day length, and ambient temperature) or social context on fertility, sex, and parental behavior. Some were studying the interactive effects of several different environmental and social factors. Physiologists were allowing their research to be guided by ecological and evolutionary considerations and vice versa. Wild, rather than laboratory species were studied in natural or seminatural environments. Some were integrating knowledge and techniques from fields never before considered in the realm of reproductive biology. Often the conclusions, interpretations, and clinical implications that came out of these studies were quite different from those that resulted from more traditional experiments that focused more narrowly on endocrine and sexual events in male and female rats, pigeons, or monkeys paired in laboratory cages. To many of us, it seems that these studies of reproduction in the environmental and social context lead to a much deeper and more complete understanding of the rich complexity of sex behavior, sexuality, and reproduction. For the twenty-fifth CRB, Tony and Lyn asked us to put together a group of speakers for two symposia, one on the environmental and one on the social context of reproduction. It turned out to be an exhilarating conference. The response to our symposia was so enthusiastic we were inspired to edit a book on the subject. Our first candidates for chapter authors were the symposia speakers from the twenty-fifth

CRB. In some cases, on the recommendation of those speakers, we also called upon their academic offspring (students or postdoctoral associates). In other cases, we chose still other outstanding representatives of the integrative approach to the study of reproduction. We hope the resulting book will be as exciting and inspiring as the twenty-fifth conference. The annual conference has now evolved into the Society for Behavior Neuroendocrinology (SBN), and has taken on a broader, more synthetic, and integrative character. At last year's SBN conference, the keynote speaker, National Academy of Sciences member Bruce McEwen, declared that behavioral neuroendocrinology is one of the few "complete" fields of biology, encompassing and integrating cellular and molecular events with higher levels of biological organization leading all the way up to complex social behavior. We hope this book will adequately represent and inspire more work in this expanding field of biology. With special thanks to Tony and Lyn for their support of and contributions to the integrative approach to reproduction in its social and environmental context.

Acknowledgments

This book started with a phone call from Lyn Clemens asking that I organize a symposium on social influences on sexual behavior for the twenty-fifth Conference on Reproduction Behavior. I am deeply indebted to Lyn for the opportunity, which led me to think more broadly about this topic. The participants in that original symposium, David Abbott, Emilie Rissman, and John Wingfield are thanked for their excellent contributions at the meeting which demonstrated the power and importance of this approach to understanding reproductive behavior. A deep debt is owed to the contributors of this volume who have made putting this book together an unexpected pleasure.

Without Jill Schneider's indefatigable support and encouragement this book would have never been finished. One could not ask for a better co-editor and I offer my deepest appreciation for her optimism and superhuman effort. Though both of us experienced difficult times during the course of creating this book, Jill always was able to muster the perspective necessary to move us forward. Michael Rutter, our editor at MIT Press, is thanked for his unflagging support, even when the editors had doubts about the sanity of this enterprise. Support from NIMH K02-MH01062 during editing of this volume is gratefully acknowledged.

Kim Wallen

First and foremost, I'd like to thank Kim Wallen for suggesting that we edit this book. The project has been one of the most intellectually refreshing experiences in my career. All of the authors contributions have been exciting, and Kim's insights and especially his steady good nature have been essential to the completion of the project. Also, I would like to thank the speakers in the symposium "Environmental Influences on Reproduction" at the twenty-fifth Conference on Reproductive Behavior: Frank Bronson, Sarah Newman, and Irving Zucker. Also, I am very grateful for the kind support and encouragement I received from Theresa Lee, Jeff French, Antonio Nunez, Emilie Rissman, John Wingfield, Ruth Wood, and my entire laboratory group at various critical junctures of this enterprise. I also thank Carol B. Lynch for past and present help with understanding and expressing ideas about evolution, and Greg Ball for enlightening discussions about the history of our field, the proper order of the chapters and also for his help in writing the introduction to the "bird chapters." Thanks to Kay Holekamp and Randy Nelson for their feedback on chapter 1, and George N. Wade for contributing figure 1.1. Finally, warm thanks to Michael Rutter and Katherine A. Almeida from MIT Press.

Jill E. Schneider

Reproduction in Context

1 Introduction to the Study of Reproduction in Context

Jill E. Schneider

Research on sex, sexuality, and fertility is of enormous, probably immeasurable, value to society. This research has an impact on a number of societal problems, such as the treatment of sexual dysfunction, the prevention of sexually transmitted disease, the development and use of birth control methods, the prevention of unwanted teen pregnancy, and the discovery of treatments for devastating diseases that adversely influence the reproductive system. More fundamentally, research on sex and sexuality can help us understand the diversity of sexual life styles we encounter and make informed, educated choices in all aspects of our lives that are related to sex and reproduction. Despite a long-standing reluctance of some segments of society to confront sex-related topics in an objective, open, rational manner, sex and reproduction are central driving forces in our lives and present questions and problems that cannot be ignored or avoided. Whether we are conscious of it or not, our happiness and satisfaction in life depend on understanding the various factors that affect sex and reproduction. Thanks to public and private recognition of these facts, a great deal of recent research has deepened and broadened our understanding of sex and reproduction. The clinical and social applications are only one part of this endeavor. The application of knowledge for the sake of human society starts with basic research, and much of this research involves the detailed study of a variety of animals. *Reproduction in Context* contains basic research that is both pregnant with potential applications, and fascinating in and of itself.

Three issues come to mind with regard to an edited volume, particularly a book on the subject of reproduction, sex, and sexuality. First, on a professional note, I address the issue of our specific aims and why we chose particular chapter authors and not others. Next, I discuss the importance and pitfalls in studying sex, sexuality, and reproduction within a biological, and especially an evolutionary framework. Third, I provide some general background material on the physiology and neuroendocrinology of reproduction as it concerns all of the subsequent chapters.

Specific Aims of This Book

The general purpose of this book is easy to explain. We wanted to bring together, in one volume, some of the most interesting work in our field of research. Defining our field of research is both challenging and gratifying, because the aspects of our work that make it hard to define and classify are the same aspects that make the science important and the scientists engaging. The authors of the upcoming chapters are a diverse lot, many of them trained in traditional fields such as reproductive science, ethology, or behavioral endocrinology. Reproductive science, in the traditional sense, is a clinically or agriculturally based study of the reproductive system that largely excludes animal behavior. Ethology is the study of animal behavior in natural habitats in relation to adaptive function. Behavioral endocrinology is the study of hormonal effects on behavior, as well as the effects of behavior on hormone systems, and is typically carried out on laboratory species, especially rats, in an experimental setting. The authors of this book, in contrast, almost never remain within the boundaries of these fields of research and none of them confine their studies to laboratory rats. Some began as physiologists and later took on the challenge of studying animal behavior. Others were trained in fields such as ethology, behavior genetics, or evolutionary ecology and later added techniques from neuroanatomy or neuroendocrinology to their repertoire. Some are studying the interactions of the reproductive system with systems that control immune response or food intake and energy balance, and thus are incorporating the fields of biochemistry or immunology into their research programs.

The operative words that describe this book are "synthesis" and "integration." The cross-fertilization among different scientific fields has reached such an extent that some of our chapter authors have experienced difficulty in finding a scientific "home," an appropriate academic department, scientific society, or professional conference that consistently provides a forum for their ideas, methodologies, and contributions. Even with these difficulties, the use of the integrative approach is growing, due to its success and the consequent acknowledgment that all biological, chemical, and physical systems are inextricably interrelated and cannot be fully understood as separate, parallel systems of knowledge. Synthesis of the artificial categories we call "the scientific disciplines" is inevitable, since life is really an interplay between myriad aspects of the organism, its environment, and its evolutionary history.

Our goal was *not* to provide a comprehensive coverage of the traditional aspects of reproductive biology: sexual differentiation, puberty, mating, gestation, and lactation. Nor did we intend to represent the complete range of reproductive behaviors or all taxonomic classes of animals, although the book happens to contain a wealth of information about many different aspects of reproduction in a variety of species of mammals and birds. The authors were chosen because each in his or her own way is an excellent example of an integrative biologist studying reproduction in an environmental or social context or in both contexts. Each of these authors has taken our understanding of reproduction to a new level by amalgamating the ideas and techniques of traditional reproductive endocrinology with those of other fields of research. The publisher's constraints precluded more than one or two chapters on any particular environmental or social factor, such as photoperiod, energy, or pheromones. Thus, we chose the authors we felt were the most synthetic in their approaches. For example, in choosing authors to write about the effects of photoperiod, we chose those who had creatively combined ideas from more than one field, and in so doing changed our perspectives and the way we do experiments. We have come to believe that in order to understand reproduction fully, the study of mechanism must be coupled with an appreciation of the natural, social, and environmental context in which reproduction occurs, as well as an appreciation of the organism as whole. This book is a forum for scientists who hold this perspective most dear.

It would have been easy to find distinguished authors who focus narrowly on neuroendocrine mechanisms. Why have we chosen authors who include complex and confusing social and ecological factors in their study of reproduction? A full understanding of reproduction requires that we look beyond the proximate factors. We must also examine the ultimate evolutionary influences that have led to the reproductive mechanisms we observe. Theodosius Dobzhansky put it this way, "Nothing in biology makes sense, except in the light of evolution" (Dobzhansky, 1973). The theory of evolution is defined here as the idea that the diversity of life we see around us is the product of a combination of natural selection and stochastic processes acting on random heritable variation. This paradigm gives meaning to what might otherwise be seen as a few general mechanistic principles accompanied by innumerable exceptions to those principles. For example, in this book we find that female mammals have spontaneous ovulatory cycles, except those that are reflex ovulators. Male-typical sex behavior is determined by circulating titers of testosterone, except in species whose sex behavior is determined by progesterone levels, or in species that show vigorous behavior in the absence of circulating gonadal steroids. Males of many species tend to be more aggressive than females, with the most pugnacious males enjoying the greatest reproductive success. In this book we find species in which the females are the highly aggressive sex. In at least one of these species, female reproductive success is determined by their ability to dominate both males and other females, while male reproductive success seems to depend on their submissive behavior and lack of aggression toward all clan members. To most it may seem obvious that male-typical sex behavior is displayed primarily by males. In this book we confront the reality of species in which all members of the species routinely display male- and female-typical sex behaviors with equal frequency. It is commonly held by the lay public and even by some biologists that phenotypic sex, that is, whether an individual is male or female, is determined by sex chromosomes. In this book, we examine species in which maleness and femaleness is determined by other factors, such as environmental temperature or social cues from conspecifics. The contributors to

Reproduction in Context do not dismiss these phenomena as problematic aberrations. Rather, they focus their interests and energies on making sense of these phenomena by casting them in the light of evolution.

Natural selection is considered by the vast majority of evolutionary biologists to be the primary force of evolution and molds phenotypes by acting on random genetic variation in populations. Since selection acts on preexisting variation within a population, we expect some continuity among populations and species. We also expect to find differences among populations and species because selection pressures differ with different habitats and over time. Accordingly, the theory of evolution by natural selection encompasses both the continuity as well as the variation among living organisms. Within this framework we can do more than form general principles and then marvel at or ignore the "exceptions." We can ask, What specific types of selection pressures may have brought about particular reproductive strategies? What particular types of selection pressures have led to feisty female and meek male spotted hyenas? Furthermore, starting with the assumption that particular mechanisms conferred an adaptive advantage in the face of some environmental selection pressure, then it follows that we should be able to find clues to environmental-neuroendocrine connections by studying the habitat and life history of the species. For example, a survey of the breeding strategies from representatives of a variety of mammalian orders shows that the selection pressures faced by most mammals mandate that reproduction in the wild occurs *without* regard to long-term body energy reserves (e.g., body fat content). Contrary to popular myth, in free-living species in natural habitats, the neuroendocrine mechanisms that control reproduction are highly responsive to acute changes in energy balance, not energy reserves. These considerations predict that the physiological signals that control reproduction arise from acute changes in energy balance, rather than in changes in long-term adipose tissue depots. Such evolutionary considerations have led to mechanistic models that emphasizes the importance of intracellular sensory signals generated by changes in energy balance, and deemphasizes signals arising from changes in body fat content. A traditionally trained immunologist might be surprised to find immune responses decreased at times when it might be advantageous to have full immune function. However, an ecological perspective emphasizes that immune responses, thermoregulation and breeding are all energetically costly processes that cannot be sustained simultaneously under energetically challenging conditions. Thus, the ecological perspective suggests that, in the face of energetic constraints, evolutionary tradeoffs have been made to balance the energy allocated to immune function with the energy allocated for reproduction and thermoregulation. This kind of thinking leads directly to speculations about physiological connections between the immune and neuroendocrine systems. The ability to cross traditional scientific boundaries was a necessary prerequisite to the discovery and elucidation of the association between immune cells and neurosecretory cells in the brain. These are just a few examples of the utility of the integrative approach. Throughout this book, the reader will find this synthesis of evolutionary and ecological thinking with mechanistic methodologies.

Advantages and Pitfalls of an Evolutionary Approach to the Study of Sex, Sexuality, and Reproduction

Having stated the assumption that "particular mechanisms conferred an adaptive advantage in the face of some environmental pressures," we must add some cautionary notes. A strictly adaptationist approach may limit our understanding of the phenomena we study because natural selection is only one of the forces postulated to result in evolution, the others being mutation, migration, and drift. Second, directional selection is only one type of natural selection. Certain mechanisms may confer an advantage under stabilizing or diversifying selection as well. Finally, it is important to realize that natural selection acts in the absence of design, and studying sex, sexuality, and reproduction within a genetic or evolutionary framework does not necessarily imply biological determinism.

It is very common among the lay public, biologists of all kinds, and especially neuroscientists, to assume that every mechanism, every hormone, every receptor is the result of directional selection for the best possible functional design. Darwin was one of the first to lament this misinterpretation of evolutionary theory. At any one moment in time, a trait under observation, such as a hormone or receptor, may represent an adaptive trait, a trait that was adaptive during some long-past evolutionary time period but that no longer confers any particular advantage, or a trait that has never conferred any particular advantage or disadvantage. Darwin and subsequent evolutionary biologists have documented countless examples of existing traits that serve no known purpose in the present, but may have either served some adaptive function in the evolutionary past, or have come about via their genetic correlation with other selected traits.

Some of the above concerns about simplistic adaptationism can be illustrated by retelling the story of the panda's thumb, popularized by Stephen Jay Gould's book of the same name (Gould, 1980). Giant pandas are unique among the other members of their order in having a functional "thumb," which they use to strip leaves from stalks of bamboo, their dietary staple. Unlike the thumb of primates, the panda's thumblike appendage arose as an extended growth of one particular wrist bone. Ordinary bears, relatives of the panda, also show an enlargement of the same wrist bone, but to a much lesser degree. Anatomists argue that the functional thumb of the panda arose by an enlargement of this wrist bone accompanied by shifts in already existing musculature. The panda's thumb is used here as an example of how a functionally adaptive mechanism can arise from preexisting anatomical structures. A related point is that whatever genetic change provided the heritable variability in the size of the wrist bone also led to correlated heritable variability (genetic covariance) in the size of the foot bone. In other words, at least some of the same genes that code for the size of the wrist bone also code for the size of a particular foot bone. As a result, giant pandas also show an enlargement of this particular foot bone, although this foot bone per se has not been shown to confer any particular advantage to the survival and reproduction of the species. The panda's foot bone is thought to be an example of a trait that is not strictly an adaptation, but is found in the population due to its genetic correlation with a trait that has been under natural selection.

A message for neuroendocrinologists is that the mere presence of receptors for a particular substance in the brain cannot be taken as strong evidence for an adaptive neuroendocrine mechanism. How often we hear the question, If these sex hormone receptors in this particular brain area do not control reproduction, then why did Mother Nature put them in the brain? One might also ask, If it has no purpose, why did Mother Nature give the panda an enlarged foot bone? Why do males of many species have mammae (nipples)? Why do we have an appendix? Adaptation for current function is not always the correct answer. Understanding the function of neural receptors and pathways requires extensive study of the effects of the hormones at naturally occurring endogenous levels, the effects of antagonists specific to those receptors, and, in the ideal case, contextual knowledge of the hormone and receptor within the life history and breeding stages of the organism. When we demonstrate the presence of hormones and their receptors or when we demonstrate that the hormone has significant effects when applied directly to the central nervous system, these are only the first steps in understanding their functional role in the present or in the evolutionary past.

Another common misinterpretation of evolutionary theory often made by neuroscientists is the idea that neural and physiological mechanisms are designed efficiently or elegantly, or in a way that makes logical sense to the neuroscientist. Selection acts on the random variability that is already present in existing populations; it cannot create *de novo* previously nonexistent traits. Evolutionary change can be constrained in ways that may be subtle or profound. Energetic constraints that disallow maximum immune capabilities during reproduction may be one example of the limitations imposed on the creative force of natural selection. Some of these constraints are simple and easy to document. Others are complex and may never be understood. The result is that the behavioral, anatomical, physiological, and biochemical mechanisms we observe do not appear as perfectly engineered devices designed with a specific purpose in mind; they are more likely to be like the panda's thumb, a sort of contraption jury-rigged from a set of preexisting components, the range of which is limited by the prior evolutionary history of the organism.

Finally, neither the editors of nor the contributors to this book claim that natural selection has preordained fixed behavior patterns that determine human social arrangements. Species comparisons are not used to imply homology with the sex behaviors and reproductive strategies found in our own species. Species comparisons illustrate interactions between genotype and environment that lead to the complex and diverse possibilities and potentialities that exist among animals. Cultural, environmental, and genetic factors may or may not lead to similar behaviors in human societies.

For example, the study of chemical communication in insects, rodents and other animals may have led the lay public to the tacit assumption that human pheromones control sexual impulses. To the contrary, the recent study of human pheromones as described in *Reproduction in Context* led to a redefinition and reclassification of pheromones as emotional state modulators rather than behavior releasers. The careful study of these pheromones in human subjects is now actually changing the study of chemical communication in nonhuman animals. In a similar fashion, the reader will find nothing in this book to fuel popular myths such as the notion that primate evolution makes human males innately aggressive, territorial, and promiscuous, and females cooperative and nurturing; the notion that sexual dimorphisms in the size of brain structures underlie gender inequality; or the notion that male homosexuality is related to "feminized" brain structures. Those who make such assumptions misunderstand the uses and limitations of the theory of evolution, genetics, and the comparative method. No biological data in this book are intended to relieve us of our fundamental responsibility for the racial, sexual, and economic inequities in our society. In fact, a growing number of studies in behavioral neuroendocrinology have turned biological determinism on its head by demonstrating the effects of complex social systems on gene transcription. Nevertheless, we caution that the types of scientific questions posed and the interpretations they favor are inevitably affected by the social and political climate and history of the investigators. Thus, it is essential that any comparisons that are drawn between the animal models in this book and our own species include a careful painstaking analysis of the cultural, economic, and political forces that are also involved in the expression of behavior and the reproductive choices of *Homo sapiens*. Our theories and hypotheses concerning the evolution of sex, sexuality, and reproduction must be evaluated and reevaluated with this latter concern in mind. With these caveats in mind, we invite all readers to take our data and evolutionary explanations with many grains of salt and to continue to question how and why reproductive strategies and neuroendocrine mechanisms have come about.

Basic Reproductive Neuroendocrinology

Reproductive endocrinologists typically think of the reproductive system as the gonads, the pituitary, and the parts of the hypothalamus that control pituitary secretions and sexual behavior (Karsch, 1984) (figure 1.1). In both males and females, gonadal function depends primarily on the secretion of two glycoproteins, the gonadotropins, synthesized in and secreted from the anterior pituitary gland (adenohypophysis). The gonadotropins, luteinizing hormone (LH) and follicle-stimulating

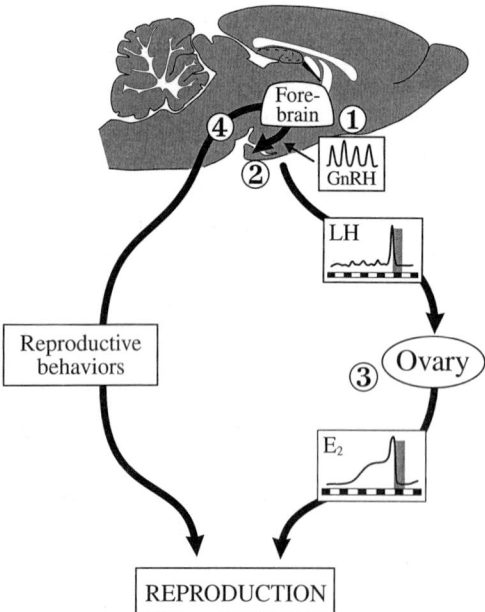

Figure 1.1
Hypothalamic-pituitary-gonadal system showing the possible loci of the effects of environmental and social stimuli on reproduction. This figure appears courtesy of George N. Wade.

hormone (FSH), are secreted by the same pituitary gonadotrophs to initiate follicle maturation and ovarian steroid secretion in females, as well as sperm production and androgenic steroid secretion in males. Secretion of LH is typically pulsatile, while secretion of FSH is less regular, depending on the species in question. Pituitary LH secretion is under control of pulsatile secretion of gonadotropin-releasing hormone (GnRH), synthesized in cells in the hypothalamus and preoptic area, secreted from terminals in the median eminence, and then transported to the anterior pituitary via the hypophyseal portal system. The origin of the pulsatile pattern of gonadotropin secretion has been traced to the pulsatile secretion of hypothalamic GnRH. Each LH pulse is preceded in time by a GnRH pulse. Without GnRH stimulation, LH pulses do not occur. The origin of the pulsatile secretion of GnRH is not well understood, but is thought to involve the intrinsic properties of the GnRH cell or volleys of synaptic activity within a neural network that includes catecholaminergic, peptidergic, and GnRH cells. Steroids secreted from the gonads tend to inhibit pulsatile LH secretion. LH pulses increase in frequency and amplitude in gonadectomized animals, and decrease in steroid-replaced animals. This empirical outcome is often referred to as "steroid negative feedback." In both males and females, the possible loci of environmental and social effects on reproduction include the (1) GnRH cells and the hypothalamic GnRH pulse generator, (2) LH and FSH synthesis and secretion from the pituitary, (3) gonadal activity including steroid secretion and steroid negative feedback at the hypothalamus, pituitary, or both, and (4) direct effects on the neural mechanisms that control sex behavior (figure 1.1). In females, another possible locus of environmental or social effects is the LH surge mechanism. This might involved effects on follicular steroid secretion, on brain mechanisms responsible for surge-level secretion of GnRH, or on pituitary mechanisms controlling the LH surge. As previously mentioned, estradiol has negative feedback effects on pulsatile LH secretion at the relatively low levels of estradiol secreted during the early follicular period. In contrast, at higher levels of estradiol secreted by the more mature follicle(s), secretion of LH is enhanced and eventually reaches levels that induce ovulation. Subsequently, the corpus luteum is formed from the ruptured follicle (initiating the luteal phase). The length of the cycle, as well as the relative lengths of the luteal and follicular phases varies greatly across species.

In some, but not all species, reproduction involves a variety of courtship and sexual behaviors that result from priming and activational effects of gonadal steroids on the neural structures and pathways that control these behaviors. Of course, not all species have spontaneous ovulatory cycles. In some species, termed "reflex ovulators" social interactions with males provide the critical stimulation that leads to behavioral, brain, pituitary, and gonadal changes in females that lead to ovulation. In both spontaneous and reflex ovulators, if fertilization takes place, increases and later decreases in estradiol levels, and increases in progesterone levels are necessary for successful gestation and parturition. Parturition and lactation also involve secretion of oxytocin and prolactin among other factors. These are important for affiliative or maternal behavior in some, but not all species.

Figure 1.1 is a simplified representation of a more complex system. The traditional field of reproductive neuroendocrinology has been concerned with elucidating the detailed physiological, neuroendocrine, cellular and molecular mechanisms that underlie this system. *Reproduction in Context* extends this view to incorporate the influences of other physiological systems into reproductive biology. Some chapters emphasize the role of such organs as adipose tissue, skeletal muscle, liver, pancreas, and the brain areas involved in control of food intake, body weight, and energy balance. Some chapters discuss the sensory systems that detect day length, temperature, and social cues from conspecifics, while other chapters discuss the interactions between the reproductive and immune systems. Still others incorporate all aspects of the organism that impinge on the individual differences that define sexuality. It can be argued that reproduction involves all physiological systems in the body.

Again, this can be understood in terms of evolutionary considerations. Selection acts on whole organisms, not on particular traits. A wonderful example is from experiments in which mice were selectively bred for high or low levels of a single behavioral trait, the size of the nest built at 22°C (Bult and Lynch, 1997). After several generations of selection for this one variable, the high- and low-selected lines differed severalfold in a large array of traits, some related to winter adaptation and some related to reproduction such as litter size and responsiveness to ovarian steroid hormones. In addition to building larger nests, having more thermogenic brown adipose tissue, thicker winter pelage and a larger body size, the high-selected lines also had larger litters, whether or not the mothers were provided with nest building material during pregnancy and lactation. In addition, selection for high levels of nest building resulted in low attack latencies in males of the high nest building lines. Thus, a portion of the genes that influence the size of the nest built also influence aspects of reproduction and social behavior. This is a dramatic example of the interrelationships among the various systems that underlie reproductive outcomes. From this perspective, the reproductive system as depicted in figure 1.1 is a primitive, two dimensional, static representation of what is in reality a highly complex, multidimensional, dynamic process. The following chapters will begin to fill in the gaps in figure 1.1 with the richness and diversity of sex and reproduction in real animals.

References

Bult A and Lynch CB (1997). Nesting and fitness: lifetime reproductive success in house mice bidirectionally selected for thermoregulatory nest-building. Behav Genet 27:231–240.

Dobzhansky T (1973). Nothing in biology makes sense except in the light of evolution. Am Biol Teacher 35:125–129.

Gould SJ (1980). The Panda's Thumb: More Reflections on Natural History. New York: WW Norton.

Karsch FJ (1984). The hypothalamus and pituitary gland. *In* Reproduction in Mammals, vol 3, Hormonal Control of Reproduction (Austin CR, Short RV, eds). Cambridge: Cambridge University press.

IA ENVIRONMENTAL CONTEXT: ENERGY BALANCE AND REPRODUCTION

Frank Bronson has provided a strong case for the idea that food availability is the environmental factor most critical for successful reproduction. Compared to other factors such as day length, ambient temperature, or social cues, food availability is critical to the largest number of species regardless of body size or geographic location. Furthermore, the adaptive advantage of reproductive responsiveness to photoperiod and temperature can be understood only in relation to food availability and its effects on energy balance. Day length provides a cue that allows animals to predict the season when food availability is most likely to support the high energetic costs of reproduction. Low ambient temperatures affect reproduction only when the energy expenditure required for thermoregulation cannot be offset by increased food intake. Other than food availability, the only factor of such importance might be water availability.

Those of us who have been influenced by the energetic perspective owe a great debt to Frank Bronson. First, he provided a comprehensive examination of the energetics of reproduction in species representing all mammalian orders in his influential book, *Mammalian Reproductive Biology*. Perhaps more important, he was one of the first and most successful mammalogists to integrate ecological with neuroendocrine approaches. He produced a large body of data that combined techniques and ideas from these two fields in a variety of ways. Historically, ecologists were concerned with measuring correlations between environmental variables and reproductive characteristics in animals living in their natural habitats. Bronson recognized two important characteristics of traditional ecologists, (1) their ability to accept the complexity of animals in their natural habitats and (2) their courage to examine many diverse factors that act on reproduction. In contrast, the early endocrinologists began with narrower, short-term goals such as elucidation of the function of the pineal gland and the gonads. Later, with the discovery that these glands were controlled by the central nervous system, the endocrinologists evolved, in a manner of speaking, into neuroendocrinologists. Bronson saw that their advantage was their ability to design experiments that could determine causation, rather than mere correlation. These types of experiments started with a falsifiable hypothesis and an experiment (e.g., removal of a gland or brain area) that resulted in an outcome that was either consistent or inconsistent with the hypothesis. Bronson working with mammals and Wingfield working with birds (chapter 4) integrated the broad vantage point of ecology with the experimental methods of endocrinology and neuroendocrinology. First, by taking the time to understand the animals in their natural habitat, scientists like Bronson and Wingfield were able to make informed predictions about the physiological mechanisms that link reproduction to environmental factors. Second, they designed experiments in the laboratory and field that tested these predictions. In some of their more creative experiments they examined reproduction in response to more than one environmental factor simultaneously. This approach is emphasized in the following two chapters and throughout this book.

In chapter 2, Bronson examines a popular notion that pervades the clinical literature in reproductive neuroendocrinology. It is widely held that reproductive maturation is dependent on a critical level of energy reserves in the form of body fat. According to this idea, a female should not attempt reproduction until she has energy reserves sufficient to raise her offspring in the event of unpredictable energetic challenges, such as a famine or cold spell. Bronson examines this idea in the context of the life history strategies of a wide range of animals living in their natural habitats. Drawing evidence from the most relevant examples, he concludes that there may be no such biological mandate for a particular level of body fat. Rather, the primary signals that control reproductive maturation are generated the moment females reach positive energy balance.

Chapter 3, by Schneider and Wade, begins with the idea that reproduction is linked to energy balance, and provides a current perspective on the primary metabolic sensory signals that lead to short-term effects on reproductive processes. The work of Schneider and Wade is based, without a doubt, on the ecology-neuroendocrinology foundation laid by scientists like Frank Bronson. In addition, they have synthesized ideas and methods commonly used in the study reproductive behavior with those methods used to study ingestive behavior and fuel homeostasis. This synthesis has led to a growing field of research aimed at elucidating metabolic and biochemical effects on the neuroendocrine system that controls reproduction.

2 Puberty and Energy Reserves: A Walk on the Wild Side

Franklin H. Bronson

Puberty, growth, and energy balance are interrelated in the female mammal but the metabolic and neuroendocrine mechanisms linking these processes are confusing and controversial. Two facts are without dispute. First, puberty as well as growth, is suppressed whenever a young female is forced into negative energy balance, that is, when energy expenditure exceeds energy available from intake and storage. This has been demonstrated in relation to food restriction, excessive exercise, and exposure to cold in a wide variety of species (reviewed in Foster et al. 1998). Second, the onset of fertility is more closely associated with body size than with chronological age in females whose growth has been retarded by energetic challenge (Kennedy and Mitra 1963). The adaptationist argument that combines these two facts suggests that young females should not ovulate and risk pregnancy until their energy reserves are sufficient to support pregnancy and lactation in case food supplies dwindle after conception. This is an ancient and often-quoted argument of uncertain origin.

The possibility that the reproductive axis of a young female might routinely monitor the accumulation of stored energy has given rise to several specific hypotheses about the source and nature of the presumed metabolic signal. Frisch and McArthur (1974) postulated that the source of the signal was body fat. Steiner and coworkers (1983) suggested that a blood-borne substance(s) such as insulin, glucose, or the amino acids might signal the brain about the size of a female's energy stores. Most recently, it has been hypothesized that puberty depends on a critical level of circulating leptin, the protein product of the obese gene expressed primarily in adipocytes (Cheung et al. 1997). Schneider and Wade (see chapter 3) argue that the body fat or "lipostatic" hypothesis has been rejected repeatedly on experimental grounds (reviewed in Bronson and Manning 1991; Wade and Schneider 1992), but the lipostatic hypothesis still finds adherents among clinically oriented researchers because fat is a good correlate of the state of energy balance in girls (Frisch 1994). This, in turn, provides a conceptual basis for the leptin hypothesis (e.g., see Matkovic et al. 1997). The search for the presumptive blood-borne signal that informs the reproductive axis about the size of a female's energy stores continues unabated today (e.g., see Cameron and Schreihofer 1995; Foster et al. 1995; 1998).

For someone with a biological background, there is a worry about these various hypotheses and, indeed, about the adaptationist argument itself. This argument and its associated hypotheses reflect the study of a handful of species in artificial environments that are far simpler and far more temporally stable than the wild habitats in which the relationships between puberty, growth, and energy balance actually evolved. Whether working with rats in small cages, larger mammals in larger pens, or well-fed humans in energetically pampered urban environments, "normal" has been defined routinely by researchers in this area as a situation in which there is no variation in food availability or quality, where there is no need to expend energy in foraging, and where there are no meaningful thermoregulatory challenges. In such energetically utopian conditions the young female grows, matures metabolically, puts on fat, and increases the secretion of leptin progressively, all at a maximum rate. Thus puberty occurs routinely at a predictable stage of metabolic development. This makes emphasis on the young female's energy stores seem quite reasonable.

In contrast, there is a more biological perspective of "normal." This perspective visualizes the relationships between puberty, growth, and energy balance as having evolved for over two hundred million years in a diversity of mammalian radiations that have exploited an equally diverse collection of highly complex and mostly unstable natural habitats. The biological perspective insists that in order to have merit, the adaptationist argument and its associated hypotheses must be compatible with the kinds of energetic challenges mammals actually face in the wild and with the kinds of responses they actually exhibit in wild habitats. When viewed from this perspective, there are many populations of mammals in which the magnitude of a young female's energy reserves would provide a particularly poor substrate on which to base a prediction about future reproductive success. This, in turn, makes one wonder about the hypotheses associated with the adaptationist argument and whether the experimentalist's emphasis on energy reserves may not be an oversimplification.

This chapter represents an intellectual exercise whose intent is to view the potential dependency of puberty on energy stores from the biologist's perspective. First, our laboratory-derived knowledge about puberty, growth, and energy balance is summarized, and in some cases reinterpreted. Second, the ecological literature is reviewed to identify the factors that might antagonize or, alternatively, promote the evolution of a direct link between energy reserves and puberty. Then, third, these different bodies of knowledge are meshed to assess the argument and hypotheses in question. Do they make sense when examined from an ecological perspective?

The View from the Laboratory

Three characteristics of the relationship between puberty, growth, and energy balance are relevant here. First, the energy-balancing process focuses directly on the immediate need to survive and only indirectly on functions like growth and puberty. Second, this is a rapid process that is designed to operate over short periods of time and

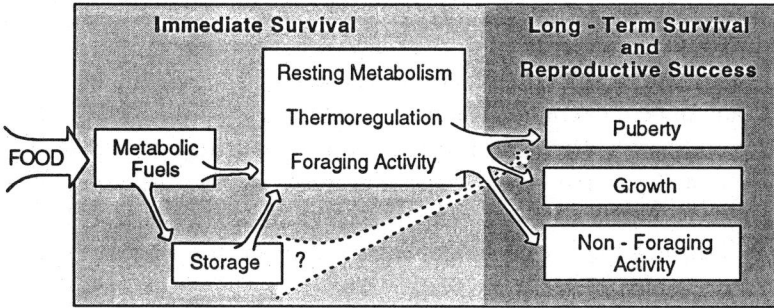

Figure 2.1
The functional relationship between energy and puberty. Foraging activity includes all energy expended while searching for food, including aggressive encounters with neighbors related to food availability and predator avoidance. Non-foraging activity includes the cost of emigration in response to food shortage, if necessary, and play, including physical exercise unrelated to foraging in humans. The concern of this chapter is with the postulated direct link between energy reserves and puberty.

it involves multiple inputs to the brain from the periphery. Third, the metabolic control of ovulation reflects these characteristics and, importantly, peripubertal and fully adult females do not differ in this regard, nor does the metabolic control of the reproductive axis differ in males and females.

Survival and Puberty

At a coarse-grained level of analysis, the reason why puberty is delayed by energetically challenging conditions is simple. Puberty has a low priority for energy relative to some other functions. As suggested in figure 2.1, the energy in ingested food is turned into oxidizable metabolic fuels—glucose or nonesterified fatty acids—which can then be stored, mostly in adipose tissue, or expended to satisfy a variety of physiological demands. In times of energetic constraint these demands compete with each other, and the onset of reproductive capability has a relatively low priority in this competition. Growth, pubertal development, and nonforaging locomotion compete for whatever energy is left over after the need for immediate survival is satisfied; that is, after the costs of resting metabolism, thermoregulation, and foraging have been paid.

As would be expected then, puberty is always suppressed when food intake is restricted beyond a critical level (e.g., see Foster et al. 1998). Likewise, puberty is suppressed in small mammals maintained at low ambient temperatures unless they can increase their food intake sufficiently to counterbalance their increased thermoregulatory costs (Barnett 1965; Manning and Bronson 1990). Similarly, puberty is delayed in high-performance athletes and ballet dancers when they cannot or will not increase their food intake sufficiently to balance these energetically costly activities (Warren 1989; Dueck et al. 1996). The negative effect of excessive exercise has been shown experimentally in peripubertal female rats also (Manning and Bronson 1991).

Importantly, figure 2.1 also shows (in dashed lines) the postulated direct link between energy reserves and puberty. It must be emphasized that the question of interest here is not whether puberty requires a female to be in positive energy balance. That is a well-established fact. *The concern of this chapter is whether puberty depends on the accumulation of a critical level of energy reserves.*

Characteristics of the Energy-Balancing Process Per Se

The central effector pathways that control foraging and eating, and that act peripherally via the autonomic nervous system to determine the fate of ingested energy, are controlled by a wide variety of metabolic signals (Kaiyala et al. 1995). Circulating levels of the oxidizable metabolic fuels or variation in the intracellular utilization of these fuels is of core importance for the moment-to-moment regulation of these pathways (e.g., see Levin and Routh 1996; Friedman 1997). The glucocorticoids and a variety of peptides produced in the gastrointestinal tract also serve this purpose. The latter act directly on the brain or via stimulation of the vagus nerve, or in both ways.

Two adipose-related signals—leptin and insulin—are also of great importance in regulating the central effector pathways. Until the discovery of leptin in 1994, heavy emphasis was placed on insulin as an adiposity signal (e.g., see Schwartz et al. 1992). Now the field seems almost totally dominated by the potential role of leptin (e.g., see Campfield et al. 1996). In either case, the degree to which these metabolic hormones act directly on the central effector pathways, as opposed to acting indirectly via another kind of signal, is not well understood.

The Energetic Control of Gonadotropin-Releasing Hormone Secretion

The characteristics noted above dovetail nicely with what is known about the reaction of the reproductive axis to energetic challenge. The rate of activity of this axis is determined by a pulse generator that controls the release of gonadotropin-releasing hormone (GnRH) and hence luteinizing hormone (LH). The pulsatile secretion of GnRH is readily suppressed by food restriction, low ambient temperature, or excessive exercise, and, at least in regard to the effect of food restriction, GnRH pulsatility returns *rapidly* when this energetic challenge is alleviated—usually in an *hour or two* at most. The available data suggest that this is true across a range of species in adult as well as peripubertal females and in both sexes (e.g., see Bronson 1986; Foster, et al. 1989; Parfitt et al. 1991). Studies in sheep and rhesus monkeys have established the fact that it indeed is calories per se that are important in this regard, whether they are obtained from fat, protein, or carbohydrate (l'Anson et al. 1991; Schreihofer et al. 1996). Interest in the neuropeptides linking energy balance and the GnRH pulse generator is intense at this time (Kalra and Kalra 1996).

The rapid responsiveness of the peripubertal female's GnRH pulse generator to perturbations in energy balance can be illustrated by studies done with rats. Pubertal ovulation is routinely blocked in young rats if they are given only enough food to maintain their body weight but not enough to grow. As shown in figure 2.2, LH pulsing is almost but not entirely blocked in rats treated in this manner. Many young females show one or two LH pulses an hour or so after their one daily meal. When abruptly returned to ad lib. feeding, LH pulsing simply continues the trend seen after the single meal rather than stopping until the next day's feeding; the females ovulate

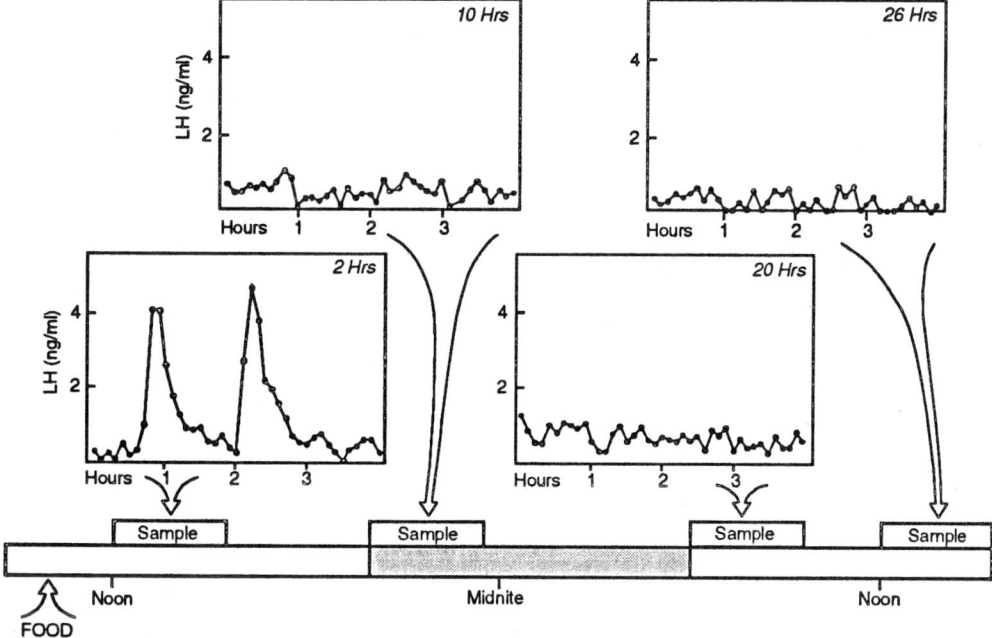

Figure 2.2
The effect of a single meal on luteinizing hormone (LH) pulsing in a food-restricted, peripubertal female rat. Not all rats always show two pulses after eating their one daily meal; many show one and some show none. (From Bronson and Heideman 1990.)

either 2 or 3 days later (Bronson and Heideman 1990). It must be emphasized here that the response of the GnRH pulse generator occurs in complete independence of any detectable change in total body fat and certainly with no dependence on a "critical level" of body fat, whatever that might be.

As might be expected from what is known about the process of energy balance per se, there is ample evidence now that the GnRH pulse generator of the young female is influenced by a variety of signals related to energy balance (Cameron 1997), including the oxidizable metabolic fuels, insulin and insulin-like growth factor I (IGF-I) (e.g., see Hiney et al. 1996; Schneider and Wade, chapter 3; see also Cagampang et al. 1990). Without diminishing the importance of the different kinds of signals, particularly the metabolic fuels for which there is so much evidence now (Wade et al. 1996), the concern in this chapter is with leptin and the hypothesis that circulating levels of this peptide serve as a metabolic gate for puberty. Administration of leptin recovers reproductive function in *ob/ob* mice that lack this peptide (Barash et al. 1996); it can stimulate LH release in median eminence/arcuate nucleus explants (Yu et al. 1997), and antisera to it can decrease LH pulsatility (Carro et al. 1997). Leptin injections can counteract the effect of food restriction on reproductive development in peripubertal mice and rats (Ahima et al. 1996; Cheung et al. 1997). Importantly, leptin gene expression shows the kind of 1- to 2-hour robust changes one would expect from the timing characteristics of the GnRH response to energy manipulation, as discussed in chapter 3 (cf. figure 2.2 and Trayhurn et al. 1995a,b).

In brief summary, the evidence obtained in the laboratory actually provides little support for the concept that puberty depends on a critical level of energy reserves in female mammals. Instead, the available data suggest the GnRH pulse generator is regulated by energy balance on a moment-to-moment basis in complete independence of the size of a young female's energy reserves. So how does the adaptationist argument and its associated hypotheses look when viewed from the perspective of wild mammals living in wild habitats?

The View from Natural Habitats

Three limitations in the literature dealing with wild populations must be noted. First, relatively few of the four thousand-odd species of mammals have been studied in any detail and thus some conclusions must be based on case studies of individual species rather than generalities drawn from large numbers of species. Second, most mammals are tropical in residence, yet most of our knowledge relates to the relatively few species that live in the energetically harsher Temperate Zone; this is where most of us live and work and where most of our experimental mammals evolved. Third, it is often difficult to tell the difference between peripubertal and fully adult females in the wild. Thus, some conclusions must be based on the control of ovulation in groups of females of all ages.

As a generality, the degree to which natural selection would favor a direct link between puberty and energy reserves undoubtedly varies from radiation to radiation. Whether it is favored in a particular radiation should depend on five factors: (1) stability of a mammal's habitat of residence, (2) life span, (3) body size, and the presence or absence of energy-conserving adaptations like (4) hibernation and (5) feeding strategy.

Habitat Stability

A direct link between puberty and energy reserves should be favored in habitats in which the energetic milieu remains constant or in which it changes seasonally in a reasonably predictable manner. If the energetic challenges offered by a habitat change rapidly and unpredictably, the result will be temporal variation in the economics of foraging (calories lost vs. calories gained) and this, in turn, will produce temporal variation in the magnitude of a mammal's energy stores. As will be noted later, in some cases the result can be great moment-to-moment variation in energy stores and thus a particularly "noisy" substrate upon which to make predictions about future reproductive success. While the brain could probably integrate even the noisiest of such signals, other factors that will be discussed shortly make this seem like an unlikely solution.

It is important to realize that most wild habitats in which mammals exist are indeed quite unstable. Energetic challenges change with season in most habitats, sometimes predictably so and sometimes not. Food availability varies seasonally, even in the tropics, and often this pattern is reasonably predictable (Bronson and Heideman 1994). On the other hand, droughts and their resulting food shortages occur unpredictably; sometimes they last for years and sometimes they disappear in a few months. On a shorter time scale, weather fronts come and go over periods of days, and foraging conditions often change dramatically and rapidly with their passage. Indeed, foraging conditions can change dramatically over periods of hours in many habitats. The energy-balancing process evolved to ensure immediate survival in the face of all such variation.

Life Span

Perhaps as many as a third of the four thousand-odd species of mammals have life spans measured in weeks in the wild, including most of the small rodents and insectivores and many marsupials (Eisenberg 1981). Most of these animals live for 2 or 3

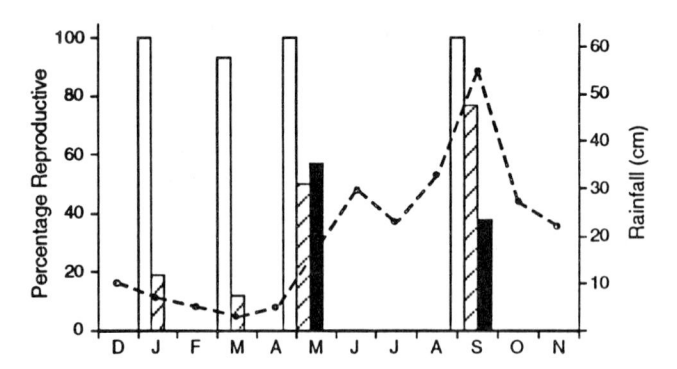

Figure 2.3
Seasonal variation in reproductive success in wild-caught collections of female cloud forest mice in Costa Rica in relation to seasonal variation in rainfall. The open bars indicate females showing recent corpora lutea; the hatched bars indicate females with embryos; the black bars indicate lactating females; and the dashed line shows variation in rainfall. (From Heideman and Bronson 1992.)

years in the laboratory, but not in the wild. Indeed, the life expectancy of some of these animals in the wild barely equals the time required for a female to mature and produce weaned offspring. A short life span dictates an opportunistic reproductive strategy in which females mature and ovulate without concern about the ultimate success of their reproductive effort. Thus a short life span should antagonize the evolution of a requirement that puberty be delayed until energy reserves are adequate to support pregnancy and lactation; a long life span should promote it.

To illustrate this principle, we can look at a representative of our most common kind of mammal, a tropical rodent, the cloud forest mouse (*Peromyscus nudipes*) of Costa Rica at 8 degrees of latitude. Like most parts of the tropics, a distinctly seasonal pattern of rainfall produces seasonal variation in food availability in Costa Rica. Cloud forest mice ovulate year-round but they only successfully produce offspring during the wet season when food is readily available (figure 2.3). Embryos resulting from a fertile mating either fail to implant or are reabsorbed after implantation during the dry season. The young born late in the wet season find enough food to mature and ovulate during the dry season but, like the adults, they fail to reproduce successfully until the rains come again. The syndrome of reproductive failure seen during the dry season—ovulation followed occasionally by implantation but never by birth—can be reproduced in the laboratory by employing mild food restriction (Heideman and Bronson 1992).

Cloud forest mice obviously mature and ovulate when their energy reserves are totally inadequate to support even the relatively low costs of pregnancy, let alone the much more expensive process of lactation. These animals must gamble routinely because of their short life span, winning during the wet season when food is readily available and losing routinely during the dry season when it is scarce. Such opportunism is by far the most common strategy seen in short-lived mammals. Only at much higher latitudes do short-lived mammals rely on a seasonal predictor like photoperiod to avoid maturation and ovulation during the winter when energetic conditions are particularly harsh, and many populations at the highen latitudes are actually genetically heterogeneous for this trait (Bronson and Heideman 1994). Opportunism is also seen in larger mammals living in unpredictable environments, such as the well-studied red kangaroo (*Macropus rufus*) of central Australia where rainfall and hence food availability vary erratically (Newsome 1966).

Body Size

Almost all mammals whose life spans are measured in weeks in the wild also weigh less than 100 gs, as do most bats, which are our second most common kind of mammal behind the rodents. Thus, the majority of mammals are quite small in stature. Small body size should antagonize the evolution of a direct link between puberty and energy reserves for two reasons. First, small mammals can store only enough energy to protect them from starvation for a few days—a minuscule amount compared to that required for a successful fertile cycle of pregnancy, lactation, and the weaning of healthy young (Thompson 1992). Second, small size can lead to immense temporal variation in the size of a mammal's energy stores in natural habitats. The large surface-to-mass ratio of small mammals makes them exceptionally vulnerable to variation in ambient temperature. As shown in figure 2.4, the fat reserves of a 20-g peripubertal house mouse can be depleted by starvation in only 24 hours at 10°C (50°F)—a relatively mild temperature for a larger mammal—and in less than 3 days at 24°C (75°C). Indeed, even at normal room temperature, 22°C (72°F), peripubertal female house mice lose almost a third of their body fat during the light phase of a light cycle if allowed to feed only during the dark phase (Bronson 1987).

To appreciate the relevance of these experimental observations, one must visualize the daily pattern of energy expenditure and gain in a young female in a natural habitat. Unless she caches food or goes into torpor (e.g., see Blank 1992), the young female must burn fat continuously when in her nest during the daylight hours, sometimes at a high rate. Then, when it is dark, she must leave her nest and forage regardless of the temperature outside. There is little choice in this matter; she needs to replenish her lost energy. The young female will lose calories continuously through her skin while foraging, more so at lower temperatures. In times of food shortage, the female must increase her outside foraging time. When severe, the combination of

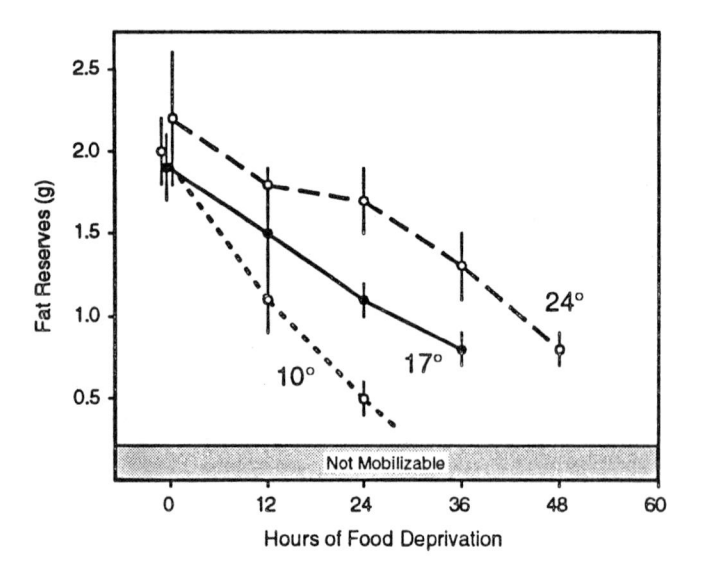

Figure 2.4
Effect of ambient temperature (°C) on the rate at which body fat is depleted in peripubertal female laboratory mice after all food is removed from their cages. (From Bronson et al. 1991.)

food shortage and low temperature can lead to uneconomical foraging—more loss than gain—and thus energetic catastrophe. This challenge is most marked in the Temperate Zone, but it would be a factor of importance even in the warm tropics (see figure 2.4; 24°C), where wet conditions should lead to rapid drain of stored energy when foraging. Over the planet as a whole, then, small size can lead to dramatic day-to-day, hour-to-hour, and even moment-to-moment variation in the level of stored energy. Such temporal instability would provide a poor substrate upon which to predict the best time to ovulate for the first time, even if the small female could store enough energy to support a major portion of a cycle of pregnancy and lactation, which it cannot.

Hibernation

Hibernation should antagonize the evolution of a direct link between puberty and energy reserves because the pubertal ovulation occurs immediately after a young female emerges from hibernation and this is when its energy reserves are at an annual low. Female golden-mantled ground squirrels (*Spermophilus saturatus*) of western Washington State, for example, expend about 8000 kJ of energy to produce their one litter of young each year (figure 2.5), but their fat stores at the time of ovulation

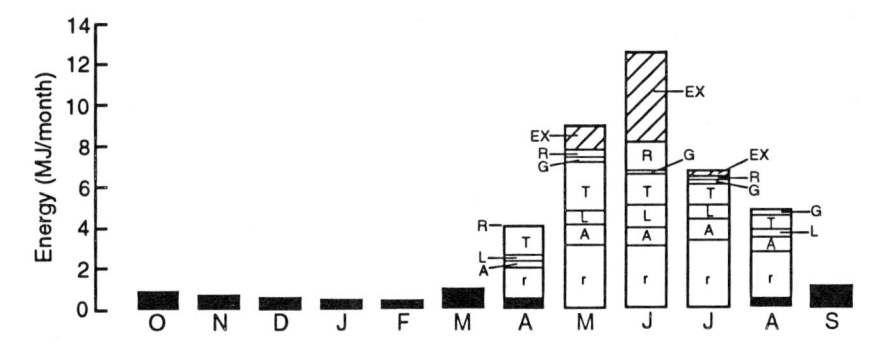

Figure 2.5
The annual cycle of energy metabolism in female golden-mantled ground squirrels. Black bars, resting metabolism during hibernation; r, resting metabolism at other times; A, alertness (the energy required to move around in the burrow and come out of it and sit at its edge); L, locomotion; T, thermoregulation; G, growth; R, reproduction; EX, energy exported as offspring or milk. (From Kenagy 1987.)

average only about 400 kJ (Kenagy 1987). This is enough to survive for only 2 or 3 days in the absence of feeding (Kenagy and Barnes 1988). The annual cycle of these animals is regulated by a photoperiod-synchronized, endogenous circannual rhythm. Thus, environmental conditions can vary immensely when they emerge from hibernation and ovulate. Sometimes food is readily available at this time; at other times it is extremely limited and buried under snow. Obviously these females totally ignore the fact that they have relatively tiny and probably highly labile fat stores when they ovulate for the first time.

Feeding Strategy

Sometimes predators are lucky and find food quickly when foraging; sometimes they find nothing, in which case they must survive by burning stored energy until their next kill. The result of poor luck can be great temporal variation in the level of stored energy and thus a predacious feeding strategy would probably antagonize selection for a direct link between puberty and energy stores. This kind of challenge is most obvious in the small insectivores, some of which are legendary for their need to find and consume their own body weight in prey each day if they are to survive (e.g., see Pearson 1944). While never measured, body energy reserves must fluctuate wildly on a moment-to-moment basis in these animals, particularly when ambient temperatures are low.

The potent influence of a predacious feeding strategy can be seen in the larger carnivores as well, at least in the Temperate Zone (e.g., see Cavallini and Santini 1996). According to Poulle et al. (1995), for example, the 10- to 15-kg coyote (*Canis latrans*) produces one litter each year on a strictly seasonal basis in Quebec. *Average* body fat levels in these animals peak just prior to the breeding season in late winter. Hunting is difficult at this time of the year, however, and coyotes sometimes go several days without making a kill. Thus, prey availability, low temperature, and luck can combine to produce great day-to-day variation in their energy stores. The annual pattern of fat deposition and depletion is the same in male and female coyotes, reemphasizing the fact that any reproductive-related function of fat is strictly of secondary importance to its survival function.

Where Does One See a Good Correlation Between Puberty and Energy Reserves?

In the wild, good correlations between the size of a female's energy stores and puberty are seen only in stable (rare) or predictably changing (more common) environments, and then only among some of the larger, longer-lived herbivores and omnivores, and perhaps some of the large carnivores. One of the most interesting such species is the human—not the well-studied, modern urban human, but those living in primitive subsistence societies in the tropics. These societies often show profound seasonal variations in birth rates which presumably reflect in part variation in ovulation in response to seasonal variation in food availability and workload (e.g., see Leslie and Fry 1989; Bailey et al. 1992; reviewed in Bronson 1995).

The agriculturist society of Kenaba in Gambia is one of the best-studied such populations (e.g., see Prentice, et al. 1996). In Gambia, there is a "hungry season" coinciding with the rainy season when females must work long hours each day in the fields and when the amount of food left over from the previous year's crop is almost depleted. The resulting seasonal variation in body weight averages about 10%, variation in body fat averages 40%, and the variation in birth rate is over twofold, with the highest conception rates coinciding with the highest levels of fat (figure 2.6). Like other such societies (Galdikis and Wood 1990), a typical fertile cycle extends for well over 2 years in the Gambian population because infants are routinely suckled for over 18 months. An average 27-month fertile cycle requires about 1500 MJ of energy beyond that required by nonpregnant, nonlactating women (Singh et al. 1989; Heini et al. 1991; Prentice et al. 1996). At the time of conception these women *average* about 11 kg of fat, which is enough stored energy to sustain pregnancy and the first few months of lactation (Lawrence et al. 1987).

On a strictly *correlational* basis then, the data for the Kenaba population seem to support the concept that ovulation, pubertal or otherwise, depends on accumulating enough stored energy to support pregnancy and lactation in times of food shortage. One well-substantiated fact argues against this inference. The relationship between

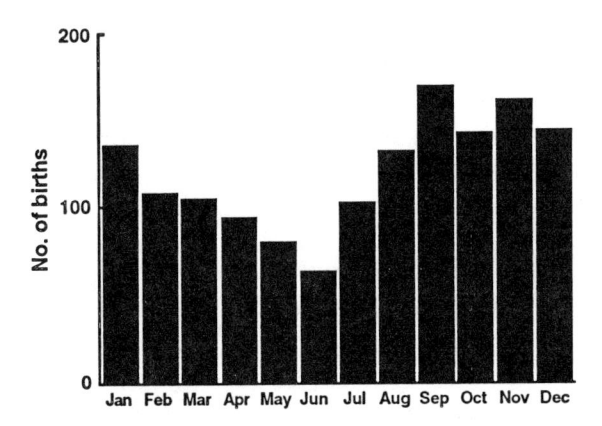

Figure 2.6
Seasonal variation in birth rate of women in an agriculturalist subsistence society in Gambia. (From Prentice and Cole 1994.)

body fat and ovulation has been examined numerous times in athletes and dancers from developed countries and the evidence overwhelmingly rejects the body fat hypothesis (e.g., see Loucks and Horvath 1985; see other references in Bronson and Manning 1991). Likewise, the body fat hypothesis has been subjected repeatedly to experimental testing in laboratory animals and been rejected over and over again (Wade and Schneider 1992). The Gambian women probably exist routinely in the gray area surrounding neutral energy balance, ovulating only when workload and food intake allow them to slide far enough into positive energy balance to do so. Overall, there is virtually no support in the literature dealing with wild populations for the postulate that puberty depends on a critical level of stored energy.

Conclusions

It is important to remember that the adaptationist argument and the hypotheses of interest here relate to a direct link between a young female's *energy reserves* and the GnRH pulse generator and not to the link between *energy balance* and the pulse generator. There is no question about the latter. Females, young or old, cannot ovulate unless they are in positive energy balance, and the fact that the GnRH pulse generator routinely monitors energy balance, perhaps by sensing a blood-borne signal of some kind, is well established. In contrast, the present analysis argues strongly against the long-held assumption that puberty depends directly on the size of a young female's energy reserves. Thus it also argues against the various hypotheses about a blood-borne signal that keeps the GnRH pulse generator informed specifically about the magnitude of a young female's energy reserves (see figure 2.1).

The strongest argument against these propositions is the fact that many mammals have short life spans and cannot afford the luxury of postponing their pubertal ovulation until their energy stores are adequate to ensure reproductive success. Like the cloud forest mouse, many short-lived mammals mature rapidly and then ovulate whenever they are in positive energy balance despite the fact that their energy reserves are hopelessly inadequate to ensure the successful production of offspring. The often-stated adaptationist argument that young females should delay ovulation until their energy reserves reach some critical level does not agree with what is known about a multitude of species in some of our most populous orders of mammals—the rodents, insectivores, and marsupials. Without support for the adaptationist argument, there is no conceptual basis for the hypotheses that focus on the source and identity of the putative blood-borne signal—leptin, insulin, and so on—that links the GnRH pulse generator to energy reserves. Of great relevance in this regard is the total lack of experimental support for the body fat hypothesis proposed by Frisch (1994).

The second argument against these propositions is the fact that most mammals are relatively small, and small stature translates as a miniscule amount of stored energy compared with that needed to fuel a full fertile cycle of pregnancy and lactation. Additionally, small size can lead to immense moment-to-moment variation in the size of a female's energy stores and hence a very "noisy" substrate upon which to base the timing of puberty. It is not inconceivable that the brain could integrate such a noisy signal and thus rely upon a long-term average level of stored energy to regulate the GnRH pulse generator, but why would natural selection act in this manner given the fact that most of these animals are also short-lived? With specific regard to the leptin hypothesis, circulating levels of this protein are even more labile than body fat levels (see figure 3.9 and Trayhurn et al. 1995a,b; Weigle et al. 1997) and thus leptin titers must fluctuate wildly in many mammals under some conditions (for review, see chapter 3). Again, the brain could integrate this signal, but why would it do so if short-lived animals indeed ignore the magnitude of their stored energy when ovulating for the first time?

Some mammals hibernate and many are predators and these factors also argue against the proposition that there is a direct link between a young female's energy reserves and the onset of fertility. Hibernators experience their first ovulation when their energy stores are at their annual minimum so this is not a question of integration of a labile blood-borne signal; both hibernation and a predacious feeding strategy should lead to temporal variation in a young female's energy stores. Of particular interest in this regard are the many Temperate Zone insectivores, all of which are very small, short-lived predators inhabiting cool to cold climates and thus their energy stores must be extremely labile.

Laboratory experimentation also argues against the proposition that puberty depends on the size of a young female's energy reserves. Of great importance is the fact that there is no evidence to date for any difference between the two sexes in the way the GnRH pulse generator is regulated by energy balance. This is in spite of the fact that the energetic requirements for reproductive success are totally different in the two sexes. Furthermore, the GnRH pulse generator responds to energetic manipulation rapidly—in an hour or two—in both sexes, which emphasizes the fact that the primary focus of the energy-balancing process is on immediate survival and only secondarily on functions associated with reproduction.

All in all, then, there seem to be two possibilities here. Either puberty does not depend on the magnitude of a young female's energy reserves in mammals and there is no blood-borne signal that links puberty directly to energy reserves or, alternatively, such a dependence and such a signal have evolved in some but not all mammals. In the latter case, this would involve a minority of the 4000-odd species and it would imply a *totally independent evolution* of these mechanisms in radiations far removed from each other evolutionarily. This seems highly unlikely.

Careful experimentation is absolutely crucial to isolate and identify the links between energy balance and puberty, but a strict reliance on this approach can lead to oversimplification and indeed probably has done so in regard to the long-postulated link between energy reserves and puberty. If one's framework of reference is limited to a rodent in a small cage, a larger mammal in a larger pen, or a modern urban woman living in an energetically pampered environment, the adaptationist argument of interest here and its associated hypotheses all seem quite reasonable. Expanding one's framework of reference to the four thousand-odd species that live in a diversity of natural habitats finds little support for this argument and these hypotheses, however. It would seem much better at this point to apply Ockham's razor and conclude that puberty is linked to the magnitude of a young female's energy reserves only in the sense that pubertal ovulation requires that a mammal be in positive energy balance, and energy reserves are an important component of the energy-balancing equation.

One final comment seems worthwhile. There is not much doubt that the onset of fertility is closely linked to body size. This phenomenon has been documented repeatedly since first reported in 1963 by Kennedy and Mitra (e.g., see Bronson and Desjardins 1974). If the energy-balancing process and specifically a female's energy reserves do not serve as the conduit through which puberty is linked to body size, what does? Ultimately, the answer must be related to the interaction between genes, that is, the timing of puberty is encoded in the genome, but the physiological entity or process through which the genes act poses a tough question. One possibility that deserves serious consideration is that the brain monitors changes in total body protein. Unfortunately, there is almost no literature available with which to assess this possibility. While sometimes mentioned within other contexts (e.g., see Webster 1993; Hirvonen and Keesey 1996), there has been little interest in determining if the brain monitors total protein developmentally as a way of regulating growth, and no interest in whether the GnRH pulse generator might monitor protein accumulation in the young female. Since a wide range of possible metabolic signals are generated when animals change from positive to negative energy balance, it is difficult to dissociate the effects of one macronutrient, such as protein, from the other correlated factors. Chapter 3 suggests an organizational framework in which we might begin to examine the primary sensory signals, the circulating fuels and macronutrients (protein, glucose), hormones (leptin, melanocortin), neurotransmitters, and aspects of body composition that link energy balance to the GnRH pulse generator.

References

Ahima RS, Prabakaran D, Mantzoros C, Qu D, Lowell B, Maratos-Flier E, Flier JS (1996). Role of leptin in the neuroendocrine response to fasting. Nature 382:250–252.

Bailey RC, Jenike MR, Ellison PT, Bentley GR, Harrigan AM, Peacock, NR (1992). The ecology of birth seasonality among agriculturalists in central Africa J Biosoc Sci 24:393–412.

Barash IA, Cheung CC, Weigle DS, Ren H, Kabigting EB, Kuiper JL, Clifton DK, Steiner RA (1996). Leptin is a metabolic signal to the reproductive system. Endocrinology 137:3144–3147.

Barnett SA (1965). Adaptation of mice to cold. Biol Rev 40:5–51.

Blank JL (1992). Phenotypic variation in physiological response to seasonal environments. In Mammalian Energetics (Tomasi T, Horton TH, eds) pp 186–212. Ithaca, NY: Comstock.

Bronson FH (1986). Food-restricted, prepubertal female rats: rapid recovery of luteinizing hormone pulsing with excess food, and full of recovery of pubertal development with gonadotropin-releasing hormone. Endocrinology 118:2483–2487.

Bronson FH (1987). Susceptibility of fat reserves to natural challenges in mice. J Comp Physiol [B] 157:551–554.

Bronson FH (1995). Seasonal variation in human reproduction: environmental factors. Rev Biol 70:141–164.

Bronson FH, Desjardins C (1974). Circulating concentrations of FSH, LH, estradiol and progesterone associated with acute, male-induced puberty in female mice. Endocrinology 94:1658–1668.

Bronson FH, Heideman PD (1990). Short-term hormonal responses to food intake in peripubertal female rats. Am J Physiol 259:R25–R31.

Bronson FH, Heideman PD (1994). Seasonal regulation of reproduction in mammals. In The Physiology of Reproduction, 2nd edi (Knobil E, Neill JD, eds), pp 541–584. New York: Raven Press.

Bronson FH, Heideman PD, Kerbeshian MC (1991). Lability of fat stores in peripubertal wild house mice. J Comp Physiol [B] 161:15–18.

Bronson FH, Manning JM (1991). Minireview—the energetic regulation of ovulation: a realistic role for body fat. Biol Reprod 44:945–950.

Cagampang RRA, Maeda K-I, Ota K (1990). Involvement of the gastric vagal nerve in the suppression of pulsatile luteinizing hormone release during acute fasting in rats. Endocrinology 130:3003–3006.

Cameron JL (1997). Search for the signal that conveys metabolic status to the reproductive axis. Curr Opin Endocrinol Diabetes 4:158–163.

Cameron JL, Schreihofer DA (1995). Metabolic signals and puberty in primates. In The Neurobiology of Puberty (Plant TM, Lee PA, eds), pp 259–270. Bristol, UK: Journal of Endocrinology.

Campfield LA, Smith FJ, Burn P (1996). The OB protein (leptin) pathway—a link between adipose tissue mass and central neural networks. Horm Metab Res 28:619–632.

Carro E, Pinilla L, Seoane LM, Considine RV, Aguilar E, Casanueva FF, Dieguez C (1997). Influence of endogenous leptin tone on the estrous cycle and luteinizing hormone pulsatility in female rats. Neuroendocrinology. 66:375–377.

Cavallini P, Santini S (1996). Reproduction of the red fox *Vulpes vulpes* in Central Italy. Ann Zool Fenn 33:267–274.

Cheung CC, Thornton JE, Kuijper KL, Weigle DS, Clifton DK, Steiner RA (1997). Leptin is a metabolic gate for the onset of puberty in the female rat. Endocrinology 138:855–858.

Dueck CA, Matt KS, Manore MM, Skinner JS (1996). Treatment of athletic amenorrhea with a diet and training intervention program. Int J Sport Nutr 6:24–40.

Eisenberg JF (1981). The Mammalian Radiations. Chicago: University of Chicago Press.

Foster DL, Ebling FJP, Micka AF, Vannerson LA, Bucholtz DC, Wood RI, Suttie JM, Fenner DE (1989). Metabolic interfaces between growth and reproduction. I. Nutritional modulation of gonadotropin, prolactin, and growth hormone secretion in the growth-limited female lamb. Endocrinology 125:342–350.

Foster DL, Bucholtz DC, Herbosa CG (1995). Metabolic signals and the timing of puberty in sheep. In The Neurobiology of Puberty (Plant TM, Lee PA, eds) pp 243–257. Bristol UK: Journal of Endocrinology.

Foster DL, Nagatani S, Bucholtz DC, Tsukamura H, Tanaka T (1998). Metabolic links between nutrition and reproduction: signals, sensors and pathways controlling GnRH secretion. In Pennington Symposium on Nutrition and Reproduction (Hansel W, Brayer G, eds). Baton Rouge: Louisiana State University Press.

Friedman MI (1997). An energy sensor for control of energy intake. Proc Nutr Soc 56:41–50.

Frisch RE (1994). The right weight: body fat, menarche and fertility. Proc Nutr Soc 53:113–129.

Frisch RE, McArthur J (1974). Menstrual cycles: fatness as a determinant of minimum weight for height necessary for their maintenance or onset. Science 185:949–951.

Galdikis MFB, Wood JW (1990). Birth spacing patterns in humans and apes. Am J Phys Anthropol 83:185–191.

Heideman PD, Bronson FH (1992). A pseudoseasonal reproductive strategy in a tropical rodent, *Peromyscus nudipes*. J Reprod Fertil 95:57–67.

Heini A, Schutz Y, Diaz E, Prentice AM, Whitehead RG, Jequier E (1991). Free-living energy expenditure measured by two independent techniques in pregnant and nonpregnant Gambian women. Am J Physiol 261:E9–14.

Heideman PD, Bronson FH (1992). A pseudo seasonal reproductive strategy in a tropical rodent, *Peromyscus nudipes*. J Reprod Fert 95:57–67.

Hiney JK, Srivastava V, Nyberg CL, Ojeda SR, Dees WL (1996). Insulin-like growth factor I of peripheral origin acts centrally to accelerate the initiation of female puberty. Endocrinology 137:3717–3728.

Hirvonen MD, Keesey RE (1996). The specific locus and time course of the body protein adjustments produced in rats by lesions of the lateral hypothalamus. Physiol Behav 60:725–731.

Kaiyala KJ, Woods SC, Schwartz MW (1995). New model for the regulation of energy balance and adiposity by the central nervous system. J Clin Nutr l62 suppl:1123S–1234S.

Kalra SP, Kalra PS (1996). Nutritional infertility: the role of the interconnected hypothalamic neuropeptide Y-galanin-opioid network. Front Neuroendocrinol 17:371–401.

Kenagy GL (1987). Energy allocation for reproduction in the golden-mantled ground squirrel. *In* Reproductive Energetics in Mammals (Racey PA, Loudon ASI, eds), pp 259–275. Oxford: Oxford Science.

Kenagy GJ, Barnes BM (1988). Seasonal reproductive patterns in four coexisting rodent species from the Cascade Mountains, Washington. J Mammal 69:274–292.

Kennedy GC, Mitra J (1963). Body weight and food intake as initiating factors for puberty in the rat. J Physiol 166:408–418.

l'Anson H, Foster DL, Foxcroft GR, Booth PJ (1991). Nutrition and reproduction. Oxford Rev Reprod Biol 13:239–311.

Lawrence M, Lawrence F, Coward WA, Cole TJ, Whitehead RG (1987). Energy requirements of pregnancy in Gambia. Lancet 2:1072.

Leslie PW, Fry PH (1989). Extreme seasonality of births among nomadic Turkana pastoralists. Am J Phys Anthropol 79:103–115.

Levin BE, Routh VH (1996). Role of the brain in energy balance and obesity. Am J Physiol 271:R491–500.

Loucks AB, Horvath SM (1985) Athletic amenorrhea: a review. Med Sci Sports Exerc 17:56–72.

Manning JM, Bronson FH (1990). The effects of low temperature and food intake on ovulation in domestic mice. Phys Zool 63:938–948.

Manning JM, Bronson FH (1991). Suppression of puberty in rats by exercise: effects on hormone levels and reversal with GnRH infusion. Am J Physiol 29:R717–723.

Matkovic V, Ilch JZ, Skugor M, Badenhop NE, Goel P, Clairmont A, Klisovic D, Nahas RW, Landoll JD (1997). Leptin is inversely related to age at menarche in human females. J Clin Endocrinol Metab 82:3239–3245.

Newsome AE (1966). The influence of food on breeding of the red kangaroo in central Australia. CSIRO Wildl Res 11:187–196.

Parfitt, DB, Church KR, Cameron JL (1991). Restoration of pulsatile luteinizing hormone secretion after fasting in rhesus monkeys (*Macaca mulatta*): dependence on size of the refed meal. Endocrinology. 129:7490–7556.

Pearson OP (1944). Reproduction in the shrew (*Blarina brevicauda* Say). Am J Anat 75:39–93.

Poulle ML, Crete M, Huot J (1995). Seasonal variation in body mass and composition of eastern coyotes. Can J Zool 73:1625–1633.

Prentice AM, Cole TJ (1994). Seasonal changes in growth and energy status in the Third World. Proc Nutr Soc 53:509–519.

Prentice AM, Spaaij CJK, Goldberg GR, Poppitt SD, van Raaij JMA, Totton M, Swann D, Blank RE (1996). Energy requirements of pregnant and lactating women. Eur J Clin Metab 50 (suppl 1): S82–S111.

Schneider JE, Wade GN (1999). Inhibition of reproduction in service of energy balance. *In* Reproduction in Context: Social and Environmental Influences on Reproduction (Schneider JE, Wallen K, eds). Cambridge, MA: MIT Press.

Schreihofer DA, Renda F, Cameron JL (1996). Feeding-induced stimulation of luteinizing hormone secretion in male rhesus monkeys is not dependent on a rise in blood glucose concentration. Endocrinology 137:3770–3776.

Schwartz MW, Figlewicz DP, Baskin DG, Woods, SC, Porte DJ (1992). Insulin in the brain: a hormonal regulator of energy balance. Endocr. Rev 13:387–414.

Singh J, Prentice AM, Diaz E, Coward WA, Ashford J Sawyer M, Whitehead RG (1989). Energy expenditure of Gambian women during peak agricultural activity measured by the doubly-labelled water method. Br J Nutr 62:315–321.

Steiner RA, Cameron JL, McNeil TH, Clifton DK, Brenner WJ (1983). Metabolic signals for the onset of puberty. *In* Neuroendocrine Aspects of Reproduction (Norman RL, ed), pp 183–227. New York: Academic Press.

Thompson SD (1992). Gestation and lactation in small mammals: basal metabolic rate and the limits of energy use. *In* Mammalian Energetics (Tomasi T, Horton TH, eds), pp 213–259. Ithaca, NY: Comstock.

Trayhurn P, Thomas ME, Duncan JS, Rayner DV (1995a). Effects of fasting and refeeding on *ob* gene expression in white adipose tissue of lean and obese (*ob/ob*) mice. FEBS Lett 368:488–490.

Trayhurn P, Duncan JS, Rayner DV (1995b). Acute cold-induced suppression of ob (obese) gene expression in white adipose tissue of mice: mediation by the sympathetic system. Biochem J 311:729–733.

Wade GN, Schneider JE (1992). Metabolic fuels and reproduction in female mammals. Neurosci Biobehav Rev 16:235–272.

Wade GN, Schneider JE, Li H-Y (1996). Control of fertility by metabolic cues. Am J Physiol 270:E1–E19.

Warren MP (1989). Reproductive function in the ballet dancer. *In* The Menstrual Cycle and Its Disorders (Pirke KM, Wuttke W, Schweiger U, eds), pp 161–170. Heidelberg: Springer-Verlag.

Webster AJF (1993). Energy partitioning, tissue growth and appetite control. Proc Nutr Soc 52:69–76.

Weigle DS, Duell PB, Connor WE, Steiner RA, Soules MR, Kuijper JL (1997). Effect of fasting, refeeding, and dietary fat restriction on plasma leptin levels. J Clin Endocrinol Metab 82:561–565.

Yu WH, Kimura M, Walczewska A, Karanth S, McCann SM (1997). Role of leptin in hypothalamic-pituitary function. Proc Natl Acad Sci U S A 94:1023–1028.

3 Inhibition of Reproduction in Service of Energy Balance

Jill E. Schneider and George N. Wade

Many of us are aware that our physical energy levels can influence our sexual desire and fertility (Wade et al. 1996; Juiter et al. 1993; Stewart 1992; White et al. 1990; Kusche et al. 1987). This association can hardly go unnoticed in modern western industrialized countries where cultural expectations for women during the past three decades have prescribed either a lean, athletic build or a waiflike, peripubertal body type (Schwartz 1986; Banner 1983; Pugliese et al. 1983; Garner et al. 1980). Reports of menstrual irregularities, amenorrhea (the cessation of menstrual cycles), diminished sexual desire and activity, and infertility are common in categories of women who restrict their food intake, for example fashion models and women with the eating disorder *anorexia nervosa* (Loucks et al. 1994; Gardiner et al. 1991; Katz and Weiner 1991). Similar symptoms are reported in athletes and dancers, particularly those who do not increase their food intake to compensate for the unusual energetic demands of their training schedules (Loucks et al. 1992; Abraham et al. 1982; Dueck et al. 1996). The feminine body type favored in prehistoric human societies appears to have been quite different from that prescribed for modern women. The link between fertility and energy balance is quite apparent in the earliest known sculptures of the human body (figure 3.1). The Venus of Willendorf and the Goddess of Lespugue are thought to be representations of the fertility goddesses of human societies that existed over twenty thousand years ago in geographic areas that are now in the countries of Austria and France. The most striking features of these fertility figures are the ample body fat depots, particularly those located in the breasts, hips, and thighs.

If we look beyond our own species, there can be no doubt that the link between energy balance and fertility is a taxonomically widespread, biologically relevant phenomenon. The link between food intake and fertility was central to the theory of evolution by natural selection; food availability was one of the first selection pressures postulated by Charles Darwin (Darwin 1859). Darwin noticed that lack of food and hard living influenced the fertility of free living relatives of domestic animals. He reasoned that wild species, unlike their domesticated counterparts, must "struggle for their food ... during some seasons" and that "individuals of the same species, having different constitutions" would reproduce differentially under what he termed "natural selection." A related observation made by Darwin is still an unsolved problem in our modern meat and dairy industry. Energetically costly activities such as milk production and rapid growth are associated with delayed puberty and longer breeding intervals in domestic cattle.

Almost a century of laboratory research has documented that fertility is impaired when food intake is restricted or when energy expenditure is increased by enforced

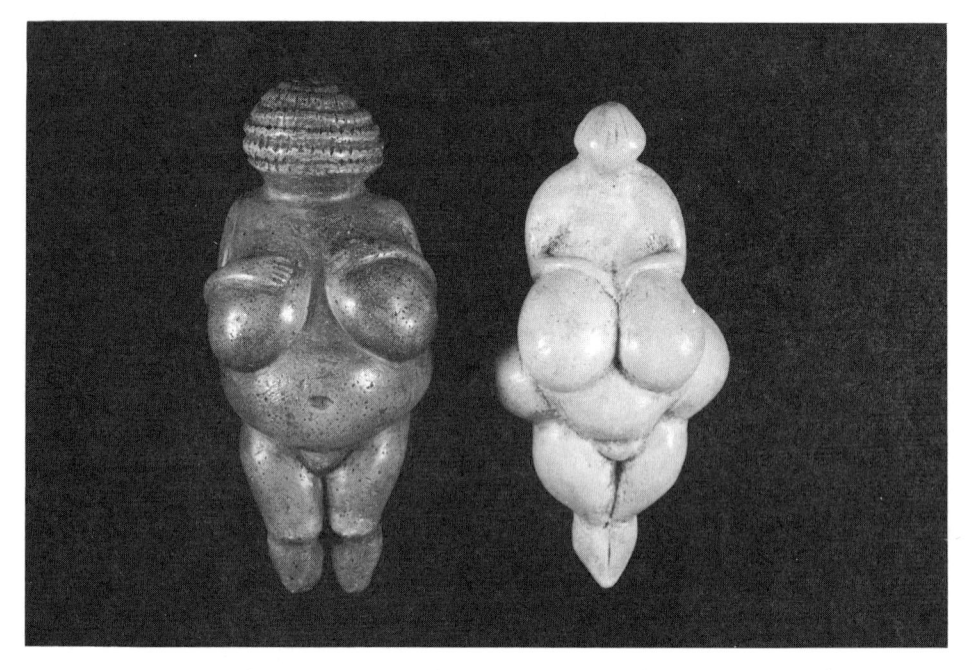

Figure 3.1
The Venus of Willendorf (left) and the Goddess of Lespugue (right) are thought to be representations of the fertility goddesses of human societies that existed over twenty thousand years ago in geographic areas that are now the countries of Austria and France. The most striking features of these fertility figures are the ample body fat deposits, particularly those located in the breasts, hips, and thighs. (Photo: S. Lichak)

exercise or housing at cold temperatures. In species representative of all mammalian orders, including primates such as ourselves, ovulatory cycles are lengthened or interrupted by inadequate food intake or by excessive energy expenditure that is not offset by compensatory increases in food intake (Bronson 1989; l'Anson et al. 1991; Wade and Schneider, 1992; Wade et al. 1996).

Despite clinical, economic, and biological interest in the phenomenon, little is known about the mechanisms that underlie nutritional infertility. In chapter 2, Bronson notes that years of research on a variety of species living in natural habitats has yielded little specific information about the signals, detectors, and pathways involved in the onset of reproductive maturity. Laboratory research has provided a larger quantity of data, but those data are often presented within an equally large degree of confusion concerning the primary sensory stimuli, sensory detectors, the metabolic and endocrine sequelae to stimulus detection, and the effector pathways. Over the past forty years, at least twenty different putative signals for nutritional infertility have been proposed. Some of these factors are aspects of body composition, such as adipocyte size and number, or whole-body protein. Hormonal candidates include glucocorticoids, insulin, and recently, leptin. Neurotransmitter and gut peptide candidates include neuropeptide Y (NPY), corticotropin-releasing factor (CRF), pro-opiomelanocortin (POMC) gene products, norepinephrine (NE), and cholecystokinin (CCK). Metabolic substrates such as glucose, free fatty acids (FFAs), and amino acids (AAs) also have been proposed as signals. For many of the factors on the list, the discovery that each could affect reproduction was often followed by a declaration that this factor might be the critical signal for nutritional infertility, creating a new bandwagon onto which investigators in the field predictably leapt. They tested the effects of the new factor on reproduction in their own animal model until the discovery of the next factor initiated the next bandwagon.

The parade of putative signals marches on as new hormones and peptides are discovered, and as technological advances provide new measures of the neuroendocrine events that underlie reproduction. It is likely that glucagon-like peptide 1 (GLP-1), urocortin, alpha-melanocyte stimulating hormone, orexin (hypocretin) and cocaine- and amphetamine-regulated transcript (CART) will appear next, if they have not already by the time this book is published. However, adding putative signals and targets to the list does not necessarily deepen our understanding of the phenomenon. We need to organize this list into those factors that are the primary sensory signals and those factors that change in response to the primary sensory signal. The primary sensory signal is generated by metabolic events that occur when animals fall into negative energy balance. Most of the factors listed above are not primary sensory signals; they change *in response to* the primary sensory signal. For example, circulating insulin levels fall when animals are food deprived. Primary sensory signals generated intracellularly during metabolic fuel oxidation link energy balance to mechanisms that control insulin secretion (MacDonald 1990; Prentki et al. 1997). Factors downstream from the primary sensory signal, that is, neurotransmitters, may be viewed as mediators, modulators, or nonfunctional correlates. While we continue to add secondary factors such as hormones and neurotransmitters to the list of putative "metabolic signals," the primary sensory signal remains largely unexamined. This analysis echoes Mark Friedman's assessment of current trends in the study of food intake.

It almost seems that there is the expectation that a clear understanding of food intake will materialize when enough controls are added to the list. The opposite, however, is probably true: the longer the list, the less we really know. The growing roster betrays a lack of integration and synthesis, of understanding and explanation. (Friedman 1990, p. 513)

The analogy to the study of food intake is particularly appropriate because both food intake and reproduction are regulated in service of energy balance. Energy balance refers to the difference between energy intake and expenditure. Regardless of whether the organism has just eaten or is starving, blood levels of metabolic fuels and cellular adenosine triphosphate (ATP) charge do not fall below certain limits. Animals continually expend energy to live, and therefore must ingest and metabolize food. Similarly, when food availability is limited, some types of energy expenditure can be curtailed in order to maintain energy balance. A myriad of metabolic adaptations occur to conserve energy during seasonal changes in climate and food availability. When food is abundant and energetic demands are minimal, energy is available for the cellular processes necessary for survival, as well as for growth, reproduction, and other activities. However, when food is scarce or energetic demands are high, activities necessary for individual survival, especially eating, will often take precedence over processes such as growth and reproduction. In fact, most of the factors that increase hunger and food intake also inhibit reproduction in a wide variety of species.

Based on these interrelations between food intake, energy balance, and reproduction, we have argued previously that research on nutritional infertility should be guided by hypotheses and experimental paradigms that have been successful in the study of ingestive behavior (Wade and Schneider 1992; Wade et al. 1996). In this chapter we reiterate this theme with the caveat that some ideas and methodologies have been more useful than others. The addition of new putative signals may be less important than a hypothetical framework and a set of experimental methodologies that allow us to categorize these factors as either primary sensory signals, secondary and tertiary intermediates, or nonfunctional correlates. A critical lesson from the literature on food intake and energy balance is that the metabolic hormones, neurotransmitters, fuels, and measures of body composition all rise and fall together during feast and famine, winter and summer, activity and inactivity. In chronically well-fed animals, body fat stores, lipogenic enzymes, levels of insulin and leptin, glucose and AAs available from the diet, and liver glycogen are all relatively high. Levels of glucagon, ketone bodies, lipolytic enzymes, circulating FFAs in circulation, and hypothalamic NPY, CRF and catecholamines in certain brain nuclei are comparably low. In fasted animals, relative levels of all of these substances are reversed. It is no surprise that all of these changes appear to be correlated with changes in reproduction and food intake. The correlations alone do not provide information about causation.

As Friedman has explained (Friedman 1990), we can make two assumptions about cause and effect relationships. First, just as food intake and reproduction are regulated in service of energy balance, so are the hormones, neurotransmitter, and glucose levels. They all respond to metabolic signals, so, by strict definition, they themselves are not the primary sensory signals. Rather than acting as primary sensory signals, these factors may convey information about primary sensory events to the central nervous system (CNS), and thus modulate neural, behavioral, and metabolic responses. Thus, our organizational framework categorizes these hormones as mediators or modulators of the primary sensory event. Second, exogenous treatment with hormones, in either brain or periphery, often affects metabolic events that serve as primary sensory stimuli. We will describe the effects of hormones such as insulin and leptin on metabolic fuel availability and oxidation. Hormones and neurotransmitters, whether they act in brain or periphery, can have measurable effects on energy metabolism, and that these changes in energy metabolism can explain most changes in food intake (Friedman 1978, 1995, 1998, 1991). These are two of the most important ideas from the study of ingestive behavior that may be applied to the study of reproduction.

In this chapter we attempt to provide a logical framework for understanding previously collected data and for designing new experiments. This framework allows the organization of factors that affect reproduction into three categories: (1) primary sensory signals, (2) intermediates that affect reproduction directly, and (3) intermediates that affect reproduction indirectly by changing the primary sensory signals.

The first three sections of this chapter provide background material necessary for understanding the experimental data. First, we provide basic information about the changes in metabolic substrates, hormones, and neurotransmitters that occur as animals fluctuate between positive and negative energy balance. Second, we give a brief overview of the neuroendocrine events that can be affected by changes in energy balance. Third, we provide a description of the utility and significance of various animal models that can be used to study nutritional infertility. After these preliminary sections, we evaluate the different hypotheses that have been proposed to explain the link between energy balance and reproductive function. The hypotheses are grouped into three categories for analysis: the lipostatic, hormonal, and metabolic hypotheses. After each of these different hypotheses is evaluated, we describe our own hypothetical framework designed to organize the factors that affect reproduction into either primary sensory signals, mediators that affect reproduction directly, and intermediates that affect reproduction indirectly by changing the primary sensory signals.

Metabolic Interrelations During Feeding and Fasting

Reproduction is inhibited in service of energy balance. Therefore, a deeper understanding of reproductive inhibition might be facilitated by developing a better understanding of the different metabolic events that occur in the well-fed compared to the fasted condition. In the "well-fed" or "storage mode" of metabolism (figure 3.2), the diet supplies a large portion of the energy requirements (Harris and Crabb 1992). Glucose and AAs pass from the intestine directly into the blood and are delivered to the liver, nervous system, muscles, and other organs. The liver is primarily concerned with the conversion of fuels, primarily lactate (indirect pathway), and to some extent

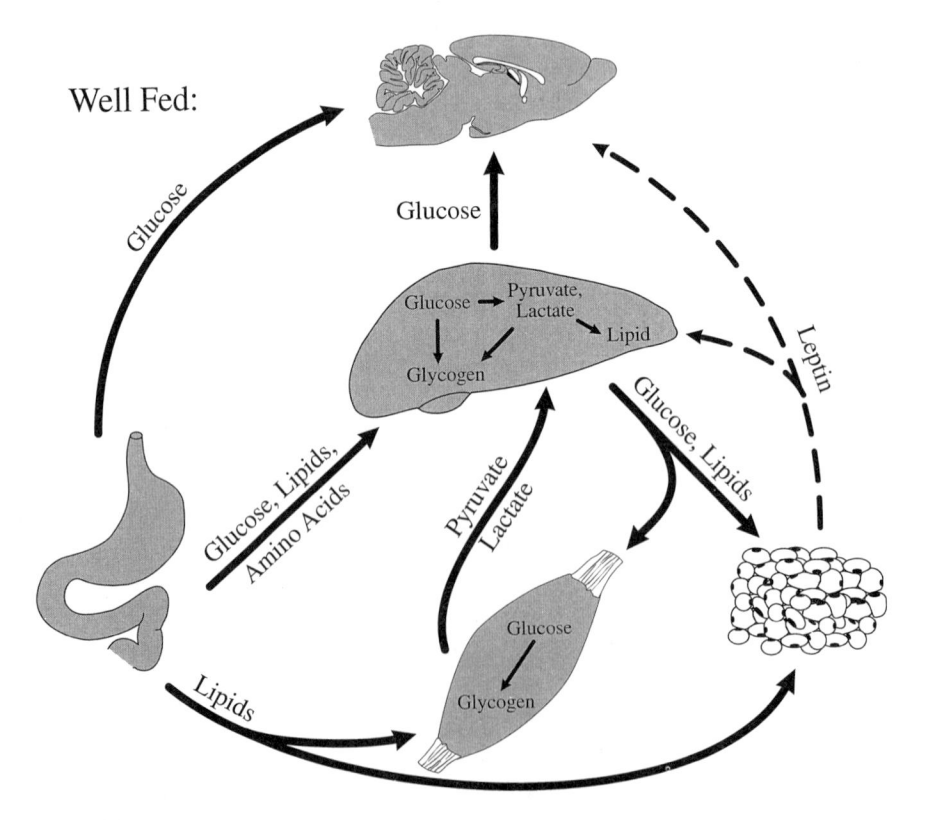

Figure 3.2
Metabolic consequences of ingested macronutrients (glucose, amino acids, and fat) by various tissues in the well-fed state.

glucose (direct pathway), into stored glycogen. Other tissues oxidize glucose and store the excess. Storage involves the conversion of glucose, lactate, and pyruvate into fat by the process known as lipogenesis in white adipose tissue, as well as in liver and other organs. Peripheral metabolism of the nutrients is aided by the release of insulin from beta cells in the pancreas. Insulin secretion is highest after meals. Insulin is important for glucose uptake in peripheral cells, and when glucose levels are high, this hormone promotes lipogenic processes. In the well-fed phase of metabolism, pancreatic secretion of glucagon is at its nadir.

In the early stages of fasting, maintenance of blood glucose is achieved by hepatic glycogenolysis, the breakdown of glycogen into glucose (Harris and Crabb, 1992). Lipogenesis is curtailed and, if fasting continues, lactate, pyruvate, and AAs are diverted into formation of glucose by the process of gluconeogenesis in liver (figure 3.3). Pancreatic insulin and adipocyte leptin secretion fall to relatively low levels, while pancreatic glucagon and adrenal glucocorticoid secretion rise. Lipolysis, the process by which triglycerides are broken down to glycerol and FFAs, may be initiated largely owing to the fall in insulin. FFAs, primarily from white adipose tissue, are released into blood, providing an oxidizable substrate for peripheral tissues. A portion of the FFAs are transported to the liver, where they can be oxidized completely, or partially oxidized and used in formation of ketone bodies. During prolonged fasting, ketone bodies can be oxidized to provide ATP for cellular processes in various tissues. During the fed condition, the brain is almost solely dependent on glucose for production of ATP. The uptake of glucose in brain can occur with or without the presence of insulin. However, during starvation, ketone bodies synthesized in liver can be transported through the blood-brain barrier to be oxidized by cells in the central nervous system (CNS), albeit after a major metabolic adjustment by these CNS cells. Other fasting-induced changes include increased NPY and CRF, secretion in brain, particularly the paraventricular nucleus of the hypothalamus and median eminence, increased adrenocorticotropic hormone (ACTH) release from pituitary, increased catecholamine release in central and autonomic neurons, and increased catecholamine release from the adrenal glands. When levels of oxidizable fuels reaching the brain become dangerously low, the cerebral metabolic emergency induces a compensatory sympathoadrenal response aimed at mobilizing oxidizable fuels. This condition is characterized by increased secretion of catecholamines and a consequent sharp rise in plasma glucose levels (hyperglycemia). All of these changes facilitate the metabolic responses necessary to maintain caloric homeostasis.

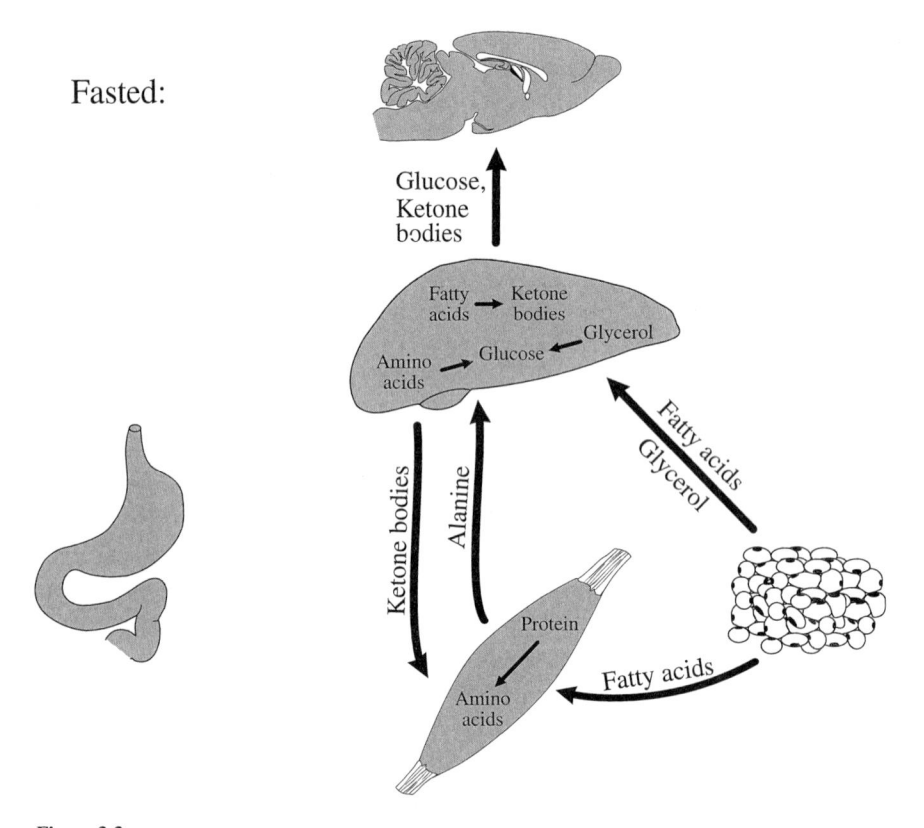

Figure 3.3
Metabolic interrelationships of the major tissues of the body in the fasting state.

The Loci of Metabolic Effects

Metabolic stimuli, hormones, neurotransmitters, and neuromodulators have multiple effects on reproductive processes. As illustrated in figure 1.1 (chapter 1), inhibition may occur in the hypothalamus by affecting the pulsatile secretion of gonadotropin-releasing hormone (GnRH). Food deprivation decreases the pulse frequency of hypothalamic GnRH secretion (Foster et al. 1995; Foster and Bucholtz 1995), and consequently decreases the frequency of pulsatile luteinizing hormone (LH) secretion (Foster et al.1998; Cameron 1997; Piacsek 1996).

Negative energy balance is commonly thought to have its primary effect on the GnRH pulse generator. Normal ovulatory cycles can be reinstated by treatment with artificial pulses of GnRH in fasted rats, sheep, pigs, cows, monkeys, and women

(Bronson 1986; Cameron 1996; Day et al.1986; Foster and Olster, 1985; Kile et al. 1991; Manning and Bronson, 1991; Cameron and Nosbisch 1991; Nillius et al. 1975; Armstrong and Britt 1987). However, as with any artificial hormone treatment, it is not known whether GnRH treatment simulates natural endogenous secretion, or results in artificially elevated levels of GnRH that might mask a decreased sensitivity to the peptide.

It is unlikely that negative energy balance affects GnRH synthesis, because in at least a few species GnRH messenger RNA (mRNA), GnRH content, and the number of GnRH immunoreactive neurons are not decreased, and in some cases may be increased by energetic challenges (l'Anson et al. 1997; Ebling et al. 1990; McShane et al. 1993). In sheep and rats, energetic challenges also fail to decrease pituitary LH and follicle stimulating hormone (FSH) content, or message (Bronson 1988; Beckett et al. 1997a).

Reproductive inhibition by energetic challenges may occur at the level of the pituitary by changing sensitivity to GnRH (Day et al. 1986; Booth 1990; Beckett et al. 1997a; l'Anson et al. 1997), particularly under severe energetic challenges. For example, estradiol treatment enhances pituitary GnRH receptor as well as mRNA for the receptor in well-fed and moderately food-restricted sheep, but this estradiol-induced enhancement is absent in severely food-restricted sheep (Beckett et al. 1997a).

In many species, it is evident that inhibition of GnRH and gonadotropin secretion may be due either to heightened sensitivity to steroid-negative feedback in brain or pituitary or both (Cagampang et al. 1990; Cagampang et al. 1991; Nagatani et al. 1994; Spillar and Piacsek, 1991; Howland and Ibrahim, 1971; Sprangers and Piacsek 1988; Piacsek 1985; Foster and Olster 1985; Beckett et al. 1997b) or to steroid-independent effects (Foster, Olster, 1985; Nagatani et al. 1996b; Bronson 1988; Murahashi et al. 1996). It has been suggested that increased sensitivity to steroids might be due to increases in receptors for those steroids. Consistent with this idea, metabolic challenges are associated with increased estrogen receptor immunoreactivity in the medial preoptic area (mPOA) in Syrian hamsters (Li et al. 1994), and in the paraventricular nucleus (PVN) of the hypothalamus in hamsters and rats (Estacio et al. 1996a). These brain areas either contain GnRH neurons or contain nuclei that project to GnRH neurons. It has been suggested that the essential estradiol-negative feedback may occur in the PVN, because microinfusion of estradiol into this area accentuates fasting-induced suppression of pulsatile LH secretion in ovariectomized rats (Nagatani et al. 1994). Changes in energy intake and expenditure seem to affect the frequency, and sometimes the amplitude of GnRH pulses and consequent gonadotropin secretion via changes in secretion of neurotransmitters such as NPY, NE, opiates, and CRF. Many of these neuropeptides increase sensitivity to steroid-negative feedback (Nagatani et al. 1996b; Cagampang et al. 1992; Maeda et al. 1994; McShane et al. 1993; Kalra and Kalra 1984; Goubillon and Thalabard 1996; Cosgrove et al. 1991). In addition, in some species, it has been demonstrated that energetically challenged animals show heightened sensitivity to estradiol-positive feedback (Sprangers and Piacsek, 1998; Mangels et al. 1996). It is conceivable that enhanced positive feedback might lead to premature LH surges that fail to induce ovulation because they occur prior to adequate follicle development. An additional or alternative reason for anovulation might be the suppression of the FSH surge that occurs in food restricted rats (Sprangers and Piacsek, 1998). Reduced or suppressed FSH surges may be at least one contributor to retarded follicular development and anestrus (Schwartz 1974; Richards and Midgley 1976; Hirschfield and Midgley 1978).

Energetic challenges also have effects on the central neural mechanisms that control courtship and sex behavior and that are independent of the effects of these challenges on GnRH, LH, FSH, follicle development, and steroid secretion (Li et al. 1994; Dickerman et al. 1993; Panicker et al. 1998; Panicker and Wade 1998; Siegel and Wade 1979). Changes in the behavioral responsiveness to steroids might be related to the decrease in estrogen receptors in areas critical for estrous behavior, such as the ventromedial hypothalamus (VMH) (Li et al. 1994; Panicker and Wade, 1998; Roemmich et al. 1997; Estacio et al. 1996b; Estacio et al. 1996a).

Details of the specific neuroendocrine loci whereby environmental factors inhibit reproduction have been covered in most of the chapters in this book and have been the subject of a number of recent reviews (Cameron 1997; Foster et al. 1998; Piacsek 1996). The aim of this chapter is to direct attention to the primary metabolic sensory stimuli and detectors. Consequently, the details of the neuroendocrine consequences of changes in energy balance are not discussed here.

Animal Models

Nutritional infertility has been studied in three main experimental models: nutritionally delayed puberty (Kennedy and Mitra 1963; Hiney et al. 1996; Foster 1994; Foster et al. 1995; Cameron 1991; Cameron and Schreihofer 1995), nutritionally prolonged lactational diestrus (Walker et al. 1995; Woodside et al. 1998b; Woodside et al. 1998a), and fasting-induced inhibition of estrous cycles (anestrus) (discussed throughout this chapter). Nutritionally delayed puberty has been studied in rats, sheep, and monkeys that have been food-restricted just prior to the onset of sexual maturity. As Lee and Gorman illustrate in chapter 7, a number of environmental variables such as food availability, photoperiod, and social influences converge to synchronize first ovulation and estrus with favorable environmental conditions, thereby increasing the likelihood of successful reproduction, weaning, and offspring survival to reproductive maturity. Nutritionally delayed lactational diestrus has been studied by subjecting lactating rats to 2 weeks of food restriction or 2 days of fasting. These manipulations result in a significantly lengthened interval between parturition and next ovulation and estrus (Walker et al. 1995; Woodside et al. 1998b). The effects of restriction on the length of lactational diestrus are not due to lack of single macronutrients in dams or to changes related to offspring intake or behavior. Rather, they can be accounted for by the overall energy balance of the dams (Woodside et al. 1998a). Fasting-induced anestrus is studied by subjecting estrous cycling females to metabolic challenges that delay the next ovulation and estrus. Food restriction or food deprivation inhibits estrous cyclicity and other aspects of fertility and behavior in adults of a number of species (for review, see Wade and Schneider 1992 and l'Anson et al. 1991). In addition, the effects of energetic challenges on sex behavior, independent of effects on the ovulatory cycle, can be studied in animals that are ovariectomized and brought into estrus by exogenous treatment with estradiol and progesterone. Exogenous hormone treatment induces sex behavior in fasted animals. At least some aspects of sex behavior, such as the duration of lordosis, are significantly compromised by fasting and other energetic challenges. From these data it might be expected that the exogenous doses of steroid hormones required for lordosis would be higher in fasted, ovariectomized hamsters than in ovariectomized hamsters fed ad lib. If so, it would be concluded that fasting decreases neural sensitivity to steroid hormones.

It might be argued that the neuroendocrine mechanisms that deserve the most attention are those that would be operating in animals faced with energetic challenges in their natural habitats. For many short-lived species, the reproductive stage that is most likely to be interrupted is either timing of first estrus or the length of lactational diestrus. This is because females of species with short life spans generally do not undergo a long series of estrous cycles in which they do not conceive. They become pregnant soon after puberty, and thereafter reproductive success is largely defined by the interval between gestation and the next conception (in addition to factors such as litter size and offspring survival to reproductive maturity). These considerations suggest that the delay of puberty by energetic challenges and the length of lactational diestrus are the more ecologically relevant measures of reproductive regulation, at least in some species.

In other species, such as human beings and other primates, sex behavior and fertility are more often dissociated. In women, sexual activity may persist when fertility is curtailed by tubal ligation, ovariectomy, hormone treatment, or various birth control methods. However, sexual desire and activity, as well as the neuroendocrine mechanisms that control menstrual cyclicity, are both adversely affected by metabolic challenges (Loucks et al. 1994; Dueck et al. 1996; Juiter et al. 1993; Rabach and Faltus 1991; Kusche et al. 1987). Modern commercial breeding of dairy cattle relies almost exclusively on artificial insemination, a process that requires robust estrous behavior from females so that the introduction of sperm coincides with ovulation. However, in stocks with high milk yield, ovulation often occurs without normal estrous behavior, making it difficult or impossible to successfully time artificial insemination. From the basic research, clinical, and commercial perspectives, all of the above considerations reveal that it is important to understand the mechanisms that control sex behavior as processes independent of the mechanisms that control ovulatory cycles. To this end, it is important to examine data and conclusions that arise from some of the more convenient animal models in a variety of species under a variety of natural and seminatural conditions.

Fasting-induced anestrus in Syrian hamsters (*Mesocricetus auratus*) is an experimental model unsurpassed in its convenience for the experimenter. Unlike other commonly studied laboratory rodents, Syrian hamsters display 4-day estrous cycles that are exceptionally predictable and easy to monitor using noninvasive external indices. For example, the first day of the cycle, immediately after the termination of estrus, is marked by the appearance of a conspicuous vaginal discharge (Orsini 1961). On the fourth day, beginning just prior to the onset of the dark phase of the photoperiod and lasting until the following morning, females show stereotypical estrous behavior (lordosis) promptly upon being introduced to a sexually experienced male. Ovulation occurs during the evening of day 4 of the cycle. Most relevant for the present discussion, Syrian hamsters' estrous cycles are highly sensitive to food deprivation and other treatments that decrease availability of metabolic fuels for oxidation, but are typically not affected by nonenergetic stressors such as routine handling, repeated injections, or gastric intubations (Morin 1975; Morin 1986). Morin developed a paradigm that opened the doors to a whole new line of research on reproduction (Morin 1986). Female hamsters were food deprived on days 1 and 2 of the estrous cycle, and fed ad lib on days 3 and 4. This 48-hour period of food deprivation blocked the next expected period of estrus and ovulation and the postovulatory vaginal discharge. Further research showed that this fasting regimen decreased concentrations of plasma estradiol, inhibited the LH surge (Morin 1986), enhanced the responsiveness to the positive effects of estradiol (Mangels et al. 1996), decreased activation of GnRH immunoreactive cells in the mediobasal hypothalamus and caudal preoptic area (Berriman et al. 1992), decreased lordosis duration in ovariectomized females treated with estradiol (Dickerman et al. 1993; Li et al. 1994), decreased VMH, and increased mPOA estrogen receptor immunoreactivity (Li et al. 1994).

Data resulting from the use of Morin's paradigm will be referred to again and again throughout this chapter. Fasting-induced anestrus is defined as an inhibition of lordosis, ovulation, and the postovulatory vaginal discharge after a 48- to 60-hour fast. Experimental treatments, such as fasting, feeding single macronutrients, or pharmacological treatments, are administered during the first 48 hours of the estrous cycle. Estrous behavior is measured on the evening of day 4, and vaginal discharge and tubal ova are monitored on day 1 of the next expected estrous cycle. To examine sex behavior independent of the effects of fasting on estrous cyclicity, ovariectomized hamsters are measured for the duration of lordosis in response to a male or to manual stimulation by the experimenter after a 48-hour fast, and 48 hours after injection of estradiol at a standard dose known to induce lordosis.

Three Types of Hypotheses

A number of ideas have been proposed over the years to explain the effects of energy balance on reproduction. We have grouped these hypotheses into three categories: adipostatic, hormone, and metabolic. Adiposity, hormones, and the availability and oxidation of metabolic fuels all influence reproduction. However, the articulation and evaluation of these distinct hypotheses might aid in distinguishing primary sensory signals from secondary mediators, modulators, and correlates.

The first type of hypothesis is the *lipostatic* or *adipostatic hypothesis*, in which the critical signal that controls reproduction is the level of energy reserves in the form of body fat. Central to this hypothesis is the concept of a signal, often proposed to be a hormone, that reports the level of body fat directly to the mechanisms that control reproduction (figure 3.4). As body fat levels rise, secretion and plasma levels of some blood-borne factor also rise. Critical levels of this factor in plasma or in cerebrospinal fluid are postulated to either facilitate or permit reproductive function. Body fat content is the putative primary signal.

The second category is the counterregulatory *hormone hypothesis*. According to this idea, levels of one or a number of hormones reflect the immediate state of energy balance (not the body fat content per se). It is usually suggested that the hormone reports the state of energy balance directly to brain mechanisms controlling reproduction by binding to neural receptors (figure 3.5). Levels of hormones are the putative primary signals.

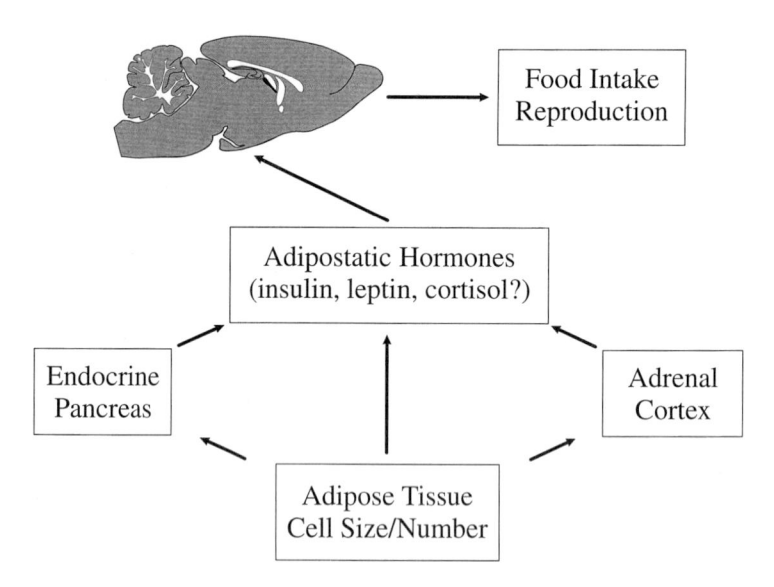

Figure 3.4
Adipostatic hypothesis: The adipostatic signal reports current levels of body fat content directly to hypothalamic mechanisms that control gonadotropin-releasing hormone secretion and food intake. Over the years, the adipostatic signal has been postulated to be a succession of hormones such as insulin and leptin.

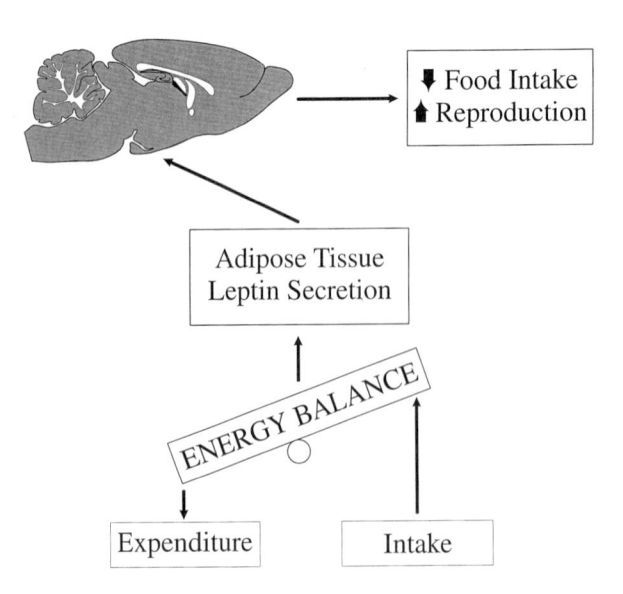

Figure 3.5
Hormonal hypothesis: In this case, the levels of the hormone report current levels of energy availability, rather than levels of energy storage. This hypothesis incorporates data showing that hormones such as insulin and leptin change rapidly with energetic challenges such as fasting and feeding.

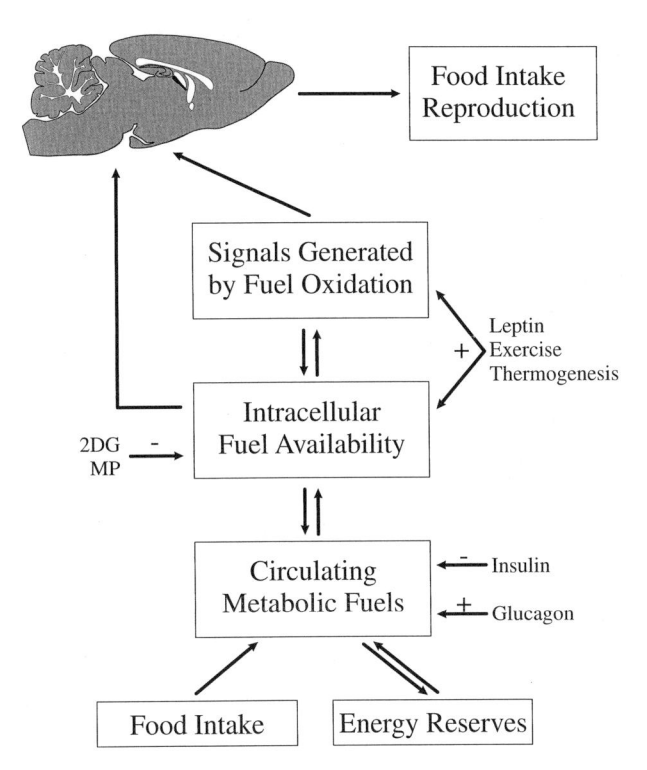

Figure 3.6
Metabolic fuels hypothesis: Reproduction and food intake are responsive to increases and decreases in metabolic fuel oxidation. Fuel oxidation depends on intracellular availability of oxidizable metabolic fuels. Increases in fuel availability and oxidation generate signals that facilitate or permit normal reproduction. These signals may be generated by increases in levels of intermediates, enzymes, or adenosine triphosphate. Hormones may have their effects on reproduction by changing intracellular fuel availability and oxidation. Hormone synthesis and secretion may be influenced by the intracellular availability and oxidation of fuels. 2DG, 2-deoxy-D-glucose; MP, methylpalmoxirate.

A third type of hypothesis is the *metabolic hypothesis* in which the putative primary signal is generated by changes in the availability and oxidation of metabolic fuels. The primary signal controls hormone levels and may also influence the hypothalamic-pituitary-gonadal axis either directly or indirectly via neural or humoral secondary signals (figure 3.6).

The hormonal and metabolic hypotheses must share in common the notion that counterregulatory hormone levels are controlled by signals generated by fuel oxidation. However, investigators studying the reproductive effects of hormones such as insulin and leptin often fail to articulate the concept that levels of these hormones are controlled by a primary metabolic sensory signal. The central tenet of the metabolic hypothesis is that both counterregulatory hormone secretion and reproductive function are controlled by a primary sensory signal generated by changes in the availability and oxidation of metabolic fuels. This is emphasized because articulation of the hypothesis is a prerequisite to testing the hypothesis.

Some investigators have suggested that the levels of circulating metabolic fuels might serve as a signal for control of reproduction. Levels of circulating fuels are subsumed under the metabolic hypothesis because signals related to energy availability and oxidation may be affected by changes in circulating metabolic fuels. However, for reasons discussed in this chapter, the primary sensory signal is probably located inside the cell. The metabolic hypothesis posits an intracellular signal generated as fuels are oxidized in the biochemical processes that lead to formation of ATP. The metabolic hypothesis is unique in that it brings the search for the metabolic signal inside the cell, and perhaps inside the mitochondria.

The Lipostatic Hypothesis

Of the great array of variables that are correlated with reproductive function, body fat content or the fat-to-lean ratio has received the most attention in clinical practice and in the popular press (see figure 3.3). The lipostatic idea stems from the observation that food intake is often correlated with body fat content, and body fat content is correlated with measures of pubertal development in women, domestic animals, laboratory rats, and most other species (Kennedy and Mitra, 1963; Reed et al. 1988; Frisch 1990). Based solely on these correlations, it was hypothesized that when levels of adiposity fall below a critical threshold, a signal is generated and sent to the hypothalamic-pituitary-gonadal system thereby inhibiting ovulatory cycles. The key feature of this hypothesis is the notion that the circulating levels of some substance, most often presumed to be a hormone, can accurately reflect the level of adiposity. For many years, it was proposed repeatedly that the signal was estrogen generated by aromatization of androgens in adipocytes (Frisch 1990). However, a coherent mechanism whereby decreased estrogen from adipocytes could interact with the normal neuroendocrine events of the ovulatory cycle was never supported. Discovery of the *ob* protein leptin has led some investigators to revisit the lipostatic hypothesis.

Numerous theoretical and empirical pitfalls in the lipostatic hypothesis have been discussed at length in several different research articles, review articles, book chapters, and letters to the editors of various journals (Scott and Johnston 1982; Bronson and Manning 1991; Bronson 1989; Schneider and Wade 1989b; Schneider and Wade 1997; Van Ittalie and Kissileff 1990; Wade and Schneider 1992). The correlations between adiposity and fertility are not consistent over all species under all energetic conditions. As more data have been accumulated, the correlations argue against the idea that a critical level of body fat is the exclusive, overriding signal that permits reproduction (Schneider and Wade 1989a; Schneider and Wade 1990b; Bucholtz et al. 1996; Bronson and Manning 1991; Cameron and Nosbisch 1991; Wade et al. 1991).

In chapter 2, Bronson points out that most species living in a wild habitat reproduce when they are in positive energy balance, in most cases without regard to any particular level of body fat. This chapter provides a number of examples of laboratory experiments designed to dissociate body fat content from its many metabolic and endocrine correlates. Bronson was one of the first to show that the normal frequency of LH pulses was decreased by food deprivation and was restored within hours of a meal (see figure 2.1). The effect of the meal was far too rapid to be related to an accumulation of body fat. Similarly, rapid effects of meals and infusion of metabolic fuels on pulsatile LH secretion have been shown in male and female cattle, sheep, swine, and monkeys (McCann and Hansel 1986; Cameron and Nosbisch 1991; Mattern et al. 1993; Schreihofer et al. 1996; Bucholtz et al. 1993). All of these experiments consistently demonstrate that a threshold level of body fat content is not necessary for normal frequency and amplitude of LH secretion. The level of body fat or the fat-to-lean ratio is not a proximate cause, but rather, one of many contributing factors to metabolic fuel availability.

As might be predicted from the well-known correlations between body fat content and reproductive function, fasting-induced anestrus occurs in hamsters with a low body fat content, but does not occur in those with a relatively high body fat content prior to fasting (Schneider and Wade 1989; Schneider and Wade 1990b). Hamsters with a relatively low body fat content were either fasted or fed a single macronutrient, either sugar or fat, instead of their regular chow. Weight loss among the groups that were fasted or fed a single macronutrient did not differ significantly. However, hamsters allowed to ingest a single metabolic fuel such as glucose (a 30% solution of glucose) or triglycerides (pure vegetable shortening) continued to show 4-day estrous cycles, while those that received no dietary supplements became anestrous (Schneider and Wade 1990b). Thus, in Syrian hamsters, as well as rats, sheep, and monkeys, a particular body weight is not necessary for normal reproductive function.

The results were extended to show that a particular body weight and fat content is not sufficient for normal estrous cyclicity. Previous work showed that systemic treatment with 2-deoxy-D-glucose (2DG) or methyl palmoxirate (MP) significantly increased food intake in rats by inhibiting oxidation of glucose or fatty acids, respectively (Ritter and Taylor 1990; Scharrer and Langhans 1986; Friedman et al. 1986; Friedman and Tordoff 1986). In the first experiment, we examined the effects of 2DG and MP on estrous cyclicity in hamsters fed ad lib. Treatment with 2DG at high doses (greater than 1750 mg/kg) every 6 hours on days 1 and 2 of the estrous cycle induced anestrus without a significant decrease in food intake or loss of body fat in Syrian hamsters fed ad lib. (Schneider et al. 1997a). In contrast, we found no dose of MP that induced anestrus without causing a hibernation-like torpor.

The effects of these drugs were also examined in hamsters with body fat contents high enough to prevent fasting-induced anestrus. Fat hamsters were fasted and treated with either 2DG or MP twice a day at low doses that did not induce anestrus in hamsters fed *ad lib*. While the fat, fasted, vehicle-treated hamsters showed normal estrous cycles, fat, fasted hamsters treated with MP or 2DG became anestrous (Schneider and Wade 1989a) (figure 3.7). These experiments demonstrated again that the effects of adiposity could be dissociated from the effects of changes in the oxidation of fuels. This idea was confirmed by a great deal of research showing that a high body fat content was not sufficient for normal reproductive function when fuel oxidation was limited in rats, sheep, cows, and monkeys (Bucholtz et al. 1996; Nagatani et al. 1996a; Murahashi et al. 1996; Funston et al. 1995; Heisler et al. 1993; Chen et al. 1992). In addition to ruling out a proximate role for critical body fat levels, these experiments led us and other investigators to examine the role of oxidizable metabolic fuels, as discussed later in this chapter.

The Hormonal Hypothesis

In the lipostatic hypothesis discussed above, the hormonal signal reports information about the level of energy reserves in the form of body fat (see figure 3.3). A second type of hypothesis holds that hormones report information about the state of energy balance, rather than the level of body fat content (see figure 3.4). In this part of the chapter we discuss the evidence that levels of hormones such as insulin and leptin are rapidly changing indicators of energy intake and expenditure, rather than slowly changing indices of adipocyte filling. In addition, we illustrate examples of experimental paradigms that have attempted to separate the direct effects of hormones from their metabolic consequences.

Insulin A number of investigators have suggested that hormones such as insulin serve as important signals controlling food intake and body weight. These investigators posit that tonic levels of central insulin serve as signals that keep body weight and food intake in check, and that obesity may result from impaired sensitivity to insulin as a central signal (M. W. Schwartz et al. 1992; Woods et al. 1996). The evi-

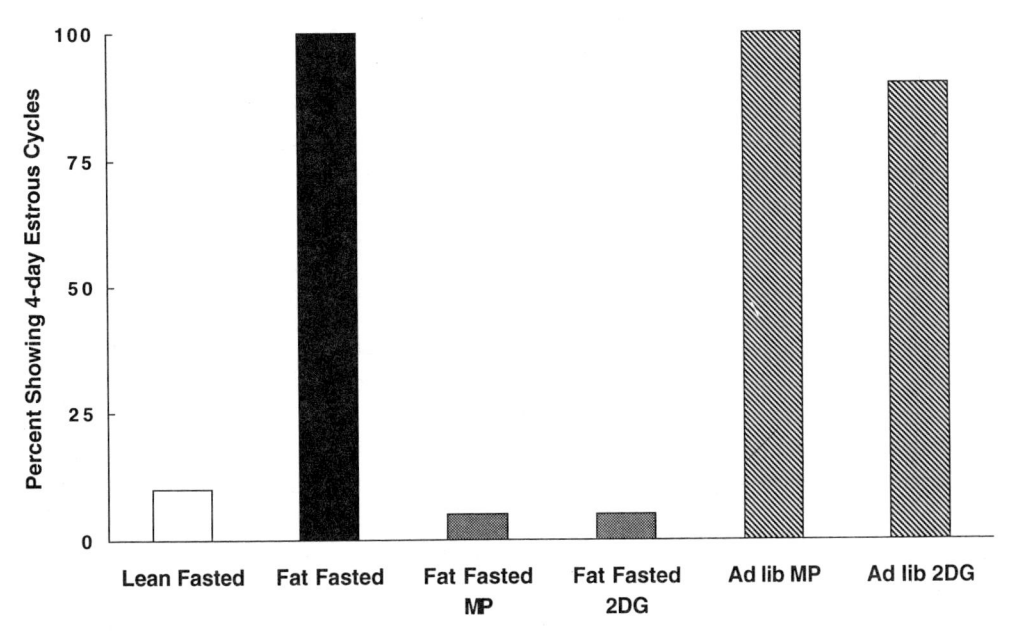

Figure 3.7
Percent of Syrian hamsters showing 4-day estrous cycles in groups that were either lean or fattened prior to being either fasted or fed during days 1 and 2 of the estrous cycle. Fattened groups were treated every 12 hours either intraperitoneally with 2-deoxy-D-glucose (2DG) at 1000 mg/kg, or intragastrically with methyl palmoxirate (MP) at 20 mg/kg, or the appropriate vehicles. (1) Lean fasted; (2) fat fasted; (3) fat, fasted plus MP treatment; (4) fat, fasted + 2DG treatment; (5) fat, fed plus MP treatment; (6) fat, fed plus 2DG treatment. Estrous behavior was examined on the evening of day 4 and vaginal discharge was examined on the morning of day 1 of the next expected cycle. Fasting-induced anestrus occurred more frequently in the fattened, fasted hamsters treated with MP or 2DG than in the fasted, vehicle-treated hamsters, or in the fed hamsters treated with MP or 2DG ($P < 0.01$) (see Schneider and Wade, 1989 and Schneider et al., 1997a.)

dence for this idea is primarily based on the notion that animals appear to defend a particular body weight "set point," that is, animals often make changes in food intake and energy expenditure that keep body weight within a certain range. Circulating levels of substances like insulin and leptin are often correlated with adiposity and therefore it is proposed that levels of these hormones might provide information to the brain about the level of stored fuels. For example, in contrast to systemic treatment with insulin, intracerebroventricular (ICV) insulin treatment causes animals to decrease food intake, and maintain their body weight at a lower level than before insulin treatment (Carlsson et al. 1997).

This hypothesis is consistent with the effects of ICV insulin treatment on food intake. However, it is not necessarily consistent with the severalfold changes in circulating levels of insulin experienced by most animals over short time periods. It is well-known that insulin secretion is highly responsive to fasting and feeding (Harris and Crabb 1992). How might these short-term fluctuations be averaged and translated into long-term signals that are hypothesized to control food intake and body weight? For insulin, a mechanism has yet to be articulated. It can be argued that whether an experimenter is studying long-term vs. short-term mechanisms is merely an artifact of the time period over which food intake or body weight change is measured (Friedman 1990). A physiological mechanism has not been elucidated to explain the mechanisms whereby these purported long-term signals are superimposed onto short-term effects. If insulin levels provide a signal to mechanisms that control food intake, it seems more likely that the signal reflects a short-term change in energy balance, rather than a change in energy reserves.

It is also important to determine whether insulin affects brain mechanisms that control food intake directly or indirectly, via secondary processes such as fuel availability and oxidation. It has been possible to dissociate the influence of insulin, and other hormones, from the effects of fuel oxidation by measuring food intake in rats with pharmacologically induced diabetes (Friedman 1978; Friedman et al. 1985; Chavez et al. 1998). Diabetic rats showed reduced plasma insulin and leptin levels and increased plasma corticosterone levels. These diabetic rats were hyperphagic, that is, they ate more food than nondiabetic control rats or diabetic rats treated with insulin, in keeping with the notion that these hormones act in brain to control food intake. To test this notion, hormone levels were dissociated from levels of fuel availability by feeding the diabetic rats two different diets. Diabetic rats were hyperphagic on a low-fat, high-carbohydrate diet, but ate normal amounts of food when fed a high-fat, low-carbohydrate diet. Although the caloric intakes of these two diet groups were significantly different, plasma levels of insulin, corticosterone, and leptin were not different (Chavez et al. 1998). The different caloric intakes resulting from the same levels of hormones indicate that the hormone level was not the critical signal for control of food intake. To the contrary, the different levels of caloric intake can be explained by differences in metabolic fuels available for oxidation in the high fat–fed vs. low fat–fed diabetics. Fats, but not carbohydrates, are readily oxidized by rats lacking insulin. Thus, the diabetics on the low-fat, high-carbohydrate diet may have been stimulated to overeat by their inability to oxidize the type of fuels they were ingesting. In contrast, the diabetics on the high-fat, low-carbohydrate diet were better able to oxidize the fuels they were ingesting and thus did not overeat (Friedman 1978; Friedman et al. 1985).

We examined the relationship between estrous cyclicity and plasma insulin levels in Syrian hamsters. We have demonstrated previously that hamsters that weigh less

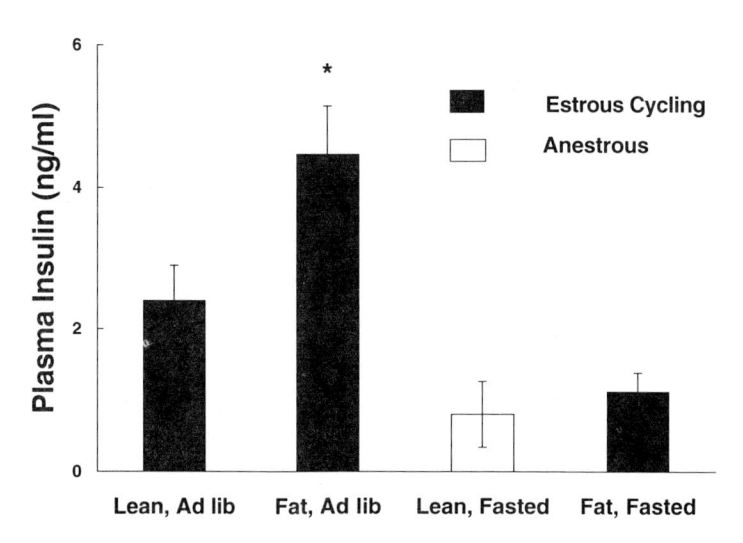

Figure 3.8
Mean and standard error of plasma insulin levels in fat or lean hamsters. Half of each group was either fed ad lib or fasted for 48 hours on days 1 and 2 of the estrous cycle. Blood was drawn on day 2 of the estrous cycle. Insulin levels were lowest in, and not significantly different between, fat and lean fasted hamsters, although only lean fasted hamsters became anestrous. Fat hamsters fed ad lib had significantly higher insulin levels than lean hamsters fed ad lib ($P < .01$).

than 110 g typically show fasting-induced anestrus, while those that weigh greater than 120 g typically do not show fasting-induced anestrus (Schneider and Wade 1990a, 1989a; Schneider et al. 1997; Schneider et al. 1997a). Thus, for this experiment, we used two groups of hamsters fed different diets so that they were significantly different in body weight and fat content at the start of the experiment. Half of the fat hamsters (body weights greater than 130 g) and half of the lean hamsters (body weights less than 105 g) were either fasted or fed ad lib on days 1 and 2 of the estrous cycle. Blood samples were drawn on the afternoon of day 2, after 36 hours of fasting. In keeping with the hormone hypothesis, fasting caused a dramatic decrease in plasma insulin in both the lean and fat hamsters (figure 3.8). Contrary to the hormone hypothesis, plasma insulin levels did not differ between the lean and fat fasted hamsters. The difference in estrous cyclicity between fasted lean (anestrus) and fasted fat (normal estrous cycles) hamsters could not be explained by differences in plasma insulin levels on day 2 of the estrous cycle.

In collaboration with Mark Friedman, we designed an experiment to distinguish between the direct effects of insulin treatment and the metabolic consequences of insulin treatment. High prolonged doses of insulin promote the redirection of metabolic

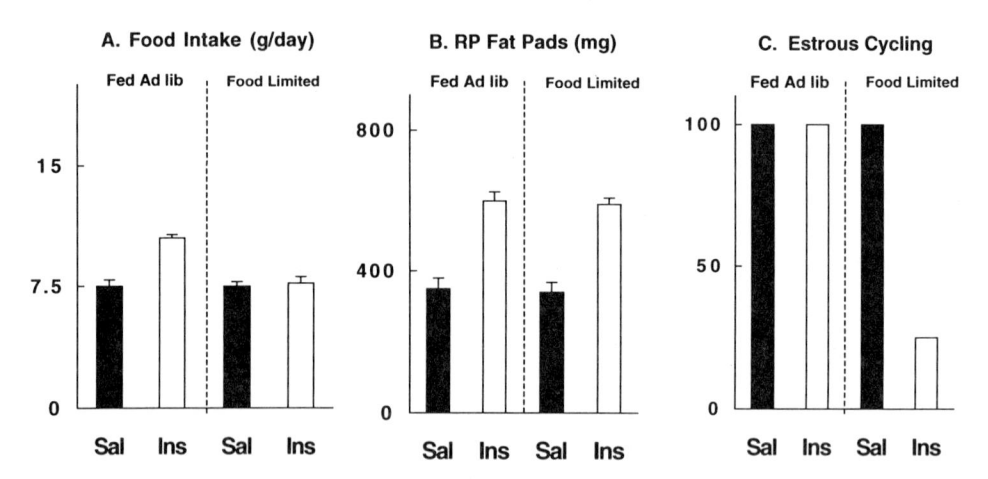

Figure 3.9
(A) Food intake, (B) retroperitoneal (RP) fat pad weight and (C) percent showing 4-day estrous cycles in Syrian hamsters. Half of each group were either treated with long-acting insulin (Ins) or saline (Sal) vehicle every 12 hours on days 1 and 2 of the estrous cycle. The two groups were further divided into groups either fed ad lib or limited to 110% of the food intake of the vehicle-treated, fed group. Estrous behavior was examined on the evening of day 4 and vaginal discharge was examined on the morning of day 1 of the next expected cycle. A significantly lower proportion of females showed normal 4-day estrous cycles when treated with insulin and with their food intake limited ($P < .05$). However, insulin had no significant effect on estrous cyclicity in hamsters allowed to overeat.

fuels from the circulation into tissues where they are stored. Syrian hamsters were treated systemically with either insulin or vehicle every 12 hours on days 1 and 2 of the estrous cycle and tested for indices of estrous cyclicity on the evening of day 4 and on day 1 of the next expected cycle. Insulin treatment at these doses typically increases food intake; thus, a third group of hamsters was treated with insulin and limited to 110% of the intake of the vehicle-treated control group. Hamsters treated with insulin and allowed to eat ad lib showed normal estrous cycles not significantly different from vehicle-treated hamsters. In contrast, insulin-treated hamsters restricted to their pre-insulin food intake became anestrous, despite the fact that they gained body fat (Wade et al. 1991) (figure 3.9). The results could not be explained by direct effects of insulin because ad lib-fed, insulin-treated hamsters showed normal estrous cycles. Insulin-treated, food-limited hamsters became anestrous despite a gain in body fat, thereby dissociating anestrus from a below-threshold level of body fat once again.

In the above experiment, we challenged the notion that the inhibitory effects of insulin are due to the direct effects of this hormone. Other investigators have examined whether the endogenous rise in insulin associated with a meal could have direct facilitating effects on reproductive processes by enhancing LH secretion. Fasted male monkeys display inhibited LH secretion that is rapidly restored after a meal and is correlated with a rise in plasma insulin concentration (Williams et al. 1996). Treatment with the insulin antagonist diazoxide blocked the meal-induced increase in insulin but failed to block the meal-induced restoration of pulsatile LH secretion. Again, the restoration of LH secretion is due to some consequence of meal ingestion other than the direct effects of insulin.

Similar results have been found in experiments using sheep as a model system to study the role of insulin in nutritional fertility. In fasted or ad lib-fed ewe lambs, ICV treatment with insulin failed to increase mean LH levels, LH pulse frequency, or LH pulse amplitude (Hileman et al. 1993). In contrast, ICV insulin or insulin plus glucose treatment significantly increased LH pulse frequency in adult rams. However, the insulin and insulin plus glucose treatments did not restore gonadotropin secretion to the extent seen in a control group in which both LH and FSH pulse frequency were restored by dietary supplements (Miller et al. 1995). These results are consistent with the idea that changes in central hormone levels can influence the reproductive system. More data are needed to determine whether insulin acts directly or via indirect effects on peripheral fuel metabolism.

Leptin Leptin is secreted predominantly, if not exclusively, by white adipose tissue cells (Zhang et al. 1994). Levels of leptin in plasma correlate positively with adiposity in rats, mice, and human beings of both sexes (Considine et al. 1995). Mutations in the mouse leptin gene lead to severe obesity and infertility. Two adult human subjects that are homozygous for the mutant *ob* allele have been identified. They are obese, infertile, and failed to undergo pubertal maturation (Strobel et al. 1998). Exogenous treatment with leptin results in decreased food intake and body weight, and reverses the effects of underfeeding on several aspects of reproduction in both lean and obese laboratory rodents (Ahima et al. 1996; Chehab et al. 1996, 1997; Barash et al. 1996; Yu et al. 1997a,b; Nagatani et al. 1998; Woodside et al. 1998b). It is not surprising, then, that there has been an intense focus on this protein and its potential role in control of reproduction.

For a lipostatic signal to be effective, it must show stable levels of transcription of the gene leading to stable synthesis and secretion of the protein under a variety of conditions in which body fat levels remain unchanged. To the contrary, both *ob* gene expression and serum leptin concentrations change rapidly during fasting and refeeding when body weight and fat content change either very little or not at all in human beings and laboratory animals. Unfortunately most of the investigators that measured *ob* mRNA did not measure leptin levels and vice versa. Thus, we will discuss the correlation between adiposity and *ob* mRNA levels separately from the correlation between adiposity and circulating leptin levels.

First, we will consider transcription of the *ob* gene that codes for leptin. In a lean strain of laboratory rats, *ob* gene mRNA levels were greater in rats that had been fasted and refed compared with rats that had been fed ad lib, despite the significantly higher levels of body fat in the fed control rats (Frederich et al. 1995). In the UCP-DTA strain, fasted for 12 hours, *ob* mRNA was significantly decreased, by 75%, despite no significant loss of body weight. Unfortunately, these investigators did not measure body fat content, but it is unlikely that a 12-hour fast produced a loss of body fat that was proportional to the 75% change in *ob* mRNA (Frederich et al. 1995). In another rat strain, *ob* gene expression, as well as expression of genes involved in various aspects of lipid uptake and synthesis, were measured immediately prior to and 3 hours after a meal. The postmeal expression of ob mRNA was 4.8-fold higher within 3 hours after the start of the meal, and the change occurred prior to any changes in expression of the key genes involved in lipid uptake and synthesis (Thompson 1996). In another study, rats were either allowed access to food ad lib or meal-fed, that is, provided three meals at specified times by gastric intubation. One meal-fed group was provided with 100%, while another was provided with 200% of the caloric intake of the group allowed to eat ad lib. Meal feeding caused rapid three-fold increases in *ob* mRNA expression in the absence of significant increases in either food intake or body fat content (Harris et al. 1996). Similar effects of fasting and feeding (Saladin et al. 1995; Levy et al. 1997), as well as cold exposure (Trayhurn et al. 1995) on *ob* gene expression have led several investigators to doubt that leptin gene expression could serve as a long term index of body fat content (Rentsch et al. 1995; Trayhurn and Rayner 1996).

While the above examples illustrate a dissociation between leptin message and body fat content, the same has been demonstrated for plasma leptin concentrations and body weight, body fat content, and fat metabolism in a variety of experimental models. In C57BL female mice, plasma leptin concentrations decreased significantly after a 48-hour fast. However, the reduction in leptin concentrations during fasting was not significantly correlated with body weight or change in body weight (Ahren et al. 1997). In men, serum leptin concentrations declined to a nadir at 36 hours after the start of a fast and returned to baseline concentrations within 24 hours after refeeding. The magnitude of the leptin decline did not correlate significantly with the percent body fat or body fat loss (Kolaczynski et al. 1996). After a 72-hour fast, serum leptin concentrations were decreased by 64%, although the subjects lost only 0.5% of their body weight (Boden et al. 1996). In subjects in whom the fasting-induced fall in plasma glucose was prevented by glucose infusion that did not prevent body fat loss, serum leptin levels did not drop from baseline values (Boden et al. 1996). Even when glucose infusions provided only 100 kcal during a 6-hour fast, leptin levels were not significantly different from leptin levels in subjects fed 1750 kcal in a balanced diet over the same time period (Kolaczynski et al. 1996).

Together, all of the above results demonstrate that levels of *ob* gene expression and serum leptin do not provide an accurate profile of long-term body fat stores. Thus, it seems unlikely that changes in circulating levels of leptin could represent a long-acting, slowly changing index of adipocyte filling. Rather, it appears that leptin levels are quite labile, fluctuating acutely with short-term energy intake and expenditure. Thus, if leptin is a hormonal signal that reports directly to the reproductive system, it is more reasonable to hypothesize that leptin levels reflect the state of short-term energy balance, rather than levels of energy reserves (see figure 3.4). Consistent with this idea, a number of investigators have found that leptin treatment in rodents reverses the effects of food restriction or food deprivation on aspects of reproduction, including the onset of puberty (Chehab et al. 1997; Ahima et al. 1997; Barash et al. 1996; Apter 1997), the length of lactational diestrus (Woodside et al. 1998b), the length of the estrous cycle (Chehab et al. 1996; Schneider et al. 1997b, 1998), gonadotropin and gonadal steroid levels (Ahima et al. 1996; Yu et al. 1997a,b), and pulsatile LH secretion (Nagatani et al. 1998).

To examine the hypothesis that leptin might be a hormonal signal that links reproduction to energy balance, we examined the relationship between estrous cyclicity and plasma leptin levels in Syrian hamsters that differed in body weight and fat content and thus differed in their susceptibility to fasting-induced anestrus. Half of the fat hamsters (body weights greater than 130 g) and half of the lean hamsters (body weights less than 105 g) were either fasted or fed ad lib on days 1 and 2 of the estrous cycle. Blood samples were drawn on the afternoon of day 2, after a 36-hour fast. Fasting caused a dramatic decrease in plasma leptin in both the lean and fat hamsters (figure 3.10). These results were more consistent with the hormonal hypothesis than the adipostatic hypothesis in that leptin levels plummeted over the 36-hour fast in fat, fasted hamsters that showed little change in body fat content. However, fat fasted hamsters showed normal estrous cycles despite dramatically decreased leptin levels. Thus, fasting-induced changes in leptin did not necessarily result in anestrous. In addition, lean fed hamsters that showed normal estrous cycles did not have significantly higher leptin levels that lean fasted hamsters.

In subsequent experiments, we examined leptin, insulin and body fat levels at earlier time points during the fasting period. Again, the rapid dramatic effects of fasting on plasma insulin and leptin concentrations occurred by 12 and 24 hours after the start of fasting, without changes in body fat content and without changes in estrous

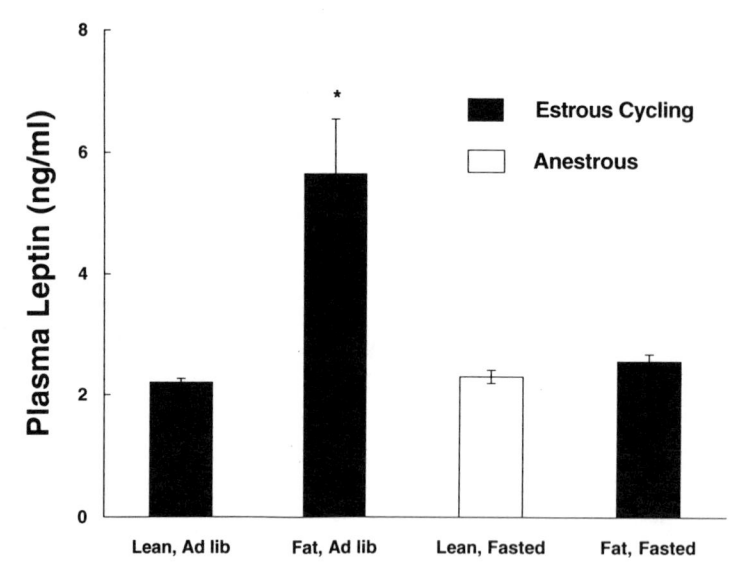

Figure 3.10
Mean and standard error of plasma leptin levels in fat or lean hamsters. Half of each group was either fed ad lib or fasted for 48 hours on days 1 and 2 of the estrous cycle. Blood was drawn on day 2 of the estrous cycle. Leptin levels were lowest in, and not significantly different between, fat and lean fasted hamsters, although only lean fasted hamsters became anestrous. Fat hamsters fed ad lib had significantly higher leptin levels than lean hamsters fed ad lib ($P < .05$).

cyclicity. However, lean fasted hamsters that became anestrous had slightly but significantly lower levels of leptin, insulin and body fat compared to lean fed hamsters at 24 hours. Hamsters fasted for only 24 hours failed to increase plasma leptin at 3, 6, 12, and 24 hours after the start of refeeding and unlike hamsters fasted for 48 hours, showed normal estrous cycles. Thus, fasting leptin levels are not sufficient to induce anestrus (Schneider et al., 1999).

As we have pointed out with regard to insulin, it is important to ask whether the effects of leptin treatment are secondary to the effects of the hormone on energy availability and expenditure. We tested whether fasting-induced anestrus could be reversed by intraperitoneal (IP) injections of murine leptin, and if so, whether these effects of leptin could be blocked by treatments known to inhibit FFA or glucose oxidation in either periphery or brain (Schneider et al. 1998). Glucose utilization was inhibited by IP injections of 2DG. Fatty acid oxidation was inhibited by intragastric treatment with MP. We used doses of MP and 2DG that inhibit glucose and FFA oxidation, but did not induce anestrus in hamsters fed ad lib (Schneider et al. 1988, 1993; Tutwiler et al. 1985). Fasting-induced anestrus was reversed and normal sex

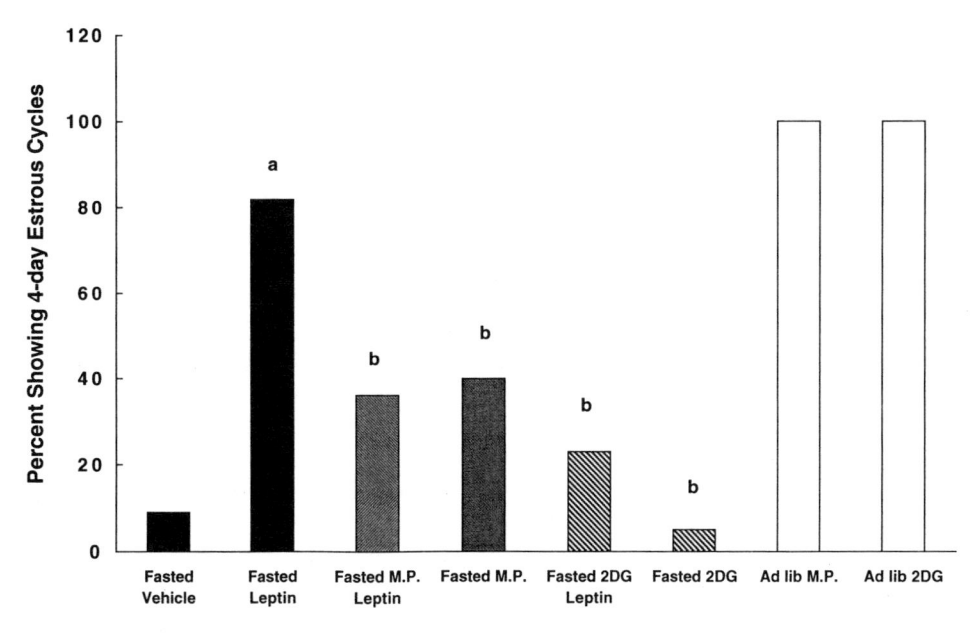

Figure 3.11
Effects of systemic leptin in combination with inhibitors of metabolic fuel availability. High levels of leptin prevent fasting-induced anestrus, except when leptin's ability to increase metabolic fuel oxidation is blocked by treatment with metabolic inhibitors. Percent of Syrian hamsters showing 4-day estrous cycles in groups either fed ad lib (white bars) or fasted (dark bars) for 60 hours on days 1 and 2 of the estrous cycle: Hamsters were fasted and either (1) treated with the appropriate vehicles, (2) pretreated with vehicles and treated with 5 mg/kg leptin, (3) pretreated with methylpalmoxirate (MP) (20 mg/kg) and treated 1 hour later with leptin, (4) pretreated with MP and treated with vehicle, (5) pretreated with 2-deoxy-D-glucose (2DG) (1000 mg/kg) and treated with leptin, or (6) pretreated with 2DG and treated with vehicle. Two additional groups were fed ad lib and treated with either MP or 2DG. Any hamsters that did not receive leptin, 2DG, or MP received the appropriate vehicle. a, significantly different from fasted, vehicle-treated at $P < .05$; b, significantly different from fasted, leptin-treated at $P < .05$.

and social behavior and ovulation rate were restored in hamsters injected with 5 mg/kg leptin every 12 hours during fasting on days 1 and 2 of the estrous cycle (figure 3.11). In contrast, when each injection of leptin was preceded by an injection of 2DG (1000 mg/kg) or MP (20 mg/kg), fasting-induced anestrus was not significantly attenuated by 5 mg/kg leptin relative to that seen in food-deprived, vehicle-treated hamsters. Estrous cyclicity was not restored when oxidation of metabolic fuels was blocked, despite high endogenous levels of leptin (Schneider et al. 1998). These results do not support a leptinostatic hypothesis for control of reproduction, because fasted-leptin-MP and fasted-leptin-2DG-treated hamsters showed anestrus despite high levels of leptin.

The Metabolic Hypothesis

A large and still growing body of data supports the idea that reproduction is controlled by the availability, and hence the oxidation of metabolic fuels (see figure 3.5). All of the data presented in this chapter are consistent with this hypothesis (see figures 3.6 through 3.12).

Metabolic Effects of 2DG Estrous cycles can be inhibited in a number of species by treatment with 2DG, an inhibitor of glucose oxidation. These effects on estrous cycles can be explained by the known metabolic effects of 2DG. 2DG is a glucose analogue that is transported into cells and phosphorylated to 2DG-6-phosphate. Because of its missing hydroxyl group, 2DG-6-phosphate cannot be metabolized further. Glycolysis is thus terminated at the phosphohexoseisomerase step, an action that may also inhibit further uptake of glucose (Brown 1962). Consistent with the idea that anestrus is induced by 2DG's competitive inhibition of glycolysis, the effects of 2DG on estrous cycles in Syrian hamsters can be overcome by pretreatment with an equal dose of glucose (Schneider et al. 1997a).

Metabolic Effects of MP Similarly, a high body fat content does not prevent fasting-induced anestrus when hamsters are treated with MP (see figure 3.6). This result can be explained by the metabolic effects of MP. MP binds irreversibly to carnitine palmitoyltransferase I (CPT-I), the enzyme necessary for transport of long-chain FFAs into mitochondria (Tutwiler et al. 1985). Treatment with MP inhibits FFA oxidation with a consequent reduction in the formation of ketone bodies (Tutwiler et al. 1985). We have confirmed that treatment with MP significantly lowers plasma ketone bodies in fasted Syrian hamsters relative to vehicle-treated, fasted controls, consistent with the intended effects of MP treatment on FFA oxidation (Schneider et al. 1988). Despite these metabolic effects of MP, even the highest doses of MP do not induce anestrus in hamsters fed ad lib (Schneider et al. 1993). The effects of MP are only revealed in fasted hamsters (see figure 3.6). During fasting, caloric homeostasis is maintained by hydrolysis of triglycerides to FFAs and glycerol. FFAs are mobilized to various peripheral tissues where they are preferentially oxidized, thereby sparing glucose for oxidation in the CNS (see figure 3.2). Fasted hamsters may be more sensitive to MP than fed hamsters because they are predisposed to FFA utilization. The metabolic signal that induces anestrus in fasted, MP-treated hamsters may be generated by either the inhibition of FFA oxidation in peripheral tissues, or by attenuation of the glucose-sparing effects of mobilizing FFAs in brain or periphery, or both signals. The latter idea is supported by data showing that anestrus occurs in fat fasted hamsters treated with low doses of 2DG (see figure 3.6).

We are not aware of data demonstrating that the effects of fasting are attenuated by a high body fat content in species other than Syrian hamsters. Furthermore, there are no published data demonstrating the effects of metabolic inhibitors of fatty acid oxidation on reproductive processes in species other than Syrian hamsters. Investigators studying species such as rats, sheep, and monkeys might find that the effects of MP and mercaptoacetate (MA) might only be revealed in animals predisposed toward FFA utilization. For example, it should be expected that MP and MA would only inhibit pulsatile LH secretion in fattened, fasted animals, or in animals fed a high-fat diet for several weeks. These have been very useful methodologies that should be tested in a variety of species.

Metabolic Effects of Insulin Concurrent with the metabolic fuels hypothesis, figure 3.8 shows that insulin-treated hamsters, when not allowed to overeat, became anestrous, whereas insulin-treated hamsters that were allowed to overeat continued to show normal estrous cycles. This result can be explained by the well-known effects of insulin treatment on metabolic fuel partitioning. Prolonged insulin treatment of the type used in this experiment promotes uptake and storage of metabolic fuels, presumably leaving less for oxidation, and thereby stimulating food intake. Hamsters that are not allowed to compensate for this lack of oxidizable fuels by overeating become anestrous. Insulin has the same effects on the neural mechanisms that control steroid-induced estrous behavior. Fasting decreases lordosis duration displayed by ovariectomized hamsters treated with estradiol, and this effect can be mimicked by treating hamsters with insulin and limiting their food intake (Panicker and Wade 1998). Furthermore, it is important to note that insulin-treated hamsters limited to the food intake of the vehicle-treated controls became anestrous and showed decreased lordosis duration despite the fact that they ate as much food as ad lib-fed, vehicle-treated hamsters and gained body fat content. Despite their ample body fat stores, the insulin-treated, food-limited hamsters became anestrous and had decreased lordosis duration because the fuels stored in body fat were not available for oxidation.

Similar results have been shown in other species. For example, in ewes infused with insulin, pulsatile LH secretion was inhibited. However, the same treatment failed to inhibit pulsatile LH secretion when both glucose and insulin were infused simultaneously (Clarke et al. 1990). Again, insulin-induced effects on availability of oxidizable fuels can account for the effects of insulin.

Metabolic Effects of Leptin The negative effects of 2DG, MP, and insulin on estrous cycles and pulsatile LH secretion can be accounted for by the ability of these pharmacological agents to decrease fuel availability and oxidation. The positive effects of exogenous leptin treatment on puberty, estrous cycles, and pulsatile LH secretion can be accounted for by the ability of this hormone to increase fuel availability and oxidation. Leptin treatment appears to have unique and dramatic effects on fuel metabolism. Leptin treatment has been shown to increase FFA oxidation and prevent lipogenesis, and increase glucose oxidation, glucose uptake, glucose turnover, and glycogen depletion in rats and mice and in various types of cells in culture (Kamohara et al. 1997; Shimabukuro et al. 1997; Hwa et al. 1996; G. Chen et al. 1996; Flier 1997; Sarmiento et al. 1997; Rossetti et al. 1997). Leptin's effects on fuel metabolism occur in situ, that is, fuels are oxidized within the same cells where they were hydrolyzed from triglycerides (Shimabukuro et al. 1997).

This is notable because it differs from the metabolic shifts that occur during fasting. In fasting, triglycerides are hydrolyzed to FFAs and glycerol, and these fuels are mobilized to a variety of peripheral tissues that can use them as fuels (see figure 3.2). In the liver, FFAs can be partially oxidized and used in the formation of ketone bodies. Elevated levels of ketone bodies, FFAs and glycerol are diagnostic of starvation. However, in leptin-treated rats, it appears that FFAs are oxidized as they are hydrolyzed, precluding mobilization of FFAs and hepatic synthesis of ketone bodies in the liver. This idea is supported by data showing that leptin treatment results in the rapid loss of body fat, without a fasting-induced increase in plasma FFAs, ketone bodies, or urine ketones in rats (Shimabukuro et al. 1997). Leptin treatment results in increased rates of FFA oxidation and reduced rates of esterification within a number of tissues such as pancreatic islets and skeletal muscle, as well as pancreatic islets in culture. Leptin treatment also increased mRNA for enzymes of FFA oxidation and decreased mRNA for an enzyme known to inhibit CPT-I activity (Zhou et al. 1997). In mice, systemic leptin treatment resulted in increased mRNA expression of thermogenic proteins, lipoprotein lipase and hormone-sensitive lipase (Sarmiento et al. 1997).

These results raise the possibility that leptin treatment does not increase the general pool of metabolic fuels; rather it appears to increase fuel availability and oxidation within cells of specific tissues. Furthermore, this increased availability of oxidizable fuels is just as rapidly depleted by increased fuel oxidation in those same tissues. The mechanisms by which leptin increases fuel oxidation are under investigation. Leptin may act directly on these tissues or leptin may bind in brain areas that control the sympathetic nervous system. A variety of intermediates may be involved in leptin's effects on peripheral metabolism. For example, leptin activates neurons in the retrochiasmatic area and lateral arcuate nucleus that innervate the thoracic spinal cord and contain CART and POMC. It is thought that this pathway may contribute to leptin-induced increases in thermogenesis and energy expenditure (Elias et al. 1998).

Leptin is known to increase synthesis and secretion of both uncoupling protein-1 (UCP-1) and UCP-2 in rats and mice (Zhou et al. 1997; Sarmiento et al. 1997; Scarpace et al. 1997). The former is a protein long known to be responsible for the uncoupling of oxidation and phosphorylation that leads to nonshivering thermogenesis in brown adipose tissue. The latter is a more recently discovered protein, ubiquitously expressed throughout the body and thought to induce thermogenesis that occurs in tissues other than brown adipose tissue. In leptin-treated rats, UCP-2 mRNA was overexpressed in pancreatic islets over 10 times the level of pair-fed controls (Zhou et al. 1997). The effects of leptin on the mechanisms that govern fuel oxidation are perhaps the most important effects of this protein. Thus, exogenous leptin treatment might stimulate fuel oxidation and availability within particular tissues, generating a signal that permits normal estrus in the face of food deprivation or other energetic manipulations. It is conceivable that this increased intracellular fuel oxidation provides a permissive signal for the onset of estrous cyclicity, similar to the way in which a minimum level of circulating thyroid hormone is necessary for the onset of reproductive function in some seasonal breeding species. Whether endogenous fluctuations in leptin have the same influence on reproductive processes is unknown.

The ability of MP and 2DG to block the effects of leptin on estrous cycles (see figure 3.11) can be explained by the metabolic effects of these agents. Leptin increases the availability and oxidation of glucose (Kamohara et al. 1997; Zhou et al. 1998); 2DG treatment has the opposite effect (Brown, 1962). Leptin increases the intracellular availability and oxidation of FFAs, possibly via inhibition of acetyl CoA decarboxylase and consequent disinhibition of CPT-I (Bai et al. 1996; Zhou et al. 1997). MP treatment has the opposite effects, irreversibly binding CPT-I and preventing transport of FFAs into mitochondria (Tutwiler et al. 1985).

These results most likely reflect an interaction between leptin and metabolic inhibitors at the level of intracellular fuel oxidation, although interactions at other levels are possible. One possible outcome of such an interaction might be that leptin levels change the animal's reproductive responsiveness to energetic challenges. When leptin levels are increased and consequently levels of fuel availability and oxidation are increased, the reproductive system may be less sensitive to energetic challenges that decrease fuel availability and oxidation. For example, increased secretion of leptin around the time of puberty might raise the level of fuel availability and oxidation, decreasing the animal's responsiveness to energetic challenges, and thereby increasing pulsatile LH secretion.

If exogenous leptin treatment acted via increases in metabolic fuel oxidation, it would not be expected to promote estrous cyclicity beyond limits of available metabolic fuels. For example, leptin can fully reverse the effects of fasting-delayed puberty in rats restricted to 80% of their ad lib food intake, but not in rats restricted to 70% of their ad lib intake (Cheung et al. 1997). In ovariectomized hamsters brought into estrus by estradiol and progesterone treatment, leptin increases the duration of lordosis in ad lib fed, but not in fasted hamsters (Wade et al. 1997). Similarly, enforced exercise, which promotes fuel oxidation and decreases energy storage, can counter the effects of various environmental inhibitors of puberty, gonadal regression, gonadotropin secretion, anestrus, and estradiol-enhanced sex behavior (Borer et al. 1983; Bronson 1987; Perrigo and Bronson, 1983; Pieper et al. 1995). In keeping with a metabolic hypothesis, these positive effects of exercise on reproduction did not occur when energy from intake and storage could not meet the demands of increased energy oxidation (Powers et al. 1994; Perrigo and Bronson 1983). Thus, leptin treatment, and possibly exercise, facilitates reproduction by increasing oxidation and intracellular availability of metabolic fuels, but this facility is constrained ultimately by the overall availability of the fuels from intake and storage, as illustrated in figure 3.6.

Figure 3.11 provides an explanation of the mechanism whereby *exogenous* leptin treatment may reverse fasting-induced anestrus, that is, by reversing the effects of fasting on fuel oxidation. Implicit in this explanation is the notion that estrous cycles are inhibited by decreased fuel oxidation and are permitted or initiated by increased fuel oxidation. However, the question as to whether *endogenous* changes in leptin secretion might mediate the effects of metabolic challenges on reproduction needs to be addressed. It is important to demonstrate that fluctuations in leptin levels within physiological ranges can have significant effects on either reproduction or metabolic fuel oxidation. Endogenous plasma levels of leptin, including pulse frequency and circadian rhythms in plasma and in cerebral spinal fluid, must be determined for the species in question under a variety of metabolic conditions. The field is progressing rapidly and these data may be available by the time this chapter is published.

From the Primary Metabolic Sensory Signal to the GnRH Pulse Generator

Peripheral Metabolic Signals

It is reasonable to conclude that metabolic fuel availability and oxidation control reproductive processes. Where do these metabolic signals occur? The characteristics of the metabolic signal and the relative importance of peripheral vs. central signals are unknown, and the data do not yet paint a clear picture. On the one hand, it might be suggested that decreased fuel oxidation in the periphery is a potent signal for anestrus. In rats, fasting-induced suppression of pulsatile LH secretion, as well as increases in PVN estrogen receptor immunoreactivity, are prevented by gastric vagotomy (Cagampang et al. 1992a; Estacio et al. 1996a). Furthermore, in hamsters, the increase in estrogen receptor immunoreactivity in the mPOA that is normally seen in response to fasting or treatment with metabolic inhibitors is abolished by vagotomy (Li et al. 1994). Treatment with MP, a pharmacological agent that does not reach the brain in appreciable quantities (Tutwiler et al. 1985; Brown 1962), induces anestrus in fat fasted hamsters (figure 3.7). Furthermore, leptin treatment reverses fasting-induced anestrus, but fails to do so in hamsters treated with MP (figure 3.11). As mentioned previously, leptin and MP have opposite effects on the activity of CPT-I, the enzyme necessary for transport of long-chain FFAs into mitochondria. These results suggest that a signal generated by changes in CPT-I activity might originate in tissues such as liver or gut and may reach the brain via the vagus nerve, or might originate in brown adipose tissue, skeletal muscle, or pancreas and these signals may reach the mPOA or other brain areas known to contain GnRH neurons, by way of peripheral afferents.

Other data on estrous cyclicity in Syrian hamsters are not consistent with a peripheral signal, at least not one mediated by the vagus nerve. For example, MP-induced anestrus in fat fasted hamsters does not require an intact vagus nerve (Schneider et al. 1997c). One possibility is that MP treatment induces anestrus because it blocks the glucose-sparing effects of mobilizing FFAs liberated from stored triglycerides. A possible explanation for the data illustrated in figure 3.10 is that leptin may increase the rate of fuel oxidation in brain or periphery or both, creating a critical need for glucose in the CNS. When MP treatment is used to block peripheral FFA oxidation, it may preclude the glucose sparing effect of oxidizing these fuels, thereby inducing anestrus. In this scenario, signals generated peripherally in the oxidization of FFAs affect reproduction indirectly via their effects on glucose availability for the CNS.

Data on the role of the vagus nerve are in direct conflict. Subdiaphragmatic vagotomy does not significantly attenuate the effects of metabolic inhibitors on estrous cyclicity, lordosis duration, or on estrogen receptor immunoreactivity in the VMH in Syrian hamsters (Li et al. 1994; Schneider et al. 1997c). In contrast, vagotomy abolishes the increase in mPOA estrogen receptor immunoreactivity typically found in metabolically challenged hamsters (Li et al. 1994), and also blocks suppression of pulsatile LH secretion and the increase in PVN estrogen receptor immunoreactivity typically found in fasted rats (Cagampang et al. 1992a; Estacio et al. 1996a). In order to fully evaluate these discrepancies, pulsatile LH secretion and PVN estrogen receptor immunoreactivity in response to vagotomy must be examined in hamsters. Vagotomy may have prevented the effects of MP (or even 2DG) on pulsatile LH secretion in hamsters, but the effects of these inhibitors on other factors such as the LH surge may have masked this effect. Finally, we must keep in mind that even though an intact vagus nerve is not necessary, vagal signals may be sufficient to affect reproductive processes under some circumstances.

Most accounts of leptin's action on food intake and reproduction emphasize direct effects on leptin in the hypothalamus. The presence of receptors in hypothalamus provides circumstantial evidence for the possibility that leptin may act directly on hypothalamic mechanisms that control reproduction (Hakansson et al. 1996; Huang et al. 1996; Mercer et al. 1996; Fei et al. 1997). Leptin receptor binding, however, has been shown in organs such as adipose tissue, pancreas, liver, kidney, lung, adrenal medulla and skeletal muscle (Kieffer et al. 1996; Lynn et al. 1996; H. Chen et al. 1996; Chua et al. 1996; Lee et al. 1966; Mercer et al. 1996; Wang et al. 1997; Hoggard et al. 1997; Zamorano et al. 1997; Fei et al. 1997; Leclereq-Meyer et al. 1996; Emilsson et al. 1997; Cao et al. 1997; Barash et al. 1996; Zamorano et al. 1997; Cioffi et al. 1996), and thus it is also possible that reproduction is controlled by leptin binding in these peripheral tissues. Furthermore, experiments should be designed to determine whether peripheral effects of leptin on fuel metabolism influence reproduction via vagal or other afferents. Whether the focus is on the positive effects of leptin or the negative effects of metabolic challenges, peripheral signals mediated by the vagus or other pathways cannot be ruled out at this time.

Metabolic Detection and/or Integration in the Caudal Brain Stem

Historically, research on food intake focused narrowly on the hypothalamus. More recently, it was demonstrated that areas in the caudal brain stem could act independently of the hypothalamus to control the fundamental aspects of food intake. These areas, in turn, project to hypothalamic areas more traditionally associated with control of food intake. Areas such as the area postrema (AP), the nucleus of the solitary tract (NTS), and the lateral parabrachial nucleus (LPBN) are part of a well-studied afferent pathway that mediates a variety of autonomic enteroreceptive functions related to food and water ingestion, caloric homeostasis, as well as other functions. Chronically decerebrate rats display metabolic, hormonal, and neurotransmitter control of the ingestive motor program and sympathoadrenal responses, even though all connections to the forebrain are severed (Grill and Kaplan 1990; Kaplan et al. 1993, 1998; Grill et al. 1997). Metabolic treatments that affect food intake (as well as reproduction) increase neural activation in the AP, NTS and LPBN, as well as forebrain areas (Ritter et al. 1992; Schneider et al. 1995; Horn and Friedman 1998; Keen et al. 1998). Effects of metabolic inhibitors on food intake are severely attenuated by lesions of the AP that include varying amounts of damage to the medial nucleus of the solitary tract NTS (Bird et al. 1983; Contreras et al. 1982; Hyde and Miselis 1983; Ritter and Taylor 1990). Infusion of 5TG (5-thioglucose, another inhibitor of glucose utilization) restricted to the fourth, but not the third ventricle increases food intake in rats (Ritter et al. 1981).

Despite all this work demonstrating the importance of the caudal brain stem, early work on the effects of leptin focused almost exclusively on the hypothalamus as a site of action. More recently, experiments indicate that effects of leptin on gastric emptying could be elicited by slow microinfusion of leptin into the fourth ventricle (Smedh et al. 1998).

Inspired by data showing that the caudal brain stem is involved in metabolic control of food intake, similar experiments have been performed with regard to metabolic control of reproduction. Well-known afferent pathways from the NTS lead to a number of hypothalamic structures that could affect GnRH secretion and sex behavior. AP lesions block the effects of metabolic inhibitors on estrous cycles in hamsters, estrogen receptor immunoreactivity in VMH and arcuate nucleus (Arc), and lordosis induced by exogenous hormone treatment in ovariectomized hamsters (Li et al. 1994; Schneider and Zhu, 1994; Panicker et al. 1998). Slow microinfusion of 2DG into the fourth ventricle in rats inhibited pulsatile LH secretion (Murahashi et al. 1996). Treatment with metabolic inhibitors increases neural activation in the AP, as well as the NTS and the well-known pathways to the hypothalamic areas involved in reproduction and food intake (Ritter et al. 1992; Schneider et al. 1995; Schneider et al. 1997a). The AP, like most circumventricular brain areas, is well located to receive stimuli of all kinds that may be present in cerebrospinal fluid. While most investigators studying the effects of leptin on reproduction have assumed that leptin acts directly on hypothalamic neurons controlling GnRH secretion, we found that leptin treatment increases neural activation in the caudal brain stem (figure 3.12). Experiments in progress will determine whether leptin infused into the fourth ventricle facilitates, and whether AP lesions block, the effects of exogenous leptin treatment on estrous cycles in Syrian hamsters. Effects of leptin treatment that involve the AP are certainly consistent with the notion that exogenous leptin treatment influences reproduction via its well known effects of fuel oxidation.

There is no doubt that reproductive processes are responsive to changes in metabolic fuel oxidation. A model whereby changes in fuel oxidation might control reproduction is shown in figure 3.13. This model emphasizes that changes in oxidation are detected peripherally or in the caudal brain stem, or both, and lead to a variety of changes in more rostral hypothalamic areas that control the GnRH pulse generator as well as food intake and fuel homeostasis.

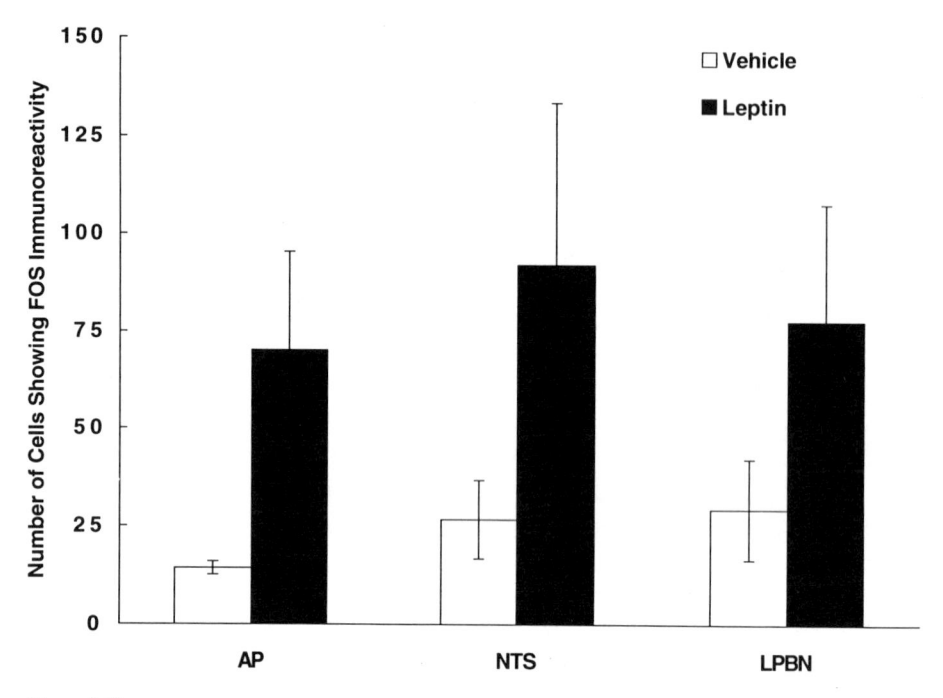

Figure 3.12
Mean and standard error of number of cells showing Fos immunrectivity in the area postrema (AP) and nucleus of the solitary tract (NTS) and lateral parabrachial nucleus (LPBN) in Syrian hamsters fed ad lib and treated with leptin (5 mg/kg) 1.5 hours prior to perfusion.

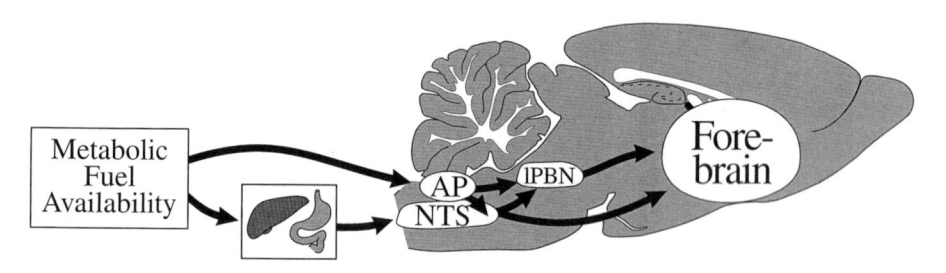

Figure 3.13
Diagram of a possible pathway whereby primary metabolic signals are detected and sent to brain areas that control the gonadotropin-releasing hormone (GnRH) pulse generator. It is hypothesized that peripheral signals are received via the vagus nerve or other afferents, or via the caudal brain stem, possibly in the area postrema (AP). From the caudal brain stem information may be passed rostrally via the nucleus of the solitary tract (NTS) and lateral parabrachial nucleus (LPBN) to brain areas such as the central nucleus of the amygdala or the paraventricular nucleus of the hypothalamus, which may in turn project to GnRH neurons.

Summary

The evidence supporting three types of hypotheses has been evaluated. The lipostatic hypothesis (see figure 3.4) is consistent with vast amounts of data demonstrating a positive correlation between body fat content and fertility. However, plasma levels of the putative lipostatic signaling hormones do not always provide an accurate profile of body fat content. For example, changes in plasma insulin and leptin occur rapidly in response to short-term changes in energy intake and expenditure and can be readily dissociated from body fat content. Furthermore, these dramatic endogenous fluctuations in hormone levels have not been associated with changes in ovulatory cycles (see figures 3.8 and 3.10). Although there is no doubt that body fat content can influence reproduction, it may do so through a mechanisms other than an adipostat. For example, adipose tissue is a storage depot for metabolic fuels, and changes in oxidation of these fuels may generate the primary sensory signal that control reproduction.

The hormonal hypothesis, in which hormone levels change as a result of changes in energy availability and oxidation, accounts for the rapid changes in insulin, leptin, and glucocorticoids that are seen in response to short-term changes in energy intake and expenditure (see figure 3.5). One possibility is that these rapid changes in hormones mediate the rapid changes in pulsatile LH secretion that are seen in response to fasting and refeeding. Prior to claiming that a particular hormone controls reproduction in a particular species, we need to determine the magnitude of endogenous fluctuations in the hormone in both brain and periphery, determine the effects of these endogenous fluctuations in the hormone on metabolism, and distinguish between the direct and indirect effects of the hormone. Most likely this hormone will be a secondary signal that is downstream from the primary sensory signal that links energy balance to secretion of the hormone.

The metabolic fuels hypothesis (see figure 3.6) postulates that reproductive processes, presumably the GnRH pulse generator, are controlled by signals generated intracellularly when there are changes in the availability and oxidation of metabolic fuels. Endocrinologists may be slow to embrace this idea because the signal is not a hormone binding to its receptor, but rather a metabolic event. However, there is precedence for a secretagogue that undergoes metabolism rather than interacting with a receptor. In the pancreatic beta-cell the proximal signal for glucose-induced insulin release occurs between the central part of the glycolytic pathway and mitochondrial metabolism (MacDonald, 1990). More recent evidence suggests that malonyl-CoA and long chain acyl-CoA esters may act as metabolic coupling factors in beta-cell signalling (Prentki et al. 1997). Thus, understanding these primary metabolic signals is important to understanding control of counterregulatory hormones.

In addition, these primary sensory signals may control a variety of peripheral and central cells involved in the hypothalamic-pituitary-gonadal axis.

Counterregulatory hormones also have effects on fuel partitioning and oxidation in brain and periphery, which in turn may influence the hypothalamic-pituitary-gonadal axis. As illustrated in figure 3.6, body fat content, food intake, circulating metabolic fuels, hormones, and energy expenditure all contribute to changes in the availability and oxidation of metabolic fuels. Sensory systems that monitor these changes in fuel oxidation must be pivotal to the control of food intake and reproduction.

It is unlikely that feeding and reproduction will be explained entirely by the rise and fall of body fat content, hormones, and neurotransmitters. Survival and reproduction cannot be served without a sensory system that links feeding and reproduction to energy balance. Our efforts should be concentrated on charcterization of this sensory sytem, the intracellular metabolic stimuli, and their detectors. Are metabolic signals generated by the increased presence of intracellular substrates for oxidation, or by some aspect or by-product of increased oxidation, or by all of these signals? Are they related to changes generated in glycolysis, the tricarboxylic acid cycle, or electron transport? Are they related to the activities of glucokinase or hexosamine, or to levels of particular intermediates? Are they related to levels of ATP, phosphorylation potential, or to ADP:ATP ratios? Characterization of these metabolic sensory properties might lead more readily to the location of their detectors as well as their effector pathways. When we identify the primary metabolic stimuli and detectors, confusion concerning their location and connection to the GnRH pulse generator may be more easily resolved.

Acknowledgments

We are grateful to Mark I. Friedman for laying much of the conceptual foundation for our research. We thank Linda Lowe-Krenz, Antonio Nunez, Theresa Lee, Kim Wallen, Laura Lubbers, Eric Corp, and Gayatri Sonti for helpful advice and discussion, and Dan Zhou and Robert Blum for contributing their unpublished data. This work was supported by Research Scientist Development Award MH01096 and Senior Scientist Award MH00321-17 from the National Institute of Mental Health, and research grants NS10873-25, HD30372-04, and DK53402 from the National Institutes of Health and IBN9723938 from the National Science Foundation.

References

Abraham SF, Beumont PJV, Fraser IS, Llewellyn-Jones D (1982). Body weight, exercise and menstrual status among ballet dancers in training. Br J Obstet Gynaecal 89:507–510.

Ahima RS, Dushay J, Flier SN, Prabakaran D, Flier JS (1997). Leptin accelerates the onset of puberty in normal female mice. J Clin Invest 99:391–395.

Ahima RS, Prabakaran D, Mantzoros C, Qu D, Lowell B, Maratos-Flier E, Flier J (1996). Role of leptin in the neuroendocrine response to fasting. Nature 382:250–252.

Ahren B, Mansson S, Gingerrich R, Havel PJ (1997). Regulation of plasma leptin in mice: influence of age, high-fat diet, and fasting. Am J Physiol 273:R113–R120.

Apter D (1997) Leptin in puberty. Clin Endocrinol 47:175–176.

Armstrong JD, Britt JH (1987). Nutritionally-induced anestrus in gilts: Metabolic and endocrine changes associated with cessation and resumption of estrous cyles. J Anim Sci 65:508–523.

Bai Y, Zhang S, Kim KS, Lee JK, Kim KH (1996). Obese gene expression alters the ability of 30A5 preadipocytes to respond to lipogenic hormones. J Biol Chem 271:13939–13942.

Banner L (1983). American Beauty: A Social History Through Two Centuries of the American Idea, Ideal, and Image of the Beautiful Woman. New York: Knopf.

Barash IA, Cheung CC, Weigle DS, Ren H, Kabigting EB, Kuijper JL, Clifton DK, Steiner RA (1996). Leptin is a metabolic signal to the reproductive system. Endocrinology 137:3144–3147.

Beckett JL, Sakurai H, Adams BM, Adams TE (1997a). Moderate and severe nutrient restriction has divergent effects on gonadotroph function in orchidectomized sheep. Biol Reprod 57:415–419.

Beckett JL, Sakurai H, Famula TR, Adams TE (1997b). Negative feedback potency of estradiol is increased in orchidectomized sheep during chronic nutrient restriction. Biol Reprod 57:408–414.

Berriman SJ, Wade GN, Blaustein JD (1992). Expression of fos-like proteins in gonadotropin-releasing hormone neurons of Syrian hamsters: effects of estrous cycles and metabolic fuels. Endocrinology 131:2222–2228.

Bird E, Cardone CC, Contreras RJ (1983). Area postrema lesions disrupt food intake induced by cerebroventricular infusions of 5-thioglucose in the rat. Brain Res 270:193–196.

Boden G, Chen X, Mozzoli M, Ryan I (1996). Effect of fasting on serum leptin in normal human subjects. J Clin Endocrinol Metab 81:3419–3423.

Booth PJ (1990). Metabolic influences on hypothalamic-pituitary-ovarian function in the pig. J Reprod Fertil 40:89–100.

Borer KT, Campbell CS, Tabor J, Jorgenson K, Kandarian S, Gordon L (1983). Exercise reverses photoperiodic anestrus in golden hamsters. Biol Reprod 29:38–47.

Bronson FH (1986). Food-restricted, prepubertal female rats: rapid recovery of luteinizing hormone pulsing with excess food, and full recovery of pubertal development with gonadotropin-releasing hormone. Endocrinology 118:2483–2487.

Bronson FH (1987). Puberty in female rats: relative effect of exercise and food restriction. Am J Physiol 252:R140–R144.

Bronson FH (1988). Effect of food manipulation on the GnRH-LH-estradiol axis of young female rats. Am J Physiol 254:R616–R621.

Bronson FH (1989). Mammalian Reproductive Biology. Chicago: University of Chicago Press.

Bronson FH, Manning JM (1991). The energetic regulation of ovulation: a realistic role for body fat. Biol Reprod 44:945–950.

Brown J (1962). Effects of 2-deoxy-D-glucose on carbohydrate metabolism: review of the literature and studies in the rat. Metabolism 11:1098–1112.

Bucholtz DC, Manning JM, Herbosa CG, Schillo KK, Foster DL (1993). The energetics of LH secretion: a temporally-focused view of sexual maturation. Soc Neurosci Abstracts 349.

Bucholtz DC, Vidwans NM, Herbosa CG, Schillo KK, Foster DL (1996). Metabolic interfaces between growth and reproduction. V. Pulsatile luteinizing hormone secretion is dependent on glucose availability. Endocrinology 137:601–607.

Cagampang FRA, Maeda K-I, Yokoyama A, Ota K (1990). Effect of food deprivation on the pulsatile LH release in cycling and ovariectomized female rats. Horm Metab Res 22:269–272.

Cagampang FRA, Maeda K-I, Tsukamura H, Ohkura S, Ota K (1991). Involvement of ovarian steroids and endogenous opioids in the fasting-induced suppression of pulsatile LH release in ovariectomized rats. J Endocrinol 129:321–328.

Cagampang FRA, Maeda K-I, Ota K (1992a). Involvement of the gastric vagal nerve in the suppression of pulsatile luteinizing hormone release during acute fasting in rats. Endocrinology 130:3003–3006.

Cagampang FRA. Ohkura S, Tsukamura H, Coen CW, Ota K, Maeda K-I (1992b). α_2-Adrenergic receptors are involved in the suppression of luteinizing hormone release during acute fasting in the ovariectomized estradiol-primed rat. Neuroendocrinology 56:724–728.

Cameron JL (1991). Metabolic cues for the onset of puberty. Horm Res 36:97–103.

Cameron JL (1996). Regulation of reproductive function in primates by short-term changes in nutrition. Rev Reprod 1:117–126.

Cameron JL (1997). Search for the signal that conveys metabolic status to the reproductive axis. Curr Opini Endocrinol Diabetes 4:158–163.

Cameron JL, Nosbisch C (1991). Suppression of pulsatile luteinizing hormone and testosterone secretion during short term food restriction in the adult male rhesus monkey (*Macaca mulatta*). Endocrinology 128:1532–1540.

Cameron JL, Schreihofer DA (1995). Metabolic signals and puberty in primates. *In* The Neurobiology of Puberty (Plant TM, Lee PA, eds), pp 259–270. Bristol, UK: Journal of Endocrinology.

Cao GY, Considine RV, Lynn RB (1997). Leptin receptors in the adrenal medulla of the rat. Am J Physiol 273:E448–E452.

Carlsson B, Lindell K, Gabrielsson B, Karlsson C, Bjarnason R, Westphal O, Karlsson U, Sjostrom L, Carlsson MS (1997). Obese (ob) gene defects are rare in human obesity. Obes Res 5:30–35.

Chavez M, Seeley RJ, Havel PJ, Friedman MI, Woods SC, Schwartz MW (1998). Effect of a high-fat diet on food intake and hypothalamic neuropeptide gene expression in streptozotocin diabetes. J Clin Invest. 102:340–346.

Chehab F, Lim M, Lu R (1996). Correction of the sterility defect in homozygous obese female mice by treatment with the human recombinant leptin. Nat Genet 12:318–320.

Chehab FF (1997). The reproductive side of leptin. Nat Med 3:952–953.

Chehab FF, Mounzih K, Lu R, Lim ME (1997). Early onset of reproductive function in normal female mice treated with leptin. Science 275:88–90.

Chen G, Koyama K, Yuan X, Lee Y, Zhou YT, O'Doherty R, Newgard CB, Unger RH (1996). Disappearance of body fat in normal rats induced by adenovirus-mediated leptin gene therapy. Proc Natl Acad Sci U S A 93:14795–14799.

Chen H, Charlat O, Tartaglia LA, Woolf EA, Weng X, Ellis SJ, Lakey ND, Culpepper J, Moore KJ, Breitbart RE, Duyk GM, Tepper RI, Morgenstern JP (1996). Evidence that the diabetes gene encodes the leptin receptor: identification of a mutation in the leptin receptor gene in db/db mice. Cell 84:491–495.

Chen M, Ordog T, O'Byrne KT, Goldsmith JR (1996). The insulin hypoglycemia-induced inhibition of gonadotropin-releasing hormone pulse generator activity in the rhesus monkey: roles of vasopressin and corticotropin-releasing factor. Endocrinology 137:2012–2021.

Chen M-D, O'Byrne KT, Chiappini SE, Hotchkiss J, Knobil E (1992). Hypoglycemic "stress" and gonadotropin-releasing hormone pulse generator activity in the rhesus monkey: role of the ovary. Neuroendocrinology 56:666–673.

Cheung CC, Thornton JE, Kuijper JL, Weigle DS, Clifton DK, Steiner RA (1997). Leptin is a metabolic gate for the onset of puberty in the female rat. Endocrinology 138:855–858.

Chua SC, Chung WK, Wu-Peng XS, Zhang Y, Liu S, Tartaglia LA, Leibel RL (1996). Phenotypes of mouse diabetes and fat fatty due to mutations in the OB (leptin) receptor. Science 271:994–996.

Cioffi JA, Shafer AW, Zupancic TJ, Smith-Gbur J, Mikhail A, Platika D, Snodgrass HR (1996). Novel B219/OB receptor isoforms: possible role of leptin in hematopoiesis and reproduction. Nat Med 2:585–589.

Clarke IJ, Horton RJE, Doughton BW (1990). Investigation of the mechanism by which insulin-induced hypoglycemia decreases luteinizing hormone secretion in ovariectomized ewes. Endocrinology 127:1470–1476.

Considine RV, Considine EL, Williams CJ, Nyce MR, Magosin SA, Bauer TL, Rosato EL, Colberg J, Caro JF (1995). Evidence against either a premature stop codon or the absence of obese gene mRNA in human obesity. J Clin Invest 95:2986–2988.

Contreras RJ, Fox E, Drugovich ML (1982). Area postrema lesions produce feeding deficits in the rat: effects of preoperative dieting and 2-deoxy-D-glucose. Physiol Behav 29:875–884.

Conway GS, Jacobs HS (1997). Leptin: a hormone of reproduction. Presented to European Society for Human Reproduction and Embryology pp 633–635.

Cosgrove JR, Booth PJ, Foxcroft GR (1991). Opioidergic control of gonadotrophin secretion in the prepubertal gilt during restricted feeding and realimentation. J Reprod Fertil 91:277–284.

Darwin C (1859/1985). The Origin of Species. London: Penguin.

Day ML, Imakawa K, Zalesky DD, Kittok RJ, Kinder JE (1986). Effects of restriction of dietary energy intake during the prepubertal period on secretion of luteinizing hormone and responsiveness of the pituitary to luteinizing hormone-releasing hormone in heifers. J Anim Sci 62:1641–1647.

Dickerman RW, Li H-Y, Wade GN (1993). Decreased availability of metabolic fuels suppresses estrous behavior in Syrian hamsters. Am J Physiol 264:R568–R572.

Dueck CA, Matt KS, Manore MM, Skinner JS (1996). Treatment of athletic amenorrhea with a diet and training intervention program. Int J Sport Nutr 6:24–40.

Ebling FJP, Wood RI, Karsch FJ, Wannerson LA, Suttie JM, Bucholtz DC, Schall RE, Foster DL (1990). Metabolic interfaces between growth and reproduction. III. Central mechanisms controlling pulsatile luteinizing hormone secretion in the nutritionally growth-limited female lamb. Endocrinology 126:2719–2727.

Elias CF, Lee C, Kelly J, Aschkenasi C, Ahima RS, Couceyro PR, Kuhar MJ, Saper CB, Elmquist JK (1998). Leptin activates hypothalamic CART neurons projecting to the spinal cord. Neuron 21:1375–1385.

Emilsson V, Liu YL, Cawthorne MA, Morton NM, Davenport M (1997). Expression of the functional leptin receptor mRNA in pancreatic islets and direct inhibitory action of leptin on insulin secretion. Diabetes 46:313–316.

Estacio MAC, Tsukamura H, Yamada S, Tsukahara S, Hirunagi K, Maeda K (1996a). Vagus nerve mediates the increase in estrogen receptors in the hypothalamic paraventricular nucleus and nucleus of the solitary tract during fasting in ovariectomized rats. Neurosci Lett 208:25–28.

Estacio MAC, Yamada S, Tsukamura H, Hirunagi K, Maeda K (1996b). Effect of fasting and immobilization stress on estrogen receptor immunoreactivity in the brain in ovariectomized female rats. Brain Res 717:55–61.

Fei H, Okano HJ, Li C, Lee GH, Zhao C, Darnell R, Friedman JM (1997). Anatomic localization of alternatively spliced leptin receptors (Ob-R) in mouse brain and other tissues. Proc Natl Acad Sci U S A 94:7001–7005.

Flier JS (1997). Leptin expression and action: new experimental paradigm. Proc Natl Acad Sci U S A 94:4242–4245.

Foster DL (1994). Puberty in Sheep. *In* The Physiology of Reproduction (Knobil E, Neill JD, eds), pp 411–451. New York: Raven Press.

Foster DL, Bucholtz DC (1995). Glucose as a possible metabolic cue timing puberty. *In* Serono International Symposium on Puberty: Basic and Clinical Aspects (Bergada C, Moguilevsky JA, eds), pp 319–332. Rome: Ares-Serono Symposia.

Foster DL, Olster DH (1985). Effect of restricted nutrition on puberty in the lamb: patterns of tonic luteinizing hormone (LH) secretion and competency of the LH surge system. Endocrinology 116:375–381.

Foster DL, Bucholtz DC, Herbosa CG (1995). Metabolic signals and the timing of puberty in sheep. In The Neurobiology of Puberty (Plant TM, Lee PA, eds), pp 243–257. Bristol, UK: Journal of Endocrinology.

Foster DL, Nagatani S, Bucholtz DC, Tsukamura H, Tanaka T (1998). Metabolic links between nutrition and reproduction: signals, sensors and pathways controlling GnRH secretion. In Pennington Symposium on Nutrition and Reproduction (Hansel W, Brayer G, eds). Baton Rouge: Louisiana State University Press.

Frederich RC, Lollmann B, Hamann A, Napolitano-Rosen A, Kahn BB, Lowell BB, Flier JS (1995). Expression of ob mRNA and its encoded protein in rodents. J Clin Invest 96:1658–1663.

Friedman MI (1978). Hyperphagia in rats with experimental diabetes mellitus: a response to a decreased supply of utilizable fuels. J Comp Physiol Psychol 92:109–117.

Friedman MI (1990). Making sense out of calories. In Handbook of Behavioral Neurobiology 10. Neurobiology of Food and Fluid Intake (Stricker EM, ed), pp 513–529. New York: Plenum Press.

Friedman MI (1991). Metabolic control of calorie intake. In Chemical Senses, vol 4: Appetite and Nutrition (Friedman MI, Tordoff MG, Kare MR, eds), pp 19–38. New York: Marcel Dekker.

Friedman MI (1995). Control of energy intake by energy metabolism. Am J Clin Nutr 62 (suppl):1096S–1100S.

Friedman MI (1998). Fuel partitioning and food intake. Am J Clin Nutr 67:s513–s518.

Friedman MI, Tordoff MG (1986). Fatty acid oxidation and glucose utilization interact to control food intake in rats. Am J Physiol 251:R840–R845.

Friedman MI, Ramirez I, Edens NK, Granneman J (1985). Food intake in diabetic rats: isolation of primary metabolic effects of fat feeding. Am J Physiol 249:R44–R51.

Friedman MI, Tordoff MG, Ramirez I (1986). Integrated metabolic control of food intake. Brain Res Bull 17:855–859.

Frisch RE (1990). Body fat, menarche, fitness and fertility. In Adipose Tissue and Reproduction (Frisch RE, ed), pp 1–26. Basel: Karger.

Funston RN, Roberts AJ, Hixon DL, Hallford DM, Sanson DW, Moss GE (1995). Effect of acute glucose antagonism on hypophyseal hormones and concentrations of insulin-like growth factor (IGF)-1 and IGF-binding proteins in serum, anterior pituitary, and hypothalamus of the ewe. Biol Reprod 52:1179–1186.

Gardiner RJ, Martin F, Jikier L (1991). Anorexia nervosa: endocrine studies of two distinct clinical subgroups. In Anorexia Nervosa: Recent Developments in Research (Darby PL, Garfinkel PE, Garner DM, Coscina DV, eds), pp 285–289. New York: Alan R Liss.

Garner D, Garfinkel P, Schwartz D, Thompson M (1980). Cultural expectations of thinness in women. Psychol Rep 47:483–491.

Goubillon M-L, Thalabard J-C (1996). Insulin-induced hypoglycemia decreases luteinizing hormone secretion in the castrated male rat: involvement of opiate peptides. Neuroendocrinology 64:49–56.

Grill HJ, Kaplan JM (1990). Caudal brainstem participates in the distributed neural control of feeding. In Handbook of Behavioral Neurobiology: Neurobiology of Food and Fluid Intake (Stricker EM, ed), pp 125–150. New York: Plenum Press.

Grill HJ, Donahey JCK, King L, Kaplan JM (1997). Contribution of caudal brainstem to D-fenfluramine anorexia. Psychopharmacology 130:375–381.

Hakansson M, Hulting A, Meister B (1996). Expression of leptin receptor mRNA in the hypothalamic arcuate nucleus-relationship with NPY neurones. Neuroreport 7:3087–3092.

Harris RA, Crabb DW (1992). Metabolic interrelationships. In Textbook of Biochemistry with Clinical Correlations (Devlin TM, ed), pp 575–606. New York: John Wiley & Sons.

Harris RBS, Ramsay TG, Smith SR, Bruch RC (1996). Early and late stimulation of ob mRNA expression in meal-fed and overfed rats. J Clin Invest 97:2020–2026.

Heisler LE, Pallotta CM, Reid RL, van Vugt DA (1993). Hypoglycemia-induced inhibition of luteinizing hormone secretion in the rhesus monkey is not mediated by endogenous opioid peptides. J Clin Endocrinol Metab 76:1280–1285.

Hileman SM, Schillo KK, Hall JB (1993). Effects of acute, intracerebroventricular administration of insulin on serum concentrations of luteinizing hormone, insulin, and glucose in ovariectomized lambs during restricted and *ad libitum* feed intake. Biol Reprod 48:117–124.

Hiney JK, Srivastava V, Nyberg CL, Ojeda SR, Dees WL (1996). Insulin-like growth factor I of peripheral origin acts centrally to accelerate the initiation of female puberty. Endocrinology 137:3717–3728.

Hirschfield AN, Midgley AR (1978). The role of FSH in the selection of large ovarian follicles in the rat. Biol Reprod 19:606–611.

Hoggard N, Mercer JG, Rayner V, Moar K (1997). Localization of leptin receptor mRNA splice variants in murine peripheral tissues by RT-PCR and *in situ* hybridization. Biochem Biophys Res Commun 232:383–387.

Horn C, Friedman MI (1998). Metabolic inhibition increases feeding and brain fos-like immunoreactivity as a function of diet. Am J Physiol, 275:R448–R459.

Howland BE (1980). Effect of glucoprivation induced by 2-deoxy-D-glucose on serum gonadotropin levels, pituitary response to GnRH and progesterone-induced release of luteinizing hormone in rats. Horm Metab Res 12:520–523.

Howland BE, Ibrahim EA (1971). Increased LH suppressing effects of oestrogen in ovariectomized rats as a result of underfeeding. J Reprod Fertil 35:1484–1485.

Huang XF, Koutcherov I, Lin S, Wang HQ, Storlien L (1996). Localization of leptin receptor mRNA expression in mouse brain. Neuroreport 7:2635–2638.

Hwa JJ, Ghibaudi L, Compton D, Fawziand AB, Strader CD (1996). Intracerebroventricular injection of leptin increases thermogenesis and mobilizes fat metabolism in ob/ob mice. Horm Metab Res 28:659–663.

Hyde TM, Miselis RR (1983). Effects of area postrema/caudal medial nucleus of solitary tract lesions on food intake and body weight. Am J Physiol 244:R577–R587.

l'Anson H, Foster DL, Foxcroft CR, Booth PJ (1991). Nutrition and reproduction. Oxf Rev Reprod Biol 13:239–311.

l'Anson H, Terry SK, Lehman MN, Foster DL (1997). Regional differences in the distribution of gonadotropin-releasing hormone cells between rapidly growing and growth-prepubertal female sheep. Endocrinology 138:230–236.

Juiter A, Panhuysen G, Everaerd W, Koppeschaar H, Krabbe P, Zebssen P (1993). The paradoxical nature of sexuality in anorexia nervosa. J Sex Marital Ther 19:259–275.

Kalra SP, Kalra PS (1984). Opioid-adrenergic-steroid connection in regulation of luteinizing hormone secretion in the rat. Neuroendocrinology 38:418–426.

Kamohara S, Burcelin R, Halaas JL, Friedman JM, Charron MJ (1997). Acute stimulation of glucose metabolism in mice by leptin treatment. Nature 389:374–377.

Kaplan JM, Seeley RJ, Grill HJ (1993). Daily caloric intake in intact and chronic decerebrate rats. Behav Neurosci 107:876–881.

Kaplan JM, Song S, Grill HJ (1998). Serotonin receptors in the caudal brainstem are necessary and sufficient for the anorectic effect of peripherally administered mCPP. Psychopharmacology, 137:43–49.

Katz JL, Weiner H (1991). The aberrant reproductive endocrinology of anorexia nervosa. *In* Brain, Behavior and Bodily Disease (Weiner H, Hofer MA, Stunkard AJ, eds), pp 165–180. New York: Raven Press.

Kennedy GC, Mitra J (1963). Body weight and food intake as initiating factors for puberty in the rat. J Physiol 166:408–418.

Kieffer TJ, Heller RS, Habener JF (1996). Leptin receptors expressed on pancreatic beta-cells. Biochem Biophys Res Commun 224:522–527.

Kile JP, Alexander BM, Moss GE, Hallford DM, Nett TM (1991). Gonadotropin-releasing hormone overrides the negative effect of reduced dietary energy on gonadotropin synthesis and secretion in ewes. Endocrinology 128:843–849.

Kolaczynski JW, Considine RV, Ohannesian J, Marco C, Opentanova I, Nyce MR, Myint M, Caro JF (1996). Responses of leptin to short-term fasting and refeeding in humans. Diabetes 45:1511–1515.

Kusche M, Henrich W, Bolte A (1987). Effect of competitive sports on the menstrual cycle and sexuality— results of a survey of the West German team for the Olympic games in Los Angeles. Geburtshilfe Frauenheilkd 47:808–811.

Leclereq-Meyer L, Considine RV, Sener A, Malaisse WJ (1996). Do leptin receptors play a functional role in the endocrine pancreas? Biochem Biophys Res Commun 229:794–798.

Lee GH, Proenca R, Montez JM, Carroll KM, Darvishzadeh JG, Lee JI, Friedman JM (1966). Abnormal splicing of the leptin receptor in diabetic mice. Nature 379:632–635.

Levy JR, LeGall-Salmon E, Santos M, Pandak WM, Stevens W (1997). Effect of enternal versus parenteral nutrition on leptin gene expression and release into the circulation. Biochem Biophys Res Commun 237:98–102.

Li H-Y, Wade GN, Blaustein JD (1994). Manipulations of metabolic fuel availability alter estrous behavior and neural estrogen-receptor immunoreactivity in Syrian hamsters. Endocrinology 135:240–247.

Loucks AB, Vaitukaitis J, Cameron JL, Rogol AD, Skrinar G, Warren MP, Kendrick J, Limacher MC (1992). The reproductive system and exercise in women. Med Sci Sport Exerc 24:S288–S293.

Loucks AB, Heath EM, Verdun M, Watts JR (1994). Dietary restriction reduces luteinizing hormone (LH) pulse frequency during waking hours and increases LH pulse amplitude during sleep in young menstruating women. J Clin Endocrinol Metab 78:910–915.

Lynn RB, Cao G, Considine RV, Hyde TM, Caro JF (1996). Autoradiographic localization of leptin binding in the choroid plexus of ob/ob and db/db mice. Biochem Biophys Res Commun 219:884–889.

MacDonald, MJ (1990). Elusive proximal signals of beta-cells for insulin secretion. Diabetes 39:1461–1466.

Maeda K-I, Cagampang FRA, Coen CW, Tsukamura H (1994). Involvement of the catecholaminergic input to the paraventricular nucleus and of corticotropin-releasing hormone in the fasting-induced suppression of luteinizing hormone release in female rats. Endocrinology 134:1718–1722.

Mangels RA, Jetton AE, Powers JB, Wade GN (1996). Food deprivation and the facilitory effects of estrogen in female hamsters: the LH surge and locomotor activity. Physiol Behav 60:837–843.

Manning JM, Bronson FH (1991). Suppression of puberty in rats by exercise: effects on hormone levels and reversal with GnRH infusion. Am J Physiol 260:R717–R723.

Mattern LG, Helmreich DL, Cameron JL (1993). Diurnal pattern of pulsatile LH and testosterone secretion in adult male rhesus monkeys (*Macaca mulatta*): influence of daily meal intake. Endocrinology 132:1044–1054.

McCann JP, Hansel W (1986). Relationship between insulin and glucose metabolism and pituitary-ovarian functions in fasted heifers. Biol Reprod 34:630–641.

McShane TM, Petersen SL, McCrane S, Keisler DH (1993). Influence of food restriction on neuropeptide-Y, proopiomelanocortin, and luteinizing hormone–releasing hormone gene expression in sheep hypothalami. Biol Reprod 49:831–839.

Mercer JG, Hoggard N, Williams LM, Lawrence CB, Hannah LT, Trayhurn P (1996). Localization of leptin receptor mRNA and the long form splice variant (Ob-Rb) in mouse hypothalamus and adjacent brain regions by *in situ* hybridization. FEBS Lett 387:113–116.

Miller DW, Blache D, Martin GB (1995). The role of intracerebral insulin in the effect of nutrition on gonadotropin secretion in mature male sheep. J Endocrinol 147:321–329.

Morin LP (1975). Effects of various feeding regimens and photoperiod or pinealectomy on ovulation in the hamster. Biol Reprod 13:99–103.

Morin LP (1986). Environment and hamster reproduction: Responses to phase-specific starvation during estrous cycle. Am J Physiol 251:R663–R669.

Murahashi K, Bucholtz DC, Nagatani S, Tsukahara S, Tsukamura H, Foster DL, Maeda K-I (1996). Suppression of luteinizing hormone pulses by restriction of glucose availability is mediated by sensors in the brain stem. Endocrinology 137:1171–1176.

Nagatani S, Tsukamura H, Maeda K-I (1994). Estrogen feedback needed at the paraventricular nucleus or A2 to suppress pulsatile luteinizing hormone release in fasting female rats. Endocrinology 135:870–875.

Nagatani S, Bucholtz DC, Murahashi K, Estacio MAC, Tsukamura H, Foster DL, Maeda K-I (1996a). Reduction of glucose availability suppresses pulsatile LH release in female and male rats. Endocrinology 137:1166–1170.

Nagatani S, Tsukamura H, Murahashi K (1996b). Paraventricular norepinephrine release mediates glucoprivic suppression of pulsatile luteinizing hormone secretion. Endocrinology 137:3183–3186.

Nagatani S, Guthikonda P, Thompson RC, Tsukamura H, Maeda K, Foster DL (1998). Evidence for GnRH regulation by leptin: Leptin administration prevents reduced pulsatile LH secretion during fasting. Neuroendocrinology 47:370–376.

Nillius SJ, Fries H, Wide L (1975). Successful induction of follicular maturation and ovulation by prolonged treatment with LH-releasing hormone in women with anorexia nervosa. Am J Obstet Gynecol 122:921–928.

Orsini M (1961). The external vaginal phenomena characterizing the stages of estrous cycles, pregnancy, pseudopregnancy, lactation and the anestrous hamsters, *Mesocricetus auratus* Waterhouse. Proc Anim Care Panel 16:193–206.

Panicker AK, Wade GN (1998). Insulin-induced repartitioning of metaboic fuels inhibits estrous behavior in Syrian hamsters: role of area postrema. Am J Physiol, 274:R1094–R1098.

Panicker AK, Mangels RA, Powers JB, Wade GN, Schneider JE (1998). AP lesions block suppression of estrous behavior, but not estrous cyclicity, in food-deprived Syrian hamsters. Am J Physiol 275:R158–R164.

Perrigo G, Bronson FH (1983). Foraging effort, food intake, fat deposition and puberty in female mice. Biol Reprod 29:455–463.

Piacsek BE (1985). Altered negative feedback response to ovariectomy and estrogen in prepubertal restricted-diet rats. Biol Reprod 32:1062–1068.

Piacsek BE (1996). Effects of nutrition on reproductive endocrine function. *In* Effects of Nutrition on Reproductive Endocrine Function (Gass GH, Kaplan HM, eds). Boca Raton, FL: CRC Press.

Pieper DR, Ali HY, Benson L, Shows M, Lobocki CA, Subramanian MG (1995). Voluntary exercise increases gonadotropin secretion in male golden hamsters. Am J Physiol 269:R179–R185.

Prentki M, Tornheim K, Corkey BE (1997). Signal transduction mechanisms in nutrient-induced insulin secretion. Diabetologia 40 Suppl 2:S32–S41.

Powers JB, Jetton AE, Wade GN (1994). Interactive effects of food deprivation and exercise on reproductive function in female hamsters. Am J Physiol 267:R185–R190.

Pugliese MT, Lifshitz F, Grad G, Fort P, Marks-Katz M (1983). Fear of obesity: A cause of short stature and delayed puberty. N Engl J Med 309:513–517.

Rabach J, Faltus F (1991). Sexuality of women with anorexia nervosa. Acta Psychiatr Scand 84:9–11.

Reed DR, Contreras RJ, Maggio C, Greenwood MRC, Rodin J (1988). Weight cycling in female rats increases dietary fat selection and adiposity. Physiol Behav 42:389–395.

Rentsch J, Levens N, Chiesi M (1995). Recombinant ob-gene product reduces food intake in fasted mice. Biochem Biophys Res Commun 214:131–136.

Richards JS, Midgley AR (1976). Protein hormone action: a key to understanding ovarian follicular and luteal cell development. Biol Reprod 14:82–94.

Ritter S, Taylor JS (1990). Vagal sensory neurons are required for lipoprivic but not glucoprivic feeding in rats. Am J Physiol 258:R1395–R1401.

Ritter RC, Slusser PG, Stone S (1981). Glucoreceptors controlling feeding and blood glucose are in the hindbrain. Science 213:451–453.

Ritter S, Calingasan NY, Hutton B, Dinh TT (1992). Cooperation of vagal and central neural systems in monitoring metabolic events controlling feeding behavior. *In* Neuroanatomy and Physiology of Abdominal Vagal Afferents (Ritter S, Ritter RC, Barnes CD, eds), pp 249–277. Boca Raton, FL: CRC Press.

Roemmich JN, Li X, Rogol AD, Rissman EF (1997). Food availability affects neural estrogen receptor immunoreactivity in prepubertal mice. Endocrinology 138:5366–5373.

Rossetti L, Massillon D, Barzilai N, Vuguin P, Chen W, Hawkins M, Wu J, Wang J (1997). Short term effects of leptin on hepatic gluconeogenesis and in vivo insulin action. J Biol Chem 272:27758–27763.

Saladin R, de Vos P, Guerre-Millo M, Leturque A, Girard J, Staels B, Auwerx J (1995). Transient increase in obese gene expression after food intake or insulin administration. Nature 377:527–529.

Sarmiento U, Benson B, Kaufman S, Ross L, Qi M, Scully S, DiPalma C (1997). Morphologic and molecular changes induced by recombinant human leptin in the white and brown adipose tissues of C57BL/6 mice. Lab Invest 77:243–256.

Scarpace PJ, Matheny M, Pollock BH, Tumer N (1997). Leptin increases uncoupling protein expression and energy expenditure. Am J Physiol 273:E226–E230.

Scharrer E, Langhans W (1986). Control of food intake by fatty acid oxidation. Am J Physiol 250:R1003–R1006.

Schneider JE, Wade GN (1989a). Availability of metabolic fuels controls estrous cyclicity of Syrian hamsters. Science 244:1326–1328.

Schneider, JE, Wade GN (1989b). Body weight and reproduction (letter). Science 246:432.

Schneider JE, Wade GN (1990a). Decreased availability of metabolic fuels induces anestrus in golden hamsters. Am J Physiol 258:R750–R755.

Schneider JE, Wade GN (1990b). Effects of diet and body fat content on cold-induced anestrus in Syrian hamsters. Am J Physiol 259:R1198–R1204.

Schneider JE, Wade GN (1997). Critical fatness hypothesis (letter). Am J Physiol 273:E231–E232.

Schneider JE, Zhou D (1999). Central and peripheral fuel oxidation interact to control estrous cycles in Syrian hamsters. Am J Physiol, in press.

Schneider JE, Zhu Y (1994). Caudal brain stem plays a role in metabolic control of estrous cycles in Syrian hamsters. Brain Res 661:70–74.

Schneider JE, Blum RM, Wade GN (1999). Control of food intake and estrous cycles in Syrian hamsters: the role of insulin and leptin. Am J Physiol, in press.

Schneider JE, Friedenson DG, Hall A, Wade GN (1993). Glucoprivation induces anestrus and lipoprivation may induce hibernation in Syrian hamsters. Am J Physiol 264:R573–R577.

Schneider JE, Finnerty BC, Swann JM, Gabriel JM (1995). Glucoprivic treatments that induce anestrus, but do not affect food intake, increase FOS-like immunoreactivity in the area postrema and nucleus of the solitary tract in Syrian hamsters. Brain Res 698:107–113.

Schneider JE, Goldman MD, Leo NA, Rosen ME (1997a). Central vs. peripheral metabolic control of estrous cycles in Syrian hamsters. II. Glucoprivation. Am J Physiol 272:R406–R412.

Schneider JE, Goldman MD, Tang S, Bean B (1997b). Leptin, metabolic fuels and reproduction in Syrian hamsters. Soc Neurosci Abstracts 23:417.

Schneider JE, Hall AJ, Wade GN (1997c). Central vs. peripheral metabolic control of estrous cycles in Syrian hamsters. I. Lipoprivation. Am J Physiol 272:R400–R405.

Schneider JE, Goldman MD, Tang S, Bean B, Ji H, Friedman MI (1998). Leptin indirectly affects estrous cycles by increasing metabolic fuel oxidation. Horm Behav 33:217–228.

Schneider JE, Lazzarini SJ, Friedman MI, Wade GN (1988). Role of fatty acid oxidation in food intake and hunger motivation in Syrian hamsters. Physiol Behav 43:617–623.

Schreihofer DA, Renda F, Cameron JL (1996). Feeding-induced stimulation of luteinizing hormone secretion in male rhesus monkeys is not dependent on a rise in blood glucose concentration. Endocrinology 137:3770–3776.

Schwartz H (1986). Never Satisfied: A Cultural History of Diets, Fantasies and Fat. New York: Free Press.

Schwartz MW, Figlewicz DP, Baskin DG, Woods SC, Porte DJr (1992). Insulin in the brain: a hormonal regulator of energy balance. Endocr Rev 13:387–414.

Schwartz NB (1974). The role of FSH and LH and their antibodies on follicle growth and ovulation. Biol Reprod 10:236–272.

Scott EC, Johnston FE (1982). Critical fat, menarche, and the maintenance of manstrual cycles: a critical review. J Adolescen Health Care 2:249–260.

Shimabukuro M, Koyama K, Chen G, Wang MY, Trieu F, Lee Y, Newgard CB, Unger RH (1997). Direct antidiabetic effect of leptin through triglyceride depletion of tissues. Proc Natl Acad Sci U S A 94:4637–4641.

Siegel LI, Wade GN (1979). Insulin withdrawal impairs sexual receptivity and retention of brain cell nuclear estrogen receptors in diabetic rats. Neuroendocrinology 29:200–206.

Smedh U, Hakansson ML, Meister B, UvnasMoberg K (1998). Leptin injected into the fourth ventricle inhibits gastric emptying . Neuroreport 9:297–301.

Spillar PA, Piacsek BE (1991). Underfeeding alters the effect of low levels of estradiol on luteinizing hormone pulsatility in ovariectomized female rats. Neuroendocrinology 53:253–260.

Sprangers SA, Piacsek BE (1988). Increased suppression of luteinizing hormone secretion by chronic and acute estradiol administration in underfed adult female rats. Biol Reprod 39:81–87.

Sprangers SA, Piacsek BE (1998). Chronic underfeeding increases the positive feedback efficacy of estrogen on gonadotropin secretion. Proc Soc Exp Biol Med 216:398–403.

Stewart DE (1992). Reproductive functions in eating disorders. Annu Med 24:287–291.

Strobel A, Issad T, Camoin L, Ozata M, Strosberg AD (1998). A leptin missense mutation associated with hypogonadism and morbid obesity. Nat Genet 18:213–215.

Thompson MP (1996). Meal-feeding specifically induces obese mRNA expression. Biochem Biophys Res Commun 224:332–337.

Trayhurn P, Rayner DV (1996). Hormones and the ob gene product (leptin) in the control of energy balance. Biochem Soc Trans 24:565–570.

Trayhurn P, Duncan JS, Rayner DV (1995). Acute cold-induced suppression of ob (obese) gene expression in white adipose tissue of mice: mediation by the sympathetic system. Biochem J 311:729–733.

Tutwiler GF, Brentzel HJ, Kiorpes TC (1985). Inhibition of mitochondrial carnitine palmitoyl transferase A in vivo with methyl-2-tetradecylglycidate (methyl palmoxirate) and its relationship to ketonemia and glycemia. Proc Soc Exp Biol Med 178:288–296.

VanItallie, TB, Kissileff HR (1990). Human obesity: a problem in body energy economics. *In* Handbook of Behavioral Neurobiology 10. Neurobiology of Food and Fluid Intake (Stricker EM, ed), pp 513–529. New York: Plenum Press.

Wade GN, Schneider JE (1992). Metabolic fuels and reproduction in female mammals. Neurosci Biobehav Rev 16:235–272.

Wade GN, Schneider JE, Friedman MI (1991). Insulin-induced anestrus in Syrian hamsters. Am J Physiol 260:R148–R152.

Wade GN, Schneider JE, Li H-Y (1996). Control of fertility by metabolic cues. Am J Physiol 270:E1–E19.

Wade GN, Lempicki RL, Panicker AK, Frisbee RM, Blaustein JD (1997). Leptin facilitates and inhibits sexual behavior in female hamsters. Am J Physiol 272:R1354–R1358.

Walker C-D, Mitchell JB, Woodside BC (1995). Suppression of LH secretion in food-restricted lactating females: effects of ovariectomy and bromocryptine treatment. J Endocrinol 146:95–104.

Wang Y, Kuropatwinski KK, White DW, Hawley TS, Hawley RG, Tartaglia LA, Baumann H (1997). Leptin receptor action in hepatic cells. J Biol Chem 272:16216–16223.

White JR, Case DA, McWhirter D, Mattison AN (1990). Enhanced sexual behavior in exercising men. Arch Sex Behav 19:193–209.

Williams NI, Lancas MJ, Cameron JL (1996). Stimulation of luteinizing hormone secretion by food intake: evidence against a role for insulin. Endocrinology 137:2565–2571.

Woods SC, Chavez M, Park CR, Riedy C, Kaiyala K, Richardson RD, Figlewicz DP, Schwartz MW, Porte D, Seeley RJ (1996). The evaluation of insulin as a metabolic signal influencing behavior via the brain. Neurosci Biobehav Rev 20:139–144.

Woodside B, Abizaid A, Caporale M (1998a). The role of specific macronutrient availability in the effect of food restriction on length of lactational diestrus in rats. Physiol Behav 64:409–414.

Woodside B, Abizaid A, Jafferali S (1998b). Effect of acute food deprivation on lactational infertility in rats is reduced by leptin administration. Am J Physiol 274:R1653–R1658.

Yu WH, Kimur M, Walczewska A, Karanth S, McCann SM (1997a). Role of leptin in hypothalamic-pituitary function. Proc Natl Acad Sci U S A 94:1023–1028.

Yu WH, Walczewska A, Karanth S, McCann SM (1997b). Nitric oxide mediates leptin-induced luteinizing hormone–releasing hormone (LHRH) and LHRH and leptin-induced LH release from the pituitary gland. Endocrinology 138:5055–5058.

Zamorano PL, Mahesh VB, De Sevilla LM, Chorich LP, Bhat GW, Brann DW (1997). Expression and localization of the leptin receptor in endocrine and neuroendocrine tissues of the rat. Neuroendocrinology 65:223–228.

Zhang Y, Proenca R, Maffei M, Barone M, Leopold L, Friedman JM (1994). Positional cloning of the mouse Obese gene and its human homologue. Nature 372:425–431.

Zhou YT, Shimabukuro M, Koyama K, Lee Y, Wang MY, Trieu F, Newgard CB, Unger RH (1997). Induction by leptin of uncoupling protein-2 and enzymes of fatty acid oxidation. Proc Natl Acad Sci U S A 94:6386–6390.

IB ENVIRONMENTAL CONTEXT: INTEGRATION OF ENVIRONMENTAL FACTORS IN BIRDS

The next chapters concern the environmental and social regulation of reproduction in birds. The integrative theme of this book originated with studies in birds, and the study of birds continues to be a source of important thoeries and methodologies relevant to other taxa (Konishi et al. 1989). A great deal of present evolutionary and ecological theory, as well as our current understanding of the behavioral endocrinology of reproduction, can be traced to the work of various ornithologists. For example, Lack has been cast as the father of evolutionary ecology, the study of behavior as it is influenced by life history and evolution. Later, Crook was instrumental in demonstrating how ecological factors shape social organization in populations of weaverbirds, and his influence spread to investigators studying other taxa. Information about the ecology of bird reproduction, especially the work of Ricklefs on resource allocation, has provided a very useful database for informing studies of the physiological mechanisms that control mating and fertility. Many "firsts" in reproductive biology came from bird studies. Berthold conducted the first study demonstrating that gonadal secretions influenced reproduction in roosters. Rowan performed the first studies of the effects of photoperiod on reproduction in dark-eyed juncos. Lehrman demonstrated the influence of behavior on hormone secretion and the reciprocal influence of hormone secretion on behavior in ringdoves.

Pioneering work at the interface of ecology and endocrinology was accomplished over the last 30 years by the distinguished first author of chapter 4, John C. Wingfield. Beginning from knowledge of the specific life-history stages of a variety of avian species, he made predictions about the endocrine and environmental control of behavior, then used the techniques of behavioral endocrinology to test these predictions on this same array of species. Wingfield, Jacobs, Tramontin, Perfito, Meddle, Maney, and Soma have graced us here with a detailed introduction to his ecological framework concerning life-history stages and breeding phases, along with an up-to-date account of the field and laboratory data that test his predictions. At this point in time, there is no more complete resource available for an integration of ecological with behavioral endocrine strategies.

Chapter 5, by Ball and Bentley starts from Wingfield's perspective and takes this work further into the neuroendocrine mechanisms whereby these life-history stages, environmental factors, and social cues are translated from sensory stimulus, to changes in the nervous system, and finally to reproductive behavior. They have used state-of-the-art neuroanatomy and neuroendocrinology techniques while still maintaing an ecological perspective.

Silver and Silverman also share an interest in environmental and social influences on reproduction. Prior to their work in chapter 6, each of these investigators made important contributions to reproductive neuroendocrinology as well as to circadian biology. In chapter 6, the reader is treated to the unusual account of a serendipitous discovery that may alter our view of the immune system. While studying the effects of courtship behavior on gonadotropin-releasing hormone (GnRH) cells in the brains of ringdoves, they discovered the appearance of immune cells that actually contain GnRH and appear during courtship. Traditionally, the fields of immunology and reproductive biology have progressed in parallel, with little crosstalk and cooperation between the two disciplines. With the work of Silver and Silverman, these two fields merged to become psychoneuroimmunology. It is interesting that in chapter 6, recognition of the link between immune function and reproduction came about through a neuroanatomical investigation, whereas in chapter 8 the same link is recognized in relation to the energetic tradeoffs between immune function and reproductive processes.

Reference

Konishi K, Emlen ST, Ricklefs RE, Wingfield J (1989). Contributions of bird studies to biology. Science 246:465–246.

4 Toward an Ecological Basis of Hormone-Behavior Interactions in Reproduction of Birds

John C. Wingfield, Jerry D. Jacobs, Anthony D. Tramontin, Nicole Perfito, Simone Meddle, Donna L. Maney, and Kiran Soma

As endocrinologists, our focus is on specific actions of hormones on reproductive function. Typically this follows a reductionist approach in the laboratory to resolve pathways for control of chemical signals and their mechanisms of action. Chapters 2 and 3 together illustrate a more integrative approach. Ecological considerations predicted the existence of a reversible mechanism that controls reproduction in service of energy balance (rather than in service of energy reserves), and laboratory experiments were guided by considerations concerning wild populations of mammals. Chapters 2 and 3 are concerned with one environmental variable, that is, energy availability. However, if we consider an individual organism in its natural environment, reproduction in context involves *multiple* levels of environmental, social, and chemical cues that interact on a perplexing scale—far beyond the scope of simple laboratory experimentation. Is it possible to unravel this Gordian knot of morphological, physiological, and behavioral interactions to determine mechanisms? Can the bewildering and apparently chaotic events in the field be rationalized and understood? Although answers to these questions may be unattainable at present, in this chapter we outline some ideas and approaches at ecological and theoretical levels that have heuristic value in addressing these critical issues.

An individual's offspring contribution to the next generation, or fitness, is a summation of all reproductive events that occur over the course of the individual's lifetime. Thus, it is necessary to consider reproductive function in the context of a life cycle, then focus on what specific processes are changing and require regulation. If a theoretical framework can be developed, it will aid the formulation of precise hypotheses and predictions that can be tested in the laboratory and field, and at organismal and molecular levels. Organisms use environmental information to time and integrate various stages in their life cycle to maximize lifetime fitness. These signals affect neural and endocrine functions which in turn orchestrate appropriate changes in morphology, physiology, and behavior. The nervous system responds very rapidly to external signals. In contrast, changes in life-history stage, such as reproductive function, in response to a fluctuating environment depend more upon neuroendocrine and endocrine secretions. These secretions, in turn, may determine the likelihood of particular neural and behavioral responses. It is here that the field of endocrine physiology intertwines with ecology and behavior, forming a continuum to cell and molecular biology.

We have developed field endocrine techniques to obtain information from free-living animals (particularly birds). Field studies allow us to work at these interfaces

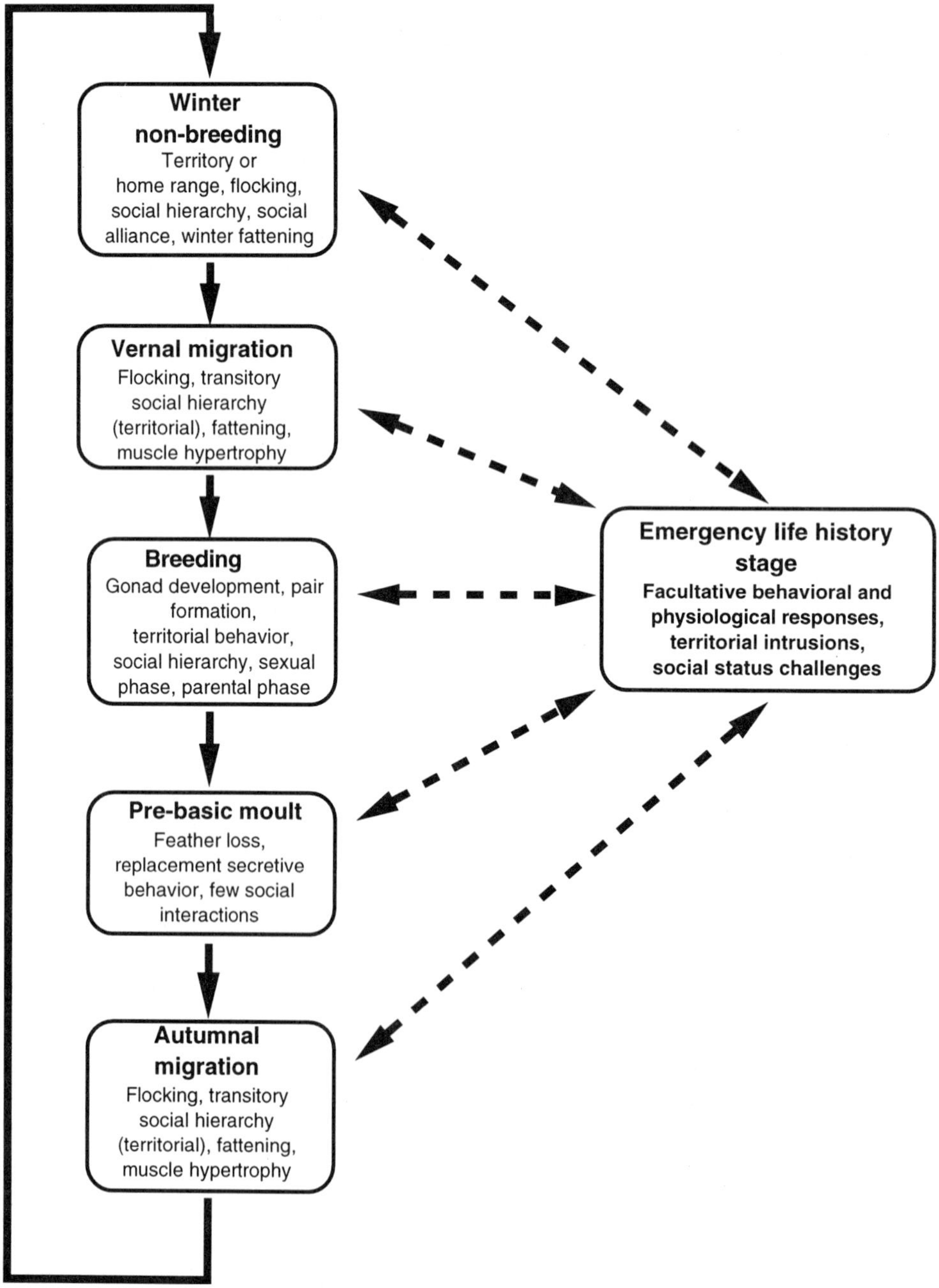
Winter non-breeding
Territory or home range, flocking, social hierarchy, social alliance, winter fattening
Vernal migration
Flocking, transitory social hierarchy (territorial), fattening, muscle hypertrophy
Breeding
Gonad development, pair formation, territorial behavior, social hierarchy, sexual phase, parental phase
Pre-basic moult
Feather loss, replacement secretive behavior, few social interactions
Autumnal migration
Flocking, transitory social hierarchy (territorial), fattening, muscle hypertrophy
Emergency life history stage
Facultative behavioral and physiological responses, territorial intrusions, social status challenges

of ecology and behavior with endocrinology and provide insight for physiology and molecular studies in the laboratory. Mathematical theory also helps us design experimental manipulation in the field and can suggest where new research directions should be initiated. It is important to note that although the highly integrated approach taken here focuses on reproductive function in birds, the principles involved also have future application to how organisms regulate the progression of life-history processes in general, and what the endocrine mechanisms may be. We also place emphasis on the class Aves because in many respects it is the most extensively studied in relation to field and laboratory investigations.

The chapter begins with a general discussion of life-history stages using the white-crowned sparrow as an example. The focus then narrows to the stage of interest, breeding. Next, the different types of environmental factors that impinge on the breeding stage are described and classified. For detailed discussion of theoretical predictions and empirical tests of these predictions, the breeding stage is subdivided into three distinct phases. Each phase of the breeding stage is associated with different environmental, social, and hormonal controls. It will become evident that the study of reproduction within the framework of the life-history stage has been critical to understanding both the mechanisms and function of reproductive behavior.

Finite-State Machine Theory and Life History

The life cycle of vertebrates is made up of series of distinct stages each independent of the other, although they may overlap to varying degrees (Jacobs 1996; Wingfield et al. 1997a,b, 1998a). A typical example of life-history stages is given in figure 4.1 for the white-crowned sparrow (*Zonotrichia leucophrys*). Each box represents a distinct life-history stage within which are several unique substages. On the left-hand side of the figure is a sequence of life-history stages that occur in a predictable progression of about 1 year. Each occurs in a set sequence that cannot be reversed. For example, it

Figure 4.1
Life-history stages of a typical passerine bird (e.g., white-crowned sparrow; modified from Jacobs 1996). Note that the right-hand box is a transitory stage (the "emergency life-history stage," ELHS) that is triggered in response to perturbations in the environment. These events are unpredictable and thus the ELHS can be initiated at any time in the life cycle as indicated by the broken lines. Each stage is represented by a box and within each box is a list of substages expressed within that stage. Finite-state machine theory (Jacobs 1996) suggests that even though some substages are expressed in several life-history stages, their context and control mechanisms are different. If true, then this model has considerable implications for environmental and endocrine control of behavior and physiology in changing environments. (Modified from Jacobs 1996; Wingfield et al. 1998a.)

is not possible to revert to vernal migration after the breeding season; the sequence must move on to the next stage, in this case prebasic molt. Substages are unique to each stage, but can be expressed in many sequences and combinations within that stage (Jacobs 1996). For each organism we can postulate a finite-state machine of several phenotypic (life-history) stages that occur in a fixed sequence and usually on a schedule determined by, for example, the changing seasons. Additionally, there is a transitory, facultative life-history stage called the "emergency life-history stage" (ELHS, right-hand side of figure 4.1), which is completely separate from the normal life-history sequence (left side of the figure). The ELHS is triggered by unpredictable perturbations of the environment. It is designed to redirect the individual away from the normal sequence of life-history stages to maximize survival in the face of unfavorable conditions. Once the perturbation passes, the individual can return to the life-history stage appropriate for that time of year (see figure 4.1; Wingfield et al. 1998b). Thus there are two distinct components to the finite-state machine, one dealing with the normal progression of life-history stages and the other with unpredictable perturbations. They represent adaptations to the two major forms of environmental variables to which an organism must adjust in order to survive and breed successfully.

The Breeding Stage

The breeding life-history stage and its characteristic set of substages are involved in the whole reproductive process. We will use the theoretical framework provided by finite-state machine theory as a template to investigate how reproductive processes are regulated by environmental and social cues, and how these influence neuroendocrine and endocrine secretions that orchestrate all aspects of reproduction. We also address the ELHS in response to perturbations that disrupt reproductive function. It is hoped that this framework may represent the beginnings of ecological bases of endocrine phenomena in general. Only in this way do we have any hope of addressing the complex interactions of an organism with its environment.

A major role of hormones is to regulate the expression of a life-history stage that includes transitions in morphology, physiological state, and behavior. There are three distinct phases of a life-history stage. First, there is a developmental phase in which morphological and physiological adjustments are made prior to the second phase—"mature capability"—in which a number of substages can be activated (figures 4.2 and 4.3). Hormones also play a major role in the activation of these substages through mechanisms that may be very different from their roles promoting development of that stage (Jacobs 1996). The third phase is termination of the stage, which

Ecological Bases of Endocrine Phenomena

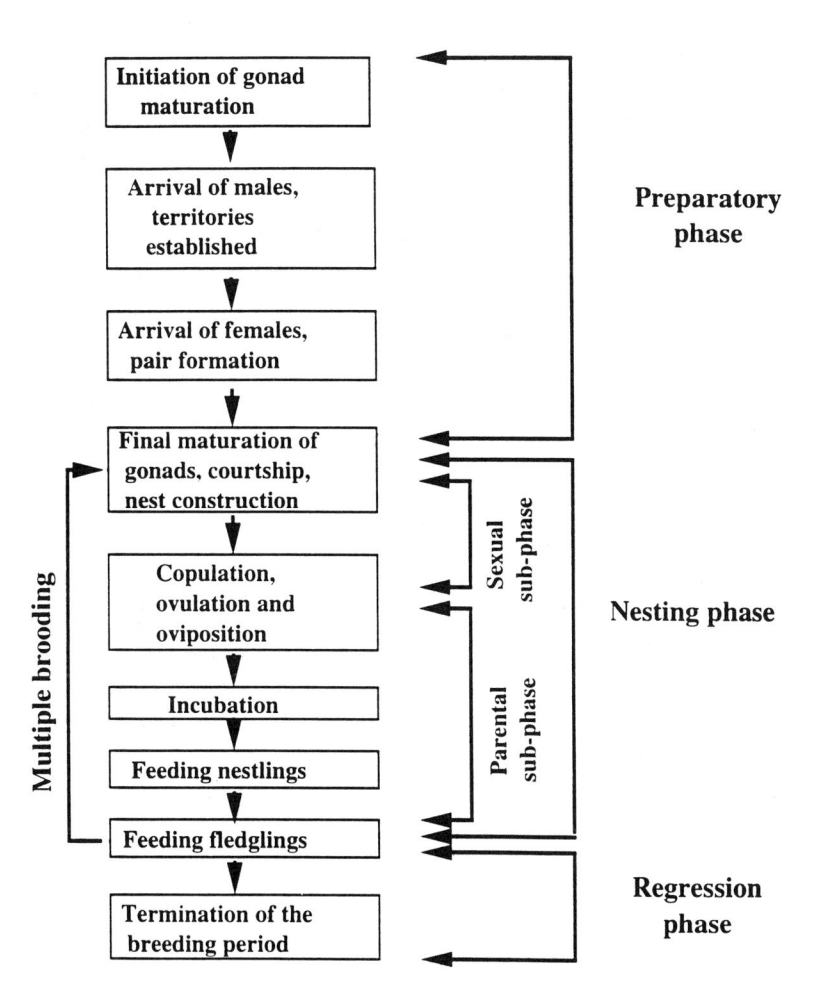

Figure 4.2
A detailed analysis of the major events and substages in the breeding life-history stage of the white-crowned sparrow (modified from Wingfield and Farner 1980; Wingfield and Moore 1987). Note that there are three distinct phases—development of the stage (preparatory phase), mature capability of the stage in which reproduction can actually begin (nesting phase), and termination of the stage (regression phase). Each phase has its own suite of substages that can be activated in various combinations or alone (see Jacobs 1996). It is the combination of these substages expressed at any point that gives the reproductive state at that time. See text for details.

INHIBITORY SUPPLEMENTARY INFORMATION

INITIAL PREDICTIVE INFORMATION

ACCELERATORY SUPPLEMENTARY INFORMATION

Development of Life History Stage
(cell division, differentiation, metamorphosis etc.)

Mature Capability of Life History Stage
(characteristic events can be expressed)

Onset of Events Characteristic of the Life History Stage

Termination of Life History Stage
(next state can now begin development)

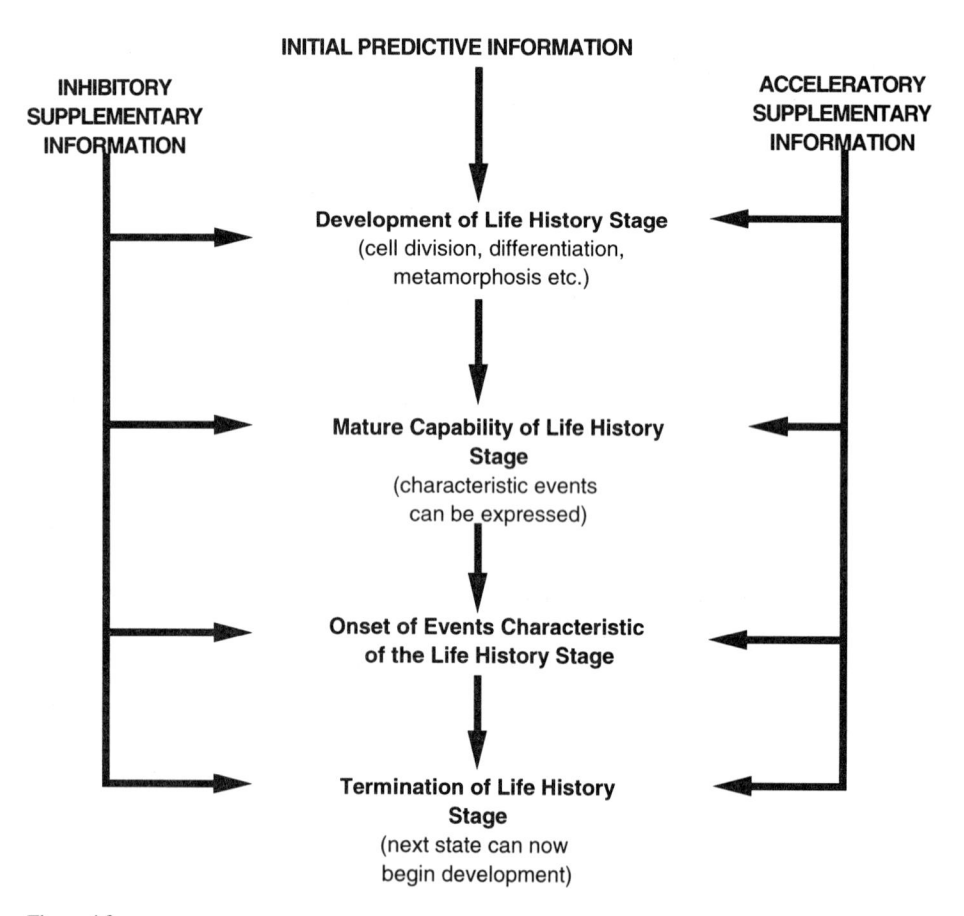

Figure 4.3
Development and termination of the reproductive stage with reference to how initial predictive and supplementary cues may regulate the process. Note that social cues (synchronizing and integrating information) may influence reproductive function at any point in this process. Additionally, unpredictable perturbations in the environment may trigger the transitory, emergency life-history stage at any time, thus interrupting reproductive function until the event passes.

then allows development of the next. There can be varying degrees of overlap of stages, however. For example, in many avian species onset of molt may occur while individuals are still feeding young from the last nesting attempt (Wingfield and Farner 1980,1993). Note again that hormonal mechanisms involved in termination of a stage may be completely different from those involved in development.

In figure 4.2, we have detailed the components of the breeding stage of white-crowned sparrows. Phase 1, the development (preparatory phase) of reproductive function, involves initiation of gonadal development, territory establishment, and pair formation. This culminates in phase 2, mature capability of the breeding stage (see figure 4.3) when onset of breeding, the nesting phase (see figures 4.2 and 4.3), can begin. The nesting phase has two distinct subphases: sexual (courtship, copulation, and egg-laying), and parental (incubation of eggs, feeding of young) that involve dramatic changes in physiology and behavior. Furthermore, in many species that have longer breeding seasons, several broods of young may be raised, requiring multiple "subcycling" within the breeding stage (see figure 4.2). Finally, in phase 3, the breeding stage is terminated (regression phase).

How are all these events regulated, timed, and synchronized between members of a breeding pair? The breeding stage can be expressed precisely in time and duration (i.e., seasonal), or can be continual or opportunistic. With the possible exception of those few species that breed continually, information from the environment is used to time gonadal maturation (development of the breeding stage) in anticipation of the breeding season, and to cease reproductive effort (termination of the breeding stage) before conditions become unfavorable. These provide "predictive" information (e.g., see Farner 1970). Within the breeding stage, environmental signals may retard or advance the onset of nesting (mature capability). A whole set of additional unpredictable factors in the habitat can trigger the ELHS. Intimately related to all these environmental cues are the social signals among individuals that can influence all phases of the life-history stage. Although there is a vast literature on the responses of vertebrates to environmental cues and social signals, there have been few attempts to analyze what these factors might do and what their interrelationships may be. About 20 years ago we began proposing a theoretical framework that allows us to postulate testable hypotheses on how this complex and seemingly intractable web of relationships might work (Wingfield 1980, 1983). This framework is described next.

Classification of Environmental Signals

Environmental information and social cues used by animals to regulate life-history stages in general can be organized into four major types based on their major biological effects (Wingfield 1980,1983; Wingfield and Kenagy 1991).

Initial Predictive Cues The first group, called "initial predictive cues," provides very reliable long-term predictive information so that an individual can begin preparing for the next life-history stage several weeks or even months in advance. An example is the seasonal change in day length (photoperiod) which can act as a signal to promote gonadal development in anticipation of the breeding season. There is considerable evidence that changes in day length also regulate the timing of molt and migration (e.g., see Farner 1970, 1985; Farner and Lewis 1971, 1973).

Supplementary Factors These long-term predictive cues are then integrated with the second group of environmental signals, called "supplementary factors." In many habitats, there are temporal variations in predictive events. For example, at midlatitudes spring may be warm, allowing early onset of nesting, or may be cool, resulting in a delay. Changes in reproductive function induced by initial predictive cues need to be adjusted to these short-term predictive changes to time onset of breeding precisely in relation to the most favorable conditions. The three phases of a life-history stage, such as breeding in response to *predictable* fluctuations of the environment, are coordinated by these two types of signals. Examples of supplementary cues are local temperature, availability of food, nest sites, rainfall, and so forth. These can act in two ways—as inhibitors that slow down development and onset of nesting or as accelerators that do the opposite (e.g., see Marshall 1959, 1970). Hormonal mechanisms by which these two subtypes of supplementary factors act may be markedly different.

Modifying Information Superimposed on the predictable changes in an environment are *unpredictable* events that require sudden adjustments, often critical to survival, until the habitat returns to "normal." This suite of unpredictable environmental events constitutes the third class of cues, called "modifying information" (e.g., see Wingfield et al. 1998b). There are two types of modifying factors, more precisely called labile perturbation factors (Jacobs 1996). Some modifying factors have indirect effects with regard to the breeding individual. For example, a predator may take the nest and young but the parents are not affected directly. Nonetheless, the breeding stage undergoes a marked readjustment as the individual and its mate may re-nest (see Wingfield and Farner 1979; Wingfield and Moore 1987). Factors that directly modify reproduction of the breeding individual include the effects of prolonged bad weather, increased predation pressure, human disturbance, and so on, which may not affect the nest and young, but result in an increase in energy expenditure by the adults as they attempt to continue breeding in the face of deteriorating conditions. In this case, direct modifying factors may trigger the ELHS. Clearly unpredictable perturbations of the environment that disrupt the breeding stage can have different mechanisms of action depending upon context and type of disturbance.

Integrating and Synchronizing Information The fourth group of environmental factors is called "integrating and synchronizing information" and comprises all of the behavioral interactions, inter- and intrasexual, among groups and between adults and young. They can have marked effects on all phases of a life-history stage and there are four major subtypes. First, in the breeding stage behavioral interactions can speed up or delay gonadal maturation and termination of the reproductive effort. Second, they have many well-known effects on onset of breeding (i.e., mature capability, onset of nesting; Wingfield and Moore 1987; Wingfield et al. 1994c). Third, interactions of parents and young (parental subphase) are essential to maximize reproductive success. Fourth, behavioral interrelationships can also have important support effects such as establishment of a territory, mate choice, and so forth (see Wingfield and Kenagy 1991; Wingfield et al. 1994c).

The classification described above provides a simple way of organizing all environmental signals that affect life-history stages within a framework that can be approached experimentally to identify possible mechanisms. This classification scheme has withstood considerable testing over the past 20 years; thus far it has been possible to place all known environmental signals into one of the described categories. With these frameworks in mind it is now possible to address the three major phases of the breeding stage in a systematic manner and in ecological context.

Expression and Control of the Breeding Stage

Environmental signals and social cues play crucial roles in regulating the timing of phases and duration of the breeding stage (see figures 4.2 and 4.3) by influencing neuroendocrine and endocrine secretions. These in turn modulate changes in morphology, physiology, and behavior that constitute the reproductive state at any point in the breeding season. Initial predictive information triggers development of the gonads and associated reproductive functions, maintains the breeding stage (i.e., mature capability) so that the individual can respond to many other environmental cues that regulate expression of substages, and then terminates the breeding stage before conditions conducive to breeding deteriorate (see figure 4.3). Inhibitory and acceleratory supplementary information can slow down or speed up this process and also have a major role in regulating onset of nesting once mature capability is reached (see figure 4.3). The interrelationship of these types of factors is explored next.

Phase 1: Development of the Breeding Stage

Much experimental work has addressed the mechanisms by which initial predictive information regulates avian gonadal cycles, particularly photoperiodic mechanisms (e.g., see Farner and Follett 1979; Follett 1984) and endogenous rhythms (e.g., see Gwinner 1987; Berthold 1988; Farner 1985), and is beyond the scope of this chapter. Here we focus on the integration of initial predictive and supplementary signals. The mechanisms by which initial predictive and supplementary cues are integrated are poorly understood, and the underlying neuroendocrine and endocrine components remain largely unknown. Photoperiodic regulation of the first phase of reproductive development involves increased secretion of chicken gonadotropin-releasing hormone I (cGnRH-I), the major GnRH in passerines (Sherwood et al. 1988). This then regulates release of the gonadotropins luteinizing hormone (LH) and follicle-stimulating hormone (FSH) which in turn orchestrate gonadal growth and secretion of sex steroid hormones. The latter then trigger development of secondary sex characters and reproductive behavior (see Wingfield and Farner 1993 for a review of wild birds). This form of GnRH and other forms may also regulate reproductive behavior (e.g., see Maney et al. 1997b). Increased levels of gonadotropins in blood are accompanied by elevated LH β-subunit messenger (mRNA) titers in the anterior pituitary, and a rise in LH and FSH receptors in the testes of Gambel's white-crowned sparrow (*Z.l. gambelii*) (Ishii and Farner 1976; Kubokawa et al. 1994). In females, photoperiodic cues trigger release of the same reproductive hormones, and ovarian maturation follows (Wingfield and Farner 1980).

For some species, especially those breeding in environments in which the onset of conditions favorable for raising young are highly predictable, initial predictive information may be the predominant type of environmental cue used to time breeding. For many others (perhaps the majority of species on a global scale), conditions favorable for breeding are not so predictable, and supplementary cues are crucial for timing. In those taxa that rely on totally unpredictable resources for reproduction (i.e., opportunistic breeders), supplementary (short-term) cues may predominate, and initial predictive (long-term) information may be absent or of little use. A few species utilize resources that are always available in quantities favorable for breeding, and in these exceptional cases environmental cues may not be essential to regulating gonadal maturation. So predictability of the breeding season, the time of year when reproduction is most successful, is an important ecological consideration and may have major importance for how much environmental information populations need to monitor to regulate development of the breeding stage.

The Concept of Predictability Availability of food for offspring is one of the most limiting factors in a reproductive cycle in birds, as well as in mammals. Natural selection would be expected to favor those individuals that produce young at a time when abundance of food can support growth of young. Cohen (1967) showed mathematically that if the timing of a future event, such as onset of breeding, is predictable, then one or a few reliable environmental cues will suffice to regulate gonadal growth. Alternatively, if onset of breeding is less predictable, then more environmental information about whether to inhibit or accelerate gonadal maturation will result in optimal synchronization of breeding with local phenological conditions. Unfortunately it is very difficult to assess and measure ultimate factors directly and analyze them mathematically as a measure of predictability. Wingfield et al. (1992a,1993) used egg-laying dates as an indirect measure of ultimate factors because they are themselves a result of natural selection and vary from year to year as a function of local conditions.

Log-linear analysis of egg-laying data from several avian taxa provides a preliminary assessment of predictability of the breeding season, and an estimate of the importance of initial predictive environmental information for timing development of the breeding stage (see Wingfield et al. 1993; figure 4.4). The analysis provides three important types of information: seasonality, the variation in egg-laying dates by month within each year (U month in figure 4.4, top); yearly variation in egg-laying dates from year to year (U year in figure 4.4); and the interaction of U month.year, which represents unpredictability of timing. The converse of unpredictability is predictability, which can then be estimated. Each component of variation is plotted along one of the sides of a triangle. The point at which these values intersect, along lines indicated in figure 4.4, is a measure of overall predictability and seasonality (Wingfield et al. 1993). In seasonal breeders (e.g., sparrows of the genera *Melospiza* and *Zonotrichia*), these points lie high on the seasonality line with low year-to-year variation and thus very high predictability (see figure 4.4, below). Continual breeders, such as rock doves (*Columba livia*) and sooty terns (*Sterna fuscata*), show low seasonality with little year-to-year variation (i.e., they can breed at all times) with resultant low predictability (see figure 4.4, below, Wingfield et al. 1993). Opportunistic breeders (e.g., zebra finches [*Poephila guttata*], Darwin's finches [*Geospiza sp.*], and crossbills [*Loxia curvirostra*]) have varying degrees of seasonality, year-to-year variation, and thus predictability (see figure 4.4).

These data suggest that for high predictability (i.e., the breeding season occurs at similar times from year to year), initial predictive information such as the annual cycle in day length is important for regulating development of the breeding stage because photoperiod is highly precise and essentially invariant from year to year. As

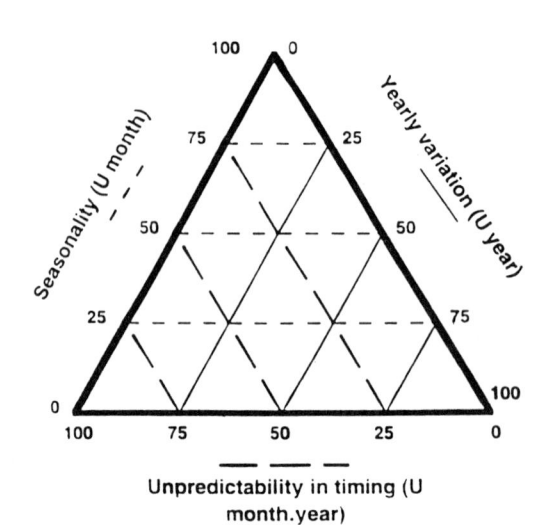

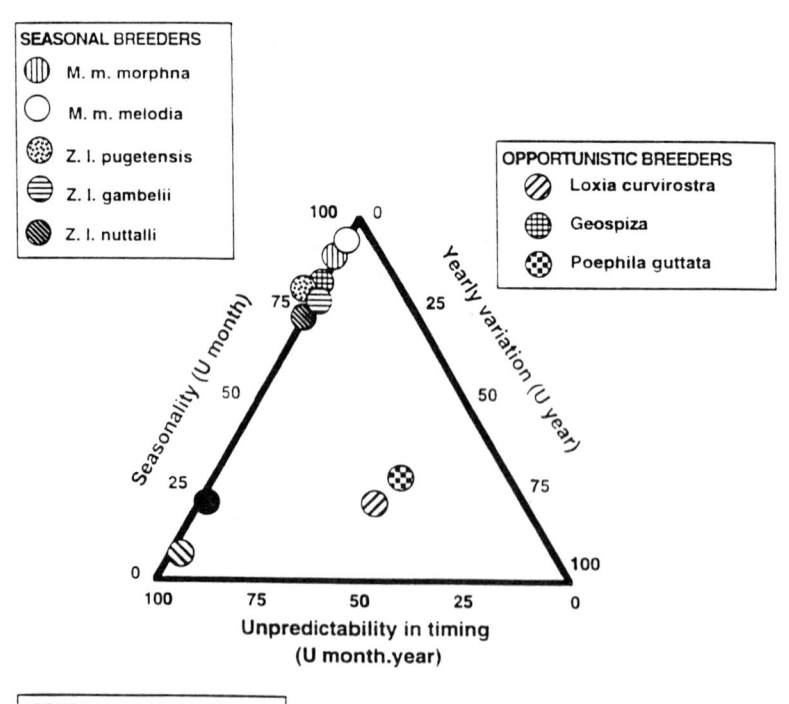

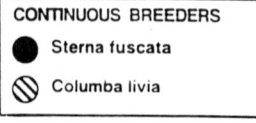

predictability decreases, then initial predictive information may be less important and other cues become important. At low predictability, the gonads may remain developed at all times (i.e., the breeding stage may be permanent). Evidence in rock doves supports this prediction (Lofts and Murton 1968; Wingfield et al. 1993).

The degree to which log-linear analysis may indicate how other cues, such as supplementary information, are involved in development of the breeding stage is limited (see how the seasonal breeders are clumped in figure 4.4, below). Alternative methods exist, however, which may resolve this issue further. Application of Colwell's (1974) model generates predictability (Pr) from egg-laying data that is different from the value generated by the log-linear method. In this case there are two components to Pr: constancy (C), the uniformity of resource levels across time intervals within a year, and contingency (M), the temporal reliability of fluctuations in resource levels across time intervals within a year. The value of Pr can vary between 0 (completely unpredictable) and 1 (highly predictable), and is equal to the sum of components C and M. The critical features of this model in relation to development of the breeding stage are the relative contributions of C and M to Pr.

How does this help us analyze patterns of gonadal development and their control by environmental signals? Total Pr indicates whether breeding should be seasonal, constant, or opportunistic, but this appears to be less useful than Pr generated by the log-linear method above (Wingfield et al. 1993). Relative contributions to Pr by C and M may be important because they indicate whether timing of breeding requires any environmental information. If C is very high, no external information will be

Figure 4.4
Log-linear analysis of egg-laying dates collated over several years provide an assessment of variation in timing of breeding. There are three major components to this variation as indicated by the three sides of a triangle (upper panel). Seasonality (U month) is the variation in egg-laying dates by months within a year; yearly variation (U year) is the difference in egg-laying dates from year to year; and the third line, U month.year, is the interaction of seasonality and yearly variation which provides a measure of unpredictability of the breeding season. Its converse is thus predictability. Each component of variation is plotted along the lines indicated in the figure. In the lower triangle, actual log-linear analyses of egg-laying dates from several avian taxa illustrate what information can be gained from this kind of analysis. Seasonal breeders clump near the apex of the triangle indicating that most of the variation in egg-laying dates is by month of the year with little year-to-year variation and thus low unpredictability (i.e., highly predictable breeding seasons). In this case we might expect that an initial predictive cue such as the annual change in day length would be a highly appropriate signal to regulate development of the breeding stage and onset of nesting. Continuous breeders (at the lower left apex of the triangle) show little variation with month of the year and little variation from year to year (i.e., they can breed at any time). This gives very high unpredictability suggesting that photoperiodic cues may be less important here and that the gonads remain in a near functional state once puberty has been reached. Opportunistic breeders, on the other hand, are spread throughout the triangle showing that some may be photoperiodic and others less so. Such predictions are eminently testable in the laboratory and may provide a powerful technique to predict how populations should respond to photoperiod cycles. (From Wingfield et al. 1993.)

required because the resource is not fluctuating. If M is high, then the resource fluctuates and external information is important for timing breeding as well as other life-history stages. In combination with Pr generated by the log-linear method, the relative contributions of C and M indicate whether initial predictive (long-term) information or primarily supplementary (short-term) information should be used to time reproductive effort. For example, when M is smaller than, or nearly equal to, C, constancy predominates and a few or one highly reliable cue would suffice to regulate development of the breeding stage. When M is greater than C, then contingency predominates, resulting in greater plasticity of breeding seasons and a need to monitor many different environmental cues so that breeding is timed appropriately (Wingfield et al. 1992a, 1993). The relative contributions of M and C can be expressed conveniently as the M/C ratio (or environmental information factor, Ie). Theoretically this is an accurate indicator of the way in which organisms may integrate initial predictive and supplementary cues which in turn provide information on whether the neuroendocrine system is responding to few or many cues. Mechanisms may thus vary markedly depending upon Ie.

We have calculated Ie values from egg-laying dates in the same avian taxa shown in figure 4.4 (Wingfield et al. 1992a, 1993). In seasonal breeders (upper panel of figure 4.5), there is almost an order of magnitude difference in Ie values. In *Z.l. gambelii*, Ie is close to 1, whereas in other related taxa Ie is greater, approaching 10 in some cases (see figure 4.5). The theoretical models suggest that predictability is high in all these seasonal breeders (see figure 4.4) and thus all should show strong responses to initial predictive information such as increasing day length in spring. However, the Ie values suggest that *Z.l. gambelii* may be less responsive to supplementary factors than the other taxa (see figure 4.5). These are clear hypotheses that can be tested under laboratory conditions. In *Z.l. gambelii*, effects of low environmental temperature and restricted food are ineffective in suppressing the response to photoperiod (e.g., see Farner and Lewis 1971,1973; Wingfield 1988; Wingfield et al. 1982b,1996), supporting the suggestion that if Ie is low then initial predictive cues should predominate. Another race of white-crowned sparrow (*Z.l. pugetensis*) breeds during the longer summers of mid-latitudes. It is a short-distance migrant and onset of the breeding season can vary by up to 3 or 4 weeks (Wingfield and Farner 1978b). In this form Ie is greater than 1 and we can expect supplementary factors to have more effect on photoperiodically induced gonadal growth than in *Z.l. gambelii*. Consistent with theoretical expectations, experimental evidence shows that temperature does modulate photoperiodically induced gonadal growth in females (Wingfield et al. 1997a). Other circumstantial evidence from the field also supports this view (Wingfield et al. 1992a, 1993). Curiously, the effects of temperature do not appear to be mediated

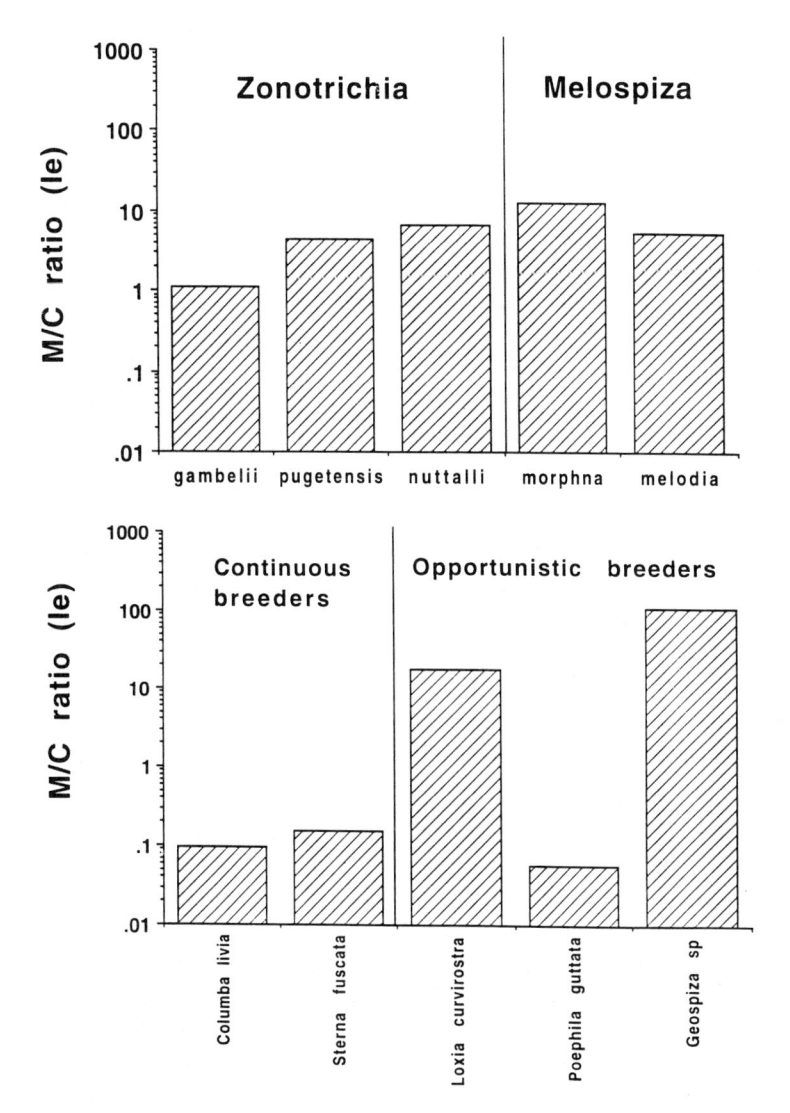

Figure 4.5
Analysis of the same egg-laying dates shown in figure 4.4 but with a different measure of predictability (the constancy-contingency model of Colwell 1974) gives a very different perspective. Here predictability has two components: constancy (C): environmental conditions are conducive to breeding (or not) for most of the time; and contingency (M): environmental conditions show major changes from favorable to unfavorable for breeding. The ratio of these two (M/C = Ie, the environmental information factor) is a measure of whether a population should be sensitive to more or fewer environmental cues. If the contribution of contingency is great, then monitoring more cues in the environment may allow for the most accurate timing of breeding. The more constancy in the environment, the less need for monitoring of many cues. Just a few, or even one, may be sufficient to time breeding successfully. Ie may vary markedly in different taxa (note the log scale of the y-axis). This method thus provides specific predictions as to how much environmental information a given population may need to assess and regulate development of the breeding stage and onset of nesting. See Wingfield et al. 1992a, 1993 and text for details.

through the secretion of GnRH because plasma concentrations of LH and FSH were not affected (Wingfield et al. 1997). Possible mechanisms by which temperature effects are mediated are currently under investigation. Comparisons of closely related taxa with different Ie values are ideal for exploring neuroendocrine and endocrine pathways for environmental signals regulating reproductive function.

In opportunistic breeders, Pr and Ie values are highly variable (Wingfield et al. 1992a; see figures 4.4 and 4.5, lower right-hand panel). Some, such as red crossbills, can respond to changes in day length (Hahn 1995); others, such as the zebra finch, do not (e.g., see Farner 1985). These species clearly offer many experimental opportunities to test theoretical calculations and the reliability of Ie factors. In red crossbills, initial predictive information, such as photoperiod, clearly stimulates gonadal development (Hahn 1995). However, food supply may also be important but in an unexpected way. Photostimulated male crossbills restricted to a level of food intake typical of short days showed similar testicular growth to males fed ad lib., and much greater development than males held on short days (Hahn 1995). After food-restricted males on long days were given food ad lib., testicular development was significantly greater than in males that had continuous access to food ad lib. Body mass and food intake were similar among groups regardless of availability of food. These exciting data suggest that perception of food availability may affect gonadal development.

Some species are able to breed continually, or have the ability to nest at any time of year. In two species for which there are data, predictability values (see figure 4.4) are very low, suggesting that initial predictive information is less important for timing gonadal development. For the rock dove, experimental data show that gonadal development is not regulated by changing day length, supporting theoretical predictions (Lofts and Murton 1968). Ie values (see figure 4.5) for continual breeders are also extremely low. This suggests that these species have reproductive systems more or less mature at all times and thus development of the breeding stage occurs only once, analogous to puberty in some mammals. Comparisons of species with very different Pr and Ie values will allow design of critical experiments to explore how vertebrates in general respond to environmental signals and regulate development of the breeding stage in widely different habitats and ecological contexts.

Social Cues and Development of the Breeding Stage The interaction of initial predictive and supplementary types of information regulates development of the breeding stage in anticipation of the breeding season. These cues emanate from the physical environment. Additionally, behavioral interactions among individuals within a population can have profound effects on all aspects of development of the breeding stage. It well-known that the presence of males can accelerate ovarian development in females and presence of females can enhance testicular growth in all the major vertebrate taxa (see Wingfield et al. 1994c for review). For example in wild birds such as *Z.l. gambelii*, broadcast of male songs enhances ovarian growth stimulated by increased day length (Morton et al. 1985). Similarly, exposure of male *Z.l. gambelii* and *Z.l. pugetensis* to females made sexually receptive by implants of estradiol resulted in an increase in circulating levels of LH and testosterone (Moore 1982, 1983). More recent evidence suggests that social cues may also affect sensitivity to supplementary information. In female *Z.l. pugetensis*, the acceleration of photoperiodically induced ovarian growth by warm temperature (30°C) was completely abolished if males were not present (Wingfield et al. 1997). How these effects of temperature are mediated in the presence of males remains unknown.

Phase 2:   Mature Capability of the Breeding Stage

Once the stage has developed, then processes characteristic of that stage can begin. This is called the "mature capability" (see figure 4.3; Jacobs 1996) and is equivalent to the nesting phase in figure 4.2. One important concept here is that activation of the various substages that make up the nesting phase probably involves an entirely separate suite of hormone control mechanisms. The mechanisms involved in development of the breeding stage may be essential to allow expression of the nesting phase, but activation mechanisms themselves may be different (Jacobs 1996; Wingfield 1983). For example, in *Zonotrichia*, the development of the ovary culminates when follicles are about 2 to 3 mm in diameter and contain white yolk only. Females generally will not progress beyond this phase unless the environment is conducive to nesting (King et al. 1966). Supplementary factors (accelerators and inhibitors) regulate the nesting phase, rapid deposition of yellow yolk (under the control of estradiol secretion), and egg-laying (Wingfield and Farner 1980). Although the regulation of ovulatory cycles has received much attention in birds and mammals, environmental regulation of onset of the nesting phase in wild birds remains largely unknown. However, production of large quantities of yolk is an obvious marker that a female has initiated the nesting phase. Ovulation and oviposition usually occur within hours to a day or so later (e.g., see Lofts and Murton 1973; Wingfield and Farner 1993). In female *Z.l. pugetensis*, yolk deposition and rapid final maturation of follicles were marked in females at 30°C *and* exposed to males by day 75 of photostimulation (Wingfield et al. 1997a). Females photostimulated at lower temperatures did not show yolk deposition even if exposed to males, suggesting that they had not initiated the nesting phase. These data indicate that social cues and temperature could act in synergy or separately. Further experiments are required to tease apart the integration of these cues.

Although it is well-known that availability of food, quality of nutrition, and endogenous reserves of fat and protein can have profound influences on reproductive function in birds (see Follett 1984), the mechanisms by which food supply acts as supplementary information for onset of the nesting phase remain equivocal (Wingfield and Kenagy 1991). There is no doubt that food is an important proximate, as well as ultimate, factor regulating breeding seasons, but how does it work? We have chosen not to review this literature in depth here, because it has been done elsewhere (see Wingfield and Kenagy 1991) and because no obvious line of approach is forthcoming for birds. However, in mammals more progress has been made (see chapters 2 and 3).

Synchronizing and Integrating Information: Social Regulation of Onset of the Nesting Phase

Onset of the nesting phase and transitions between its two major subphases—sexual and parental—are regulated to a great extent by social cues. There are several components here: first, support functions such as establishment of a territory, pair bond, or position in a hierarchy that allows access to mates; second, courtship, copulation, nest-building, and egg-laying which constitute actual onset of the sexual subphase; and third, the transition to parental subphase in which the eggs are incubated and young are fed until they reach independence. In those species that breed in environments in which more than one brood of young can be raised, then after fledging one brood there is a transition back to the sexual subphase to begin the nesting cycle anew (see figure 4.2). There are marked changes in hormone regulatory mechanisms which have been intensively studied in birds and other vertebrate taxa (see Wingfield and Moore 1987; Wingfield and Kenagy 1991; Wingfield et al. 1994c for reviews). The interrelationships of hormones and behavior involved in the nesting phase are extremely complex and we will adhere to the simple components described above. Initially it is important to address the functions of the steroid hormone testosterone and its interrelationship with aggression, and then go on to look at the implications for mating systems. Then we address the transition to the parental phase and ecological constraints that have a profound effect on patterns of hormone secretion.

Testosterone and Territorial Aggression Testosterone has a wide spectrum of biological actions (table 4.1), including promotion of spermatogenesis, growth of accessory organs and some secondary sex characters (e.g., wattles, nuptial plumages, vas deferens), and activation and organization of reproductive behaviors, including ter-

Table 4.1
Biological actions of testosterone

Physiological effects	Morphological effects	Behavioral effects	Biological "costs" of testosterone
Negative feedback on gonadotropins Miscellaneous secretions, e.g., in accessory organs, secretions of skin	Accessory organs Secondary sex characteristics Muscle hypertrophy Spermatogenesis	Sexual behavior Aggressive behavior in reproductive contexts	Increased potential for predation Increased injury Energetic costs Conflict with pair formation and courtship Interference with parental care Suppression of the immune system Possible oncogenic effects

ritorial aggression (e.g., see Arnold and Breedlove 1985; Wingfield et al. 1997b, 1998b). Given such a broad spectrum of actions, how can specific changes in behavior, such as territorial and sexual, be regulated independently of morphological and physiological actions? The concept of state levels of secretion of hormones has been introduced as a model to begin to explain how these multiple actions, often on different schedules, can be regulated by one hormone. Wingfield et al. (1997b) postulated that there are three levels of secretion of testosterone (figure 4.6, top). An increase in circulating concentrations of testosterone from a nonbreeding baseline (level A) to a breeding baseline (level B) is sufficient for complete spermatogenesis and all androgen-sensitive secondary sex characters; accessory organs develop normally and the full spectrum of reproductive behaviors can be expressed (Wingfield et al. 1990,1997a,b). However, during the breeding season, plasma levels of testosterone may show complex patterns with transitory elevations above the breeding baseline to level C (see figure 4.6). Level A represents a constitutive secretion of testosterone, probably to maintain negative feedback regulation of GnRH and gonadotropins even in the nonbreeding season, whereas level B represents regulated, periodic secretion of testosterone during the normal breeding stage in the predictable annual cycle of life-history stages (see figure 4.1). The complex patterns of testosterone seen in free-living birds appear to be generated entirely by surges of circulating concentrations above level B toward level C. This has been termed the regulated, facultative level (see figure 4.6; Wingfield et al. 1990,1998b). It is not similar to the ELHS in which facultative responses also occur (see figure 4.1 and below), because reproduction is usually not disrupted.

Maturation of the breeding stage is completed when plasma levels of testosterone are at the breeding baseline (level B), so why are there such complex patterns of testosterone among populations and species, and what is the function of increased

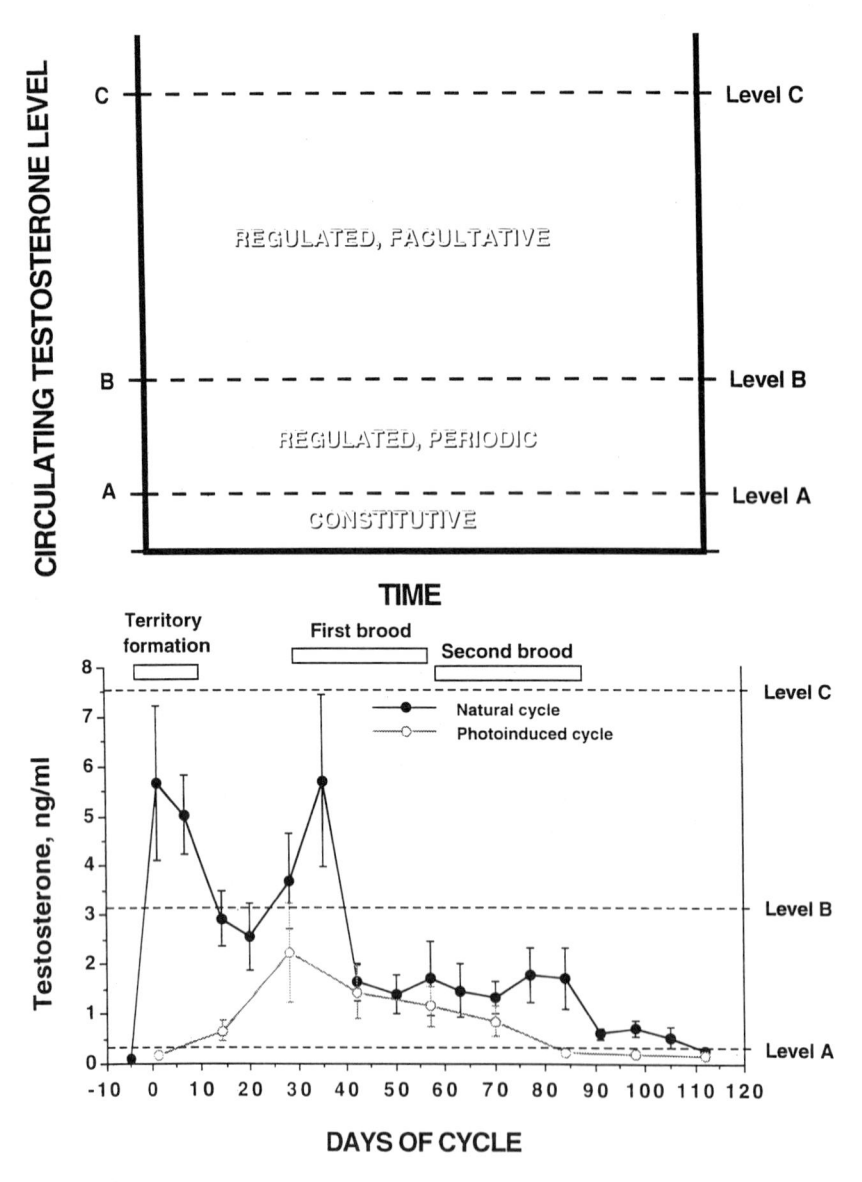

Figure 4.6
Concept of levels of testosterone secretion in relation to actions during the breeding life-history stage. Top panel shows theoretical level as a function of time (day, season, etc.). Lower panel illustrates the pattern of testosterone levels during the reproductive cycle of the song sparrow with the concept of state levels indicated. Note that surges of testosterone secretion above the breeding baseline (level B to level C) are transitory and soon return to level B. See text for details. (From Wingfield 1984b; Wingfield et al. 1990, 1998b.)

testosterone above level B? It has been suggested that increases of testosterone concentrations from level B to C are solely involved with increased frequency and intensity of male-male aggression in territorial and mate-guarding contexts. Although the full repertoire of aggressive behaviors can be expressed with testosterone concentrations at level B, further increases to level C may be required to support very high levels of aggression (Wingfield et al. 1990, 1998b). Thus variations in the patterns of testosterone levels in the blood of populations or individuals may reflect parallel changes in male-male competition over territories or mates, or both. In figure 4.6 (lower panel) the pattern of circulating testosterone levels in free-living male song sparrows is compared with that pattern obtained in captive males exposed to long days that induce development of the breeding stage, but do not result in onset of the nesting phase. In the field the pattern of testosterone secretion is very different from that in captive males and gives an indication of how testosterone levels may fluctuate in relation to levels A to C. If initial predictive cues such as increased day length drive regulated periodic secretion of testosterone (level B), what regulates increases of testosterone secretion above level B?

The "Challenge Hypothesis" Elevations of male-male aggression during periods of social instability, or when females are receptive, are often accompanied by increased secretion of testosterone in males (Sapolsky 1987; Wingfield et al. 1987). These changes in secretion of testosterone appear to be induced by behavioral cues emanating from a challenging male or the sexual behavior of a female (the "challenge hypothesis"; e.g., see Wingfield et al. 1987, 1998b). For example, if male song sparrows were challenged by placing a decoy male on the territory and playing back tape-recorded songs through a speaker placed alongside, then the resident male attacked the intruder and the interaction resulted in an increase in circulating testosterone level (Wingfield 1985). Similarly, in *Z.l. gambelii*, males paired with sexually receptive females had significantly increased testosterone levels over those of males paired with nonreceptive females or males caged alone (Moore 1983). It is important to note that increased testosterone secretion in response to these challenges does not activate territorial aggression, but appears to support a period of heightened aggression as the territory is established or defended, or during mate-guarding (Wingfield et al. 1987; Wingfield 1994b). Once the stimulus is removed, then testosterone levels decline rapidly to level B (see figure 4.6).

Temporal Patterns of Testosterone, Mating Systems, and Breeding Strategies
Seasonal patterns of testosterone concentrations in blood have now been determined for over fifty free-living avian taxa. One striking characteristic of all these studies is the high degree of variation in the temporal pattern of testosterone during the

breeding stage. Does this variation have any meaning? Is there an ecological basis for the phenomenon? Several years ago it became clear that those species in which males provide parental care, the level of testosterone remained at level B or lower when attending eggs or young (e.g., see figure 4.6). However, these males were able to increase testosterone to level C if challenged by another male (see Wingfield 1985; Wingfield et al. 1987). A lower level of testosterone, at or below the breeding baseline B, appeared to be critical for expression of male parental behavior. If male pied flycatchers (*Ficedula hypoleuca*), house sparrows (*Passer domesticus*), spotted sandpipers (*Actitis macularia*), yellow-headed blackbirds (*Xanthocephalus xanthocephalus*), and song sparrows were given implants of testosterone to maintain plasma concentrations at the seasonal maximum (i.e., level C), then parental behavior was significantly lowered, resulting in reduced reproductive success (Silverin 1980; Hegner and Wingfield 1987; Oring et al. 1989; Wingfield et al. 1989; Beletsky et al. 1995). These data suggest strongly that high levels of testosterone above level B are incompatible with male parental behavior. Thus the temporal pattern of testosterone in these species may be regulated by the degree of male-male interaction (which tends to elevate testosterone) and the extent to which the male provides parental care (which requires circulating testosterone to decrease).

In contrast, those males that show no parental care would have no such restrictions on high levels of testosterone and maximum concentrations would be expected throughout the breeding stage. This relationship was supported by the observations that if normally monogamous male song and white-crowned sparrows were given implants of testosterone to maintain high levels similar to the pattern seen in polygynous species, then these males also became polygynous (Wingfield 1984a). However, the "costs" of such prolonged high levels of testosterone resulted in reduced reproductive success despite some males having up to three mates. These data also point to the concept that there is not only selection for mechanisms that increase hormone secretion but also for mechanisms that turn off hormone secretion at inappropriate times. In table 4.1 we have summarized the potential deleterious "costs" of prolonged high levels of testosterone which stand in contrast to the known biological actions (see also Wingfield et al. 1997b, 1998b). Such classic tradeoffs of benefits and costs of hormone secretion may provide a way in future of determining ecological bases of patterns of hormone secretion in general.

A simple correlation of mating system (monogamy vs. polygyny) and the temporal pattern of testosterone level in blood may be misleading because some monogamous males show little or no parental care, and some polygynous males feed young extensively. Instead, the interrelationship of male-male aggression (increasing testosterone to level C) and the degree of male parental care (requiring testosterone titers at level

B or lower) may be a major determinant of the temporal pattern of testosterone secretion in the breeding stage. Males that show high male-male aggression and low parental care would tend to have higher testosterone for long periods throughout the breeding stage, whereas males that show low male-male aggression and high parental care would have much lower levels of testosterone. By comparing the degrees of male-male competition and parental care we can generate many intermediate theoretical patterns of testosterone secretion during a breeding season (Wingfield et al. 1990). The temporal patterns of testosterone measured in males of all avian taxa studied to date fit the theoretical patterns predicted by the degrees of male-male aggression and male parental care expressed by each species (Wingfield et al. 1990).

Some further ecological bases for temporal patterns of testosterone secretion can be made from this hypothesis. Males that show little or no parental care may have fewer restrictions on high levels of testosterone compared to males that do show parental care. To test this requires a comparison of testosterone levels in males of species showing different degrees of aggression and parental behavior. However, absolute levels of testosterone can be misleading because of species differences in testosterone receptor levels. To circumvent this, Wingfield et al. (1990) compared the ratio of the seasonal peak of testosterone corrected for the nonbreeding baseline (level C–level A) to the breeding baseline (level B–level A) with the ratio of male-male aggression to male parental care. The testosterone level ratio was estimated in those species for which a complete cycle of testosterone levels through all stages of the reproductive cycle were available. The behavioral ratio was obtained by assigning numbers to the degree of male-male aggression (low aggression = 1, moderate aggression = 2, high aggression = 3), and male parental care (low parental care = 1, high parental care = 2). Thus when the ratio is high (e.g., 3) for a species or population under investigation, then male-male aggression is high and male parental care is low. If the ratio is low (e.g., 0.5), then male-male aggression is low and male parental care is high.

Using these ratios, we have reevaluated these relationships by expanding the number of species represented from twenty in Wingfield et al. (1990) to sixty (figure 4.7). The same relationship appears to hold with those males showing lower male-male aggression and higher paternal care having a greater difference between levels C and B than those males which have high male-male aggression and little or no parental care (see figure 4.7). The important point here is that the latter males appear unable to modulate their testosterone secretion during the breeding stage (i.e., levels B and C are equal), whereas the former males showing high paternal care show a decline in testosterone level when in the parental subphase (see figures 4.2 and 4.6) but retain the ability to increase testosterone secretion (i.e., level C $\gg$ level B) should

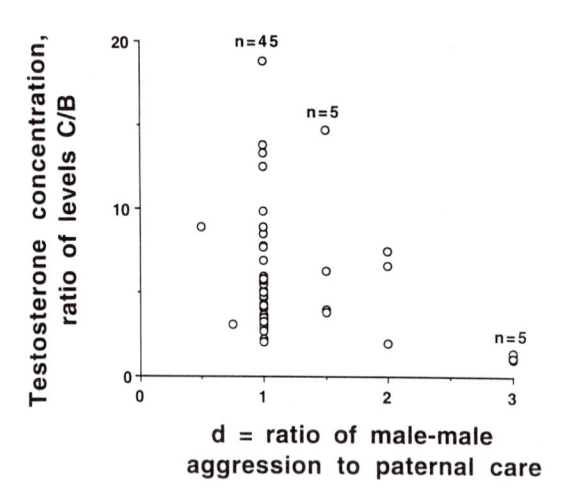

Figure 4.7
Relationship of the flexibility in testosterone secretion expressed as the ratio of peak testosterone levels measured during social instability (level C) to the breeding baseline level of testosterone (level B). The symbol d is a measure of male-male aggression vs. paternal care (see Wingfield et al. 1990 and text for details). This figure was reconstructed from data on sixty species (see source list in references). Note that these data have not been subjected to a phylogenetic analysis (see text).

they be challenged by an intruding conspecific male. It should be emphasized here that there is considerable potential for these relationships to be confounded by phylogenetic relationships (e.g., Harvey and Pagel 1991). Nevertheless these results point to some exciting experimental comparisons of closely related species which can or cannot modulate testosterone by social interactions and thus determine neural pathways for these signals that affect neuroendocrine secretion. This phenomenon may be highly adaptive in regulating the fine balance of aggression and parental care in highly complex social situations. The data also point to considerable neural plasticity in transducing information from the social environment into neuroendocrine and endocrine cascades that make up the hypothalamic-pituitary-gonadal axis. An in-depth phylogenetic analysis (e.g., Garland et al. 1993) to determine the most appropriate comparisons that could be made experimentally is in progress.

Sexual Dimorphism and Patterns of Testosterone in Male and Female Birds Another intriguing observation in the reproductive endocrinology of wild birds is the marked variation in testosterone patterns in females (Wingfield 1994a). Females are generally regarded as not aggressive and testosterone is regarded as the "male sex hormone." These views are incorrect (Wingfield 1994a) because female vertebrates have significant quantities of testosterone in blood, and may be highly aggressive, at least toward conspecific females. Females compete to defend resources such as food and nest sites for breeding, to space nests and avoid predation, to reduce intraspecific brood parasitism, and to gain exclusive access to male investment such as parental care. This aggression appears similar to that of males, and females of some species may sing at least occasionally. Otherwise, threats and similar postures are essentially identical to those of males (Wingfield 1994a). Females also show aggression in many other contexts, but the endocrine basis of female-female conflict in the breeding stage has been neglected. Maternal aggression (defense of young from a predator) has received more attention but in a different context from the ones discussed here (Wingfield 1994a).

Activation of aggression, including singing (or analogous vocalizations), by testosterone in females is well-known, although not ubiquitous. But do females have patterns of circulating testosterone that are synchronized with expression of reproductive aggression under natural conditions? Studies show that female birds with plumage and rates of aggression similar to those of males, or that are even more brightly colored (e.g., polyandrous sandpipers, Scolopacidae), have concentrations of testosterone that are correlated with aggression (reviewed in Wingfield 1994a). Females that are aggressive, but less so than males, have cycles of testosterone that are of low amplitude or are not detectable. It thus appears that some females activate aggression with high testosterone levels, whereas others do not. The mechanisms underlying these differences remain unresolved.

Is there a possible ecological basis for testosterone and aggression in relation to sexual dimorphism? Do females that are similar to males in plumage and behavior tend to have higher levels of testosterone relative to males than species with greater dimorphism? To test this, Wingfield (1994a) developed a dimorphism index to standardize sex differences in all species (table 4.2). The index comprises a score for (A) body size, (B) plumage and other integumentary structures, and (C) territorial aggression. The dimorphism index is the average of scores for traits A, B, and C (see table 4.2; Wingfield 1994a). Comparison of the dimorphism index with the ratio of peak plasma testosterone levels in males vs. females (to control for differences in absolute levels of testosterone among species that may have no direct bearing on expression of behavior) has been expanded from about thirty species reviewed by Wingfield (1994a) to forty-six in figure 4.8. When the dimorphism index is low (males and females are similar), then levels of testosterone in females tend to be high relative to those of males (top panel of figure 4.8). This is not, however, a linear relationship although at low dimorphism indices testosterone levels are uniformly similar in males and females. At high dimorphism indices, some females have high levels of testosterone; others do not (see figure 4.8). If the same relationship is plotted for socially monogamous and polygamous species, the results are more striking. In socially

Table 4.2
Dimorphism index

A. Body size
1 = 80%–100% overlap
2 = 5%–80% overlap
3 = No overlap

B. Plumage
1 = Monomorphic nuptial plumage
2 = Moderate dimorphism (e.g., eye, beak, leg color, size of feather plumes, etc.)
3 = Great dimorphism (one sex has bright nuptial plumage, other sex is cryptic)

C. Territorial aggression
1 = Both sexes defend territory or compete equally
2 = One sex defends territory or competes more than the other
3 = One sex defends territory or competes and the other does not

From Wingfield 1994a.

monogamous species there is a more linear relationship of dimorphism index and testosterone in females, but in contrast there is no relationship of dimorphism index and testosterone in polygamous species (see figure 4.8, middle and lower panels respectively). As mentioned earlier, this relationship should only be used to identity potential experimental comparisons. Phylogenetic confounds are possible (Harvey and Pagel 1991) and an analysis (cf. Garland et al. 1993) is currently underway.

These expanded investigations support the original findings of Wingfield (1994a). Whether or not this suggests sexual selection for a behavioral role of testosterone in socially monogamous females who compete for male investment requires further study. Additionally, control of aggression in females of polygamous species apparently involves mechanisms independent of testosterone, such as central actions of arginine vasotocin on song (e.g., see Maney et al. 1997a). Analyses of this type have heuristic value and provide a template for future investigations that will test hypotheses of hormonal mechanisms based on ecological theory.

Phase 3: Termination of the Breeding Stage

The final phase of a life-history stage is its termination. This allows other stages, appropriate for that time of year, to predominate or develop anew (see figure 4.1). The breeding stage is generally timed to coincide with periods of favorable trophic resources for feeding young. Equally important are control systems that terminate the breeding stage before those trophic resources decline so that other life-history stages such as prebasic molt and autumnal migration can be expressed (Marshall 1959; Lofts and Murton 1973; Farner and Lewis 1971; Farner and Follett 1979; Wingfield and Farner 1980). In avian species that nest at mid-latitudes termination of

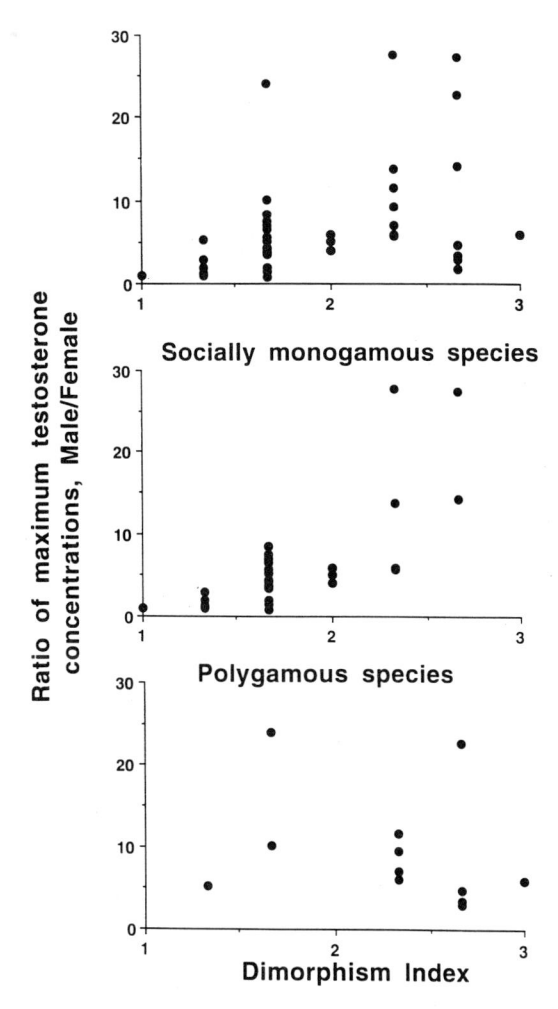

Figure 4.8
Ratio of maximum levels of testosterone (level C) in males and females of the same species as a function of the sexual dimorphism index (see Wingfield 1994a and text). Data are presented in total (upper panel), for socially monogamous species (middle panel), and for polygamous species (lower panel). These data are drawn from a total of forty-six avian species (see source list in references).

the breeding stage occurs in July and August, sometimes earlier, so that prebasic molt and preparations for migration are completed long before the first inclement weather in October and November. In most cases, termination of the breeding stage involves a refractory state in which individuals are no longer able to respond to the environmental stimuli that initiated development of the stage and onset of nesting in spring.

Photoperiodic Regulation of Termination of Reproduction The control of refractory periods by initial predictive information (photoperiod) has been investigated extensively (e.g., see Nicholls et al. 1988). Vernal increase in day length initiates development of the breeding stage (see above), but many species undergo spontaneous gonadal regression in midsummer despite continuing long days. This photorefractory condition may also be induced by artificial long days, and when in this state no known photoregimen will stimulate gonadal recrudescence. Recovery of photosensitivity usually occurs only after birds have been exposed to short days for 40 to 60 days (Lofts and Murton 1973; Farner and Follett 1979; Farner 1985). Under natural conditions, photosensitivity in many species of the North Temperate Zone is regained in late October and early November when day length is still decreasing. Development of the next breeding stage is thus prevented until days lengthen in the following spring (Farner and Follett 1979).

Circannual rhythms of gonadal growth and regression are another way in which life-history stages may be developed and then terminated. Evidence exists for several avian species (e.g., several taxa of *Phylloscopus* and *Sylvia, Sturnus vulgaris*, and others) that these rhythms regulate the timing of winter (nonbreeding), migration, breeding, and molt stages which appear to be entrained by annual photoperiod into cycles of exactly 1 year (e.g., see Berthold 1988; Gwinner 1981,1987). This mechanism is particularly attractive for migratory species that winter in equatorial regions where changes in photoperiod are minimal, or for transequatorial migrants that are exposed to long days on their wintering grounds.

Changes in the endocrine system accompany photoperiodically induced development of photorefractoriness. A role for the thyroid gland has been implicated since thyroidectomy appears to block the onset of photorefractoriness in the European starling. Starlings exposed to 11 hours of light per day underwent testicular recrudescence and remained sexually mature indefinitely. Treatment with thyroxine (T_4) resulted in rapid testicular involution and onset of a prebasic-type molt, the life-history stage that typically follows natural onset of photorefractoriness and termination of the breeding stage. Furthermore, these birds failed to show gonadal recrudescence following transfer to even longer days (18 hours of light) indicating that they were indced photorefractory (Goldsmith and Nicholls 1984). Thyroidectomized starlings maintained high hypothalamic contents of GnRH compared with controls (Dawson et al. 1985; Goldsmith and Nicholls 1984). Only short periods of T_4 injections or even a single injection is needed just after photostimulation to provoke a spontaneous regression in thyroidectomized birds several weeks later (reviewed in Nicholls et al. 1988).

Nonphotoperiodic Regulation of Termination of Reproduction: Supplementary Factors and Social Cues There appears to be no clear demonstration that birds become refractory to environmental stimuli other than day length, although there is evidence that the onset of photorefractoriness can be delayed by nonphotoperiodic (possibly supplementary) information. Investigations of free-living *Z.l. gambelii* reveal that timing of termination of the breeding stage varies. Unmated males showed gonadal regression by the fourth week of June, whereas nesting males and females did not terminate the breeding stage until the first week of July. Re-nesting birds that had lost their nest and initiated a second sexual subphase in mid-June (re-nested; see above), were able to delay gonadal regression and onset of the molt stage until the young were fledged in the third week of July (Wingfield and Farner 1978a,1979,1980). It would clearly be maladaptive for the parental subphase to be terminated abruptly by photorefractoriness because the young would die and reproductive success decrease. Natural selection may have favored mechanisms that prevent initiation of clutches (i.e., onset of new sexual subphases) beyond a certain date, but allow completion of the parental subphase already underway. In the case of *Z.l. gambelii*, this period appears to be the third and fourth weeks of June (Wingfield and Farner 1979).

The mechanisms by which photorefractoriness and termination of the breeding stage are delayed remain obscure. Elevated circulating levels of sex steriod hormones during the initiation of a second sexual subphase may delay gonadal regression. Treatment with testosterone has been shown to maintain a functional testis in Japanese quail (*Coturnix japonica*) (Brown and Follett 1977), prevent spontaneous gonadal involution in photostimulated house sparrows (*Passer domesticus*) (Turek et al. 1976), and delay onset of the molt stage in canaries (*Serinus canarius*) (Kobayashi 1952) and European starlings (Schleussner and Gwinner 1988). Social interactions between mates may also play a role since free-living female song sparrows given subcutaneous implants of estradiol remained sexually receptive and delayed onset of prebasic molt until at least the beginning of October (figure 4.9; Runfeldt and Wingfield 1985). Control females given empty implants became photorefractory and began the molt stage by mid-August, as did their male mates. All but two had vacated the breeding area by the end of August (see figure 4.9). In contrast, males

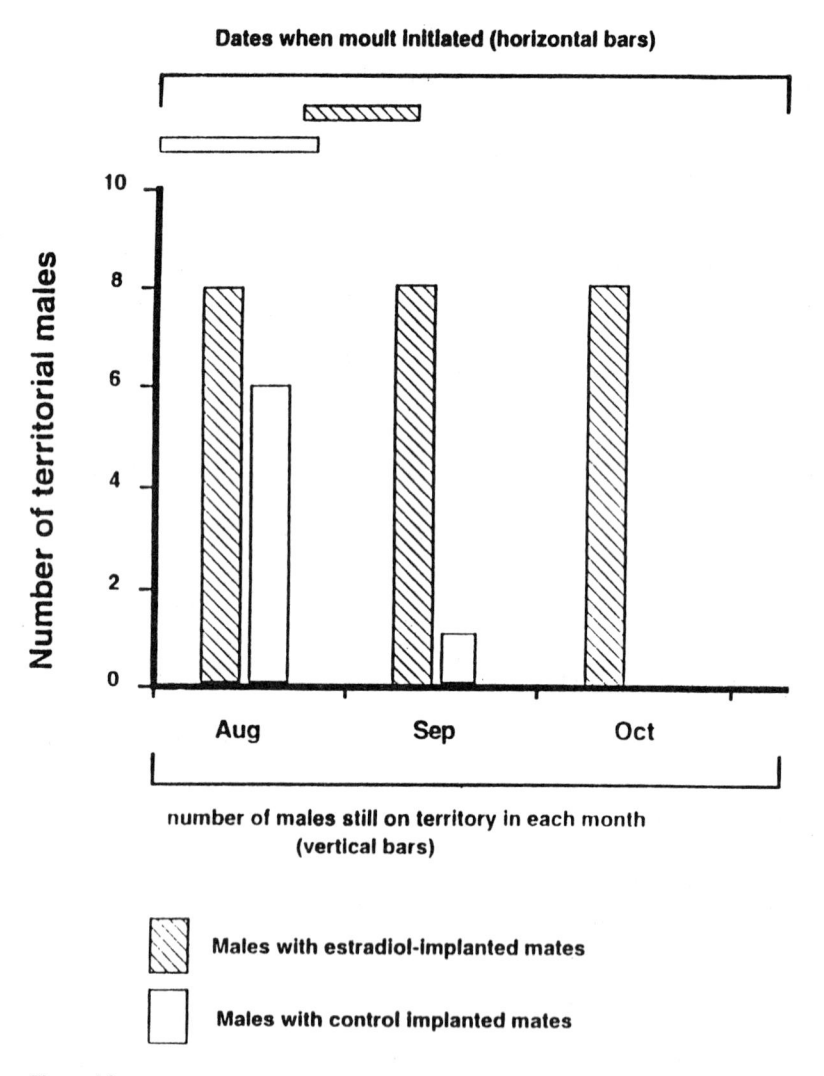

Figure 4.9
Left, Effects of estradiol-implanted female song sparrows on the termination of territorial behavior and onset of the prebasic molt life-history stage in free-living males. Right, Effects of testosterone-implanted males on termination of territorial behavior and onset of the prebasic molt life-history stage in free-living females. (Adapted from Runfeldt and Wingfield 1985.)

Ecological Bases of Endocrine Phenomena

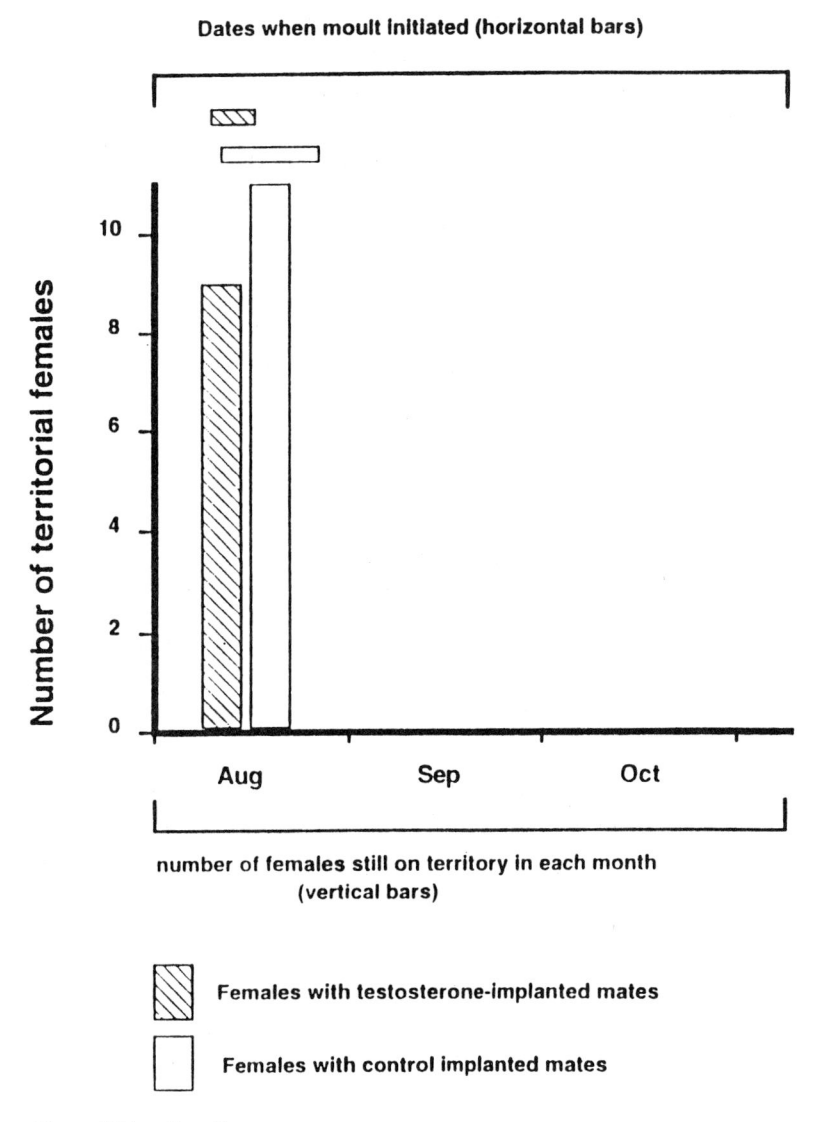

Dates when moult initiated (horizontal bars)

Number of territorial females (y-axis: 0, 2, 4, 6, 8, 10)

Aug — Sep — Oct

number of females still on territory in each month
(vertical bars)

▨ Females with testosterone-implanted mates
☐ Females with control implanted mates

Figure 4.9 (continued)

mated to estradiol-implanted females did not become photorefractory and failed to begin the molt stage until late-August to mid-September—1 full month later than controls (see figure 4.9). These males also had higher plasma levels of testosterone (Runfeldt and Wingfield 1985). Conversely, untreated females mated to males that had been given implants of testosterone became photorefractory and began the molt stage at the same time as females mated to control males (mid-August) even though testosterone-implanted males remained on territory and delayed onset of molt until at least the beginning of October (see figure 4.9). Presumably, social cues emanating from estradiol-treated females delayed onset of photorefractoriness in their mates. On the other hand, termination of the breeding stage in females does not appear to be influenced by males (Runfeldt and Wingfield 1985), unlike the developmental and mature capability phases (see above).

Emergency Life-History Stage and Perturbation of Reproductive Function

Development of the breeding stage, onset of nesting, and termination follow roughly predictable schedules depending on the type of habitat. However, unpredictable perturbations of the environment can occur at any time in the life cycle, including the breeding stage (see figure 4.1). Typically, reproduction is interrupted while the individual responds to the perturbation, but the reproductive system remains in a near functional state so that breeding can begin again once the perturbation passes (see Wingfield 1988, 1994c). We use the term *labile perturbation factors* (LPFs) to classify modifying factors because they are transitory, unpredictable, and always disruptive to the current life-history stage (Jacobs 1996). This differs from the regulated facultative responses of testosterone to unpredictable social cues (see above) which do *not* result in reproductive failure. LPFs trigger facultative behavioral and physiological responses (ELHS) that appear to be mediated by increases in corticosterone secretion (Wingfield et al. 1998a). There are several components to the ELHS (Wingfield and Ramenofsky 1997) as follows:

1. Deactivation of the current life-history stage (e.g., territorial behavior, abandonment of current nesting effort).

2. There are two, possibly three options here:

a. movements away from the source of the LPF ("leave-it" strategy);

b. if the individual remains it will seek a refuge ("take-it" strategy);

c. seek a refuge first and then move away if conditions do not improve ("take-it" at first and then "leave-it").

3. Mobilization of stored energy sources such as fat and perhaps protein to fuel the movement, or to provide energy while sheltering in a refuge.

4. Continued movement until suitable habitat is discovered or the perturbation passes.

5. Settlement in alternate habitat once an appropriate site is identified, or return to the original site and resumption of the normal sequence of life-history stages.

Rapid responses to some LPFs, for example, the fight-or-flight response to predators, result in immediate avoidance behavior and possibly an increase in corticosterone after a few minutes. These responses are generally too short-lived to activate an ELHS (Wingfield and Ramenofsky 1997). On the other hand, loss of the nest and young to a predator does disrupt the reproductive process and re-nesting follows. This is called a response to *indirect* LPFs and is independent of the ELHS. On the other hand, more chronic LPFs such as periods of inclement weather, human disturbance, and so forth may not affect the nest and young directly but result in a decrease of available food, temperature, and so on, so that the adults may become energetically stressed. It is this *direct* effect of LPFs that trigger the ELHS via increased secretion of corticosteroids from adrenocortical tissue (Wingfield and Ramenofsky 1997; Wingfield et al. 1998a). The emergency stage is temporary (hours to days) and maximizes the likelihood of survival in the face of direct LPFs. Once the perturbation passes the individual will return to the same life-history stage; in this case it will re-nest. Next we outline some of the events during re-nesting, particularly in response to an indirect LPF, and then go on to discuss mechanisms underlying the ELHS.

Hormonal Changes During Renesting After an Indirect LPF

Breeding pairs of birds that have lost the nest and young to a storm or predator frequently will re-nest. Loss of the nest is a considerable disruption of the normal temporal progression of the nesting phase and a marked reorganization of endocrine function is necessary to coordinate the re-nest attempt. In *Z.l. gambelii* there were marked resurgences of LH, testosterone, and testis mass after loss of the nest, coincident with elevated LH and estrogens in females culminating in production of a replacement clutch of eggs (Wingfield and Farner 1979). Essentially similar results have been obtained in female mallards (*Anas platyrhynchos*) after experimental removal of the eggs (Donham et al. 1976) and in song sparrows that lost nests to extensive flooding (Wingfield and Farner 1993). In white-crowned sparrows there were also increased concentrations of testosterone when producing a replacement clutch, unlike multiple brooding after *successfully* raising young (see Wingfield and Moore 1987; Wingfield and Farner 1993). Mechanisms underlying these differences in multiple brooding and re-nesting await further investigation (also reviewed in Wingfield and Farner 1993). Why should testosterone concentrations rise during the re-nesting period but not during the egg-laying period of a normal second brood? The typical second egg-laying often occurs in the presence of fledglings that still benefit from paternal care. It is possible that high levels of testosterone accompanied by increased territorial and "mate-guarding" aggression would interfere with parental behavior when males are feeding fledglings (see above and Silverin 1980; Hegner and Wingfield 1987). Presumably, fitness of the male is greater if he feeds fledglings from the first brood to independence (their chances of survival are greater than for young from later broods; e.g., see Perrins 1970) rather than neglecting offspring to mate-guard the female and ensure paternity of later clutches. On the other hand, if the nest and eggs are lost, it would be advantageous to the male to mate-guard and protect paternity of the replacement clutch (Wingfield and Moore 1987). These explanations could easily be tested in the field.

Hormonal Changes after Exposure to a Direct LPF: The Emergency Life-History Stage

A number of field investigations in birds have shown that plasma levels of corticosterone rise while responding to direct LPFs such as storms during the breeding season (see Wingfield 1998a). The behavioral and physiological responses were consistent with components of the ELHS outlined above. In breeding male *Z.l. pugetensis*, prolonged rain and wind storms in May and early June 1980 resulted in abandonment of the nest and territory (i.e., an ELHS was triggered). Plasma levels of corticosterone were up to threefold higher than the normal levels expected at that time of year (Wingfield et al. 1983). As corticosterone levels in blood rose, behavioral and physiological changes characteristic of the ELHS occurred (Wingfield 1998a). These were the short-term (minutes to hours) effects distinct from the more well-known, and detrimental long-term (chronic) effects resulting from days to weeks of sustained high levels of circulating glucocorticosteroids (table 4.3; see Wingfield, 1994c, Wingfield et al. 1998a for reviews).

There is now extensive experimental evidence to support the rapid effects of corticosterone, including suppression of reproductive and territorial behavior without inhibiting the reproductive system. This is in contrast to the chronically stressed state characterized by sustained high levels of corticosteroids that result in marked atrophy of the gonads. Other rapid effects of corticosterone include increased gluconeo-

Table 4.3
Effect of corticosterone in an emergency life-history stage

Rapid (i.e., short term, minutes to hours)	Chronic (i.e., long term, days to weeks)
Suppresses reproductive behavior	Inhibits reproductive system
Regulates immune system	Suppresses immune system
Increases gluconeogenesis	Promotes severe protein loss
Increases foraging behavior	Disrupts second messenger systems
Promotes escape (irruptive) behavior during day	Neuronal cell death
Promotes night restfulness by lowering standard metabolic rate	Suppresses growth and metamorphosis
Promotes recovery on return to normal life-history stage	

Modified and expanded from Wingfield 1994c.

genesis, enhanced foraging behavior, promotion of escape-like behavior (consistent with moving away from the LPF), and energy conservation by reduced standard metabolic rate at night (Wingfield 1994c; Wingfield et al. 1998a). The short-term effects of corticosterone comprise many facets of the facultative behavioral and physiological responses typical of an ELHS. Neural peptides such as corticotropin-releasing factor (CRF) and beta endorphin also appear to be involved (e.g., see Maney and Wingfield 1998a,b; Wingfield et al. 1998a). The sum of these is to redirect the individual away from the normal life-history stage for that time of year to maximize survival.

Glucocorticosteroids such as corticosterone act through classic intracellular receptors that bind to the genome and regulate gene expression (e.g., see McEwen et al. 1988). There are three possible roles that glucocorticosteroids may play in mammalian nervous systems: activation of neural activity during diurnal changes in pituitary-adrenal function; adaptation of the central nervous system to "stress"; and, in cases of chronically high levels of glucocorticosteroids, loss of neurons, especially in the hippocampus. Downregulation of corticosteroid receptors in response to chronic high levels of corticosteroids may, however, act as a protective mechanism to reduce deleterious effects such as neuron loss (McEwen et al. 1988). A characteristic of genomic actions of steroid hormones is that it takes time (at least 30 minutes, usually hours) for the responses to be manifest. However, some actions of glucocorticosteroids may occur within minutes. In amphibians there is evidence for a membrane receptor for corticosterone (Orchinik et al. 1991) that may be able to mediate glucocorticosteroid effects within minutes. If such membrane receptors for steroid hormones prove to be more widespread, then it is possible that they may also play a role in mediating the ELHS in response to LPFs in conjunction with genomic actions.

Conclusions

The entire reproductive process, from initiation of gonadal growth to production of independent young and termination of breeding, involves a tremendously complex, interconnecting network of morphological, physiological, and behavioral traits. For obvious reasons, most investigations tend to focus on one small aspect of reproductive function to determine mechanisms. However, it is then easy to lose sight of the overarching process itself. In this chapter we have attempted to develop a theoretical approach that enables us to place all of the potentially overwhelming processes involved in reproduction on a framework useful for all vertebrate species. Reproduction can be regarded as a stage in a finite-state machine of several life-cycle stages each of which have three subphases—development, mature capability (allows onset of substages characteristic of that stage), and termination. Immediately we can then postulate that the endocrine mechanisms involved will not necessarily be the same at each subphase—a point that needs to be borne in mind when focusing on the control of a single hormone or action of that hormone.

It is also possible to take a theoretical approach when addressing problems associated with how organisms respond to environmental signals that influence the breeding stage, and how these signals are transduced into neuroendocrine and endocrine secretions. This area has been a morass historically. We present a mathematical approach that generates testable hypotheses about how much environmental information a population should need to regulate breeding in the context of the habitat in which that populations lives. Clearly some populations respond to far more environmental cues than others—a point that raises some fascinating questions about evolution of brain mechanisms that perceive these signals and transduce them.

All reproductive processes appear to be profoundly influenced by social cues, whether they be intra- or intersexual or between adults and offspring. Effects of social cues are intimately interwoven with those of the physical environment, but a theoretical approach allows a level of analysis that identifies specific hypotheses. These can be tested with appropriate populations that enable experimental comparisons similar to those using genetic mutants or gene knockout models in mice. For example, brain pathways for transduction of social cues that influence GnRH and gonadal hormone secretions could be compared in closely related species that show strong modulation of testosterone secretion (i.e., level C $\gg$ level B) with those that do not (i.e., level C = level B). Such comparisons are particularly powerful because they involve viable, functioning populations and all of the problems associated with the biology of genetic mutants, or other effects of knocking out a gene throughout the organism, are avoided.

The reproductive process is usually regarded, for research purposes, as a predictable series of events occurring in an orderly fashion. For populations in their natural habitat, unpredictable perturbations of the environment are a critical component with the potential not only to reduce reproductive success for that year but also to increase mortality of adults unless appropriate adjustments are made. There has been much research on the effects of chronic stress on reproduction, and this has tremendous importance for medicine and agriculture. However, for an individual in the field, chronic stress is likely to be fatal or so debilitating that survival over winter is greatly reduced. A combination of field and laboratory studies revealed that initial responses to increased corticosterone secretion (following acute "stressors" that mimic environmental perturbations) redirect an individual away from breeding, but leave the reproductive system intact in a near functional state. Neural peptides associated with stress are also involved. When the perturbation passes the individual can then return to the nesting phase immediately. We have termed this the *emergency life-history stage* (*ELHS*). It promotes survival in the best condition possible so that breeding can be attempted again. Without such an "emergency" stage, individuals would attempt to continue breeding even when environmental conditions deteriorate and are no longer conducive to breeding. Redirection to a survival mode, and then re-nesting when conditions improve, may increase overall reproductive success in the long run.

We want to emphasize here that many of the approaches we have taken are theoretical. Several have been tested in depth and thus far some of these ideas are supported. Nonetheless, we feel that the theory in general is still in its infancy but has heuristic value. Broader testing, especially with other vertebrate taxa, will doubtless result in modifications of these ideas.

Acknowledgments

Much of the research described and the theoretical ideas developed here were supported by a series of grants from the Division of Integrative and Biology and Neuroscience, and Office of Polar Programs, National Science Foundation, to J.C.W. He is also grateful for a John Simon Guggenheim Fellowship, a Benjamin Meaker Fellowship from the University of Bristol, the Russell F. Stark University Professorship, and a grant from the University of Washington, Royalties Research Fund. Scott Edwards provided valuable advice on phylogenetic aspects of multiple species comparisons.

References

Arnold AP, Breedlove SM (1985). Organizational and activational effects of sex steroids on brain and behavior: a reanalysis. Horm Behav 19:469–498.

Astheimer LB, Buttemer WA, Wingfield JC (1994). Gender and seasonal differences in the adrenocortical response to ACTH challenge in an Arctic passerine, *Zonotrichia leucophrys gambelii*. Gen Comp Endocrinol 94:33–43.

Astheimer LB, Buttemer WA, Wingfield JC (1995). Seasonal and acute changes in adrenocortical responsiveness in an Arctic-breeding bird. Horm Behav 29:442–457.

Beletsky LD, Gori DF, Freeman S, Wingfield JC (1995). Testosterone and polygyny in birds. Curr Ornithol 12:1–41.

Berthold, P (1988). The control of migration in European warblers. *In* Acta 19th Congress of International Ornithology (Ouellet H, ed), pp 215–249. Ottawa: University of Ottawa Press.

Brown NL, Follett BK (1977). Effects of androgen on the testis of intact and hypophysectomized Japanese quail. Gen Comp Endocrinol 33:267–277.

Cohen D (1967). Optimizing reproduction in a varying environment. J Theor Biol 16:1–14.

Colwell RK (1974). Predictability, constancy, and contingency of periodic phenomena. Ecology 55:1148–1153.

Dawson A, Follett BK, Goldsmith AR, Nicholls TJ (1985). Hypothalamic gonadotrophin-releasing hormone and pituitary and plasma FSH and prolactin during photostimulation and photorefractoriness in intact and thyroidectomized starlings (*Sturnus vulgaris*). J Endocrinol 105:71–77.

Donham RS, Dane CW, Farner DS (1976). Plasma luteinizing hormone and the development of ovarian follicles after loss of clutch in female mallards (*Anas platyrhynchos*). Gen Comp Endocrinol 29:152–155.

Farner DS (1970). Predictive functions in the control of annual cycles. Environ Res 3:119–131.

Farner DS (1985). Annual rhythms. Annu Rev Physiol 47:65–82.

Farner DS, Follett BK (1979). Reproductive periodicity in birds. *In* Hormones and Evolution. (Barrington EJW, ed), pp 829–872. New York: Academic Press.

Farner DS, Lewis RA (1971). Photoperiodism and reproductive cycles in birds. *In* Photophysiology, vol 6 (Giese AC, ed), pp 325–370, New York: Academic Press.

Farner DS, Lewis RA (1973). Field and experimental studies of the annual cycles of white-crowned sparrows. J Reprod Fertil 19 (Suppl):35–50.

Follett BK (1984). Birds. *In* Marshall's Physiology of Reproduction, vol 1. Reproductive Cycles of Vertebrates (Lamming GE, ed), pp 283–350. Edinburgh: Churchill Livingstone.

Garland T Jr, Dickerman AW, Janis CM and Jones JA (1993). Phylogenetic analysis of covariance by computer simulation. System Biol 42:265–292.

Goldsmith AR, Nicholls TJ (1984). Thyroxine induces photorefractoriness and stimulates prolactin secretion in European starlings (*Sturnus vulgaris*). J Endocrinol 101:R1–R3.

Gwinner E (1981). Circannual systems. *In* Handbook of Behavioral Neurobiology 4. Biological Rhythms (Aschof J, ed), pp 391–410. New York: Plenum Press.

Gwinner E (1987). Circannual Rhythms. Berlin: Springer-Verlag.

Hahn TP (1995). Integration of photoperiodic and food cues to time changes in reproductive physiology by an opportunistic breeder, the red crossbill, *Loxia curvirostra* (Aves; Carduelinae). J Exp Zool 272:213–226.

Harvey PH, Pagel MD (1991). The Comparative Method in Evolutionary Biology. Oxford: Oxford University Press.

Harvey S, Phillips JG, Rees A, Hall TR (1984). Stress and adrenal function. J Exp Zool 232:633–646.

Hegner RE, Wingfield JC (1987). Effects of experimental manipulation of testosterone levels on parental investment and breeding success in male house sparrows. Auk 104:462–469.

Ishii S, Farner DS (1976). Binding of follicle-stimulating hormone by homogenates of testes of photostimulated white-crowned sparrows, *Zonotrichia leucophrys gambelii.* Gen Comp Endocrinol 30:443–450.

Jacobs J (1996). Regulation of Life History Strategies Within Individuals in Predictable and Unpredictable Environments, Ph.D. thesis, University of Washington, Seattle.

King JR, Follett BK, Farner DS, Morton ML (1966). Annual gonadal cycles and pituitary gonadotropins in *Zonotrichia leucophrys gambelii.* Condor 68:476–487.

Kobayashi H (1952). Effects of hormonic steroids on molting and broodiness in the canary. Annot Zool Jpn. 25:128–134.

Kubokawa K, Ishii S, Wingfield JC (1994). Effect of day length on luteinizing hormone β-subunit mRNA and subsequent gonadal growth in the white-crowned sparrow, *Zonotrichia leucophrys gambelii.* Gen Comp Endocrinol 95:42–51.

Lofts B, Murton RK (1968). Photoperiodic and physiological adaptations regulating avian breeding cycles and their ecological significance. J Zool (Lond) 155:327–394.

Lofts B, Murton RK (1973). Reproduction in birds. *In* Avian Biology, vol 3. Farner DS, King JR, eds., (pp 1–107). New York: Academic Press.

Maney DL, Wingfield JC (1998a). Central opioid control of feeding behavior in the white-crowned sparrow, *Zonotrichia leucophrys gambelii.* Horm Behav 33:16–22.

Maney DL, Wingfield JC (1998b). Neuroendocrine suppression of female courtship in a wild passerine: corticotropin-releasing factor and endogenous opioids. J Neuroendocrinol 10:593–599.

Maney DL, Goode CT, Wingfield JC (1997a). Intraventricular infusion of arginine vasotocin induces singing in a female songbird. J. Neuroendocrinol 9:487–491.

Maney DL, Richardson RD, Wingfield JC (1997b). Central administration of chicken gonadotropin-releasing hormone II enhances courtship behavior in a female sparrow. Horm Behav 32:11–18.

Marshall AJ (1959). Internal and environmental control of breeding. Ibis 101:456–478.

Marshall AJ (1970). Environmental factors other than light involved in the control of sexual cycles in birds and mammals. *In* La Photorégulation de la reproduction chez les oiseaux et les mammifères (Benoit J, Assenmacher I, eds), pp 53–64. Paris: Centre National. Recherche Scientifique.

McEwan B, Brinton RE, Sapolsky RM (1988). Glucocorticoid receptors and behavior: implications for the stress response. Adv Exp Med Biol 245:35–45.

Meaney MJ, Viau V, Bhatnagar S, Betito K, Iny LJ, O'Donnell D, Mitchell JB (1991). Cellular mechanisms underlying the development and expression of individual differences in the hypothalamic-pituitary-adrenal stress response. J Steroid Biochem Mol Biol 39:265–274.

Moore MC (1982). Hormonal responses of free-living male white-crowned sparrows to experimental manipulation of female sexual behavior. Horm Behav 16:323–329.

Moore MC (1983). Effect of female sexual displays on the endocrine physiology and behavior of male white-crowned sparrows, *Zonotrichia leucophrys gambelii.* J Zool Lond 199:137–148.

Morton ML, Pereya ME, Baptista LF (1985). Photoperiodically induced ovarian growth in the white-crowned sparrow (*Zonotrichia leucophrys gambelii*) and its augmentation by song. Comp Biochem Physiol 80A:93–97.

Nicholls TJ, Goldsmith AR, Dawson A (1988). Photorefractoriness in birds and comparison with mammals. Physiol Rev 68:133–176.

Orchinik M, Murray TF, Moore FL (1991). A corticosteroid receptor in neuronal membranes. Science 252:1848–1851.

Oring LW, Fivizzani AJ, El Halawani ME (1989). Testosterone-induced inhibition of incubation in the spotted sandpiper (*Actitis macularia*). Horm Behav 23:412–423.

Perrins CM (1970). The timing of bird's breeding seasons. Ibis 112:242–255.

Runfeldt S, Wingfield JC (1985). Experimentally prolonged sexual activity in female sparrows delays termination of reproductive activity in their untreated mates. Anim Behav 33:403–410.

Sapolsky R (1987). Stress, social status, and reproductive physiology in free-living baboons. *In* Psychobiology of Reproductive Behavior. (Crews D, ed), pp 291–322. Englewood Cliffs, NJ: Prentice-Hall.

Schleussner G, Gwinner E (1988). Photoperiodic time measurement during the termination of photoerfractoriness in the starling (*Sturnus vulgaris*). Gen Comp Endocrinol 75:54–61.

Sherwood N, Wingfield JC, Ball GF, Dufty AM Jr (1988). Identity of GnRH in passerine birds: comparison of GnRH in song sparrow (*Melospiza melodia*) and starling (*Sturnus vulgaris*) with 5 vertebrate GnRHs. Gen Comp Endocrinol 69:341–351.

Silverin B (1980). Effects of long acting testosterone treatment on free-living pied flycatchers *Ficedula hypoleuca*. Anim. Behav. 28:906–912.

Silverin B (1986). Corticosterone binding proteins and behavioral effects of high plasma levels of corticosterone during the breeding period in the pied flycatcher. Gen Comp Endocrinol 64:67–74.

Turek FW, Desjardins C, Menaker M (1976). Antigonadal and progonadal effects of testosterone in male house sparrows. Gen Comp Endocrinol 28:395–402.

Wingfield JC (1980). Fine temporal adjustment of reproductive functions. *In* Avian Endocrinology (Epple A, Stetson MH, eds), pp 367–389. New York: Academic Press.

Wingfield JC (1983). Environmental and endocrine control of reproduction: an ecological approach. *In* Avian Endocrinology: Environmental and Ecological Aspects (Mikami SI, Wada M, eds), pp 205–288. Tokyo: Japanese Scientific Societies Press; and Berlin: Springer-Verlag.

Wingfield JC (1984a). Androgens and mating systems: testosterone-induced polygyny in normally monogamous birds. Auk 101:665–671.

Wingfield JC (1984b). Environmental and endocrine control of reproduction in the song sparrow, *Melospiza melodia*. I. Temporal organization of the breeding cycle. Gen Comp Endocrinol 56:406–416.

Wingfield JC (1985). Short-term changes in plasma levels of hormones during establishment and defense of a breeding territory in male song sparrows, *Melospiza melodia*. Horm Behav 19:174–187.

Wingfield JC (1988). Changes in reproductive function of free-living birds in direct response to environmental perturbations. *In* Processing of Environmental Information in Vertebrates (Stetson MH, ed), pp 121–148. Berlin: Springer-Verlag.

Wingfield JC (1994a). Hormone-behavior interactions and mating systems in male and female birds. *In* The Difference Between the Sexes (Short RV, Balaban E, eds), pp 303–330. London: Cambridge University Press.

Wingfield JC (1994b). Regulation of territorial behavior in the sedentary song sparrow, *Melospiza melodia morphna*. Horm Behav 28:1–15.

Wingfield JC (1994c). Modulation of the adrenocortical response to stress in birds. *In* Perspectives in Comparative Endocrinology (Davey KG, Peter RE, Tobe SS, eds), pp 520–528. Ottawa: National Research Council of Canada.

Wingfield JC, Farner DS (1978a). The annual cycle in plasma irLH and steroid hormones in feral populations of the white-crowned sparrow, *Zonotrichia leucophrys gambelii*. Biol Reprod 19:1046–1056.

Wingfield JC Farner DS (1978b). The endocrinology of a naturally breeding population of the white-crowned sparrow (*Zonotrichia leucophrys pugetensis*). Physiol Zool 51:188–205.

Wingfield JC, Farner DS (1979). Some endocrine correlates of renesting after loss of clutch or brood in the white-crowned sparrow, *Zonotrichia leucophrys gambelii*. Gen Comp Endocrinol 38:322–331.

Wingfield JC, Farner DS (1980). Environmental and endocrine control of seasonal reproduction in temperate zone birds. Prog Rep Biol 5:62–101.

Wingfield JC, Farner DS (1993). The endocrinology of wild species. *In* Avian Biology vol 9 (Farner DS, King JR, Parkes KC, eds), pp 163–327. New York: Academic Press.

Wingfield JC, Moore MC (1987). Hormonal, social, and environmental factors in the reproductive biology of free-living male birds. *In* Psychobiology of Reproductive Behavior: An Evolutionary Perspective (Crews D, ed), pp. 149–175. Englewood Cliffs, NJ: Prentice-Hall.

Wingfield JC, Ramenofsky M (1997). Corticosterone and facultative dispersal in response to unpredictable events. Ardea 85:155–166.

Wingfield JC, Smith JP, Farner DS (1982b). Endocrine responses of white-crowned sparrows to environmental stress. Condor 84:399–409.

Wingfield, JC, Kenagy GJ (1991). Natural regulation of reproductive cycles. *In*: Vertebrate Endocrinology: Fundamentals and Biomedical Implications" (Schreibman M, Jones RE eds), vol 4, Part B, pp. 181–241, New York: Academic Press.

Wingfield JC, Hegner RE, Dufty AM Jr, Ball GF (1990). The "challenge hypothesis": theoretical implications for patterns of testosterone secretion, mating systems, and breeding strategies. Am Nat 136:829–846.

Wingfield JC, Moore MC, Farner DS (1983). Endocrine responses to inclement weather in naturally breeding populations of white-crowned sparrows. Auk 100:56–62.

Wingfield JC, Ball GF, Dufty AM Jr, Hegner RE, Ramenofsky M. (1987). Testosterone and aggression in birds: tests of the "challenge hypothesis". Am Scientist 75:602–608.

Wingfield JC, Ronchi E, Goldsmith AR, Marler C (1989). Interactions of sex steroid hormones and prolactin in male and female song sparrows, *Melospiza melodia*. Physiol Zool 62:11–24.

Wingfield JC, Hahn TP, Levin R, Honey P (1992a). Environmental predictability and control of gonadal cycles in birds. J Exp Zool 261:214–231.

Wingfield JC, Vleck CM, Moore MC (1992b). Seasonal changes in the adrenocortical response to stress in birds of the Sonoran Desert. J Exp Zool 264:419–428.

Wingfield JC, Doak D, Hahn TP (1993). Integration of environmental cues regulating transitions of physiological state, morphology and behavior. *In* Avian Endocrinology (Sharp PJ, ed), pp 111–122. Bristol, UK: Journal of Endocrinology.

Wingfield JC, Deviche P, Sharbaugh S, Astheimer LB, Holberton R, Suydam R, Hunt K (1994a). Seasonal changes of the adrenocortical responses to stress in redpolls, *Acanthis flammea*, in Alaska. J Exp Zool 270:372–380.

Wingfield JC, Suydam R, Hunt K (1994b). Adrenocortical responses to stress in snow buntings and Lapland longspurs at Barrow, Alaska. Comp Biochem Physiol 108:299–306.

Wingfield JC, Whaling CS, Marler PR (1994c). Communication in vertebrate aggression and reproduction: The role of hormones. *In* Physiology of Reproduction, 2nd ed (Knobil E, Neill JD, eds), pp 303–342. New York: Raven Press.

Wingfield JC, O'Reilly KM, Astheimer LB (1995). Ecological bases of the modulation of adrenocortical responses to stress in Arctic birds. Am Zool 35:285–294.

Wingfield JC, Hahn TP, Wada M, Astheimer LB, Schoech S (1996). Interrelationship of day length and temperature on the control of gonadal development, body mass and fat depots in white-crowned sparrows, *Zonotrichia leucophrys gambelii*. Gen Comp Endocrinol 101:242–255.

Wingfield JC, Hahn TP, Wada M, Schoech S (1997a). Effects of day length and temperature on gonadal development, body mass and fat depots in white-crowned sparrows, *Zonotrichia leucophrys pugetensis*. Gen Comp Endocrinol 107:44–62.

Wingfield JC, Jacobs J, Hillgarth N, (1997b). Ecological constraints and the evolution of hormone-behavior interrelationships. Ann N Y Acad Sci 807:22–41.

Wingfield JC, Breuner C, Jacobs J, Lynn S, Maney D, Ramenofsky M, Richardson R (1998a). Ecological bases of hormone-behavior interactions: the "emergency life history stage." Am Zool 38:191–206.

Wingfield JC, Jacobs J, Soma K, Maney DL, Hunt K, Wisti-Peterson D, Meddle S, Ramenofsky M, Sullivan K (1998b). Testosterone, aggression and communication: ecological bases of endocrine phenomena. *In* Symposium in Honor of Peter R Marler. (Konishi M, Hauser M, eds). Cambridge, MA.: MIT Press (in press).

Sources for Figures 4.7 and 4.8

Akesson TR, Raveling DG (1981). Endocrine and body weight changes of nesting and non-nesting Canada geese. Biol Reprod 25:792–804.

Ball GF, Wingfield JC (1987). Changes in plasma levels of sex steroids in relation to multiple broodedness and nest site density in male starlings. Physiol Zool 60:191–199.

Beani L, Cervo R, Lodi L, Lupo C, Dessi-Fulgheri F (1988). Circulating levels of sex steroids and sociosexual behavior in the grey partridge (*Perdix perdix L.*). Monit Zool Ital (NS) 22:145–160.

Beletsky L, Wingfield JC, Orians GH (1989). Relationships of hormones and polygyny to territorial status, breeding experience and reproductive success in male red-winged blackbirds. Auk 106:107–117.

Beletsky L, Orians GH, Wingfield JC (1990). Steroid hormones, polygyny and parental behavior in male yellow-headed blackbirds. Auk 107:60–68.

Berry HH, Millar RP, Louw GN (1979). Environmental cues influencing the breeding biology and circulating levels of various hormones and triglycerides in the Cape cormorant. Comp Biochem Physiol 62A:879–884.

Borgia G, Wingfield JC (1991). Hormonal correlates of bower decoration and sexual display in the satin bowerbird (*Ptilonorhynchus violaceus*). Condor 93:935–942.

Burger AE, Millar RP (1980). Seasonal changes of sexual and territorial behavior and plasma testosterone levels in male lesser sheathbills (*Chionis minor*). Z Tierpsychol 52:397–406.

Cherel Y, Mauget R, Lacroix A, Gilles J (1994). Seasonal and fasting related changes in circulating gonadal steroids and prolactin in king penguins, *Aptenodytes patagonicus*. Physiol Zool 67:1154–1173.

Cockrem JF, Potter MA (1991). Reproductive endocrinology of the North Island brown kiwi, *Apteryx australis mantelli*. In Acta 20th Congress of International Ornithology (Bell BD, ed), pp 2092–2101. Wellington, NZ: New Zealand Ornithological Congress Trust Board.

Dawson A (1983). Plasma gonadal steroid levels in wild starlings (*Sturnus vulgaris*) during the annual cycle and in relation to the stages of breeding. Gen Comp Endocrinol 49:286–294.

Dittami JP (1981). Seasonal changes in the behavior and plasma titers of various hormones in bar-headed geese, *Anser indicus*. Z Tierpsychol 55:289–324.

Donham RS (1979). The annual cycle of plasma luteinizing hormone and sex hormones in male and female mallards (*Anas platyrhynchos*). Biol Reprod 21:1273–1285.

Dufty AM Jr, Wingfield JC (1986a). Temporal patterns of circulating LH and steroid hormones in a brood parasite, the brown-headed cowbird, *Molothrus ater*. I. Males. J Zool Lond 208:191–203.

Dufty AM Jr, Wingfield JC (1986b), Temporal patterns of circulating LH and steroid hormones in a brood parasite, the brown-headed cowbird, *Molothrus ater*. II. Females. J Zool Lond 208:205–214.

Feder HH, Storey A, Goodwin D, Reboulleau C, Silver R (1977). Testosterone and 5-alpha-dihydrotestosterone levels in peripheral plasma of male and female ring doves (*Streptopelia risoria*). Biol Reprod 16:666–677.

Fivizzani AJ, Oring LW (1986). Plasma steroid hormones in the polyandrous spotted sandpiper, *Actitis macularia*. Biol Reprod 35:1195–1201.

Fivizzani AJ, Colwell MA, Oring LW (1986). Plasma steroid hormone levels in free-living Wilson's phalaropes, *Phalaropus tricolor*. Gen Comp Endocrinol 62:137–144.

Fivizzani AJ, Oring LW, El-Halawani ME, Schlinger BA (1990). Hormonal basis of parental care and female intersexual competition in sex-role reversed birds. In Endocrinology of Birds: Molecular to Behavioral (Wada M, Ishii S, Scanes CG, eds), pp 273–286. Tokyo: Japanese Scientific Societies Press; and Berlin: Springer-Verlag.

Fornasari L, Bottoni L, Schwabl H, Massa R (1992). Testosterone in the breeding cycle of the male red-backed shrike, *Lanius collurio*. Ethol Ecol Evol 4:193–196.

Fowler GS, Wingfield JC, Boersma PD, Sosa RA (1994). Reproductive endocrinology and weight change in relation to reproductive success in the Magellanic penguin (*Spheniscus magellanicus*). Gen Comp Endocrinol 94:305–315.

Fraissinet M, Varriale B, Pierantoni R, Caliendo MF, Di Matteo L, Bottoni L, Milone M (1987). Annual testicular activity in the grey partridge (*Perdix perdix L.*). Gen Comp Endocrinol 68:28–32.

Gratto-Trevor CL, Fivizzani AJ, Oring LW, Cooke F (1990). Seasonal changes in gonadal steroids of a monogamous versus a polyandrous shorebird. Gen Comp Endocrinol 80:407–418.

Groscolas R, Jallageas M, Goldsmith AR, Leloup J, Assenmacher I (1985). Changes in plasma LH, gonadal and thyroid hormones in breeding and molting emperor penguins. *In* Current Trends in Comparative Endocrinology (Lofts B, Holmes WN, eds), pp. 261–264. Hong Kong: University of Hong Kong Press.

Groscolas R, Jallageas M, Goldsmith AR, Assenmacher I (1986). The endocrine control of reproduction and molt in male and female emperor (*Aptenodytes forsteri*) and adelie (*Pygoscelis adeliae*) penguins. I. Annual changes in plasma levels of gonadal steroids and LH. Gen Comp Endocrinol 62:43–53.

Groscolas R, Jallageas M, Leloup J, Goldsmith AR (1988). The endocrine control of reproduction in male and female emperor penguins (*Aptenodytes forsteri*). *In* Acta 19th congress of International. Ornithology (Ouellet H, ed), pp 1692–1701. Ottawa: University of Ottawa Press.

Hall MR (1986). Plasma concentrations of prolactin during the breeding cycle in the Cape gannet (*Sula capensis*): a foot incubator. Gen Comp Endocrinol 64:112–121.

Hall MR, Gwinner E, Bloesch M (1987). Annual cycles in moult, body mass, luteinizing hormone, prolactin, and gonadal steroids during the development of sexual maturity in the white stork (*Ciconia ciconia*). J Zool Lond 211:467–486.

Hannon SJ, Wingfield JC (1990). Endocrine correlates of territoriality, breeding stage, and body moult in free-living willow ptarmigan of both sexes. Can J Zool 68:2130–2134.

Harding CF, Follett BK (1979). Hormone changes triggered by aggression in a natural population of blackbirds. Science 203:918–920.

Hector JAL (1988). Reproductive endocrinology of albatrosses. *In* Acta 19th Congress of International Ornithology (Ouellet H, ed), pp 1702–1709. Ottawa: University of Ottawa Press.

Hector JAL, Croxall JP, Follett BK (1986a), Reproductive endocrinology of the wandering albatross, *Diomedia exulans*, in relation to biennial breeding and deferred sexual maturity. Ibis 128:9–22.

Hector JAL, Follett BK, Prince PA (1986b). Reproductive endocrinology of the black-browed albatross, *Diomedia melanophris*, and the grey-headed albatross, *D. chrysostoma*. J Zool Lond 208:237–253.

Hegner RE, Wingfield JC (1986a). Behavioral and endocrine correlates of multiple brooding in the semi-colonial house sparrow, *Passer domesticus*. I. Males. Horm Behav 20:294–312.

Hegner RE, Wingfield JC (1986b). Behavioral and endocrine correlates of multiple brooding in the semi-colonial house sparrow, *Passer domesticus*. II. Females. Horm Behav 20:313–326.

Hunt K, Wingfield JC, Astheimer LB, Buttemer WA, Hahn TP (1995). Temporal patterns of territorial behavior and circulating testosterone in the Lapland longspur and other Arctic passerines. Am Zool 35:274–284.

Jouventin P, Mauget R (1996). The endocrine basis of the reproductive cycle in the king penguin (*Aptenodytes patagonicus*). J Zool Lond 238:665–678.

Kerlan JT, Jaffe RB (1974). Plasma testosterone levels during the testicular cycle of the red-winged blackbird (*Aegelaius phoeniceus*). Gen Comp Endocrinol 22:428–432.

Ketterson ED, Nolan V Jr (1992). Hormones and life histories: an integrative approach. Am Naturalist 140:S33–S62.

Lincoln GA, Racey PA, Sharp PJ, Klandorf H (1980). Endocrine changes associated with spring and autumn sexuality of the rook, *Corvus frugilegus*. J Zool Lond 190:137–153.

Lisano ME, Kennamer JE (1977). Seasonal variations in plasma testosterone level in male eastern wild turkeys. J Wildl Manage 41:184–188.

Logan CA, Wingfield JC (1995). Hormonal correlates of breeding status, nest construction and parental care in multiple-brooded northern mockingbirds, *Mimus polyglottos.* Horm Behav 29:12–30.

Lupo C, Beani L, Cervo R, Lodi L, Dessi-Fulgheri F (1990). Steroid hormones and reproductive success in the grey partridge (*Perdix perdix*). Boll Zool 57:247–252.

Mauget R, Jouventin P, Lacroix A, Ishii S (1994). Plasma LH and steroid hormones in king penguin (*Aptenodytes patagonicus*) during the onset of the breeding cycle. Gen Comp Endocrinol 93:36–43.

Mays NA, Vleck CM, Dawson J (1991). Plasma luteinizing hormone, steroid hormones, behavioral role and nest stage in cooperatively breeding Harris' Hawks (*Parabuteo unicinctus*). Auk 108:619–637.

Meijer T, Schwabl H (1989). Hormonal patterns in breeding and non-breeding kestrels *Falco tinnunculus*: field and laboratory studies. Gen Comp Endocrinol 74:148–160.

Paulke E, Haase E (1978). A comparison of seasonal changes in the concentrations of androgens in the peripheral blood of wild and domestic ducks. Gen Comp Endocrinol 34:381–390.

Péczely P, Pethes G (1979). Alterations in plasma sexual steroid concentrations in the collared dove (*Streptopelia decaocto*) during the sexual maturation and reproduction cycle. Acta Physiol Acad Sci Hung 54:161–170.

Péczely P, Pethes G (1982). Seasonal cycle of gonadal, thyroid, and adrenocortical function in the rook, (*Corvus frugilegus*). Acta Physiol Acad Sci Hung 59:59–73.

Ramenofsky M (1984). Agonistic behavior and endogenous plasma hormones in male Japanese quail. Anim Behav 32:698–708.

Rissman EF, Wingfield JC (1984). Hormonal correlates of polyandry in the spotted sandpiper, *Actitis macularia.* Gen Comp Endocrinol 56:401–405.

Röhss M, Silverin B (1983). Seasonal variation in the ultrastructure of the Leydig cells and plasma levels of luteinizing hormone and steroid hormones in juvenile and adult male great tits, *Parus major*. Ornis Scand 14:202—212.

Saino N, Møller AP (1994). Secondary sexual characters, parasites and testosterone in the barn swallow, *Hirundo rustica.* Anim Behav 48:1325–1333.

Sakai H, Ishii S (1986). Annual cycles of gonadotropins and sex steroids in Japanese common pheasants, *Phasianus colchicus versicolor*. Gen Comp Endocrinol 63:275–283.

Schoech SJ, Mumme RL, Moore MC (1991). Reproductive endocrinology and mechanisms of breeding inhibition in cooperatively breeding Florida scrub jays (*Aphelocoma c. coerulescens*). Condor 93:354–362.

Schwabl H, Wingfield JC, Farner DS (1980). Seasonal variations in plasma levels of luteinizing hormone and steroid hormones in the European blackbird, *Turdus merula.* Vogelwarte 30:283–294.

Silverin B, Wingfield JC (1982). Patterns of breeding behavior and plasma levels of hormones in a free-living population of pied flycatchers, *Ficedula hypoleuca.* J Zool Lond 198:117–129.

Silverin B, Viebke PA, Westin J (1986). Seasonal changes in plasma LH and gonadal steroids in free-living willow tits, *Parus montanus.* Ornis Scand 17:230–236.

Stokkan K-A, Sharp PJ (1980a). Seasonal changes in the concentrations of plasma luteinizing hormone and testosterone in willow ptarmigan (*Lagopus lagopus lagopus*) with observations on the effects of permanent short days. Gen Comp Endocrinol 40:109–115.

Vleck CM (1993). Hormones, reproduction, and behavior in birds of the Sonoran desert. *In* Avian Endocrinology. (Sharp PJ, ed), pp 73–86. Bristol, UK: Journal of Endocrinology.

Williams TD (1992). Reproductive endocrinology of Macaroni (*Eudytptes chrysolophus*) and Gentoo (*Pygoscelis papua*) penguins. Gen Comp Endocrinol 85:230–240.

Wingfield JC, Newman A, Hunt GL Jr, Farner DS (1980b). Androgen high in concentration in the blood of female western gulls, *Larus occidentalis wymani.* Naturwissenschaften 67:S514.

Wingfield JC, Newman A, Hunt GL Jr, Farner DS (1982a). Endocrine aspects of female-female pairing in the western gull (*Larus occidentalis wymani*). Anim Behav 30:9–22.

5 Neuroendocrine Mechanisms Mediating the Photoperiodic and Social Regulation of Seasonal Reproduction in Birds

Gregory F. Ball and George E. Bentley

Many Temperate Zone species tend to breed seasonally so that reproductive attempts coincide with environmental conditions favorable for successful breeding, as documented in chapters 2 and 4, and as will be discussed again in chapters 7 and 8 (for review, see Bronson 1989; Wingfield and Kenagy 1991). Favorable breeding conditions result from the interactions of a number of variables that include adequate food availability, the absence of predators, and mild ambient temperatures (Perrins 1970; Bronson 1989). This chapter focuses on neuroendocrine mechanisms that mediate the environmental regulation of seasonal reproduction in avian species. This relatively well-studied vertebrate class has been a source of many of the most important general principles to emerge from the study of environmental effects on reproductive behavior and physiology (Konishi et al. 1989). Many of these principles can be applied across the vertebrate classes. However, there are several peculiarities of the life-history strategies adopted by birds that influence the timing of reproduction. These must be kept in mind in order to understand the control of reproduction in these species fully. For example, in most avian species, food for the young at the time of hatching is particularly important because the successful survival of the progeny is closely tied to the types of food ingested during development (Baker 1938; Perrins and Birkhead 1983). In the case of altricial species, food is provided by the parents to the young, whereas, in the case of precocial species, the young are guided to food sources where they feed themselves (Ricklefs 1983).

In general, avian parents are unable to store energy as body fat and then provide it to their young at a later time. This is in contrast to certain mammalian species, in which the mother will overeat and store excess calories in the form of fat during a time of abundant food availability, then convert her fat to milk during lactation, and feed her young on this milk at a later time when food availabilty may be low (Bronson 1989). This pattern is very common among pinnipeds. There are of course exceptions among the over nine thousand extant species of birds. Some species (especially in the corvid family) store food so that they can initiate breeding before appropriate quantities and types of food are available (e.g., gray jays, *Perisoreus candensis*).

Another life-history trait that greatly influences the patterns of breeding in many birds is related to their volant life style. The advantage conferred by the ability to fly favors birds of smaller size. This makes it especially important to reduce organ systems when they are not needed. A tendency to reduce mass in response to environmental changes is exemplified in most species by a mechanism that preferentially activates only the sinistral ovary (only ground-dwelling species such as kiwis [*Aptery-*

giformes] are exceptions to this rule). Another example is the fluctuating pattern of breeding activity and gonadal recrudescence that is timed so that breeding is strictly limited to a relatively short time window in the annual cycle (Murton and Westwood 1977). Among male birds, there is remarkable seasonal variation in the size of the testes, with increases as large as 1000-fold in breeding compared to nonbreeding males (Follett 1984).

It is therefore a common pattern among nontropical avian species to limit reproductive activity to the time of year when temperatures are relatively mild and the necessary food resources are present. In order to coordinate gonadal recrudescence and the associated increases in endocrine activity, birds often use specific cues in the environment, such as changes in photoperiod, that is, the length of the daylight period. In chapter 4, Wingfield and co-authors outline an ecological framework that has proved very useful in organizing these different environmental variables according to their ecological relevance. Factors such as food availability that limit reproductive success are usually referred to as ultimate causes (or factors) because their presence or absence directly affects the survival of the young. Factors such as photoperiod that initiate reproductive processes in time for the breeding season are usually referred to as the proximate causes (or cues) for successful breeding (Baker 1938). Proximate cues can then be usefully subdivided into different categories. For example, photoperiod is a powerful "initial predictive cue" that initiates or terminates the period of reproduction in many nontropical species. Other cues provide "essential supplementary information" and "synchronizing and integrating information." This category includes such factors as social interactions and the availability of nest sites (Wingfield and Kenagy 1991).

This chapter focuses on the mechanisms by which various proximate cues influence the timing of breeding in birds. In particular, we focus on the neuroendocrine mechanisms that respond to and integrate environmental and social factors. It has become apparent that a variety of different types of cues are integrated by birds to time breeding and that the gonadotropin-releasing hormone (GnRH) neuronal system in birds is a key "common final pathway" in the integration of these cues (Ball 1993; Wingfield and Kenagy 1991). First, the basics of the photoperiodic response in birds are discussed, using European starlings (*Sturnus vulgaris*) as an example. We describe the sensory detection of the photoperiodic signal, the clock mechanism, and the neuroendocrinology of reproduction in starlings. Next, we examine, for a variety of species, the importance of the hypothalamic decapeptide GnRH in photoperiodic control of reproduction. Then, in relation to GnRH, we discuss species differences in breeding flexibility and the roles played by supplementary and synchronizing cues

Environmental Regulation of Reproduction in Birds 131

such as social interactions. We discuss visual and auditory cues; however, given the importance of song in the social life of birds, we focus primarily on the role of the song system in coordinating reproduction. There is a long tradition of the investigation of the effects of social cues (e.g., see Lehrman 1965; Hinde 1965) and photoperiod (Farner and Follett 1966) on reproduction in birds.

Finally, we examine converging neural pathways that mediate the effects of photoperiod and male song on the neuroendocrine system that controls reproduction. An important question in this regard is whether initial predictive cues act independently of the effects of supplementary signals on the reproductive system. We argue that these different types of cues can act either independently or in an interactive manner, depending on the species and its life-history strategy. In this chapter we try to illustrate that studies of birds continue to be useful for the elucidation of general principles underlying the environmental regulation of reproduction, and particularly for combining ideas and paradigms from ecology with those of behavioral neuroendocrinology.

Neuroendocrine Basis of Seasonal Breeding in European Starlings

In this section we focus on European starlings (*Sturnus vulgaris*) as a model for avian seasonal breeding. Starlings are one of the most studied seasonally breeding avian species, and their robust responses to changing photoperiod have been extensively characterized. Although the basic mechanisms controlling responses to photoperiod are essentially the same in all seasonally breeding birds, there are species variations in the degree of responsiveness to photoperiod and the importance of supplementary environmental cues during the breeding season. In this respect, starlings are at one end of the spectrum; they are highly photoperiodic (Burger 1947) and the effects of supplementary cues on their reproductive system may be strictly limited by the photoperiod the birds experience.

Photoperiodism in birds can be envisaged as a physiological system that has a fluctuating responsiveness to day length (Nicholls et al. 1988; Wilson and Donham 1988). At specific times of the year (late winter and spring) the hypothalamo-pituitary-gonadal axis responds to long day lengths with a large increase in gonadotropin secretion, gonadal growth, and a complete range of hormone-dependent processes, including behavioral changes. This physiological responsiveness to long days is known as photosensitivity. However, Temperate Zone birds become insensitive to the stimulatory effects of long days following prolonged long-day exposure, so that later on in the year (late summer and early autumn) the same day length has none of the physiological effects described (for reviews, see Nicholls et al. 1988; Wilson and Donham 1988). In birds, this state is known as photorefractoriness. We would like to note here that investigators studying birds use the term *photorefractoriness* to describe different physiological phenomena than those described by investigators studying mammals using the same term. For example, photorefractoriness in photoperiodic rodents is used in reference to a loss of response to the inhibitory effects of short days. Hamsters moved from a long-day to a short-day photoperiod show gonadal regression, but after several weeks of short days their gonads spontaneously regrow. In sheep, photorefractoriness can be applied to either short or long day lengths, indicating a lack of response to the prevailing photoperiod (Robinson and Karsch 1984; Robinson et al. 1985).

A feature of photorefractoriness in birds is that even though it does not become apparent for several weeks after exposure to long days, it is initiated rapidly and the reproductive system continues to proceed toward a future refractory state regardless of subsequent photoperiod once initiation is complete. If nonphotorefractory (i.e., photosensitive) starlings are exposed to only 7 "long" days, then this is often sufficient to provide the photoperiodic drive for photorefractoriness to ensue some weeks later (Dawson et al. 1985a). The rate of onset of photorefractoriness is also proportional to the length of the photoperiod (Hamner 1971; Dawson and Goldsmith 1983). A long day is that in which the duration of the light period is over and above a "critical day length," that being the length of day below which photorefractoriness cannot be induced.

Photorefractoriness is gradually dissipated by short days (i.e., day length below the critical photoperiod), and "photosensitivity" is acquired. That is, the bird's reproductive system is able to respond to long days again. In the wild, photosensitivity is acquired in the fall, when day length falls below approximately 11.5 hours. This reacquisition of photosensitivity in the fall while on short days can result in an increase in reproductive activity, including mating behavior in some species. This has been termed "autumnal sexuality" (Dawson 1983; Lincoln et al. 1980; Murton and Westwood 1977).

Neural Basis of the Photoperiodic Response

Extraretinal Photoreceptor Seasonally breeding mammals use a combination of their eyes (retina) and pineal gland for photoreception and transduction of the light signal (see Foster et al. 1989 for review). For example, if sexually mature hamsters are blinded while housed under long days, then their gonads regress—even though they are still receiving the long-day light stimulus (Reiter 1978). This indicates that ocular photoreceptors are responsible for light detection (Foster et al. 1991). The reliance on ocular photoreceptors appears to be widespread among mammals (Nelson and Zucker 1981). The light signal is transduced into a signal that is carried along the retinohypothalamic tract (RHT) to the suprachiasmatic nucleus (SCN). From the SCN, the light information leaves the brain via the paraventricular nuclei (PVN), medial forebrain bundle (MFB), and synapses in the superior cervical ganglion (SCG) in the spinal cord. Postganglionic fibers then project back into the brain and innervate the pineal gland. It is here that the light information is transformed into the melatonin signal (melatonin being released from the pineal at night).

Birds, on the other hand, transduce seasonal changes in photoperiod via photoreceptors that are extraretinal and extrapineal (see Groos 1982 for a comparative review of extraocular photoreception). The first evidence for avian extraretinal photoreceptors came from Benoit (1935a,b), who demonstrated that simultaneous photostimulation of blinded and sighted ducks resulted in equal testicular growth rates upon transfer to long days. Furthermore, the testicular response could be abolished by covering the ducks' heads with black caps. Since then, further experiments involving ducks (Benoit and Ott 1944), house sparrows (Menaker and Keatts 1968), and tree sparrows (Wilson 1991) have demonstrated convincingly that an extraretinal photoreceptor participates in the reproductive responses to changes in photoperiod. In these examples there is no retinal involvement whatsoever (Underwood and Menaker 1970; see also Foster and Follett 1985). Thus, in birds, the detection of light for reproductive purposes occurs via an extraretinal, hypothalamic pathway (Menaker 1971).

Despite numerous studies, it is still unclear as to where precisely the photoreceptor(s) lies in the avian brain. There is some evidence that extraretinal photoreceptors may reside in the infundibular region of the hypothalamus (Oliver and Baylé 1982) and in the parolfactory lobe (avian homologue of the caudate) of quail (Sicard et al. 1983). More recent research involving immunostaining of opsin (a protein involved in signal transduction by light activation) has added weight to the idea that the deep brain photoreceptors are located in the septum or hypothalamus, or both (in this case, the lateral septum and the tuberal hypothalamus of ringdoves, ducks, and quail—Silver et al. 1988). Similar staining revealing immunoreactive opsin in the septum and tuberal hypothalamus has now been reported in songbird species, including starlings (Saldanha et al. 1994a). Support for the importance of the septum has been provided by a recent study describing rhodopsin gene expression in the lateral septum of pigeons (Wada et al. 1998). There is strong evidence that the tuberal hypothalamus is at least involved in the transduction of the light signal in quail, whether or not the photoreceptors are located there. Meddle and Follett (1995) demonstrated that rapid and transient activation of the immediate-early gene, *c-fos*, occurred in the quail tuberal hypothalamus soon after photostimulation, and was followed by a rapid rise in plasma luteinizing hormone (LH). Thus, it appears that, in birds, light must pass through the skull and into the hypothalamus for transduction of its signal to occur.

The Clock for Measurement of Day Length For a "long-day" signal to result in photostimulation, day length must be measured with a high degree of accuracy. Birds, unlike mammals, do not seem to require the melatonin time signal (from the pineal gland) to control reproductive changes in response to day length. Indeed, the function of the nightly melatonin signal from the avian pineal gland remains controversial, although melatonin mediates entrainment of circadian activity rhythms in some avian species (Menaker and Keatts 1968; Wilson 1991; Juss et al. 1993) and is involved in seasonal changes in immune function (Bentley et al. 1998a). The physiological mechanisms underlying measurement of day length are still unknown, but there are two main models for which there are some supporting experimental data. The first is the "hourglass" model, in which it is hypothesized that an "hourglass-like timer is set in motion by the onset of dusk or dawn, and a photochemical accumulates during either the light or the dark phase. If a sufficient amount of the photochemical accumulates, then a photoperiodic response is initiated. The fundamental property of the hourglass model is that it requires constant resetting by the light-dark cycle. Thus, it would not operate under conditions of constant light or constant dark. There is evidence that such a system operates in some insects (for examples, see Beck 1968; Lees 1973; Vaz Nunes and Veerman 1984, 1986; Veerman et al. 1988).

The second theory, which has been shown to hold true for birds, was first proposed by Bünning in 1936 (see also Pittendrigh and Minis 1964, 1971; Follett 1973; Elliott 1976). It assumes that there is a circadian rhythm of responsiveness to light, that is, for the first part of the cycle (subjective day) the organism is insensitive to light, whereas during the second part (subjective night) it becomes "photoinducible." Should light fall during the subjective night, then a "long-day" photoperiodic response is initiated. Evidence that this "external coincidence model" applies to avian photoperiodicity was first supplied by Hamner's night-interruption experiments on the house finch (*Carpodacus mexicanus*) (1963, 1964), and has been reinforced by experiments on other bird species (for examples, see Follett and Sharp 1969; Follett et al. 1992; Turek 1974).

The Gonadotropin-Releasing Hormone Neuronal System In the early 1980s, GnRH was purified from chicken (*Gallus domesticus*) pituitaries and the structure characterized as varying from the mammalian form at residue 8 where arginine is substituted by glutamine (see Millar and King 1984 for a review of these initial studies). At this time, a second form of GnRH was found in the chicken brain by Miyamoto and colleagues (Miyamoto et al. 1984). This form differed from the mammalian form by three amino acid substitutions. These two forms were then named chicken GnRH-I (cGnRH-I) and chicken GnRH-II (cGnRH-II). The primary structure cGnRH-I is thus [Gln8]mGnRH and the primary structure of cGnRH-II is [His5, Trp7, Tyr8] mGnRH. Both these forms of GnRH have been found to be effective releasers of the gonadotropins LH and follicle-stimulating hormone (FSH) in vivo (Millar et al. 1986; Sharp et al. 1990; Wingfield and Farner 1993).

The presence of these two forms in the avian brain is not restricted to gallinaceous birds. For example, both forms of GnRH are apparently present in European starlings, a species that is a member of the most recently evolved avian taxon, the order Passeriformes (Sherwood et al. 1988). Extracts containing GnRH from starling brain exhibit molecular heterogeneity and appear to contain equal amounts of peptides similar in form to both cGnRH-I and cGnRH-II based on the high-performance liquid chromatography (HPLC) elution pattern and cross-reactivity with four different antisera (Sherwood et al. 1988). There is no evidence that the starling brain contains a form of GnRH that could be identified as mammalian, salmon, or lamprey. The primary structure of these two peptides has yet to be determined in any avian species outside the order Galliformes so it is not known definitively if the exact forms of cGnRH-I and cGnRH-II are conserved throughout the avian class. The main population of cGnRH-I neurons is clustered bilaterally in the preoptic area (POA) of the avian brain, extending dorsally from the anterior commissure down to the supracommissural division of the organum vasculosum of the lamina terminalis (OVLT). Neurons containing cGnRH-II tend to be situated caudally to this population in the midbrain (Juss et al. 1992; Millam et al. 1993; see Ball and Hahn 1997 for a review).

In starlings that are in full breeding condition (i.e., photostimulated), cGnRH-I fibers project from the cell bodies to the median eminence. GnRH is secreted into the hypothalamo-hypophysial portal system and via this to the anterior pituitary. The pituitary is thus stimulated to secrete the gonadotropins LH and FSH, which cause the gonads to grow and mature. The main action of LH is on the Leydig cells, causing them to synthesize and secrete steroid hormones; FSH stimulates the Sertoli cells and spermatogenesis (Murton and Westwood 1977). The steroid hormones (progesterone, 17β-estradiol, and testosterone) that are thus released in turn inhibit LH and FSH release via negative feedback on the hypothalamo-pituitary axis (Wilson and Follett 1974; Mattocks et al. 1976; Nicholls and Storey 1976). Maximum gonadal size and full reproductive capability is reached in about 4 weeks in this way, but after a period of time (the amount of time depending on the species and on the day length experienced) spontaneous gonadal regression occurs, along with the other symptoms of photorefractoriness. The onset of photorefractoriness is independent of steroid hormones (see Nicholls et al. 1988).

The physiological changes involved in photorefractoriness are:

1. Levels of GnRH precursor peptide (pro-GnRH) and of cGnRH-I content in the POA and median eminence in the brain decrease as shown by radioimmunoassay and by immunocytochemical staining of cGnRH-I cell bodies in starlings (Dawson et al. 1985b; Foster et al. 1987; Parry et al. 1997).

2. The gonadotropins LH and FSH decrease to minimal levels due to a termination of cGnRH-I release.

3. The gonads regress.

4. Plasma prolactin concentrations increase.

5. Postnuptial molt occurs (which generally coincides with the peak in plasma prolactin).

6. Bill color changes from yellow to black, as androgens disappear from the circulation (Witschi and Miller 1938; Dawson 1983; Ball and Wingfield 1987).

7. Cell-mediated immune function is elevated (Bentley et al. 1998a).

Endocrine Basis of Photorefractoriness

The Role of Prolactin Vasoactive intestinal polypeptide (VIP) is the neuropeptide responsible for the release of prolactin in birds (MacNamee et al. 1986; Mauro et al. 1992; Saldanha et al. 1994b; Youngren et al. 1994; El Halawani et al. 1995; Sun and El Halawani 1995). The release of prolactin is temporally coincident with the onset of photorefractoriness in birds (see Goldsmith 1985 for review) and there is no rise in prolactin in thyroidectomized starlings, which do not become photorefractory (Goldsmith and Nicholls 1984), as will be discussed later. Thus, it might be argued that VIP or prolactin, or both, might be causally involved in the termination of the breeding season. However, immunization of starlings against VIP prevents prolactin release, but does not prevent the onset of photorefractoriness (Dawson and Sharp 1998), so it appears that increased VIP and prolactin release are simply temporally related to the onset of photorefractoriness, and do not have a causal role. In addition, intracerebroventricular infusion of prolactin has been found to be potently gonadoinhibitory but it does not induce photorefractoriness in starlings (Juss and Goldsmith 1992).

The Role of Thyroid Hormones Circulating thyroxine is necessary for the initiation and maintenance of photorefractoriness, and the condition is prevented by thyroidectomy (Woitkewitsch 1940; Wieselthier and van Tienhoven 1972; Goldsmith and Nicholls 1984). Thyroidectomy of other seasonally breeding species causes them to remain in breeding condition indefinitely. This is true of tree sparrows (Wilson and Reinert 1993, 1995a,b, 1996; Reinert and Wilson 1996a,b), sheep (Moenter et al. 1991; Dahl et al. 1994; Parkinson and Follett 1994; Parkinson et al. 1995), and red deer (Shi and Barrell 1992). In addition, starlings hatch in a prepubertal condition and develop directly into a photorefractory state. Thyroidectomy of nestling starlings causes neoteny (i.e., sexual maturation in the absence of full somatic maturation) during exposure to long days, underlining the involvement of thyroid hormones in the starling's reproductive responses to photoperiod (Dawson et al. 1987, 1994; see figure 5.1).

Plasma thyroxine concentrations increase when starlings are transferred from short to long days (Dawson 1984; Bentley et al. 1998b). This is also seen in quail (Sharp and Klandorf 1981) and tree sparrows (Reinert and Wilson 1996b). Despite the long day–induced rise in plasma thyroxine concentrations, it is now apparent that thyroid hormones have a permissive role in the termination of the breeding season of starl-

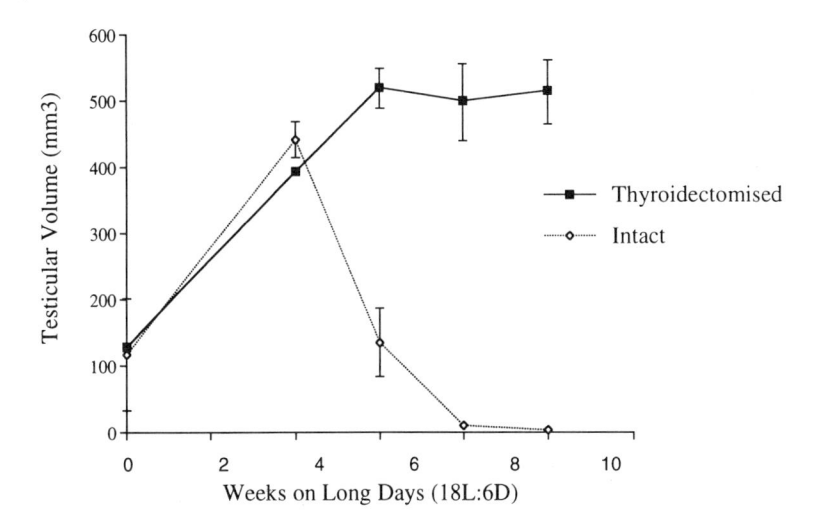

Figure 5.1
Effects of radiothyroidectomy on the onset of photorefractoriness in European starlings. Photosensitive, intact starlings maintained on long days (18L : 6D) become refractory within 6 weeks after long-day exposure (dotted line). In contrast, starlings that had been previously thyroidectomized maintain large gonads and do not become refractory (solid line).

ings (Bentley et al. 1997a), as in sheep (Dahl et al. 1995; Thrun et al. 1996, 1997). Thyroxine seems to be acting as a permissive factor for the onset of photorefractoriness in starlings in that it needs to be present in the circulation in at least minimal physiological concentrations to terminate breeding. However, when thyroxine is present in physiological concentrations it is not the rate-limiting factor in the onset of photorefractoriness. In this way thyroxine can be envisaged as a link in the chain of events that begins with the perception and transduction of a long-day signal and ends with gonadal regression and change in reproductive state. Recent data suggest that thyroid hormones may be permissive in the sense that they allow the production of thyroid-dependent neurotrophin(s) that are involved in the neuronal plasticity that is so characteristic of photorefractoriness in birds (Bentley et al. 1997b).

Changes in Brain Content and Release of cGnRH-I in the Starling Breeding Cycle

Dynamics of Cellular Changes in cGnRH-I In many seasonally breeding avian species, including starlings, the occurrence of refractoriness is associated with a marked decline in the brain content of cGnRH-I (see Ball and Hahn 1997 for review). When antibodies against the cGnRH-I peptide are used in immunocytochemical studies of the starling brain as it undergoes gonadal regression and becomes photorefractory, levels of cGnRH-I do not begin to fall until after testicular regression has begun (Goldsmith et al. 1989). This would seem counterintuitive unless the synthesis and release of the peptide are both terminated at the onset of photorefractoriness, causing gonadal regression. It is now evident that synthesis of the preprocessed GnRH (pro-GnRH) is halted at the time of gonadal regression (Parry et al. 1997), and that release of the processed peptide (i.e., cGnRH-I) is also terminated or reduced at the same time (Dawson and Goldsmith 1997). The long-term photorefractory state is characterized by an increase in the number of axosomatic terminals in contact with the cGnRH-I cell bodies, coincident with a greater number of synaptic modifications within those terminals (Parry and Goldsmith 1993). This is consistent with the idea that a neurotrophin is involved in the onset and maintenance of photorefractoriness.

As stated earlier, the steroid hormones that are released as a result of slow gonadal growth during the short-day photosensitive phase of the annual cycle inhibit LH and FSH release via negative feedback on the hypothalamo-pituitary axis (Wilson and Follett 1974; Mattocks et al. 1976; Nicholls and Storey 1976). This negative feedback does not have direct effects on the cGnRH-I system. The slow gonadal growth observed under short days is merely a manifestation of the fact that the reproductive system is in a "switched-on" state. The cGnRH-I system is priming itself for an increase in day length, which signals the onset of the breeding season. Because of the need for the cGnRH-I system to be able to respond to subsequent long days when in this photosensitive condition, small amounts of cGnRH-I are released from the cGnRH-I cell bodies onto the pituitary gland with the downstream effect of slow gonadal growth. Photostimulation subsequently overrides any negative feedback effects of the gonadal steroids by inducing a large increase in the amount of cGnRH-I that is released onto the pituitary gland.

Importance of Molecular Studies Even though GnRH-I precursor immunoreactivity in cell bodies is reduced at the time of testicular regression (Parry et al. 1997), it is not known exactly when or where the transcription of the pro-GnRH gene is initiated or terminated. Information on where pro-GnRH gene expression is initiated and on where postranslational processing occurs can only be gathered when the gene for songbird pro-GnRH is cloned. The role of thyroid hormones in the molecular mechanisms involved in seasonal GnRH production and the timing of the onset of photorefractoriness will be further elucidated when in situ hybridization studies of songbird pro-GnRH messenger RNA (mRNA) can be carried out. In addition, studies on the social regulation of GnRH in songbirds are limited by the levels of analysis of changes in GnRH that are available at present. For example, it would be very helpful to identify precisely when expression of the pro-GnRH gene starts to decline. It might be possible to investigate the endocrine and neural factors that regulate this change in gene expression.

The Role of GnRH in the Photoperiodic Control of Reproduction in Birds

A number of different species, in addition to European starlings, have been shown (using a variety of techniques; see Ball and Hahn 1997 for review) to display marked reduction in the activity of the GnRH system during absolute photorefractoriness. These species include white-crowned sparrows (*Zonotrichia leucophrys gambelii*) (Hudson and Wingfield unpublished data presented in Wingfield and Farner 1993; see also Wingfield 1985), dark-eyed juncos (*Junco hyemalis*) (Saldanha et al. 1994b), garden warblers (*Sylvia borin*) (Bluhm et al. 1991), house sparrows (*passer domesticus*) (Hahn and Ball 1995), American tree sparrows (*Spizella arborea*) (Reinert and Wilson 1996b; Wilson and Reinert 1996), and canaries (*Serinus canaria*) (Ball and Hahn 1997).

All of the species described above become absolutely refractory to the stimulatory effects of long days in a manner similar to that of starlings (Farner et al. 1983; Nicholls et al. 1988; Wilson and Donham 1988). However, there is species variation in the photoperiodic response. For example, photorefractoriness can be "absolute," as in the aforementioned starling, or "relative," as in Japanese quail (*Coturnix japonica*) (see figure 5.2). In this latter species, reproductive development occurs in the spring but relatively longer days are required to maintain reproductive competence by late summer (Robinson and Follett 1982). At least some decline in day length is necessary for the gonads to begin to regress, and the birds retain the capacity to regrow their short-day regressed gonads immediately upon reexposure to long days. Thus in species such as quail, the same day length is interpreted differently depending on the time of the year at which it is encountered. Early in the season a photoperiod of 14L:11D may stimulate gonadal growth (because it was preceded by a photoperiod of 13L:12D) while later in the season a photoperiod of 14L:11D may lead to gonadal regression (because it was preceded by a photoperiod of 15L:9D). These data point to the importance of context and experience as being factors that influence the physiological response to a given photoperiod exhibited by an individual. This work in quail is similar in many respects to more recent work on hamsters where it has also been shown that the time of year or the photoperiodic context in which an individual encounters a particular photoperiod influences the physiological response that is exhibited (see chapter 7).

To date, quail are the only species known to become relatively refractory without also becoming absolutely refractory (see Robinson and Follett 1982). There may be other examples (e.g., crossbills [*Loxia* spp.]; see Ball and Hahn 1997), and many other species may also be relatively refractory prior to becoming absolutely refractory, or during dissipation of absolute refractoriness (Hamner 1968; Dawson 1991). In contrast to the species that become absolutely refractory, Japanese quail show no decline in hypothalamic GnRH when relatively refractory (Foster et al. 1988). In fact, male quail transferred to short days after prolonged long-day stimulation actually show an increase in hypothalamic GnRH, as if GnRH production continues but secretion ceases and consequently the peptide accumulates in the brain (Foster et al. 1988). These findings suggest that there may be subtle but important species differences among certain aspects of the neuroendocrine mechanisms mediating photorefractoriness that reflect species differences in the photoperiodic response.

Photorefractoriness, either relative or absolute, is dependent upon the presence of circulating thyroid hormones for its initiation and maintenance (Follett and Nicholls 1984,1985; Goldsmith and Nicholls 1984; Woitkewitsch 1940; Wieselthier and van Tienhoven 1972). The same is true for the termination of seasonal breeding in mammals (Moenter et al. 1991; Shi and Barrell 1992; Dahl et al. 1994; Parkinson and Follett 1994; Parkinson et al. 1995). In light of this, it seems likely that there are many similarities between species in the mechanisms that control the termination of the breeding season; they are thyroid-dependent, but thyroid hormones need only be

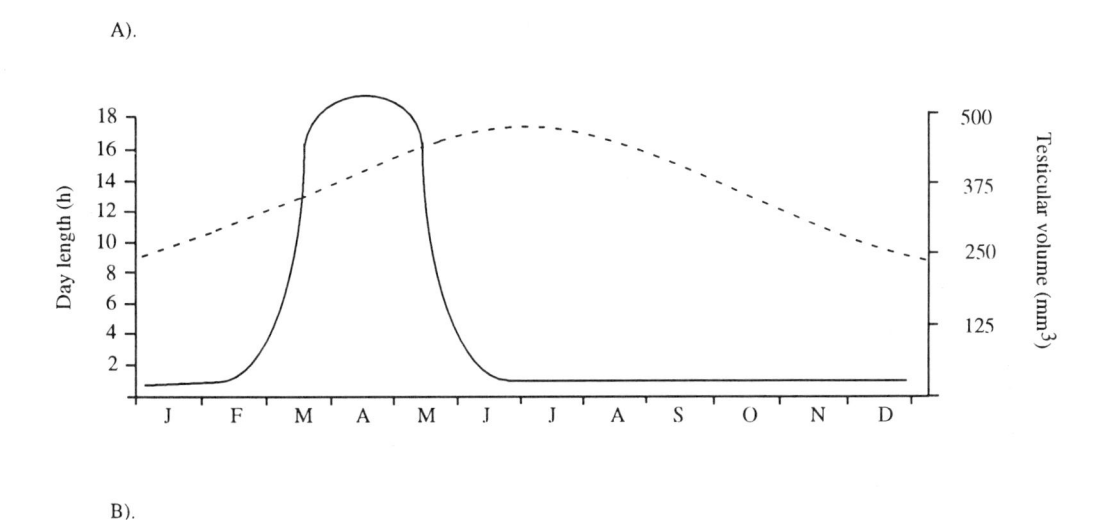

Environmental Regulation of Reproduction in Birds 141

Figure 5.2
(A) Change in testicular volume in European starlings throughout the annual cycle (solid line) and change in day length (dashed line) at 52°N. Note that in this species, which exhibits absolute photorefractoriness, gonadal regression occurs while day lengths are still increasing. (B) Change in testicular volume in Japanese quail throughout the annual cycle (solid line) and change in day length (dashed line) at 52°N. Unlike starlings, the gonads of quail change in size in proportion to the ambient day length all year-round.

present in a "permissive" fashion (Dawson 1984; Dahl et al. 1995; Bentley et al. 1997a,b; Bentley 1997). It is probable that the basic controlling mechanism for the onset of photorefractoriness is the same in all bird species, but that downstream adaptations of that mechanism have evolved to suit a particular species' ecological niche. Because quail are the only birds yet studied that only become relatively photorefractory, this idea requires further exploration.

Possible Relationship Between Species Differences in GnRH Plasticity and Breeding Flexibility

The studies just reviewed suggest that relative photorefractoriness and absolute photorefractoriness may differ with respect to changes in the GnRH system. Specifically, the GnRH system appears to be switched to an inactive state during absolute photorefractoriness, in that the peptide completely disappears. However, during relative photorefractoriness, the peptide is still present in the brain at high levels, but rather the release of GnRH into the portal vessels appears to be inhibited. This difference, if generally true, may be of significance with regard to the species' potential for temporal reproductive flexibility. Individuals that maintain high levels of GnRH in the brain and therefore potentially active neuroendocrine transduction systems may retain the capacity to respond to environmental cues, even if relative photorefractoriness has greatly reduced the net drive due to photoperiod. In contrast, individuals that have switched off GnRH production under absolute photorefractoriness tend to be unresponsive to all manner of cues (Goldsmith et al. 1992; Ball et al. 1994a). In other words, species that become absolutely photorefractory appear to display periods of reproductive quiescence that are more rigid, with little capacity to adjust breeding duration or timing in a flexible manner.

Species that only become relatively photorefractory would be most flexible, particularly if they also have evolved strong sensitivity to supplementary (i.e., nonphotic) cues whose inputs could be integrated with one another and with photoperiod to permit maintenance or reattainment of reproductive competence even when photoperiod alone is not stimulatory. As Lee and Gorman point out in chapter 7 with regard to mammalian species, day lengths shorter than the so-called critical photoperiod (see Hamner 1968; Nicholls et al. 1988; Sharp 1996) should not be viewed as inhibitory. Relative photorefractoriness may be a product of a shift in the circadian position of the photoinducible phase as a function of photoperiodic history (Robinson and Follett 1982; Dawson 1987; Follett and Nicholls 1984,1985). In quail, therefore, relatively long days near the end of the breeding summer are no longer effective in stimulating the reproductive system in relatively photorefractory individuals. The net "reproductive drive" (vis-à-vis the "photoperiodic drive" of Nicholls et al. 1988), as measured by production and release of GnRH from hypothalamic centers, might nevertheless remain strongly positive if supplementary cues remain favorable. Apparently no such drive is possible during absolute photorefractoriness.

Future studies should explore the relationships between temporal plasticity in the GnRH system and both the type of photorefractoriness (if any) that develops and the temporal flexibility of the reproductive schedule. In addition, studies specifically designed to test responsiveness to supplementary cues during relative and absolute refractoriness are needed. These studies should help to clarify the role of neuroendocrine plasticity in regulating a diversity of breeding strategies (see Hahn et al. 1997 for discussion).

The Effects of Cues Supplementary to Photoperiod on GnRH and Reproductive Physiology

Visual and Auditory Cuess

In contrast to the response to photoperiod, there is every indication that the endocrine response to supplementary information is mediated by the sensory receptors in the visual system (with the use of photoreceptors in the eyes) and in the auditory system. For example, evidence from many avian species indicates that visual and auditory cues are required to induce endocrine responses to social events such as a displaying conspecific (Lehrman 1965; Hinde and Steel 1978; Cheng 1979; Moore 1983; Wingfield 1980,1983; Erickson 1985; Crews and Silver 1985; Wingfield et al. 1994; Ball and Dufty 1998). It should also be remembered that these supplementary cues might not always be stimulatory. For example, Yokoyama and Farner (1976) found that enucleation leads to an increase in gonadal activity in white-crowned sparrows. This suggests that stimuli most commonly processed by the retinal pathway, that is, social and nonphotoperiodic physical information, may also exert an inhibitory influence. It has also been suggested by Cheng et al. (1998) that many apparently stimulatory effects of auditory inputs also involve the removal of an inhibitory input on GnRH neurons.

Endocrine responses to social supplementary cues may require rather complex neural processing to occur. For example, Friedman (1977) demonstrated that a female dove exposed to a courting male, who is directing his behavior toward her, is more apt to show an enhanced endocrine response than a female who sees the same male but from a view that indicates that he is not directing his courtship toward that specific female. Also, Cheng (1986, 1992) has suggested, from her work with ring-doves, that the female's endocrine response to male courtship is the result of the female behaviorally stimulating herself by nest cooing. Studies such as these argue that at least some types of supplementary cues undergo complex neural processing before they exert their modulatory effects on the cGnRH-I system and, through cGnRH-I, on the endocrine system. The manner in which supplementary cues are perceived can be profoundly influenced by experience. For example, Baptista and Petrinovich (1986) describe how female white-crowned sparrows, collected as nestlings in the wild and hand-reared in the laboratory, readily ovulated in captivity when housed in a situation under which ovulation has never been reported in wild-caught adult females. They suggest that young white-crowned females undergo an imprinting process during their first year with regard to the supplementary stimuli that are required for ovulation. When hand-reared in captivity they "imprinted" on substitute supplementary stimuli and therefore would ovulate under these conditions.

All these studies suggest that, in contrast to the apparently circumscribed locale of the neural pathway mediating the photoperiodic response, the neural pathways processing the endocrine response to supplementary cues will involve many more processing steps. These cues are almost assuredly processed and interpreted first by the visual and auditory systems before the information is conveyed to the hypothalamus. Nonetheless, it seems essential that supplementary information exert whatever effect it may finally have on endocrine functioning via the cGnRH-I neuronal system. Thus, both types of information appear to converge on this GnRH system. Therefore, the group of neurons in the preoptic-septal area that project to the median eminence and that are immunoreactive for cGnRH-I are one possible starting point for the elucidation of the neural pathways processing both initial predictive and supplementary cues. One of the best-studied examples of a behavioral stimulus that affects endocrine development involves the effect of vocal behavior on the development of the endocrine system. We will therefore discuss what is known about the neural basis of the stimulation of the reproductive system by vocalizations.

The Effects of Male Vocalizations on the Development of the Female Reproductive System in Birds

Early Studies Demonstrating the Effects of Vocal Behavior on Endocrine Activity

In many photoperiodic avian species, it appears that females are more sensitive to supplementary cues than males, in that photoperiod alone can induce full gonadal growth and spermatogenesis in males but females require supplementary cues to attain full gonadal development and eventually to ovulate (Wingfield and Farner 1980). Male courtship behavior is a particularly salient social cue that has been shown to enhance the reproductive development of female birds as well as many other vertebrate species (see Crews and Silver 1985; Wingfield et al. 1994; Ball and Dufty 1998 for reviews). In birds, a particularly potent aspect of male courtship behavior that stimulates female reproductive development is male vocal behavior. This was first demonstrated experimentally by pioneering studies conducted by Brockway (1965) on budgerigars (*Melopsittacus undulatus*). She found that male budgerigar "warbles" stimulate ovarian development and egg-laying in female budgerigars. Contemporaneous work by Lehrman (1965) on ringdoves (*Streptopelia risoria*) and Hinde (1965) on canaries (*Serinus canaria*) confirmed that Brockway's findings could be generalized to other species. Furthermore, females are not passive recipients of social stimuli from males, but rather the presence of a male stimulates the females to vocalize and thereby to stimulate themselves (Cheng 1992). If the females are unable to vocalize, the effect of the male courtship cooing is greatly attenuated. Thus female self-stimulation appears to mediate the enhancing effects of the male in this species.

Song Behavior and the Song System in Songbirds Although vocal behavior is widespread among the over nine thousand extant species within the class Aves (Nottebohm 1975), it is most elaborate among songbirds. In this review, the term "songbird" will be used to refer to species in the suborder Passeres (oscines). In songbirds, vocalizations are often divided into two general categories: calls and songs. Calls refer to simpler vocalizations that are used in a variety of contexts. The term "bird song" is often limited to the complex vocalizations generally produced by male members of the songbird (or perching bird) suborder, that is, the oscines (Konishi 1985; Catchpole and Slater 1995; Ball and Hulse 1998). It appears that song is learned by all members of this suborder studied to date (Kroodsma 1982) and song is generally a male-typical vocalization, though species differences in the degree of the sexual dimorphism do exist (e.g., see Brenowitz 1997). In some species only the male produces the complex song vocalizations, with the female producing only simpler call-like vocalizations. In other species both sexes will produce song but the male will sing a more complex song at a higher rate than the female.

Song primarily serves two evolutionary functions: territorial defense and mate attraction or stimulation (Kroodsma and Byers 1991). The greater use of song by males as compared to females is thought to be related to the effects of sexual selection. Both intrasexual (i.e., song is used to repel competing males from the territory) and intersexual selection (i.e., females choose males based on their songs) are thought to be involved (Catchpole 1982; Kroodsma and Byers 1991). This dimorphism in behavior is reflected by prominent sexual dimorphisms within the neural circuit that mediates song learning and production. In canaries and zebra finches (*Taeniopygia guttata*), five nuclei in the vocal control circuit have been found to be dimorphic in volume. These are the HVc (sometimes referred to as the high vocal center), the robust nucleus of the archistriatum (RA), area X, nucleus interface (Nif), and the tracheosyringeal division of the hypoglossal nucleus (nXIIts) (Nottebohm and Arnold 1976; Ball et al. 1995). Four of these nuclei are part of a pathway that is essential to vocal production (Nif → HVC → RA → nXIIts). One of these dimorphic areas, area X, is part of a rostral pathway that is important in song learning as well as production and perhaps perception (see Nottebohm et al. 1990; Wild 1997; Margoliash 1997 for reviews). These sexual dimorphisms can be quite prominent. In some cases the volume of the nucleus in the male is five times greater than in the female. In duetting species from the tropics, in which males and females sing in roughly equal amounts, such dimorphisms are not as apparent (Brenowitz et al. 1985, Brenowitz and Arnold 1986; Brenowitz 1997, but see Gahr et al. 1998). In starlings, the vocal control system exhibits a sex difference in nuclear volume to a degree that is between these two extremes (Bernard et al. 1993; Ball et al. 1994b).

The Interplay Between Photoperiod and Courtship Song In several songbird species, most notably canaries and white-crowned sparrows (*Zonotrichia leucophrys gambelii*), male song has been shown to stimulate various aspects of endocrine development in conspecific females (Kroodsma 1976; Hinde and Steel 1978; Morton et al. 1985). Although it is clear that male song can stimulate the endocrine system of conspecific females, the interplay of song and the photoperiodic condition of the female requires further investigation. The neural mechanisms by which song is perceived, and the information conveyed to the reproductive axis remain to be elucidated in oscines.

For example, both Morton et al. (1985) and Hinde and Steel (1978) found that song was not effective in enhancing reproductive development under all photoperiodic conditions. In the case of Gambel's white-crowned sparrows there seems to be a photoperiodic threshold effect. Song played back to females on photoperiods of 11L:13D or 6L:18D did not enhance ovarian growth, while it did enhance ovarian growth of females on photoperiods of 12.5L:11.5D and 14L:10D (Morton et al. 1985). In the case of canaries, Hinde and Steel (1978) report that playing back tape-recorded male song with female canaries on a photoperiod of 11L:13D significantly increased follicular growth and circulating levels of LH as compared with females on 11L:13D who did not hear male song. Females on photoperiods of 14L:10D did not exhibit enhanced reproductive development by exposure to song. Hinde and Steel (1978) suggested that this happened because a photoperiod of 14L:10D stimulated the reproductive system to a maximal extent and supplementary cues such as song could not stimulate it further.

The Role of the Song System in Song Perception Several lines of evidence have recently converged that indicate that the oscine vocal control circuit is important for song perception as well as for song production and learning (Nottebohm et al. 1990). Neurons described as "song-specific" have been identified in a variety of vocal control areas, including a key nucleus in the circuit, the HVc (Margoliash 1983; see Margoliash 1997 for a review). These neurons are specifically tuned to fire in response only to an entire song or a large fragment of a song. The presence of such complex feature detectors strongly suggest that they might be involved in the recognition or processing of song. In addition to HVc, these song-specific neurons are especially prominent in the rostral part of the song control circuit (e.g., see Doupe and Konishi 1991) which contains nuclei such as the lateral part of the magnocellular nucleus of the anterior neostriatum (lMAN) that is essential to the learning of song. Although it is known that this rostral part of the vocal control circuit is necessary for vocal learning, its role in adult activities such as song perception and production is still not well understood. Direct evidence that the song system may be essential to song perception was provided by Brenowitz (1991a) who demonstrated that electrolytic lesions to HVc would prevent female canaries from showing a preference for male conspecific song as compared to heterospecific song. These lesions did not produce any general auditory deficit but rather seemed to remove specifically the strong preference female canaries have for male canary song as compared to other species' songs such as that of the white-crowned sparrow. Similar results have been reported in female canaries that received partial excitotoxic lesions to the HVc (Del Negro et al. 1998).

Converging Neural Pathways Mediating the Effects of Photoperiod and Male Song on Reproductive Endocrine Activity

Studies such as these indicate that females use nuclei such as HVc in their vocal control circuit to perceive male vocalizations. This would suggest that the endocrine enhancement of gonadal growth by male vocalizations in female songbirds would be mediated via the song control circuit.

Cheng and Zuo (1994) have provided evidence for pathways in doves that could mediate the effects of vocal behavior on endocrine secretion in this species. They suggest that direct projections from the midbrain vocal control nucleus, intercollicularis (ICo), to areas containing GnRH-immunoreactive cell bodies in the anterior hypothalamus and other projections from ICo via the auditory thalamus to the ventromedial nucleus of the hypothalamus could form the neuroanatomical basis of the self-feedback phenomenon she has described in doves. Similar pathways have been identified in pigeons by Wild et al. (1997). Recent work has demonstrated that hypothalamic nuclei that receive these projections from the auditory thalamus in doves exhibit elevated firing rates specifically in response to playback of female nest coos (Cheng et al. 1998). The activation of these feature detectors for nest coos is associated with a significant increase in the release of LH (Cheng et al. 1998). Thus, in the dove, "coo-specific" neurons may be involved in the auditory stimulation of GnRH-stimulated LH release (Cheng et al. 1998).

Midbrain and brain stem areas controlling vocal behavior are similar in songbird species and nonsongbird species, such as ringdoves (Brenowitz 1991b; Ball 1994). Therefore a similar pathway from the midbrain to the anterior hypothalamus where GnRH-immunoreactive cells are found could be present in the oscine brain. Tract-tracing studies in zebra finches suggest that such projections from the midbrain nucleus, ICo, to the hypothalamus are present in zebra finches (Striedter and Vu 1998).

However, nonsongbird species such as ringdoves and pigeons do not have a discrete telencephalic network that connects to this midbrain and brain stem unit (Brenowitz 1991b; Ball 1994). The telencephalic portion of the vocal control system is a neural specialization limited to oscines and possibly to members of the parrot order, such as budgerigars (Brenowitz 1991b; Ball 1994; Striedter 1994). The involvement of HVc in the perception of song is unique to oscines and the evolution of specialized telencephalic structures such as HVc could well have led to the occurrence of new and different connections between vocal control areas and the hypothalamus in songbirds as compared to other avian taxa. It would therefore be useful to compare in songbirds the importance of the neural pathway from the midbrain to the hypothalamus in mediating neuroendocrine responses to vocal behavior with the role played by the telencephalic songbird control system. A connection between the hypothalamus and the telencephalic vocal control system in zebra finches has been identified by Foster et al. (1997). They report that a subregion of the lateral hypothalamus projects to a nucleus in the thalamus that in turn projects to the medial part of nucleus MAN (mMAN). Nucleus mMAN projects to HVc (Foster et al. 1997). This pathway suggests a way in which hypothalamic inputs can modulate the song system, but unless similar backprojections are identified it does not explain how information processed by the song system gets back to the GnRH system in the hypothalamus. However, the clear role played by HVc in the perception of male song by female songbirds strongly suggests that its role should be investigated, as well as other nuclei within the rostral part of the vocal control circuit that are involved in song learning and perhaps song perception.

Do Initial Predictive and Supplementary Information Act Independently of One Another or Additively?

Although both initial predictive and supplementary information appear to converge on the GnRH system, their ability to modulate these neurons, and therefore to affect endocrine activity, may not be equal. In particular, it may be the case that there is a stimulus "hierarchy" in some sense between initial predictive cues, such as photoperiod, and supplementary cues. Seasonal changes in the response of the reproductive system to photoperiod may play a role in preparing the neuroendocrine system to respond to other factors supplementary to photoperiod such as behavioral interactions, nest site, nest material, temperature, and food availability. For example, when photorefractory in the late summer, male white-crowned sparrows do not show a rise in testosterone and LH in response to a displaying female which they will show in the spring when they are being photostimulated (Moore 1983). Similar results have been reported in starlings. Female starlings do not show enhanced ovarian growth in response to the presence of a male when they are photorefractory but they do when they are photosensitive (Ball et al. 1994a). As noted previously, Morton, et al. (1985) found that playing songs to female photostimulated white-crowned sparrows (i.e., longer than a photoperiod of 12.5L:11.5D) produced a significant augmentation in ovarian growth. Playing songs to photosensitive females maintained on shorter photoperiods had no effect on ovarian growth. However, Ball et al. (1994a) did find that the presence of a male could enhance ovarian growth in female starlings when they were photosensitive and maintained on a photoperiod as short as 8L:16D. It appears that in the photorefractory state, the GnRH system is at such a low level of activity that it cannot respond to environmental stimulation of any sort, including supplementary cues. This would explain why Moore (1983) and Ball et al. (1994a) found no enhancing endocrine effects of a conspecific of the opposite sex in photorefractory birds. Similarly, it may be the case in some species that even photosensitive birds that are on short days and have not been photostimulated may be unable to respond fully to supplementary cues. In the natural situation, and in captive studies, supplementary cues are usually presented in conjunction with long-day stimulation. One possibility is that increased gonadal sex steroid secretion stimulated by long days somehow "primes" the brain to be able to respond fully to supplementary cues, which would explain why the Morton, et al. (1985) study described above found no effect of song on short day lengths. Such results suggest that there are rules of stimulus integration among the types of cues regulating reproduction in birds. Species variation in the propensity of the reproductive system to respond to supplementary cues independently of photoperiodic condition may relate to species variation in breeding flexibility

(Hahn et al. 1997). Thus, so-called opportunistic breeders, that can apparently breed in response to favorable environmental cues independent of the ambient photoperiod, differ from strictly seasonal breeders in that certain extraphotoperiodic cues are able to stimulate GnRH secretion under certain photoperiodic conditions when they cannot in seasonal breeders (Hahn et al. 1997). Detailed mechanistic studies in model species such as European starlings may reveal the key regulatory factors of the GnRH system that should be investigated in other species to ascertain how such species variability in breeding flexibility can be explained in neuroendocrine terms.

Acknowledgments

Our work on seasonal reproduction in birds is supported by grants from the National Science Foundation (IBN 951425) and the National Institutes of Health (NS 35467). We thank Jill Schneider for helpful comments on the manuscript.

References

Baker JR (1938). The evolution of breeding seasons. *In* Evolution: Essays on Aspects of Evolutionary Biology (DeBeer GB, ed), pp 161–177. Oxford: Clarendon Press.

Ball GF (1993). The neural integration of environmental information by seasonally breeding birds. Am Zool 33:185–199.

Ball GF (1994). Neurochemical specializations associated with vocal learning and production in songbirds and budgerigars. Brain Behav Evol 44:234–46.

Ball GF, Dufty Jr. AM (1998). Assessing hormonal responses to acoustic stimulation. *In* Animal Acoustic Communication (Hopp SL, Owren MJ, Evans CS, eds), pp 380–398. Berlin: Springer-Verlag.

Ball GF, Hahn TP (1997). GnRH neuronal systems in birds and their relation to the control of seasonal reproduction. *In* GnRH Neurons: Gene to Behavior (Parhar IS, Sakuma Y, eds), pp 325–342. Tokyo: Brain Shuppan.

Ball GF, Hulse SH (1998). Birdsong. Am Psychologist 53:37–58.

Ball GF, Wingfield JC (1987). Changes in plasma levels of luteinizing hormone and sex steroid hormones in relation to multiple-broodedness and nest site density in male starlings. Physiol Zool 60:191–197.

Ball GF, Besmer HR, Li Q, Ottinger MA (1994a). Effects of social stimuli on gonadal growth and brain content of cGnRH-I in female starlings on different photoperiods. Soc Neurosci Abstracts 20:159.

Ball GF, Casto JM, Bernard DJ (1994b). Sex differences in the volume of avian song control nuclei: Comparative studies and the issue of brain nucleus delineation Psychoneuroendocrinology 19:485–504.

Ball GF, Absil P, Balthazart J (1995). Peptidergic delineations of nucleus interface reveal a sex difference in volume. Neuroreport 6:957–960.

Baptista LF, Petrinovich L (1986). Egg production in hand-raised white-crowned sparrows. Condor 88:379–380.

Beck SD (1968). Insect Photoperiodism. New York: Academic Press.

Benoit J (1935a). Le rôle des yeux dans l'action stimulante de la lumière sur le développement testiculaire chez le canard. C R Soc Biol (Paris) 118:669–671.

Benoit J (1935b). Stimulation par la lumière artificielle du développement testiculaire chez les canards aveuglés par section du nerf optique. C R Soc Biol (Paris) 120:133–136.

Benoit J, Ott L (1944). External and internal factors in sexual activity. Effect of irradiation with different wavelengths on the mechanisms of photostimulation of the hypophysis and on testicular growth in the immature duck. Yale J Biol Med 17:27–46.

Bentley GE (1997). Thyroxine and photorefractoriness in starlings. Poult Avian Biol Rev 8:123–139.

Bentley GE, Goldsmith AR, Dawson A, Glennie LM, Talbot RT, Sharp PJ (1997a). Photorefractoriness in European starlings (*Sturnus vulgaris*) is not dependent upon the long-day–induced rise in plasma thyroxine. Gen Comp Endocrinol 107:428–438.

Bentley GE, Goldsmith AR, Juss TS, Dawson A (1997b). The effects of nerve growth factor and anti-nerve growth factor antibody on the neuroendocrine reproductive system in the European starling, *Sturnus vulgaris*. J Comp Physiol [A] 181:133–141.

Bentley GE, Demas GE, Nelson RJ, Ball GF (1998a). Melatonin, immunity and cost of reproductive state in male European starlings. Proc Soc Lond Biol Sci, 265:1191–1195.

Bentley GE, Goldsmith AR, Dawson A, Briggs C, Pemberton M (1998b). Decreased light intensity alters the perception of day length by male European starlings (*Sturnus vulgaris*). J Biol Rhythms 13:148–158.

Bernard DJ, Casto JM, Ball GF (1993). Sexual dimorphism in the volume of song control nuclei in European starlings: Assessment by a Nissl stain and autoradiography for muscarinic cholinergic receptors. J Comp Neurol 334:559–570.

Bluhm CK, Schwabl H, Perera A, Follett BK, Goldsmith AR, Gwinner E (1991). Variation in hypothalamic gonadotrophin-releasing hormone content, plasma and pituitary LH, and in-vitro testosterone release in a long-distance migratory bird, the garden warbler (*Sylvia borin*), under constant photoperiods. J Endocrinol 128:339–345.

Brenowitz EA (1991a). Altered perception of species-specific song by female birds after lesions of a forebrain nucleus. Science 251:303–305.

Brenowitz EA (1991b). Evolution of the vocal control system in the avian brain. Semin Neurosci 3:399–407.

Brenowitz EA (1997). Comparative approaches to the song system. J Neurobiol 33:517–531.

Brenowitz EA, Arnold AP (1986). Interspecific comparisons of the size of neural song control regions and song complexity in duetting birds: evolutionary implications. J Neurosci 6:2875–2879.

Brenowitz EA, Arnold AP, Levin RN (1985). Neural correlates of female song in tropical duetting birds. Brain Res 343:104–112.

Brockway BF (1965). Stimulation of ovarian development and egg laying by male courtship vocalizations in budgerigars (*Melopsittacus undulatus*) Anim Behav 13:575–578.

Bronson FH (1989). Mammalian Reproductive Biology. Chicago: University of Chicago Press.

Bünning E (1936). Die endogene Tagesrhythmik als Grundlage der photoperiodischen Reaktion. Berl Dtsch Bot Ges 54:590–607.

Burger JW (1947). On the relation of day length to the phases of testicular involution and inactivity of the spermatogenic cycle of the starling. J Exp Zool 105:259–268.

Catchpole CK (1982). The evolution of bird sounds in relation to mating and spacing behavior. *In* Acoustic Communication in Birds, vol 1 (Kroodsma DE, Miller EH, eds), pp 297–319. New York: Academic Press.

Catchpole CK, Slater PJB (1995). Bird Song: Biological Themes and Variations. Cambridge: Cambridge University Press.

Cheng MF (1979). Progress and prospect in ring dove research: a personal view. *In* Advances in the Study of Behavior 9 (Rosenblatt JS, Hinde RA, Beer CG, Busnel M-C, eds), pp 97–129. New York: Academic Press.

Cheng MF (1986). Individual behavioral response mediates endocrine changes induced by social inteaction. Ann N Y Acad Sci 474:4–12.

Cheng MF (1992). For whom does the female dove coo? A case for the role of vocal self-stimulation. Anim Behav 43:1035–1044.

Cheng MF, Zuo M (1994). Proposed pathways for vocal self-stimulation: met-enkephalinergic projections linking the midbrain vocal nucleus, auditory-responsive thalamic regions and neurosecretory hypothalamus. J Neurobiol 25:365–379.

Cheng MF, Peng JP, Johnson P (1998). Hypothalamic neurons preferentially respond to female nest coo stimulation: demonstration of direct acoustic stimulation of luteinizing hormone release. J Neurosci 18:5477–5489.

Crews D, Silver R, (1985). Reproductive physiology and behavior interactions in nonmammalian vertebrates. *In* Handbook of Behavioral Neurobiology, vol. 7 (Adler N, Pfaff D, Goy R, eds), pp 101–182. New York: Plenum Press.

Dahl GE, Evans NP, Moenter SM, Karsch FJ (1994). The thyroid gland is required for reproductive responses to photoperiod in the ewe. Endocrinology 135:10–15.

Dahl GE, Evans NP, Thrun LA, Karsch FJ (1995). Thyroxine is permissive to seasonal transitions in reproductive neuroendocrine activity in the ewe. Biol Reprod 52:690–696.

Dawson A (1983). Plasma gonadal steroid levels in wild starlings (*Sturnus vulgaris*) during the annual cycle and in relation to the stages of breeding. Gen Comp Endocrinol 49:286–294.

Dawson A (1984). Changes in plasma thyroxine concentration in male and female starlings (*Sturnus vulgaris*) during a photo-induced gonadal cycle. Gen Comp Endocrinol 56:193–197.

Dawson A (1987). Photorefractoriness in European starlings: Critical daylength is not affected by photoperiodic history. Physiol Zool 60:722–729.

Dawson A (1991). Effect of daylength on the rate of recovery of photosensitivity in male starlings (*Sturnus vulgaris*). J Reprod Fertil 93:521–524.

Dawson A, Goldsmith AR (1983). Plasma prolactin and gonadotrophins during gonadal development and the onset of photorefractoriness in male and female starlings (*Sturnus vulgaris*) on artificial photoperiods. J Endocrinol 97:253–260.

Dawson A, Goldsmith AR (1997). Changes in gonadotropin-releasing hormone (GnRH-I) in the pre-optic area and median eminence of starlings (*Sturnus vulgaris*) during recovery of photosensitivity and during photostimulation. J Reprod Fertil 111:1–6.

Dawson A, Sharp PJ (1998). The role of prolactin in the development of reproductive photorefractoriness and postnuptial molt in the European starling (*Sturnus vulgaris*). Endocrinology 139:485–490.

Dawson A, Goldsmith AR, Nicholls TJ (1985a). Development of photorefractoriness in intact and castrated male starlings (*Sturnus vulgaris*) exposed to different periods of long daylengths. Physiol Zool 58:253–261.

Dawson A, Follett BK, Goldsmith AR, Nicholls TJ (1985b). Hypothalamic gonadotrophin-releasing hormone and pituitary and plasma FSH and prolactin during photostimulation and photorefractoriness in intact and thyroidectomized starlings (*Sturnus vulgaris*). J Endocrinol 105:71–77.

Dawson A, Williams TD, Nicholls TJ (1987). Thyroidectomy of nestling starlings appears to cause neotenous sexual maturation. J Endocrinol 112:R5–R6.

Dawson A, McNaughton FJ, Goldsmith AR, Degen AA (1994). Ratite-like neoteny induced by neonatal thyroidectomy of European starlings, *Sturnus vulgaris*. J Zool (Lond) 232:633–639.

Del Negro C, Gahr M, Leboucher G, Kreutzer M (1998). The selectivity of sexual responses to song displays: effects of partial chemical lesion of the HVC in female canaries. Behav Brain Res 96:151–159.

Doupe AJ, Konishi M (1991). Song-selective auditory circuits in the vocal control system of the zebra finch. Proc Natl Acad Sci U S A 88:11339–11343.

El Halawani ME, Youngren OM, Rozenboim I, Pitts GR, Silsby JL, Phillips RE (1995). Serotonergic stimulation of prolactin secretion is inhibited by vasoactive intestinal peptide immunoneutralization in the turkey. Gen Comp Endocrinol 99:69–74.

Elliott JA (1976). Circadian rhythms and photoperiodic time measurement in mammals. Fed Proc 35:2339–2346.

Erickson CJ (1985). Mrs. Harvey's parrot and some problems of socioendocrine response. *In* Perspectives in Ethology 6: Mechanisms (Batson PPG, Klopfer P, eds), pp 261–285. New York: Plenum Press.

Farner DS, Follett BK (1996). Light and other environmental factors affecting avian reprodction. J Anim Sci 25 (suppl):90–118.

Farner DS, Donham RS, Matt KS, Mattocks Jr PW, Moore MC, Wingfield JC (1983). The nature of photorefractoriness. *In* Avian Endocrinology: Environmental and Ecological Perspectives (Mikami S, Homma K, Wada M, eds), pp 149–166. Berlin: Springer-Verlag.

Follett BK (1973). Circadian rhythms and photoperiodic time measurement in birds. J Reprod Fertil. 19 (suppl): 5–18.

Follett BK (1984). Birds. *In* Marshall's Physiology of Reproduction, vol 1 (Lamming GE, ed), pp 283–350. Edinburgh: Longman Green.

Follett BK, Nicholls TJ (1984). Photorefractoriness in Japanese quail: Possible involvement of the thyroid gland. J Exp Zool 232:573–580.

Follett BK, Nicholls TJ (1985). Influences of thyroidectomy and thyroxine replacement on photoperiodically controlled reproduction in quail. J Endocrinol 107:211–221.

Follett BK, Sharp PJ (1969). Circadian rhythmicity in photoperiodically induced gonadotrophin secretion and gonadal growth in quail. Nature 223:968–971.

Follett BK, Kumar V, Juss TS (1992). Circadian natue of the photoperiodic clock in Japanese quail. J Comp Physiol [A] 171:533–540.

Foster RG, Follett BK (1985). The involvement of a rhodopsin-like photopigment in the photoperiodic response of the Japanese quail. J Comp Physiol [A] 157:519–528.

Foster RG, Plowman G, Goldsmith AR, Follett BK (1987). Immunohistochemical demonstration of marked changes in the LHRH system of photosensitive and photorefractory European starlings (*Sturnus vulgaris*). J Endocrinol 115:211–220.

Foster RG, Panzica GC, Parry DM, Viglietti-Panzica C (1988). Immunocytochemical studies on the LHRH system of the Japanese quail: influence by photoperiod and aspects of sexual differentiation. Cell Tissue Res 253:327–335.

Foster RG, Timmers AM, Schalken JJ, DeGrip WJ (1989). A comparison of some photoreceptor characteristics in the pineal and retina: II. The Djungarian hamster (*Phodopus sungorus*). J Comp physiol [A] 165:565–572.

Foster RG, Provencio I, Hudson D, Fiske S, DeGrip W, Menaker M (1991). Circadian photoreception in the retinally degenerate mouse (rd/rd). J Comp Physiol [A] 169:39–50.

Foster EF, Mehta RP, Bottjer SW (1997). Axonal connections of the medial magnocellular nucleus of the anterior neostriatum in zebra finches. J Comp Neurol 382:364–281.

Friedman MB (1977). Interactions between visual and vocal courtship stimuli in the neuroendocrine response of female doves. J Comp Physiol Psychol 91:1408–1416.

Gahr, M, Sonnenschein E, Wickler W (1998). Sex difference in the size of the neural song control regions in a dueting songbird with similar song repertoire size of males and females. J Neurosci 18:1124–1131.

Goldsmith AR (1985). Prolactin in avian reproduction: incubation and the control of seasonal breeding. *In* Prolactin: Fidia Research Series, vol. 1. (Macleod RM, Scapagnini U, Thorner MO, eds), pp 411–426. Padova-Liviana Press; and Berlin: Springer-Verlag.

Goldsmith AR, Nicholls TJ (1984). Thyroidectomy prevents the development of photorefractoriness and the associated rise in plasma prolactin in starlings. Gen Comp Endocrinol 54:256–263.

Goldsmith AR, Ivings WE, Pearce-Kelly AS, Parry DM, Plowman G, Nicholls TJ, Follett BK (1989). Photoperiodic control of the development of the LHRH neurosecretory system of European starlings (*Sturnus vulgaris*) during puberty and the onset of photorefractoriness. J Endocrinol 122:255–268.

Goldsmith AR, Beakes H, Glennie L, Cuthill IC, Witter MS, Ball GF (1992). Hypothalamic LHRH in female starlings exposed to photoperiodic and non-photoperiodic stimulation. *In* Proceedings of the Fifth International Symposium on Avian Endocrinology, Edinburgh, abstract no. P85. Roslin, Scotland: AFRC Institute of Animal Physiology and Genetics Research.

Groos G (1982). The comparative physiology of extraocular photoreception. Experientia Generalia 38:989–1128.

Hahn TP, Ball GF (1995). Changes in brain GnRH associated with photorefractoriness in house sparrows (*Passer domesticus*). Gen Comp Endocrinol 99:349–363.

Hahn TP, Boswell T, Wingfield JC, Ball GF (1997). Temporal flexibility in avian reproduction: Patterns and mechanisms. *In* Current Ornithology 14 (Nolan V Jr, Ketterson ED, Thompson CF, eds), pp 39–80. New York: Plenum Press.

Hamner WM (1963). Diurnal rhythm and photoperiodism in testicular recrudescence of the house finch. Science 142:1294–1295.

Hamner WM (1964). Circadian control of photoperiodism in the house finch demonstrated by interrupted-night experiments. Nature 203:1400–1401.

Hamner WM (1968). The photorefractory period of the house finch. Ecology 49:211–227.

Hamner WM (1971). On seeking an alternative to the endogenous reproductive rhythm hypothesis in birds. *In* Biochronometry (Menaker, M, ed), pp 448–461. Washington, DC: National Academy of Sciences.

Hinde RA (1965). Interaction of internal and external factors in integration of canary reproduction. *In* Sex and Behavior (Beach F, ed), pp 381–415. New York: John Wiley & Sons.

Hinde RA, Steel E (1978). The influence of day length and male vocalizations on the estrogen-dependent behavior of female canaries and budgerigars, with discussion of data from other species. Adv Study Behav 8:39–73.

Juss TS, Goldsmith AR (1992). Intracerebroventricular prolactin is potently gonado-inhibitory but does not induce photorefractoriness. *In* Proceedings of the Fifth International Symposium on Avian Endocrinology, Edinburgh, (p 95). Roslin, Scotland: AFRC Institute of Animal Psychology and Genetics Research.

Juss TS, Ball GF, Parry DM (1992). Immunocytochemical localisation of cGnRH I and cGnRH II in the brains of photosensitive and photorefractory European starlings and Japanese quail. *In* Proceedings of the Fifth International Symposium on Avian Endocrinology, Edinburgh (p 87). Roslin, Scotland: AFRC Institute of Animal Psychology and Genetics Research.

Juss TJ, Meddle SL, Servant RS, King VM (1993). Melatonin and photoperiodic time measurement in Japanese quail. Proc R Soc Lond B Biol Sci 254:21–28.

Konishi M (1985). Birdsong: From behavior to neuron. Annu Rev Neurosci 8:125–170.

Konishi M, Emlen ST, Ricklefs RE, Wingfield JC (1989). Contributions of bird studies to biology. Science 246:465–472.

Kroodsma DE (1976). Reproductive development in a female songbird: differential stimulation by quality of male song. Science 192:574–575.

Kroodsma DE (1982). Learning and the ontogeny of sound signals in birds. *In* Acoustic Communication in Birds, vol 2 (Kroodsma DE, Miller EH, eds), pp 1–23. New York: Academic Press.

Kroodsma D, Byers BE (1991). The functions of bird song. Am Zool 31:318–328.

Lees AD (1973). Photoperiodic time measurement in the aphid *Megoura viciae*. J Insect Physiol 19:2279–2316.

Lehrman DS (1965). Interaction between internal and external environments in the regulation of the reproductive cycle of the ring dove. *In* Sex and Behavior (Beach F, ed), pp 355–380. New York: John Wiley & Sons.

Lincoln GA, Racey PA, Sharp PJ, Klandorf H (1980). Endocrine changes associated with spring and autumn sexuality of the rook (*Corvus frugilegus*). J Zool 190:137–153.

MacNamee MC, Sharp PJ, Lea RW, Sterling RJ, Harvey S (1986). Evidence that vasoactive intestinal polypeptide is a physiological prolactin-releasing factor in the bantam hen. Gen Comp Endocrinol 62:470–478.

Margoliash D (1983). Acoustic parameters underlying the responses of song-specific neurons in the white-crowned sparrow. J Neurosci 3:1039–1057.

Margoliash D (1997). Functional organization of forebrain pathways for song production and perception. J Neurobiol 33:671–693.

Mattocks PW Jr, Farner DS, Follett BK (1976). The annual cycle in luteinizing hormone in the plasma of intact and castrated white-crowned sparrows, *Zonotrichia leucophrys gambelii*. Gen Comp Endocrinol 30:156–161.

Mauro LJ, Youngren OM, Proudman JA, Phillips RE, El Halawani ME (1992). Effects of reproductive status, ovariectomy, and photoperiod on vasoactive intestinal peptide in the female turkey hypothalamus. Gen Comp Endocrinol 87:481–493.

Meddle SL, Follett BK (1995). Photoperiodic activation of Fos-like immunoreactive protein in neurones within the tuberal hypothalamus of Japanese quail. J Comp Physiol [A] 176:79–89.

Menaker M (1971). Synchronization with the photic environment via extraretinal receptors in the avian brain. *In* Biochronometry. (Menaker M, ed), pp 315–332. Washington, DC: National Academy of Sciences.

Menaker M, Keatts (1968). Extraretinal light perception in the sparrow, II. Photoperiodic stimulation of testis growth. Proc Natl Acad Sci U S A 60:146–151.

Millam JR, Faris PL, Youngren OM, El Halawani ME, Hartman BK (1993). Immunohistochemical localization of chicken goadotropin-releasing hormones I and II (cGnRH-I and II) in turkey hen brain. J Comp Neurol 333:68–82.

Millar RP, King JA (1984). Structure-activity relations of LHRH in birds. J Exper Zool 232:425–430.

Millar RP, deL. Milton RC, Follett BK, King JA (1986). Receptor binding and gonadotroping-activity of a novel chicken gonadotropin-releasing hormone ([His5, Trp7, Tyr8]GnRH) and a D-Arg6 analog. Endocrinology 119:224–231.

Miyamoto K, Hasegawa Y, Nomura M, Igarashi M, Kanagawa K, Matsuo M (1984). Identification of the second gonadotropin-releasing hormone in chicken hypothalamus: Evidence that gonadotropin secretion is probably controlled by two distinct gonadotropin-releasing hormones in avian species. Proc Natl Acad Sci U S A 81:1341–1347.

Moenter SM, Woodfill CJI, Karsch FJ (1991). Role of the thyroid gland in seasonal reproduction: Thyroidectomy blocks seasonal suppression of reproductive neuroendocrine activity in ewes. Endocrinology 128:1337–1344.

Moore MC (1983). Effect of female sexual displays on the endocrine physiology and behaviour of male white-crowned sparrows, *Zonotrichia leucophrys.* J Zool (Lond) 199:137–148.

Morton ML, Pereyra ME, Baptista LF (1985). Photoperiodically induced ovarian growth in the white-crowned sparrow (*Zonotrichia leucophrys gambelii*) and its augmentation by song. Comp Biochem Physiol 80A:93–97.

Murton RK, Westwood NJ (1977). Avian Breeding Cycles. Oxford: Clarendon.

Nelson RJ, Zucker I (1981). Absence of extrocular photoreception in diurnal and noturnal rodents exposed to direct sunlight. Comp Biochem Physiol 69A:145–148.

Nicholls TJ, Storey CR (1976). The effects of castration on plasma LH levels in photosensitive and photorefractory canaries (*Serinus canarius*). Gen Comp Endocrinol 29:170–174.

Nicholls TJ, Goldsmith AR, Dawson A (1988). Photorefractoriness in birds and comparison with mammals. Physiol Rev 68:133–176.

Nottebohm F (1975). Vocal behavior in birds. *In* Avian Biology, vol 5 (Farner DS, King JR, eds), pp 287–332. New York: Academic Press.

Nottebohm F, Arnold AP (1976). Sexual dimorphism in vocal control areas of the songbird brain. Science 194:211–213.

Nottebohm F, Alvarez-Buylla A, Cynx J, Kirn J, Ling CY, Nottebohm M, Suter R, Tolles A, Williams H (1990). Song learning in birds: the relation between perception and production. Philos Trans R Soc Lond B Biol Sci 329:115–124.

Oliver J, Baylé JD (1982). Brain photoreceptors for the photoinduced testicular response in birds. Experientia 38:1021–1029.

Parkinson TJ, Follett BK (1994). Effect of thyroidectomy upon seasonality in rams. J Reprod Fertil 101:51–58.

Parkinson TJ, Douthwaite JA, Follett BK (1995). Responses of prepubertal and mature rams to thyroidectomy. J Reprod Fertil 104:51–56.

Parry DM, Goldsmith AR (1993). Ultrastructural evidence for changes in synaptic input to the hypothalamic luteinizing hormone–releasing hormone neurons in photosensitive and photorefractory starlings. J Neuroendocrinol 5:387–395.

Parry DM, Goldsmith AR, Millar RP, Glennie LM (1997). Immunocytochemical localization of GnRH precursor in the hypothalamus of European starlings during sexual maturation and photorefractoriness. J Neuroendocrinol 9:235–243.

Perrins CM (1970). The timing of birds' breeding seasons. Ibis 112:242–255.

Perrins CM, Birkhead TR (1983). Avian Ecology. London: Blackie & Sons.

Pittendrigh CS, Minis DH (1964). The entrainment of circadian oscillators by light and their role as photoperiodic clocks. Am Naturalist 98:261–294.

Pittendrigh CS, Minis DH (1971). The photoperiodic time measurement in *Pectinophora gossypiella* and its relation to the circadian system in that species. *In* Biochronometry (Menaker M, ed). Washington, DC, National Academy of Sciences.

Reinert BD, Wilson FE (1996a). Thyroid dysfunction and thyroxine-dependent programming of photoinduced ovarian growth in American tree sparrows (*Spizella arborea*). Gen Comp Endocrinol 103:71–81.

Reinert BD, Wilson FE (1996b). The thyroid and the hypothalamus-pituitary-ovarian axis in American tree sparrows (*Spizella arborea*). Gen Comp Endocrinol 103:60–70.

Reiter RJ (1978). Interaction of photoperiod, pineal and seasonal reproduction as exemplified by findings in the hamster. Prog Reprod Biol 4:169–190.

Ricklefs RE (1983). Avian postnatal development. *In* Avian Biology, vol 7 (Farner DS, King JR, Parkes KC, eds), pp 1–83. New York: Academic Press.

Robinson JE, Karsch FJ (1984). Refractoriness to inductive day lengths terminates the breeding season of the Suffolk ewe. Biol Reprod 31:656–663.

Robinson JE, Follett BK (1982). Photoperiodism in Japanese quail: the termination of seasonal breeding by photorefractoriness. Proc R Soc Lond B Biol Sci 215:95–116.

Robinson JE, Wayne NL, Karsch FJ (1985). Refractoriness to inhibitory day length initiates the breeding season of the Suffolk ewe. Biol Reprod 32:1024–1030.

Saldanha CJ, Leak RK, Silver R (1994a). Detection and transduction of daylength in birds. Psychoneuroendocrinology 19:641–656.

Saldanha CJ, Deviche PJ, Silver R (1994b). Increased VIP and decreased GnRH expression in photorefractory dark-eyed juncos (*Junco hyemalis*). Gen Comp Endocrinol 93:128–136.

Sharp PJ (1996). Strategies in avian breeding cycles. Anim Reprod Sci 42:505–513.

Sharp PJ, Klandorf H (1981). The interaction between day length and the gonads in the regulation of levels of plasma thyroxine and triiodothyronine in the Japanese quail. Gen Comp Endocrinol 45:504–512.

Sharp PJ, Talbot RT, Main GM, Dunn IC, Fraser HM, Huskisson NS (1990). Physiological roles of chicken LHRH-I and -II in the control of gonadotrophin release in the domestic chicken. J Endocrinol 124:291–299.

Sherwood NM, Wingfield JC, Ball GF, Dufty AM (1988). Identity of GnRH in passerine birds: comparison of GnRH in song sparrow (*Melospiza melodia*) and starling (*Sturnus vulgaris*) with 5 vertebrate GnRHs. Gen Comp Endocrinol 69:341–351.

Shi ZD, Barrell GK (1992). Requirement of thyroid function for the expression of seasonal reproductive and related changes in red deer (*Cervus elaphus*) stags. J Reprod Fertil 94:251–259.

Sicard V, Oliver J, Baylé JD (1983). Gonadotrophic and photosensitive abilities of the lobus parolfactorius: electrophysiological study in quail. Neuroendocrinology 36:81–87.

Silver R, Witkovsky P, Horvath P, Alones V, Barnstable CJ, Lehman MN (1988). Coexpression of opsin-like and VIP-like immunoreactivity in CSF-contacting neurons of the avian brain. Cell Tissue Res 253:189–198.

Striedter GF (1994). The vocal control pathways in budgerigars differ from those in songbirds. J Comp Neurol 343:35–56.

Striedter GF, Vu ET (1998). Bilateral feedback projections to the forebrain in the premotor network for singing in zebra finches. J Neurobiol 34:27–40.

Sun SS, El Halawani ME (1995). Protein kinase-C mediates chicken vasoactive intestinal peptide-stimulated prolactin secretion and gene expression in turkey primary pituitary cells. Gen Comp Endocrinol 99:289–297.

Thrun LA, Dahl GE, Evans NP, Karsch FJ (1996). Time-course of thyroid-hormone involvement in the development of anestrus in the ewe. Biol Reprod 55:833–837.

Thrun LA, Dahl GE, Evans NP, Karsch FJ (1997). A critical period for thyroid hormone action on seasonal changes in reproductive neuroendocrine function in the ewe. Endocrinology 138:3402–3409.

Turek FW (1974). Circadian rhythmicity and the initiation of gonadal growth in sparrows. J Comp Physiol 92:59–64.

Underwood H, Menaker M (1970). Photoperiodically significant photoreception in sparrows: is the retina involved? Science 167:299–301.

Vaz Nunes M, Veerman A (1984). Light-break experiments and photoperiodic time measurement in the spider mite *Tetranychus urticae*. J Insect Physiol 30:891–897.

Vaz Nunes M, Veerman A (1986). A "dusk" oscillator affects photoperiodic induction of diapause in the spider mite *Tetranychus urticae*. J Insect Physiol 32:605–614.

Veerman A, Beckman M, Veenedaal RL (1988). Photoperiodic induction of diapause in the large white butterfly, *Pieris brassicae*—evidence for hour glass time measurement. J Insect Physiol 34:1063–1069.

Wada Y, Okano T, Adachi A, Ebihara S, Fukada Y (1998). Identification of rhodopsin in the pigeon deep brain. FEBS Lett 424:j53–56.

Wieselthier AS, van Tienhoven A (1972). The effect of thyroidectomy on testicular size and on the photorefractory period in the starling (*Sturnus vulgaris* L.). J Exp Zool 179:331–338.

Wild JM (1997). Neural pathways for the control of birdsong production. J Neurobiol 33:653–670.

Wild JM, Li D, Eagleton C (1997). Projections of the dorsomedial nucleus of the intercollicular complex (DM) in relation to respiratory-vocal nuclei in the brainstem of pigeon (*Columba livia*) and zebra finch (*Taeniopygia guttata*). J Comp Neurol 377:392–413.

Wilson FE (1991). Neither retinal nor pineal photoreceptors mediate photoperiodic control of seasonal reproduction in American tree sparrows (*Spizella arborea*). J Exp Zool 259:117–127.

Wilson FE, Donham RS (1988). Daylength and control of seasonal reproduction in male birds. In Processing of Environmental Information in Vertebrates (Stetson MH, ed), pp 101–119. New York: Springer-Verlag.

Wilson FE, Follett BK (1974). Plasma and pituitary luteinizing hormone in intact and castrated tree sparrows (*Spizella arborea*). Gen Comp Endocrinol 23:82–93.

Wilson FE, Reinert BD (1993). The thyroid and photoperiodic control of seasonal reproduction in American tree sparrows (*Spizella arborea*). J Comp Physiol [B] 163:563–573.

Wilson FE, Reinert BD (1995a). A one-time injection of thyroxine programmed seasonal reproduction and postnuptial moult in chronically thyroidectomised male American tree sparrows (*Spizella arborea*) exposed to long days. J Avian Biol 26:225-233.

Wilson FE, Reinert BD (1995b). The photoperiodic control circuit in euthyroid American tree sparrows (*Spizella arborea*) is already programmed for photorefractoriness by week 4 under long days. J Reprod Fertil 103:279–284.

Wilson FE, Reinert BD (1996). The timing of thyroid-dependent programming in seasonally breeding male American tree sparrows (*Spizella arborea*). Gen Comp Endocrinol 103:82–92.

Wingfield JC (1980). Fine temporal adjustments of reproductive function. *In* Avian Endocrinology (Epple A, Stetson MH, eds), pp 367–389. New York: Academic Press.

Wingfield JC (1983). Environmental and endocrine control of reproduction: an ecological approach. *In* Avian Endocrinology. Environmental and ecological Perspectives (Mikami SI, Homma K, Wada M, eds), pp 265–288. Berlin: Springer-Verlag.

Wingfield JC (1985). Environmental factors influencing the termination of reproduction in finches. *In* Acta 18th Congressus Internationlalis Ornithologici (Ilyichev VD, Gavrilov VM, eds), pp 478–487. Moscow: Nauka.

Wingfield JC, Farner DS (1980). Control of seasonal reproduction in temperate-zone birds. Prog Reprod Biol 5:62–101.

Wingfield JC, Farner DS (1993). Endocrinology of reproduction in wild species. *In* Avian Biology, vol 9 (Farner DS, King J, Parkes K, eds), pp 163–327. New York: Academic Press.

Wingfield JC, Kenagy (1991). Natural regulation of reproductive cycles. *In* Vertebrate Endocrinology: Fundamentals and Biomedical Implications, vol 4, pt B (Pang P, Schreibman M, eds), pp 181–241. New York: Academic Press.

Wingfield JC, Whaling CS, Marler PR (1994). Communication in vertebrate aggression and reproduction: the role of hormones *In* The Physiology of Reproduction (Knobil E, Neil JD, eds), pp 303–342. New York: Raven Press.

Witschi E, Miller RA (1938). Ambisexuality in the female starling. J Exp Zool 79:475–507.

Woitkewitsch AA (1940). Dependence of seasonal periodicity in gonadal changes on the thyroid gland in *Sturnus vulgaris* L. C R (Doklady) Acad Sci URSS 27:741–745.

Yokoyama K, Farner DS (1976). Photoperiodic responses in bilaterally enucleated female white-crowned sparrows, *Zonotrichia leucophrys gambelii*. Gen Comp Endocrinol 30:528–533.

Youngren OM, Silsby JL, Rozenboim I, Phillips RE, El Halawani, ME (1994). Active immunization with vasoactive intestinal peptide prevents the secretion of prolactin induced by electrical stimulation of the turkey hypothalamus. Gen Comp Endocrinol 95:330–336.

6 Sex and Brain Secretions

Rae Silver and Ann-Judith Silverman

This chapter is the story of a serendipitous discovery. Viewed from our perspective, it began with our studies of the neuroendocrine control of reproductive behavior in the ringdove. Viewed from a historical perspective, it began several years ago when the science of endocrinology was something far removed from the science of immunology. Studies of the brain, behavior, endocrinology, and immunology began as distinct fields, then merged at their borders. In the earlier half of the century, organismal biology was approached by different groups of researchers who used particular strategies within their subdisciplines. Those subdisciplines that pertain to this chapter are behavioral endocrinology, a scientific discipline concerned with hormonal effects on behavior, as well as behavioral influences on hormone secretion; neurophysiology, a discipline focused on the electrical activity of neurons; neuroendocrinology, a subdiscipline that dealt with the interaction of hormones with the nervous system; and immunology, a quite separate science concerned with the systems that defend the body against infection. More recently, we and other investigators have combined strategies and methodologies that came from these these distinct fields to form psychoneuroimmunology, or neuroendocrinimmunology, a synthetic field of research concerned with the ways in which the brain, the endocrine system, behavior, and the immune system interact.

In this chapter, we describe how our work on reproductive and courtship behavior has forced us into thinking about the immune system. We present this scientific tale within the context of the ways in which previously separate and now merging scientific disciplines have forced us to confront interrelationships between the brain and behavior, the endocrine and neuroendocrine systems, and the immune and neuro-immune systems.

To provide an organizing framework, we review the history of scientific inquiry that existed at the beginning of this story. First, we will outline studies of the brain, behavior, endocrinology, and immunology. Next, we explain how our research on the neuroendocrine regulation of reproductive behavior in ringdoves forced us to acknowledge "psycho-neuro-endocrine-immunology." Third, we review evidence that cells of hematopoietic origin (immune system) in the brain are altered during normal physiology, and change the relationships between endocrine and immune systems. Finally, we end by providing some informed speculation on the possible significance of immune interactions with the nervous system and on the types of roles that might be played by immune cells in the brain.

Studies of the Brain, Hormones, and Defense against Disease Began as Distinct Fields, then Merged at Their Interfaces

The Brain Influences Endocrine Secretions

Traditionally, *hormones* were defined as substances that are produced by specialized glands, are delivered to their targets via the vascular system, and act at sites distant from the glands they are secreted from. The term is now applied to substances not produced by specialized glands, but having similar actions at both local and anatomically remote sites. Rowan was one of the first to demonstrate experimentally that external stimuli can influence hormone secretions (Rowan 1925). He captured migratory juncos as they traveled through Edmonton in northern Canada, heading south in the fall. The birds were held in outdoor cages through an exceptionally cold winter (temperatures reached $-50°F$). One group of birds was exposed to long days by placing a light bulb in their cage; the other group was exposed to the natural light-dark cycle. When he examined the birds in midwinter, he noted that those exposed to extra hours of light had large gonads. Those exposed to short days of winter had regressed gonads—appropriate to the season. These studies showed that information from photic cues reached the reproductive system and regulated gonadal state, irrespective of thermal conditions. While Rowan initially thought that photic cues traveled through the feet, later studies showed that the brain was the route by which photic cues were translated into signals that regulated endocrine secretions. The identification of the pathway whereby sensory stimuli might influence secretions of the gonads awaited the demonstration that the central nervous system controlled secretory activity of the reproductive axis, including the pituitary and gonads.

The road to understanding that the brain itself could produce secretions had its beginnings in the work of Scharrer in the late 1920s and early 1930s. Unequivocal evidence was provided by the work of Scharrer and Bargmann in the 1940s and 1950s (Scharrer 1970). The glandular neurons, which synthesized neurohormones or neurosecretory material, were neurosecretory cells. The neurohormone traveled in neurosecretory granules by rapid axoplasmic flow, down the axons to the nerve terminals (in this case, in the neurohypophysis or posterior pituitary). Here the neurosecretions were released into the general bloodstream to act on distant targets. Because the neurons that project to the posterior pituitary were large, they were named the magnocellular neurons (Latin *magno*, large).

A connection between the brain and the adenohypophysis (anterior pituitary) was suggested as early as 1936 (see McCann et al. 1968 for review). However, it took another decade to establish that the direction of blood flow was from the brain (from the median eminence of the basal hypothalamus) to the anterior pituitary via the hypophysial portal veins (Green and Harris 1947). Subsequently, Halasz and his collaborators transplanted pituitaries either into the hypothalamus or under the kidney capsule and demonstrated that only those glands transplanted near the hypothalamus showed secretory activity. This work demonstrated the dependence of pituitary secretions on the brain (Halasz and Papp 1965). Since the areas in the hypothalamus that stimulate anterior pituitary function contained small neurons, these cells were called parvicellular neurosecretory cells (*parvus*, small). It was hypothesized that they projected to the median eminence and terminated on medium eminence blood vessels, which were drained by the portal vasculature, for direct and rapid delivery to the anterior pituitary. With knowledge of the neural and vascular pathways from the brain to the pituitary, studies of brain-behavior relationships left the realm of phenomenology and entered the domain of physiology.

Advances in understanding brain secretions have continued over the decades. The late 1960s brought the isolation and identification of brain neuropeptides that control pituitary secretions, including thyroid-releasing hormone (TRH) and gonadotropin-releasing hormone (GnRH) (Boler et al. 1969; Amoss 1971; Schally et al. 1971). In 1983, Vale and his collaborators identified corticotropin-releasing hormone (CRH), the brain hormone that regulates the adrenal system and the stress response (Vale et al. 1983). It is likely that other brain hormones remain to be identified. Taken together, this work solidifies the concept of the brain as the "master gland."

The Brain Influences Immune System Secretions

While there was a long history of anecdotal evidence that the brain might influence immune system reponses, experimental proof emerged more recently. Formal analysis of the brain–immune system link gained widespread attention with the demonstration that immune responses could be conditioned (Ader and Cohen 1982; MacQueen et al. 1989; Djuric and Bienenstock 1993). Ader and Cohen's dramatic studies showed that a conditioned stimulus could be paired with an immunosuppressant drug in a strain of mice that spontaneously develop disease because of an overactive immune system. In this conditioning paradigm, cyclophosphamide was used to suppress immune function and served as the unconditioned stimulus, that is, the response to this stimulus was innate, not learned. The presentation of saccharin was used as the conditioned stimulus; saccharin alone did not suppress immune function. By pairing saccharin with cyclophosphamide, an association was formed between the conditioned (saccharin) and unconditioned stimulus (cyclophosphamide), so that eventually the conditioned stimulus itself produced the conditioned response (immune system suppression). This landmark experiment showed that the development of an autoimmune disease was modified by classical conditioning of the immunosuppression response.

In subsequent years, the mechanisms of this enigmatic process have been examined, with the demonstration of the effectiveness of conditioning in a number of immune system responses (discussed below). In the 1990s it became increasingly clear that the brain synthesizes other proteins, including cytokines, which have long been associated with cells of the immune system (Steinman 1993; Patterson 1994). Perhaps even more surprising, it has also become clear that immune system cells contain peptides classically associated with neurosecretory cells. Examples of some of these immune system cells are the leukocytes, white blood cells or corpuscles of the immune system. The two main categories of leukocytes are granular and nongranular. Eosinophils are granular leukocytes with a nucleus that has two lobes, connected by a thin thread of chromatin (table 6.1). Within the last decade, leukocytes have been shown to be GnRH-positive (Emanuele et al. 1990), and eosinophils have been shown to contain estrogen receptors (Tchermitchin et al. 1989).

Table 6.1
Hematopoietic cells

Cell	Function
Red blood cells	Transport of oxygen to tissues.
Granulocytes	
Neutrophil	Premier defensive phagocyte. First line of defense against invaders.
Eosinophil	Contains substances highly toxic to helminthic and other parasites. Phagocytoses antigen-antibody complexes.
Basophil	Rare. Contains heparin and histamine (and serotonin in some species). Expresses receptors for IgE and shows allergic reaction to antigen recognized by IgE.
Mononuclear cells	
B cells	Differentiate in the bone marrow. Express a specific antibody on their surface. When activated by antigen they multiply and convert to a plasma cell which secretes high levels of the same antibody. This is humoral immunity.
T cells	Lymphocytes that differentiate in the thymus. The three major classes are T helper, which participates in activating B lymphocytes; T suppressor, a downregulator of the immune response; and T cytotoxic, which carry out phagocytosis of foreign antigen when presented by an antigen-presenting cell in conjunction with the major histocompatibility complex. T cells have special receptors to recognize these antigens.
Macrophages	Circulate as monocytes and differentiate in site-specific ways in tissues. They are also antigen-presenting cells.
Mast cells	Circulate as primitive but committed precursors. Enter tissues where they complete their differentiation, again in a site-specific manner. Express receptors for IgE and respond to allergen with explosive release of granular contents. Products discussed in detail in text.

Though external stimuli and behavior alter both the endocrine and immune systems, endocrinology and immunology are generally considered unrelated realms of research, as will be discussed again in chapter 8. Those who study endocrinology are typically found in departments of medicine, physiology, and biology, while those with expertise in immunology are found in pathology, rheumatology, and sometimes in medicine or biochemistry. These departments are often located far apart in different buildings of hospitals and universities. Recognition of the interactions between the nervous, endocrine, and immune systems has blossomed since the 1980s. Most work, however, has been restricted to studies of disease. The notion that these systems interact in normal physiology is virtually unexplored. In our own work, we have been confronted with the finding that a cellular element of the hematopoietic system undergoes rapid and dramatic changes following a brief period of sexual behavior. Pursuit of the causes and consequences of this event has led us into considerations of the "psychoimmunoendocrinology" of normal physiology.

Studies of Sex Behavior in Ringdoves Forced Us from Behavioral Endocrinology to Psychoneuroendocrinimmunology

Studies of Reproduction in Ringdoves

Because their behavior is highly stereotyped, and because they adapt well to domestication, ringdoves have long served as a favorite model system in behavioral endocrinology. Perhaps more important is the fact that often neuroendocrine responses of many avian forms are highly exaggerated, making it easy to uncover what might otherwise be a subtle phenomenon (Wingfield et al. 1994; Konishi et al. 1989).

Reproductively experienced doves placed together in a breeding cage engage in a stereotyped series of behaviors (Silver 1978; Cheng 1979). For a period of 7 to 9 days, they court each other. Toward the end of that interval, they copulate. The female lays a two-egg clutch, and both mates participate in incubating the eggs for about 14 days. When the eggs hatch, the parents brood and feed the young for about 21 days, at which time the young can care for themselves. The parents then start a new cycle of courtship, incubation, and rearing young.

We and many others have shown that within a few hours of placing a male and female dove together (and many days before they copulate) there is an increase in plasma androgens in males (Silver 1978). This appears to be also true for other male vertebrates, including humans (Wingfield et al. 1994; Anonymous 1970). The secretion of gonadotropins and consequently gonadal steroids is dependent on the release of GnRH into the hypophysial portal circulation (see figure 1.1). To explore the

Table 6.2
GnRH terminal peptide sequences

	1 2 3 4 5 6 7 8 9 10
Mammal	pGlu–His–Trp–Ser–Tyr–Gly–Leu–Arg–Pro–Gly–NH$_2$
Chicken I	pGlu–His–Trp–Ser–Tyr–Gly–Leu–Gln–Pro–Gly–NH$_2$
Catfish	pGlu–His–Trp–Ser–His–Gly–Leu–Asn–Pro–Gly–NH$_2$
Chicken II	pGlu–His–Trp–Ser–His–Gly–Trp–Tyr–Pro–Gly–NH$_2$
Dogfish	pGlu–His–Trp–Ser–His–Gly–Trp–Leu–Pro–Gly–NH$_2$
Salmon	pGlu–His–Trp–Ser–Tyr–Gly–Trp–Leu–Pro–Gly–NH$_2$
Lamprey	pGlu–His–Tyr–Ser–Leu–Glu–Trp–Lys–Pro–Gly–NH$_2$

neural loci where changes might occur with the onset of sexual behavior, we looked for alterations in GnRH immunoreactivity throughout the brains of courting animals, compared with controls (Silver et al. 1992). We did this by using an antibody named LR-1 directed against mammalian GnRH that recognizes amino acids 3, 4, and 7 through 10 (table 6.2). Contrary to expectation, there were no detectable changes in GnRH neurons. There was, however, the highly consistent appearance of labeled cells in the medial habenula of courted animals. These small, immunoreactive cells were clearly nonneural as they lacked neuritic or glial processes. They were always present in the courted animals, and were absent or much lower in number in animals housed in isolation (figure 6.1).

Restated, the study that launched our foray into psychoimmunoendocrinology initially sought to determine which GnRH neurons were "activated" in courting birds at the time these changes in behavior and steroid synthesis were first observed. Contrary to our expectation, there were no detectable changes in GnRH neurons. Instead, we noted the appearance of a large number of GnRH-positive, non-neuronal cells in the brain (Silver et al. 1992).

The Hunt Began to Identify the NonNeuronal Cells That Appear Within Hours of Courtship

Following our initial observations a series of studies were undertaken to characterize newly appearing cells (Silverman et al. 1994). Our first guess was that these cells

Figure 6.1
Low-magnification photomicrographs of the medial habenula from an adult male dove kept in visual isolation from conspecifics (above) and from a male that had courted a female for 2 hours (below). Note the presence of numerous mast cells (arrowheads) in the medial habenula of this bird, which are absent in the bird kept in isolation.

![Isolated — III V, Medial Habenula]

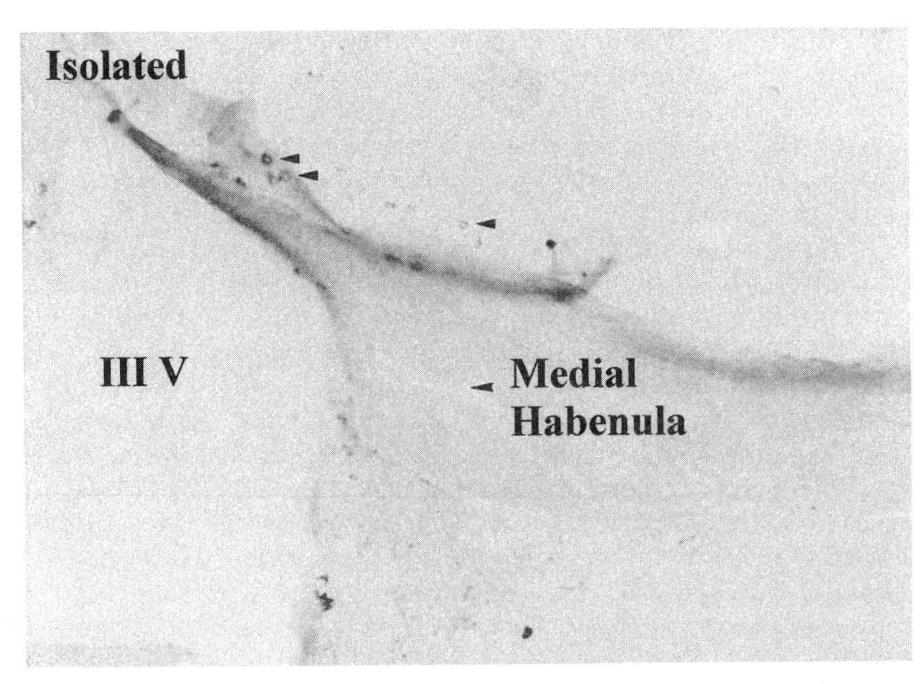

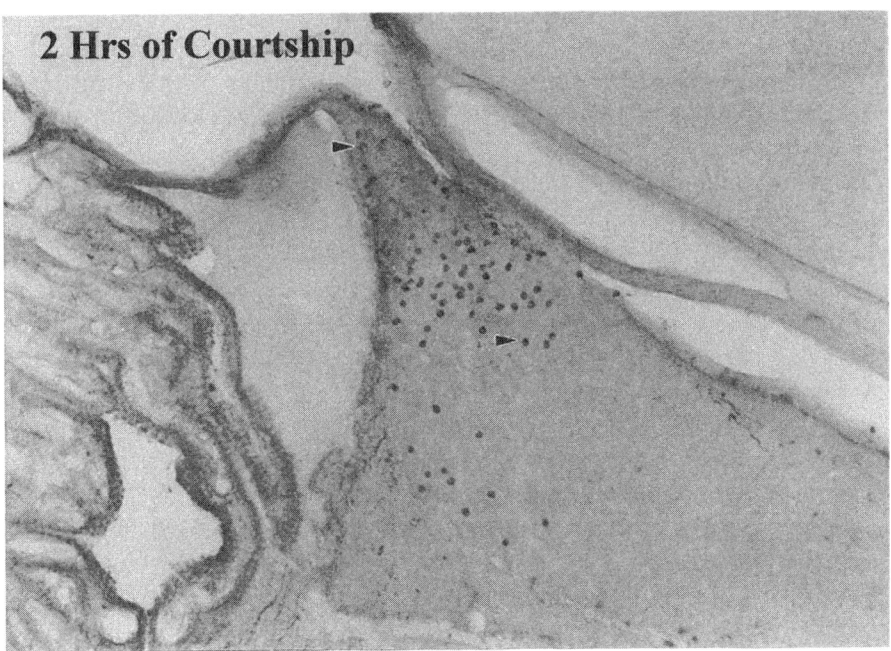

![2 Hrs of Courtship]

were microglia, a hematopoietic cell derivative of the monocyte, known to reside in the brain, and to become activated under specific local conditions (Flaris et al. 1993). To our surprise, the GnRH-immunoreactive cells of the medial habenula did not react with either of two reagents specific for microglia: a monoclonal antibody, OX42, or *Bandeiraea simplicifolia* isolectin B4, indicating that they were of a different lineage (Silver et al. 1993).

The remaining possibilities were that these were either mast cells or basophils, both of which are granulated cells of hematopoietic origin (see table 6.2). We knew that mast cells were usually located in tissues, while basophils were thought to circulate in the blood, so we set out to determine whether the non-neuronal brain cells were mast cells. We also knew mast cells played a role in allergic reactions, but that their occurrence in the brain was not widely known. We learned that mast cells could be identified by their tendency to exhibit metachromasia, a change in color, to certain aniline dyes. We used both acidic toluidine blue and toluidine blue dissolved in water and butanediol. The latter dye is specific for avian mast cells vs. avian basophils (Carlson and Hacking 1972). Use of these dyes clearly revealed metachromatic cells, approximately 8 to 10 µm in diameter, with a distribution and morphology similar to that obtained by immunostaining with the GnRH antiserum in the medial habenula. The cells had pink to deep-purple cytoplasmic granules (depending on the fixation conditions) and deep-blue, heterochromatic nuclei. They were distinct from nearby neurons that had pale-blue, nongranulated cytoplasm, prominent blue nucleoli, and euchromatic nuclei. Some of these cells also contained granules with an affinity for safranin in the presence of alcian blue, indicative of the presence of a highly sulfated proteoglycan of the heparan family.

We also used antibodies for another marker for mast cells, histamine. Immunocytochemical studies revealed that cells in the medial habenula with a similar distribution to those stained with toluidine blue or those stained with the GnRH antiserum also contained histamine. Double-label immunocytochemical experiments revealed that all medial habenular mast cells containing GnRH-like immunoreactivity also contained histamine. Cells of similar morphology, metachromasia, and immunological properties were also present in the lung and liver, organs where mast cells are known to be abundant. Recently, we have determined that mast cells in the dove brain also contain serotonin, as they do in rodents (Marathias et al. 1991).

Ultrastructural observations of LR-1 immunoreactive cells within the medial habenula revealed that they had a unilobar, ovoid, heterochromatic nucleus with the heterochromatin concentrated at the nuclear envelope, which further supports their designation as mast cells. Circulating basophils, the other potential candidate for non-neuronal cells exhibiting metachromasia, have a bilobed nucleus in domestic

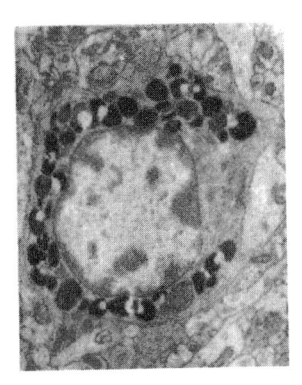

Figure 6.2
Electron micrograph of a mast cell in the medial habenula of a male dove that had courted for 2 hours. The cell has a heterochromatic nucleus and its cytoplasm is filled with many granules. There are numerous fine filamentous process on its surface (arrows).

fowl (Chand and Eyre 1978) distinct from our tissue cells. In sections from brains fixed with an aldehyde mixture for conventional electron microscopy, the granules were electron dense and displayed a variety of internal structures characteristic of both mammalian and avian mast cells (Valsala et al. 1985). Furthermore, fine filamentous processes (characteristic of mast cells but not of basophils) appeared evenly distributed on the cell surface whether cells were on the pial surface or within the brain parenchyma (figure 6.2).

There are two types of epitopes of GnRH in birds. They were first discovered in chickens, and are therefore called cGnRH-I and cGnRH-II to distinguish them from the mammalian form, called mGnRH (see table 6.2). An epitope is defined as an antigenic determinant, that is, a site which has the property of inducing a specific immune response. A more detailed discussion of the discovery of these epitopes is provided in chapter 5. In our initial experiments, the mast cells were recognized by their reactivity to a specific antiserum (LR-1) raised against mGnRH (Silverman et al. 1994). Subsequent absorption experiments suggested that the epitope within these mast cells, recognized by the LR-1 antiserum, differs from that in neuronal elements. Staining in brain mast cells was abolished when the LR-1 was absorbed with either of the two forms of avian GnRH (cGnRH-I or cGnRH-II), while the staining in neurons and axons was abolished only by absorption with cGnRH-I (figure 6.3). We have also used a battery of other GnRH polyclonal and monoclonal antibodies (Ronekliev, EL-14; Millar, Rb1076; Benoit, LR1-II; Urbanski, HU4; Jennes, monoclonal AB; monoclonal antibody from QED, Inc.) and only one, SW1 (Wray et al. 1994), stains the mast cells in the medial habenula.

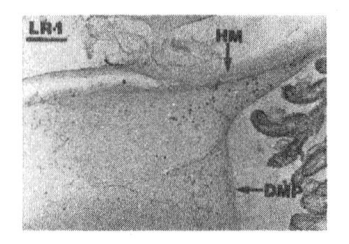

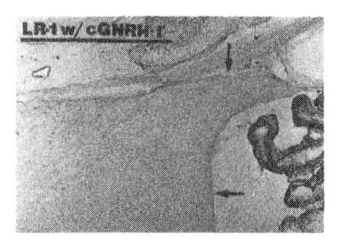

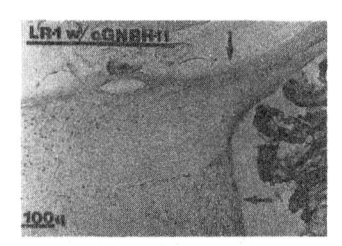

Figure 6.3
GnRH absorption.

Overview of the Hematopoietic System

Hematopoiesis is defined as the production in the bone marrow of formed elements of the blood. All of these elements are progeny of the pluripotential stem cell (Kitamura et al. 1978). Mature cells found in the circulation include red blood cells, platelets (cellular bits of the megakaryocyte), granulocytes (neutrophils, eosinophils, basophils), and lymphocytes (see table 6.1). In addition, two committed precursor populations are also in the blood. The first are the monocytes, which give rise to tissue macrophages, and the second are mast cell precursors.

Hematopoietic cells of the immune system protect us in several ways. Some of the immune cells act in an immediate fashion to defend the body against viral, bacterial, or parasitic invaders. For example, neutrophils attack bacteria by phagocytosis. Eosinophils kill the larvae of parasites by surrounding them and secreting proteases, thus digesting them. Other cells, such as lymphocytes, are designed to distinguish between self and nonself, whether the nonself is an invading organism or a tumorigenic cell. There are two types of immunity provided by lymphocytes: humoral (in which immunoglobulins are secreted by transformed B cells) and cellular (provided by activated T cells). Quiescent T cells become activated when processed antigen appears on the cell surface of an antigen-presenting cell (usually a macrophage) in conjunction with the major histocompatibility complex. B cells, on the other hand, express on their surface the immunoglobulin that they will secrete if activated. They do so if the specific antigen binds to the surface molecule.

Mast cells participate in the immune response by virtue of their capacity to bind IgE. This immunoglobulin is made by B cells, circulates in the blood, and upon entering tissues is bound to the surface of mast cells by a specific receptor (FcERCI). When the allergen specific for a particular IgE molecule is present (e.g., tree pollen, house dust), the IgE molecules dimerize and initiate a cascade of events leading to sudden or anaphylactic degranulation. The vasoactive substances dilate local blood vessels and the secreted mast cell cytokines help attract other immune system cells to the site of inflammation. Mast cells also participate in the induction of B-cell synthesis of IgE (Gauchat et al. 1993).

It is also well established that mast cells can show regulated release. For example, in the skin, substance P (SP)–positive nerve terminals can cause mast cell secretion (Yano et al. 1989). Our interests, as discussed below, relate to this regulated form of secretion and the role that the mast cells might have in normal brain.

Overview of Mast Cells There are several lines of evidence that mast cell progenitors are present in the blood. A specific set of cells defined by a high levels of the receptor, c-kit, and low levels of the surface protein, Thy^1, can be isolated by fluorescent-activated cell sorting. These cells do not yet express the receptor for IgE, which is characteristic of mature mast cells. When grown in culture these precursors give rise only to mast cells (Rodewald et al. 1996). The in vivo evidence arises from studies using specific mouse mutants that lack mast cells. When these mice are linked in a parabiosis experiment to normal mice, they acquire mast cells from the donor (Kitamura et al. 1979).

It is generally believed that mast cell precursors enter tissues and organs. They are particularly numerous in regions of the body which are at an interface with the outside world such as skin and gut. Relatively little is known about mast cell entrance into the brain. In rats (Lambracht-Hall et al. 1990) and doves (Zhuang et al. 1993a; 1999) during the neonatal period, mast cells enter the brain with the penetrating blood vessels. In the adult mammalian brain they are present in the leptomeninges, around the third ventricle, and in the parenchyma of the thalamus and hypothalamus (Ibrahim 1974; Dropp 1972; Theoharides 1990; Manning et al. 1994). In the thalamus there is some evidence that sex differences occur in the distribution of mast cells (Goldschmidt et al. 1984). How the number of mature mast cells is rapidly altered in adult brains is not known (see below for further discussion of this issue).

The final mature phenotype of a mast cell, whether it contains an amine, such as histamine, a proteoglycan, such as heparin sulfate proteoglycan, or a peptide, such as GnRH, is determined by signals within the microenvironment of the particular tissue (Galli 1990). Mast cells are generally divided into serosal and mucosal subtypes (Enerback et al. 1986). The presence of brain mast cells has been described in the older literature (e.g., see Ibrahim 1974) but only recently have mast cells in this organ (and in dura mater of the brain) been considered as a possibly distinct subtype and a focus of study in both health and disease (Theoharides 1990).

Mast cells can produce and secrete numerous mediators. They can be conceived of as migratory multifunctional glandular cells because they contain an enormous variety of substances that they either store or make when activated. Life history (quiescent or stimulated), in addition to local microenvironment, has a considerable role to play in mast cell production and release of mediators. Among those secretogogues of mast cells that are of interest to neurobiologists are neuropeptides (Yano et al. 1989; Cutz et al. 1978); neurotransmitters, such as nitric oxide (Bacci et al. 1994; Salvemini et al. 1990); adenosine triphosphate (ATP) (Osipchuk and Lahalan 1992); and neuromodulators such as prostaglandins (Levi-Schaffer and Shalit 1989; Urade et al. 1990) (for reviews, see Purcell and Atterwill 1995; Johnson and Krenger 1992; Silver et al. 1996). Mammalian mast cells contain vasoactive intestinal polypeptide (VIP) and SP. SP is also one of the most potent secretogogues of mast cells (Fewtrell et al. 1982). Because they contain many neuroactive molecules, it is logical to assume that mast cells in the brain alter neuronal functioning. Furthermore, as we shall show, gonadal steroid hormones act on mast cells, either directly or indirectly, to activate them, and induce them to release their contents.

Before we can move to this part of the tale, we must determine how and when mast cell numbers in the brain change in different behavioral and endocrine states.

Are Changes in Mast Cell Numbers in the Brain Attributable to Immigration or Changes in Detectability?

Our studies to date have shown that mast cells in the adult medial habenula can change under normal physiological conditions. These cells are defined as mature by numerous criteria: metachromasia to aniline dyes; safranin reactivity in the presence of alcian blue; histamine; secretory granules with complex subgranular patterns; and immunostaining with LR-1, an antibody to GnRH (Silverman et al. 1994). They are present in small numbers in the medial habenula of adult birds housed alone ($n \sim 400$), in smaller numbers in castrated adults ($n \sim 200$), and in dramatically higher numbers ($n \sim 1200$) when birds are allowed to court for 2 hours (Silver et al. 1992; Zhuang et al. 1993b). One of the major questions, then, is how these alterations occur.

In considering the rapid increase in cell number with courtship we have to ask whether these new, mature mast cells migrate into the central nervous system (CNS) de novo or rapidly produce the GnRH-like epitope used in our initial study for identification. While this question cannot be answered definitively until individual cells are observed over time, independent lines of evidence support the migration hypothesis. First, if mast cells were resident in the brain, but did not express GnRH until courtship began, it would be expected that the number of cells stained with general cytochemical markers would not change after courtship. To the contrary, the results demonstrate that the same increase in mast cell number is observed using either GnRH-like immunoreactivity or the general cytochemical markers that are not specific for GnRH content. Second, we can identify immature mast cells in the habenular region in the brain of young animals, even though they have very few secretory granules compared to those seen in the adult. Finally, ultrastructural images show mast cells between the adjacent pia and medial habenula with numerous filamentous surface processes oriented toward the gray matter, suggesting migration. One to two cell diameters into the medial habenula, these processes enclose considerable extracellular space, suggesting an alteration in the local environment to create room for movement. This is particularly informative as mast cells synthesize proteases, including several serine proteases such as trypase, chymase, and carboxypeptidase (Goldstein and Wintrovb 1993; Irani et al. 1989; Stevens et al. 1994; Woodbury et al. 1989). These proteases could alter the brain's extracellular matrix, perhaps increasing the space through which the mast cell can move. Finally, the rate of movement of mast cells within the medial habenula or with the transmigration of mast cells from the ventricular surface or pia into the brain parenchyma is consistent with known rates of mast cell migration (180 μm per hour) in vitro (Thompson et al. 1990).

Just as courtship increases the number of mast cells, gonadectomy results in a significant decline (Zhuang et al. 1993b). As this phenomenon was studied only in long-term orchidectomized animals, we do not yet know when the decrease occurs, relative to the withdrawal of gonadal steroid. We also do not know if this alteration is due to a loss of granular material, to emigration, or to cell death. All of these possibilities must be considered.

The medial habenula of doves attracts mast cells. We next tried to understand what quality of the medial habenula contributes to its being populated by mast cells following a period of courtship behavior, while other regions of the brain do not seem to aquire mast cells. We implanted embryonic (E15) medial habenula on one side and, as a control, embryonic optic lobe into the other side of the lateral ventricle of 4-month-old hosts (Zhuang et al. 1997). At the time of implantation, there were no detectable mast cells in the medial habenula of the embryo. Each host received both types of tissues and was sacrificed 2 to 3 months later. Both the host and transplanted medial habenula had numerous mast cells. In contrast, the transplanted optic lobe did not contain any mast cells. These data suggested that medial habenula can "attract" mast cells regardless of its location and its neural or vascular connectivity. Furthermore, access to the cerebrospinal fluid (CSF) is not the determining factor in mast cell recruitment into CNS tissue. Some factor(s) intrinsic to the medial habenula appears to play a role in attracting and sustaining mast cells. While we do not yet know what this special property of the medial habenula might be, we note that this

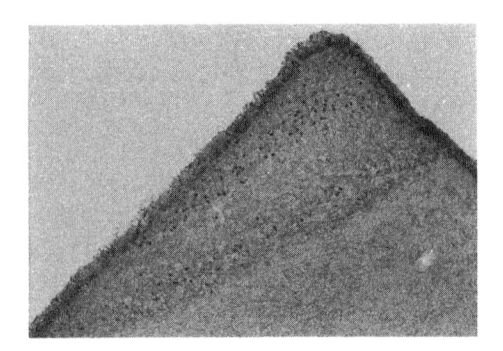

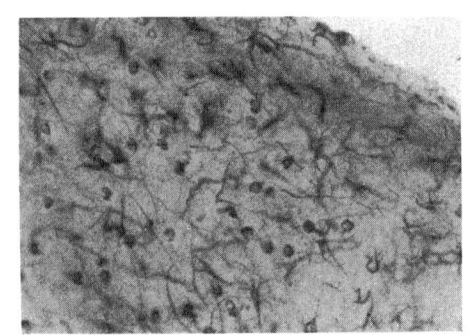

Figure 6.4
Mast cells and glia of the medial habenula.

region is very rich in glia, and that the mast cells are restricted to an area delimited by these glia (figure 6.4). Presumably, cells in this part of the brain have a unique chemoattractant.

Mast Cells and the Nervous System

Blood-Brain Barrier

One of the challenges facing current research is to determine the functional significance of the sudden, behaviorally triggered appearance of mast cells in the medial habenula. This is a question we have begun to tackle. In the periphery, mast cells are well-known for their production and secretion of vasoactive substances (Kiernan 1972; Szolcsanyi 1988) and there have been suggestions that changes in vascular permeability occur in the CNS (Belova and Jonsson 1982; Belova et al. 1991). We asked whether mast cell activation in doves could cause alterations in the permeability of the blood-brain barrier (BBB), specifically in those regions which contain mast cells (Zhuang et al. 1996). We experimentally activated mast cells and examined the effects on the BBB. In experimental animals, the mast cells were activated by injection of a degranulator, a polyamine called compound 48/80 (C48/80). Mast cell degranulators are pharmacological agents that activate mast cells, causing them to release the contents of their granules. To examine the integrity of the BBB we injected the fluorescent protein Evans blue 2 hours after injection of either C48/80 or vehicle. Under normal circumstances, large molecules, such as the tracer Evans blue, cannot leave the blood to enter the brain. Spread of the tracer Evans blue from the blood into the brain means that the BBB is compromised or "open." Control age-matched

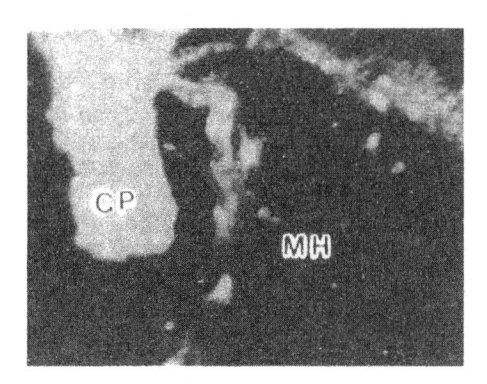

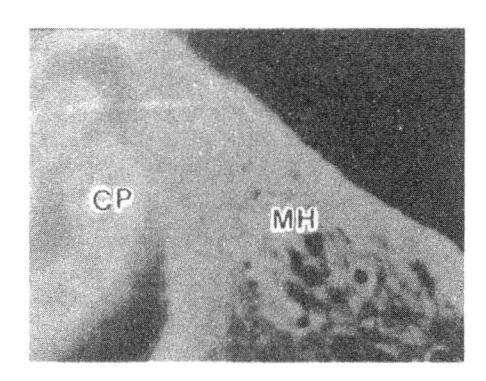

Figure 6.5
The spread of a tracer, Evan's blue, in the medial habenula following mast cell degranulation by compound 48/80 (right) vs. a control, saline-injected bird (left). Note the increased vascular permeability of the protein in the medial habenula following degranulation. The choroid plexus (CP) is labeled in both subjects.

animals were injected with saline followed by Evans blue. All birds were perfused 30 minutes after Evans blue injection.

Measurement of the optical density of the fluorescent signal indicated that C48/80 treatment significantly increased the spread of the tracer in the medial habenula (figure 6.5). In contrast, there was no leakage of tracer into the paraventricular nucleus (PVN), which is located near the ventricle, in either the experimental or control group. There was leakage of Evans blue into the brain around the lateral septal organ, a circumventricular organ that is always leaky, but the extent of leakage did not differ between groups. Degranulation of mast cells after C48/80 treatment was confirmed histochemically and ultrastructurally. The result supports the hypothesis that mast cell activation can open the BBB. These experiments form the background for studies of potential alterations in the BBB associated with activation of mast cells during normal physiological changes occurring during courtship or stress.

Mast cells respond to neurotransmitters, and can thereby act on the brain. In many regions, including the CNS, mast cells appear to be innervated or in close approximation to nerve terminals (Muller and Weihe 1991; Newson et al. 1983; Stead et al. 1990; Stead 1992); they form close contacts with sympathetic neurons in vitro and in vivo (Blennerhassett et al. 1991; Coderre et al. 1989). Mast cells are found within the cerebral vasculature and in nerves during regeneration (Olsson 1968; Dimitriadou et al. 1991), as well as in skin (Arizono et al. 1990; Olsson 1971). Mast cells are in close proximity to SP nerve terminals in the diaphragm and mesentery (Crivellato et al. 1991; Skofitsch et al. 1985). Mast cells can be stimulated to release their granular contents, including histamine, by neuropeptides (Church et al. 1991; Jansco et al. 1968; Kiernan 1972; Ratzlaff et al. 1992; Fewtrell et al. 1982).

Of particular interest to the study of mast cell and neuroimmune function and reproduction is the clinical observation that histamine secretion from mast cells, followed by cutaneous anaphylaxis, can be induced with GnRH, GnRH agonists and antagonists, although the most profound effects are seen with the latter (Karten and Rivier 1986; Rivier et al. 1986; Sundaram et al. 1988). Whether this is true for avian mast cells is not known. If true, however, the peptide could be a component of a positive feedback loop wherein mast cell secretion is hyperstimulated by its own mediator release. This scenario has been demonstrated for connective tissue mast cells which can make and release nerve growth factor (NGF) (Leon et al. 1994) and NGF stimulates mast cell degranulation (Horigome et al. 1993). In addition, activated mast cells are known to secrete molecules such as ATP, which results in the recruitment of additional mast cells into a specific region (Osipchuk and Cahalan 1992). One might envision that the initiation of courtship stimulates the release of GnRH from axon terminals near the medial habenula and nearby CSF, resulting in increasing mast cell number, and possibly more GnRH release from mast cells.

Mast Cell Secretion Can Alter Neurons

We have found clues in the literature to a number of ways that mast cells might have signaling properties that alter neurons. These, of course, are of interest to us, as we have evidence that exposure to gonadal steroids increases brain mast cell numbers, and results in activation of mast cells. As mentioned above, mast cells can secrete ATP as a self-signaling molecule. ATP can also be a neurotransmitter via its ligand-gated channel and these channels occur in the medial habenula (Edwards et al. 1992; Edwards and Gibb 1993). The ligand receptor (channel) is an excitatory one for medial habenular neurons.

Another hint comes from histamine which, via the H_1 receptor, plays a significant role in reproduction by directly or indirectly stimulating GnRH release (Miyake et al. 1987; Noris et al. 1995). Whether this involves mast cells or only histaminergic (neuronal) synapses is not yet known.

A direct role for mast cells in altering neuronal properties at a cellular level has been documented extensively in the periphery in response to allergen. Similar processes may reasonably be expected to occur in the brain. A fine example of such studies involves examination of the nodose ganglia taken from animals that have been made allergic to ovalbumin, compared with the same ganglia taken from control animals and studied in vitro. The nodose ganglion after allergen challenge contains mast cells. Upon exposure to allergen after excision and placement in a dish, these mast cells degranulate. Consequently, nodose ganglion neurons show a multitude of excitability changes, including membrane depolarization and changes in resting membrane conductance (Undem 1993). In this series of studies, Weinrich and colleagues looked at alterations in SP sensitivity (Weinrich et al. 1997). In control ganglia there is no discernible change in membrane properties or Ca^{2+} influx when SP is introduced. However, when ganglia are obtained from immunized animals and then challenged with allergen in vitro, 83% of the neurons respond to SP by depolarization and Ca^{2+} influx. The response to SP is very rapid, being maximal 30 minutes after allergen application, and long-lasting, being sustained for 8 hours in intact ganglia and 3.5 days in dispersed cells.

This mast cell–induced alteration in neuronal sensitivity is mediated by the unmasking of NK-2 tachykinin receptors via a post-translational modification of existing, silent receptor molecules. What is the functional significance of this change? Many of the nodose ganglion cells produce SP and NKA (Fischer et al. 1996) and it is known that tachykinins are released from vagal afferents in the airway (Canning and Undem 1993; Lundberg 1995). The authors postulate that if functional NK-2 receptors are unmasked at these terminals, they could subserve an excitatory autoreceptor function (Weinrich et al. 1997). This work gives us clues about the possible long-lasting consequences of mast cell activation in the brain of the type we observe following a period of sexual behavior and secretion of gonadal steroids.

Mast Cells and Hormones

Hormones Influence Mast Cell Secretion

Hormonal activation of mast cell function is important because in vitro experiments have demonstrated that mast cell secretion can be altered by the hormonal milieu. Amine secretion can be stimulated by progesterone in the rodent (Vliagoftis et al. 1990) and SP induced release is enhanced by estrogen and inhibited by tamoxifen (Vliagoftis et al. 1992). Estrogen receptors have been reported in mast cells isolated from the peritoneum of rats (Vliagoftis et al. 1992).

We have evidence that gonadal steroids play a role in mast cell function (Zhuang et al. 1993). The number of mast cells in the medial habenula (and therefore the potential to alter neuronal outflow) is increased by those behavioral (courtship) and physiological (intact vs. castrate) states that result in altered gonadal steroid levels. Light microscopic images of mast cells in the medial habenula of a female after 7 days of courtship indicate that substantial degranulation occurs in the CNS (figure 6.6). Experiments using electron microscopic visualization for direct evaluation of mast

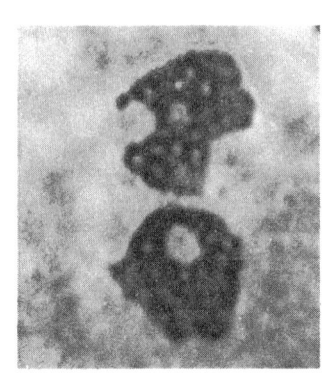

Figure 6.6
Mast cells in the medial habenula of a female dove that had courted for 7 days. The sections was processed for immunocytochemistry using the LR-1 antiserum. Note the presence of many empty appearing granules (∗). N, nucleus.

cell degranulation (as evidenced by morphological criteria) support the notion of the direct effects of gonadal steroids (Wilhelm et al., in submission). Restated, mast cells observed in animals treated with physiological doses of estrogen or testosterone show morphological signs of degranulation and refilling of emptied granules.

Hormonal Influences on Mast Cell Number and Distribution

There are many examples in the literature of alteration in mast cell number in peripheral organs in various endocrine states. The Harderian gland is sexually dimorphic in some species and gonadal steroid treatments can alter the number of mast cells in this gland. For example, estradiol treatment results in an increase in connective tissue mast cells in the Harderian gland of the female toad *Bufo viridis* (Minucci et al. 1994). Gonadal steroids also have important effects on mast cells of the Harderian gland of the Syrian hamster. Menendez-Pelaez and colleagues (Menendez-Pelaez et al. 1992) reported that testosterone implantation decreased mast cell number within 4 hours in female Syrian hamster harderian glands and at least some of this drop in detectability is due to mast cell degranulation (Mayo et al. 1997). In many of these studies, the assumption is made that an increase in mast cell number is due to proliferation, and that a decrease in mast cell number is attributable, axiomatically, to degranulation. (Upon degranulation, the loss of their granules makes mast cells hard to find).

Mast cell location also varies within reproductive organs. Changes in the number of mast cells within ovarian components has been described for the rat (Gaytan et al. 1991). There is an increase in the number of cells in the medulla, cortex, and bursal cavity on proestrus followed by a rapid decline by estrus. The latter compartment provides communication between the periovarial space and the peritoneal cavity. An increase in mast cell number in the ovary has also been reported for humans at the time of ovulation (Balboni 1976). In this case the cells are described as having the appearance and staining properties of mature mast cells rather than immature or degranulated ones. Balboni suggests that this increase might be due to migration of cells from other sources, for example, from the region of the hilum where there are numerous mast cells associated with the large blood vessels (for review, see Stern and Coulam 1992).

Mast cells have been described in the testes of rat (Gaytan et al. 1986, 1989) where they enter during postnatal life as immature (alcian blue–positive) cells and complete their maturation in situ (Gaytan et al. 1990). These authors also noted (Gaytan et al. 1986) that the testis of neonatally estrogen-treated male rats contained more mast cells than that of control rats. In hamsters kept on short or long days, mast cells were found in the testicular capsule and intratesticular blood vessels. Histamine could stimulate the production of testosterone via an H_1 receptor in the regressed, short-day animals but not in the long-day animals. However, in both groups histamine enhanced the effect of human chorionic gonadotropin (hCG) on testosterone production (Mayerhofer et al. 1989).

How do changes in mast cell numbers occur? The possibilities include immigration of mature or immature cells followed by their subsequent emigration or death. Mast cells could also proliferate locally. These possibilities are not mutually exclusive and the mechanism(s) may be tissue-specific or specific to the physiological or pathological change that elicited the alteration in mast cell number. There are now experimental methods available to test each of these possibilities. Studies must now proceed to examine how these immune system cells are changed by gonadal steroid secretion, and how mast cell secretions in turn act on their local environments in the brain and in gonads.

Hematopoeitic Cells in Normal Brain Physiology

Brain Secretions and Brain Drain

The work to date has laid out the framework for a new direction of inquiry. There is now a firm experimental basis for studying how the immune system cell—the mast cell—alters and is altered by gonadal steroid secretions. The brain has long been considered an "immune-privileged" site. This implied that effector cells and molecules of the immune system could not enter the brain. It was assumed that immune-privileged status was beneficial to the organism in that an inflammatory response in the brain might be more damaging than the antigen that produced the insult. This was attributed to the fact that the brain lacks a conventional lymphatic system. Its absence was believed to prevent the possibility of draining immunological signals from the brain to reach regional lymph nodes and spleen. On the afferent side, the presence of the BBB was assumed to prevent the entry of immune signals. This passive view of immune privilege was widely held until more recent studies indicated that both of these communication routes (to and from the brain) are in fact available to the hematopoietic system.

In most tissues of the body, interstitial fluid is produced from plasma by the process of passive filtration across the capillary endothelium. It is cleared from the interstitial spaces either by filtration (osmotic pressure) back into capillaries or by drainage into the lymphatic vessels. Lymphatic fluids contain proteins, ions, and small molecules such as amino acids, but lack the formed elements of blood (red blood cells, platelets, etc.). CSF is not the same as interstitial fluid (ISF). Its production is not passive but is an active energy requiring secretion of the epithelial cells of the choroid plexus. Its ionic composition differs from either blood or ISF and under normal circumstances the level of protein is also very low.

In contrast to previous beliefs, the afferent connection of the immune system entails the constant surveillance of the brain by the passage of cells of hematopoietic origin from the blood into the brain. Activated T cells and B cells traveling in the cerebral or spinal blood supply move across the endothelium and other components of the BBB. Not finding antigen they return to the circulation (Hickey, 1990, 1991; Hickey et al. 1991). This indicates that the brain is not a protected organ in which communication with lymphocytes is prevented. The brain actively participates in communication with the lymphatic and hematopoietic system. The efferent connection consists of the drainage from brain tissue to perivascular spaces (termed Virchow-Robin spaces in the older literature) and thence to the CSF, where it reaches cervical lymphatics and spleen (figure 6.7). The ISF of the brain differs from that of the rest of the body. The brain and spinal cord are surrounded by the CSF, and there are no conventional lymphatics in the brain. The ISF represents 10% to 20% of brain weight. It is produced by regulated flow of ISF from brain capillaries (and not through arteries or veins) to the interstitial spaces between neural elements. In addition, there is (passive) bulk flow from the CSF into brain, as ependymal cells lining the ventricle do not form a barrier to fluid flow. This fluid then drains via the perivascular spaces (described below) (Cserr and Knopf 1997).

The perivascular space is a zone that surrounds cerebral arterioles and small veins through which ISF flows, either to the CSF or to the arachnoid space. Flow through the perivascular space can be experimentally demonstrated. Tracers (e.g., Evans

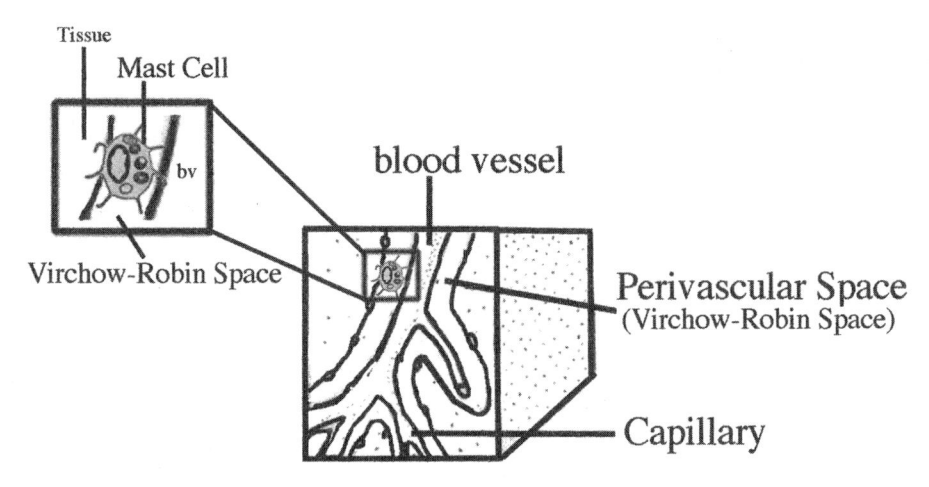

Figure 6.7
Virchow-Robins spaces: flow of interstital fluid in the brain.

blue), injected either into the ventricular system or within the brain proper, rapidly gain access to the perivascular space, while such proteins are excluded from the brain if injected into the blood. Fourteen percent to 47% of proteins injected into the brain can be collected from cervical lymphatics, revealing substantial drainage of brain ISF (see review in Cserr and Knopf 1997).

In mammalian and nonmammalian species, mast cells are located in perivascular spaces on the brain side of the BBB. This is especially true of thalamic mast cells. In some brain regions, mast cells are not associated with blood vessels, but lie directly in the parenchyma (Silverman et al. 1994; Zhuang et al. 1993a; 1999). Thus, mast cells are poised to signal the vascular and lymphatic and nervous tissues. On the sensory side, they can serve to detect antigens. Release of mast cell mediators can influence the local environment, including effects on neurons. There is also evidence of efferent modulation of mast cell function by the brain in that neurotransmitters can activate mast cells.

Current observations suggest a novel relationship between a component of the immune network and specific regions of the CNS. To visualize what the mast cells might be doing, the following image is offered: mast cells appear to function as a unicellular gland acting as a "mobile miniwarehouse cum manufacturing plant." A just-in-time delivery system results from the cell's ability to alter synthetic strategies upon activation, and by its mobility. A warehouse function is permitted by numerous tissue-specific mast cell mediators. A recruiter function derives from reciprocal mast cell interactions (direct or indirect) with other elements of the hematopoietic and nervous systems. In sum, mast cells are poised to amplify a wide variety of tissue responses. They are activated by gonadal steroids—increasing in numbers in certain locations (brain and gonadal site), and releasing their granular contents locally. The mandate for the future is to clarify the consequence of this neuro-immune-endocrine interaction.

Acknowledgments

We thank Anne K. Sutherland for her help in the preparation of the manuscript and figures. The research was supported by NSF (R.S.), NIMH 23980 (R.S.), and NIMH/NINDS 54088 (A.J.S.).

References

Ader R, Cohen N (1982). Behaviorally conditioned immunosuppression and murine systemic lupus erythematosus. Science 215:1534–1536.

Amoss M, Burges R, Blackwell R, Vale W, Fellows R, Guillemin R (1971). Purification, amino acid composition and N-terminus of the hypothalamic luteinizing hormone releasing hormone factor (LRF) of ovine origin. Biochem Biophys Res Commun 44:205–210.

Anonymous (1970). Effects of sexual activity on beard growth on man. Nature 226:869–870.

Arizono N, Matsuda S, Hattori T, Kojima Y, Maeda T, Galli SJ (1990). Anatomical variation in mast cell nerve associations in the rat small intestine, heart, lung, and skin: similarities of distances between neural processes and mast cells, eosinophils, or plasma cells in the jejunal lamina propria. Lab Invest 62:626–634.

Bacci S, Arbi-Riccardi R, Mayer B, Rumio C, Borghi-Cirri MB (1994). Localization of nitric oxide synthase immunoreactivity in mast cells of human nasal mucosa. Histochemistry 102:89–92.

Balboni GC (1976). Histology of the Ovary. London: Academic Press.

Belova I, Jonsson G (1982). Blood-brain barrier permeability and immobilization stress. Acta Physiol Scand 116:21–29.

Belova I, Sharabi Y, Danon Y, Berkenstadt H, Almog S, Mimouni-Bloch A, Zisman A, Dani S, Atsmon J (1991). Survey of symptoms following intake of pyridostigmine during the Persian Gulf War. Isr J Med Sci 27:656–658.

Blennerhassett MG, Tomioka M, Bienenstock J (1991). Formation of contacts between mast cells and sympathetic neurons in vitro. Cell Tissue Res 265:121–128.

Boler J, Enzmann F, Folkers C, Bowers Y, Schally AV (1969). The chemical identity and hormonal properties of the thyrotropin releasing hormone and pyroglutamyl-histidyl-proline amide. Biochem Biophys Res Commun 37:705–710.

Canning BJ, Undem BJ (1993). Relaxant innervation of the guinea pig trachealis: demonstration of capsaicin-sensitive and -insensitive vagal pathways. J Physiol (Lond) 460:719–739.

Carlson HC, Hacking MA (1972). Distribution of mast cells in chicken, turkeys, pheasant and quail and their differentiation from basophils. Avian Dis 16:574–577.

Chand N, Eyre P (1978). Rapid method for basophil count in domestic fowl. Avian Dis 22:639–645.

Cheng MF (1979). Progress and prospects in ringdove research: A personal view. *In* Advances in the Study of Behavior (Rosenblatt JS, Hinde RA, Beer C, Busnel M-C eds), vol 9, pp. 97–130. New York: Academic Press.

Church MK, el-Lati S, Caulfield JP (1991). Neuropeptide-induced secretion from human skin mast cells. Int Arch Allergy Appl Immunol 94:310–318.

Coderre TJ, Basbaum I, Levine JD (1989). Neural control of vascular permeability: interactions between primary afferents, mast cells, and sympathetic efferents. J Neurophysiol 62:48–58.

Crivellato E, Damiani D, Mallardi F, Travan L (1991). Suggestive evidence for a microanatomical relationship between mast cells and nerve fibres containing substance P, calcitonin gene related peptide, vasoactive intestinal polypeptide, and somatostatin in the rat mesentery. Acta Anat (Basel) 141:127–131.

Cserr H, Knopf PM (1997). Cervical lymphatics, the blood-brain barrier and immunoreactivity of the brain. *In* Immunology of the Nervous System (Keane RW, Hickey WF eds.) Oxford: Oxford University Press.

Cutz E, Chan W, Track NS, Goth A, Said SI (1978). Release of VIP in mast cells by histamine liberators. Nature 275:661–662.

Dimitriadou V, Buzzi MG, Moskowitz MA, Theoharides TC (1991). Trigeminal sensory fiber stimulation induces morphological changes reflecting secretion in rat dura mater mast cells. Neuroscience 44:97–112.

Djuric VJ, Bienenstock J (1993). Learned sensitivity. Ann Allergy Asthma Immuno 71:7833–7835.

Dropp JJ (1972). Mast cells in the central nervous system of several rodents. Anat Rec 174:227–238.

Edwards FA, Gibb AJ (1993). ATP—a fast neurotransmitter. FEBS Lett 325:86–89.

Edwards FA, Gibb AJ, Colquhoun D (1992). ATP receptor–mediated synaptic currents in the central nervous system. Nature 359:144–147.

Emanuele NV, Emanuele MA, Tentler J, Kirsteins L, Azad N, Lawrence AM (1990). Rat spleen lymphocytes contain an immunoactive and bioactive luteinizing hormone–releasing hormone. Endocrinology 126:2482–2486.

Enerback L, Miller HRP, Mayrhofer G (1986). Methods for the identification and characterization of mast cells by light microscopy. *In* Mast Cell Differentiation and Heterogeneity (Befus AD, Bienenstock J, Denburg JA, eds), pp 405–417. New York: Raven Press.

Fewtrell CMS, Foreman JC, Jordan CC, Oehme P, Renner H, Stewart JM (1982). The effects of substance P on histamine and 5-hydroxytryptamine release in the rat. J Physiol 330:393–411.

Fischer A, McGregor GP, Saria A, Philippin B, Kummer W (1996). Induction of tachykinin gene and peptide expression in guinea pig nodose primary afferent neurons by allergic airway inflammation. J Clin Invest 98:2284–2291.

Flaris NA, Kensmore TL, Molleston MC, Hickey WF (1993). Characterization of microglia and macrophages in the central nervous system of rats: definition of the differential expression of molecules using standard and novel monoclonal antibodies in normal CNS and in four models of parenchymal reaction. Glia 7:34–40.

Galli SJ (1990). New insights into "the riddle of the mast cell": micro-environmental regulation of mast cell development and phenotypic heterogeneity. Lab Invest 62:5–33.

Gauchat JF, Henchoz S, Mazzei G, Aubry JP, Brunner T, Blasey H, Life P, Talabot D, Flores-Romo L, Thompson J, Kishi K, Butterfield J, Dahinden C, Bonnefoy J-Y. (1993). Induction of human IgE synthesis in B cells by mast cells and basophils. Nature 365:340–343.

Gaytan F, Bellido C, Lucena MC, Paniagua R (1986). Increased number of mast cells in the testis of neonatally estrogenized rats. Arch Androl 16:175–182.

Gaytan F, Carrera G, Pinilla L, Aguilar R, Bellido C (1989). Mast cells in the testis, epididymis and accessory glands of the rat: effects of neonatal steroid treatment. J Androl: 351–358.

Gaytan F, Bellido C, Aceitero J, Aguilar C, Sanchez-Criado JC (1990). Leydig cell involvement in the paracrine regulation of mast cells in the testicular interstitium of the rat. Biol Reprod 43:665–671.

Gaytan F, Aceitero J, Bellido C, Sanchez-Criado JE, Aguilar E (1991). Estrous cycle–related changes in mast cell numbers in several ovarian compartments in the rat. Biol Reprod 45:27–33.

Goldschmidt RC, Hough LB, Glick SD, Padawer J (1984). Mast cells in rat thalamus: nuclear localization, sex difference and left-right asymmetry. Brain Res 323:209–217.

Goldstein SM, Wintroub BU (1993). Mast cell proteases. *In* The Mast Cell in Health and Disease. (Kaliner MA, Metcalfe DD, eds), pp 343–380. New York: Marcel Dekker.

Green JD, Harris GW (1947). Observation of the hypophysio-portal vessels of the living rat. J Physiol (Lond) 108:359–361.

Halasz B, Pupp L (1965). Hormone secretion of the anterior pituitary after physical interruption of all nervous pathways to the hypophysiotropic area. Endocrinology 77:533–562.

Hickey WF (1990). T-lymphocyte entry and antigen recognition in the central nervous system. *In* Psychoneuroimmunology, 2nd ed (Ader RA, Feletn D, Cohen N, eds). New York: Academic Press.

Hickey WF, (1991). Migration of hematogenous cells through the blood-brain barrier and initiation of CNS inflammation. Brain Pathol 1:97–105.

Hickey WF, Hsu BL, Kimura H (1991). T-lymphocyte entry into the central nervous system. J Neurosci Res 28:254–260.

Horigome K, Pryor JC, Bullock ED, Johnson EM. Jr (1993). Mediator release from mast cells by nerve growth factor. Neurotrophin specificity and receptor mediation. J Biol Chem 268:14881–14887.

Ibrahim MZM (1974). The mast cells of the mammalian central nervous system. Part I. Morphology, distribution, and histochemistry. J Neurol Sci 21:431–478.

Irani AMA, Bradford TR, Kepley CL, Schechter NM, Schwartz LB (1989). Detection of MCT and MCTC types of human mast cells by immunohistochemistry using new monoclonal anti-tryptase and anti-chymase antibodies. J Histochem Cytochem 37:1509–1515.

Jansco N, Jansco-Gabor A, Szolcsanyi J (1968). The role of sensory nerve endings in neurogenic inflammation induced in human skin and in the eye and paw of the rat. Br J Pharmacol Chemother 32:32–41.

Johnson D, Krenger W (1992). Interactions of mast cells with the nervous system—recent advances. Neurochem Res 17:939–951.

Karten HJ, Rivier JE (1986). Gonadotropin releasing hormone analog design. Structure-function studies toward the development of agonists and antagonists: rationale and perspective. Endocr Rev 7:44–65.

Kiernan JA (1972). The involvement of mast cells in vasodilation due to axon reflex in injured skin. Q J Exp Physiol 57:311–317.

Kitamura Y, Go S, Hatanaka K (1978). Decrease of mast cells in W/Wv mice and their increase by bone marrow transplantation. Blood 52:447–452.

Kitamura Y, Hatanaka K, Murakami M, Shibata H (1979). Presence of mast cell precursors in peripheral blood of mice demonstrated by parabiosis. Blood 53:1085–1088.

Konishi M, Emlen ST, Ricklefs RE, Wingfield JC (1989). Contributions of bird studies to biology. Science 246:465–472.

Lambracht-Hall M, Dimitriadou V, Theoharides TC (1990). Migration of mast cells in the developing rat brain. Brain Res Dev 56:151–159.

Leon A, Buriani A, Dal Toso R, Fabris M, Romanello S, Aloe L, Levi-Montalcini R (1994). Mast cells synthesize, store, and release nerve growth factor. Proc Natl Acad Sci U S A 91:3739–3743.

Levi-Schaffer F, Shalit M (1989). Differential release of histamine and prostaglandin D_2 in rat peritoneal mast cells activated with peptides. Int Arch Allergy Appl Immunol 90:352–357.

Lundberg JM (1995). Tachykinins, sensory nerves and asthma—an overview. Can J Physiol Pharmacol 73:908–914.

MacQueen G, Marshall J, Perdue M, Siegel S, Bienenstock J (1989). Pavlovian conditioning of rat mucosal mast cells to secrete rat mast cell protease II. Science 243:83–85.

Manning KA, Pienkowski TP, Uhlrich DJ (1994). Histaminergic and non-histamine–immunoreactive mast cells within the cat lateral geniculate complex examined with light and electron microscopy. Neuroscience 63:191–206.

Marathias K, Lambracht-Hall M, Savala J, Theoharides TC (1991). Endogenous regulation of rat brain mast cell serotonin release. Int Arch Allergy Appl Immunol 95:332–340.

Mayerhofer A, Bartke A, Amador AG, Began T (1989). Histamine affects testicular steroid production in the golden hamster. Endocrinology 125:2212–2214.

Mayo JC, Sainz RM, Antolin I, Uria H, Menendez-Pelaez A, Rodriguez C (1997). Androgen-dependent mast cell degranulation in the Harderian gland of female Syrian hamsters: in vivo and organ culture evidence. Anat Embryol 196:133–140.

Mayrhofer G (1980). Fixation and staining of granules in mucosal mast cells and intraepithelial lymphocytes in the rat jejunum, with special reference to the relationship between the acid glycosaminoglycans in the two cell types. J Histochem 12:513–526.

McCann SM, Dhariwal PS, Porter JC (1968). Regulation of the adenohypophysis. Annu Rev Physiol 30:589–640.

Menendez-Pelaez A, Mayo JC, Sainz RM, Perez M, Antolin I, Tolivia D (1992). Development and hormonal regulation of mast cells in the Harderian gland of Syrian hamsters. Anat Embryol 186:91–97.

Minucci S, Chieffi Baccari G, Di Matteo L (1994). The effect of sex hormones on lipid content and mast cell number in the Harderian gland of the female toad, *Bufo viridis*. Tissue Cell 26:797–805.

Miyake A, Ohtsuka S, Nishizake J, Tasaka K, Aono T, Tanizawa O, Yamatodani A, Watanabe T, Wada H (1987). Involvment of H_1 histamine receptor in basal and estrogen-stimulated luteinizing hormone-releasing hormone secretion in rats in vivo. Neuroendocrinology 45:191–196.

Muller S, Weihe E (1991). Interrelation of peptidergic innervation with mast cells and ED1-positive cells in rat thymus. Brain Behav Immun 5:55–72.

Newson B, Dahlstrom A, Enerback L, Ahlman H (1983). Suggestive evidence for a direct innervation of mucosal mast cells. an electron microscopic study. Neuroscience 10:565–570.

Noris G, Hol D, Clapp C, Martinez de la Escalera G (1995). Histamine directly stimulates gonadotropin-releasing hormone secretion from GT1-1 cells via H_1 receptors coupled to phosphoinositide hydrolysis. Endocrinology 136:2967–2974.

Olsson Y (1968). Mast cells in the nervous system. Int Rev Cytol 24:27–70.

Olsson Y (1971). Mast cells in human peripheral nerve. Acta Neurol Scand 47:357–368.

Osipchuk Y, Cahalan M (1992). Cell-to-cell spread of calcium signals mediated by ATP receptors in mast cells. Nature 359:241–244.

Patterson PH (1994). Leukemia inhibitory factor, a cytokine at the interface between neurobiology and immunology. Proc Natl Acad Sci U S A 91:7833–7835.

Purcell WM, Atterwill CK (1995). Mast cells in neuroimmune function: neurotoxicological and neuropharmacological perspectives. Neurochem Res 20:521–532.

Ratzlaff RE, Cavanaugh VJ, Miller GW, Oakes SG (1992). Evidence of a neurogenic component during IgE-mediated inflammation in mouse skin. J Neuroimmunol 41:89–96.

Rivier JE, Porter J, Rivier CL (1986). New effective gonadotropin releasing hormone antagonists with minimal potency for histamine release in vitro. J Med Chem 29:1846–1851.

Rodewald H-R, Dessing M, Dvorak AM, Galli SJ (1996). Identification of a committed precursor for the mast cell lineage. Science 271:818–822.

Rowan W (1925). Relationship of light to bird migration and development. Nature 115:494–495.

Salvemini D, Masini E, Anggard E, Mannaioni P, Vane J (1990). Synthesis of a nitric oxide–like factor from L-arginine by rat serosal mast cells. Biophys Biochem Res Commun 169:596–601.

Schally AV, Arimura A, Baker Y, Nair RMG, Matsuo H, Redding, TW, Debeljuk L, White WF (1971). Isolation and properties of the FSH and LH-releasing hormone. Biochem Biophys Res Commun 43:393–399.

Scharrer B (1970). General Principles of Neuroendocrine Communication. New York: Rockefeller University Press.

Silver R (1978). The parental behavior of doves. Am Scientist 66:209–213.

Silver R, Ramos CL, Silverman A-J (1992). Sexual behavior triggers the appearance of non-neuronal cells containing gonadotropin-releasing hormone-like immunoreactivity. J Neuroendocrinol 4:207–210.

Silver R, Zhuang X, Millar RP, Silverman AJ (1993). Mast cells containing GnRH-like immunoreactivity in the CNS of doves. In Avian Endocrinology P (Sharp PJ, ed.), pp 87–89. Bristol, UK: Journal of Endocrinology.

Silver R, Silveman A-J, Vitkovic L, Lederhendler II (1996). Mast cells in the brain: evidence and functional significance. Trends Neurosci 19:25–31.

Silverman A-J, Millar RP, King JA, Zhuang X, Silver R (1994). Mast cells with gonadotropin-releasing hormone-like immunoreactivity in the brain of doves. Proc Natl Acad Sci U S A 91:3695–3699.

Skofitsch G, Savitt JM, Jacobowitz DM (1985). Suggestive evidence for a functional unit between mast cells and substance P fibers in the rat diaphragm and mesentery. Histochemistry 82:5–8.

Stead RH (1992). Innervation of mucosal immune cells in the gastrointestinal tract. Reg Immunol 4:91–99.

Stead RH, Perdue MH, Blennerhassett MG, Kakuta Y, Sestini P, Bienenstock J (1990). The innervation of mast cells. In The Neuroendocrine-Immune Network (Freier S, ed.), pp 19–37. Boca Raton, FL: CRC Press.

Steinman L (1993). Connections between the immune system and the nervous system. Proc Natl Acad Sci U S A 90:7912–7914.

Stern J, Coulam CB (1992). New concepts in ovarian regulation: an immune insight. Am J Reprod Immunol 27:136–144.

Stevens RL, Friend DS, McNeil PH, Schiller V, Ghildyal N, Austen FK (1994). Strain-specific and tissue-specific expression of mouse mast cell secretory granule proteases. Immunology 91:128–132.

Sundaram K, Thau R, Chaudhuri M, Schmidt F (1988). Antagonists of LHRH bind to rat mast cells and induce histamine release. Agents Actions Suppl 25:307–313.

Szolcsanyi J (1988). Antidromic vasodilation and neurogenic inflammation. Agents Actions Suppl 25:4–11.

Tchermitchin AN, Mena MA, Soto J, Unda C (1989). The role of eosinophils in the action of estrogens and other hormones. Med Sci Res 152:265–279.

Theoharides TC (1990). Mast cells: the immune gate to the brain. Life Sci 46:607–617.

Thompson HL, Burbelo PD, Metcalfe DD (1990). Regulation of adhesion of mouse bone marrow–derived mast cells to laminin. J Immunol 145:3425–3431.

Undem BJ, Hubbard W, Weinreich D (1993). Immunologically induced neuromodulation of guinea pig nodose ganglion neurons. J Auton Nerv Syst 44:35–44.

Urade Y, Ujihara M, Horiguchi Y, Igarashi M, Nagata A, Ikai K, Hayaishi O (1990). Mast cells contain spleen-type prostaglandin D synthetase. J Biol Chem 265:371–375.

Vale W, Spiess J, Rivier C, Rivier J (1983). Characterization of a 41-residue ovine hypothalamic peptide that stimulates the secretion of corticotrophin and beta-endorphin. Science 213:1394.

Valsala KV, Jarplid B, Hansen HJ (1985). Distribution and ultrastructure of mast cells in the duck. Avian Dis 30:653–657.

Vliagoftis H, Dimitriadoum V, Theoharides TC (1990). Progesterone triggers selective secretion of 5-hydroxytryptamine. Int Arch Allergy Appl Immunol 93:113–119.

Vliagoftis H, Dimitriadou V, Boucher W, Rozniecki JJ, Correia I, Raam S, Theoharides TC (1992). Estradiol augments while tamoxifen inhibits rat mast cell secretion. Int Arch Allergy Immunol 98:398–409.

Weinrich D, Moore KA, Taylor GE (1997). Allergic inflammation in isolated vagal sensory ganglia unmasks silent NK-2 tachykinin receptors. J Neurosci 17:7683–7693.

Wilhelm M, King B, Silverman A-J, Silver R (1999). Gonadal steroids regulate the number and activational state of mast cells in the medial habenula. (in submission).

Wingfield JC, Whaling CS, Marler PR (1994). Communication in Vertebrate Aggression and Reproduction: The Role of Hormones. New York: Raven Press.

Woodbury RG, Le Trong H, Cole K, Neurath H, Miller HRP (1989). Rat mast cell proteases. *In* Mast Cell and Basophil Differentiation and Function in Health and Disease (Galli SJ, Austen KF, eds), pp 71–79. New York: Raven Press.

Wray S, Key S, Qualls R, Fueshko SM (1994). A subset of peripherin positive olfactory axons delineates the luteinizing hormone releasing hormone neuronal migratory pathway in developing mouse. Dev Biol 166:349–354.

Yano H, Wershil BK, Arizono N, Galli SJ (1989). Substance P–induced augmentation of cutaneous vascular permeability and granulocyte infiltration in mice is mast cell dependent. J Clin Invest 84:1276–1286.

Zhuang X, Machuca H, Silverman AJ, Silver R (1993a). Changes in brain mast cells during development in doves. Soc Neurosci Abstracts 19:949.

Zhuang X, Silverman A-J, Silver R (1993b). Reproductive behavior, endocrine state, and the distribution of GnRH-like immunoreactive mast cells in dove brains. Horm Behav 27:283–295.

Zhuang S, Silverman A, Silver R (1996). Brain mast cell degranulation regulates the blood brain barrier. J Neurobiol 31: 393–403.

Zhuang X, Silverman A-J, Silver R (1997). Mast cell number and maturation in the central nervous system: influence of tissue type, location and exposure to steroid hormones. Neuroscience 80:1237–1245.

Zhuang X, Silverman A-J, Silver R (1999). Distribution and local differentiation of mast cells in the parenchyma of the forebrain. J Comp Neurol 408:477–488.

IC ENVIRONMENTAL CONTEXT: INTEGRATION OF ENVIRONMENTAL FACTORS IN MAMMALS

The next three chapters are concerned with environmental and social influences on reproduction in mammals. In part IIA, on energy balance, we noted the contribution of F. H. Bronson to the integration of ecology with endocrinology. Almost every contributor to this book cites his work. The following three chapters attest to the influence of another mammalogist, Irving Zucker. Of the scientists using an integrative approach to the study of reproduction, Zucker's influence may reach the furthest because it comes from the importance of his scientific contributions, as well as from the impressive number of students and postdoctoral research associates that benefited from his mentorship and went on to successfully employ the integrative approach to the study of reproduction. No fewer than five chapter authors of *Reproduction in Context* are first-generation academic offspring of Irving Zucker: George Wade, Theresa Lee, Michael Gorman, Randy Nelson, Laura Smale, and Emilie Rissman. Several other authors, including Kay Holekamp, acknowledge, with affection, Zucker's influence as a dissertation committee member, collaborator, consultant, and friend. Most of the rest of us can proudly trace, through our own mentors, our academic lineage to Irving Zucker. It is difficult to think of another reproductive biologist of whom this is true.

Zucker's influence is obvious, as Irv himself has been unrestrained by the boundaries of particular scientific disciplines. For example, after he codiscovered that the circadian clock was located in the suprachiasmatic nucleus, he and his students moved on to make still other new discoveries such as the importance of other, separate entrainable oscillators (e.g., those that govern meal-entrainable rhythms and those that control circannual rhythms). Prior to his contributions to the field of circadian biology, he worked in classic hormones and behavior. Subsequent to his early contributions to circadian biology he and his students made many diverse and important contributions to the understanding of winter adaptations such as hibernation and daily torpor, responsiveness to short day, and regulation of body weight and energy balance. A common theme throughout most of his work is an integrative approach to understanding mechanistic processes in the context of the whole organism, its environment, and its evolution.

Theresa Lee and Michael Gorman (chapter 7) have contributed a definitive chapter on the importance of hypotheses and experimental manipulations that are informed by knowledge of naturally occurring changes in environmental stimuli experienced by free-living species. For example, prior to their work, the concept of the "critical photoperiod" served as a central guiding concept in mammalian photoperiod research. The critical photoperiod was a hypothetical species-specific length of daylight and melatonin elevation above which reproductive processes were supposedly inhibited. However, in the experiments that led to the critical day length idea, animals were housed for extended periods under long-day photoperiods and then transferred once to short-day photoperiods. Using more natural decreases in day length and duration of melatonin infusion, Lee and Gorman have overturned the critical day length hypothesis by demonstrating that reproductive processes are responsive to relative decreases even when exposed to absolute day length and melatonin durations that are above the "critical" threshold. They have gone on to show a remarkable plasticity in photoperiodic responsiveness that allows animals to integrate other predictive and proximate cues from the environment.

Randy Nelson and Sabra Klein (chapter 8) are pioneering a new synthetic field of research that links reproductive neuroendocrinology with immunology. Silver and Silverman in chapter 6 came upon this link by their own neuroanatomical investigations. Nelson and Klein, in contrast, have demonstrated that the unifying factor that links reproduction and immune function is energy balance. Immunity is compromised during breeding, pregnancy, and lactation, and immune function can be influenced by many of the same environmental, social, and intrinsic factors that affect reproductive function. Nelson and his laboratory group have documented that immune function is energetically expensive, and draws upon energy stores that could otherwise be allocated to other functions. They suggest that individuals may partition resources between the immune function and other biological processes, such as reproduction, growth, or thermogenesis. Some seasonally breeding species display what may be immunological adaptations in an effort to survive harsh environmental conditions. True to the integrative theme of this book, Nelson and Klein have incorporated methodologies and ideas covered in chapter 3 into their exploration of reproductive-immune interactions. Thus, a deeper understanding of chapter 8 might be facilitated by also reading chapter 3.

Chapter 9, by Holekamp and Smale, concerns the factors that determine reproductive success in females. The ecological tradition pioneered by investigators such as Trivers, Emlen, and Wingfield, to name a few, have led to a large body of literature on the factors that determine reproductive success in males (mate choice, male-male competition, etc.). Meanwhile, variance in female reproductive success has received comparatively little attention, and thus the work by Holekamp and Smale begins to add some much needed balance to a male-biased literature. They have examined contextual influences on reproduction in one plural breeder that exhibits extreme variance in female reproductive success, the spotted hyena (*Crocuta crocuta*). True to the integrative theme of this book, they work at the ecological and evolutionary levels, and also at the level of physiological mechanisms mediating variation in reproductive success. From impressive amounts of field data collected over several years, they demonstrate that an interplay between patterns of food availability and agonistic social interactions leads to natural selection that favors large, aggressive choosy females and, in turn, sexual selection has promoted the evolution of male characteristics that differ markedly from those found in other plural breeders. In other words, hyenas are an example of a species in which natural and sexual selection appears to have favored feisty females and meek males.

7 Timing of Reproduction by the Integration of Photoperiod with Other Seasonal Signals

Theresa M. Lee and Michael R. Gorman

The world contains climates that are challenging, inhospitable, and even hostile, and yet animals manage to inhabit almost all of them. The physiological mechanisms that allow animals to live, breed, and raise their offspring to maturity in these habitats are complex and often remarkable. For small mammals, the problem of reproduction in adverse climates is exacerbated by their high surface area-to-volume ratio which makes them particularly susceptible to heat loss. This chapter examines the intricate physiological mechanisms that allow small mammals to maximize reproductive success in environments that are periodically inhospitable because of fluctuations in temperature, humidity, food and water availability, or some unfriendly combination of these factors.

In some parts of the globe, permissive breeding conditions (adequate food and water, good thermogenic conditions, and the presence of a suitable mating cohort) occur at unpredictable intervals. Species that live in these habitats typically mount a rapid reproductive response when conditions become favorable. In desert environments, for example, species such as budgerigars and lungfish initiate reproductive activity within a few days of significant rainfall. As a result they are able to take advantage of the available water and subsequent increase in food for their developing young. The production of offspring in response to rain is not as rapid in mammals as in the above-mentioned species. However, a number of mammalian species, such as California voles and degus, use rainfall as a major, if not the only, environmental cue to time the beginning of the reproductive season (Nelson et al. 1983; Meserve et al. 1995). At the other end of the spectrum are species that live in reliably mild climates or areas with relatively constant energy, nutrient, and water availability. For example, South American cane mice of the Torrid Zone enjoy a warm climate with abundant food and water essentially year-round (Bronson and Heideman 1992). Such conditions allow continual production of offspring.

Most mammalian species living in temperate and arctic climates, as well as some in tropical areas, experience regular seasonal environmental variation in food, water, or temperature that influences reproductive success. These species respond to environmental cues that predict the coming change in conditions that will either support or cease to support reproduction. Thus, at nonequatorial latitudes, the majority of mammals become pregnant at a time of year that will ensure that offspring will be born during a season with mild temperatures and abundant food and water to support rapid growth and maturation (Bronson 1989).

Within these seasonal environments several variables may reliably predict, season after season, the best breeding period of the year. Annual changes in photoperiod are correlated, to varying degrees, with annual changes in temperature, rainfall, and food availability. Photoperiod-driven physiological changes control the onset and cessation of breeding effort in many nonequatorial mammals. Photoperiod is essentially constant from year to year. However, while photoperiod is a good "average" predictor of seasonal change, the exact timing of snow melt, first frost, and appearance of key food resources can vary from year to year by several weeks in a given location, thereby increasing or decreasing the prime reproductive period for the local population. Figure 7.1 provides an example from southern Michigan of the variability in food availability, temperature, and snowfall against a background of annual photoperiod change that does not vary between years. In some years, plants are green and growing with mean daily temperatures well above freezing in November or March. In other years, the temperatures drop below freezing with significant snowfall and little green, growing vegetation available in those same months.

We might therefore expect to find many species that respond flexibly to specific conditions arising from year to year, perhaps using photoperiod to open a window of responsiveness to other immediate environmental variables such as the presence of adequate food, water, temperature, and social group. This flexibility would be more likely to be adaptive for short-lived species such as rodents, which breed only within a single breeding season and produce, on average, only one or two litters. As expected, for species such as mice, voles, and deer mice, the year-to-year variation in the onset or end of suitable reproductive conditions may exceed 6 weeks, long enough to produce an additional litter (see figure 7.1). The variation in the onset of winter conditions may be a selection pressure that favors small animals able to take advantage of an unexpectedly extended breeding season.

It might also be predicted that longer-lived species with long gestations (ungulates) or those with short gestations that produce a single pregnancy each year (ground squirrels) would show sensitivity to fewer immediate environmental factors that allow fine-tuning the timing of reproduction. For example, most ungulates have long fixed-duration gestation periods that begin well before the onset of the birth season. As predicted, the young of such species will, in some years, be born earlier or later than is ideal for weather or food conditions. In species in which the gestation period is lengthy, the immediate conditions at the time of mating or implantation have little value in predicting conditions months later at the end of pregnancy. For these species, the potential usefulness of multiple environmental cues for fine-tuning the timing of birth is limited.

It is well established that many environmental and physiological variables related to energetics, including calories and water availability, nutritional quality of food, exercise level, and ambient temperature, interact with season, age, sex, and population level to determine the likelihood of reproduction (for review, see Bronson 1989; Bronson and Rissman 1986; Negus and Berger 1987; Nelson 1987). Factors such as photoperiod, ambient temperature, and some non-nutritive components of the diet are classified as predictive variables, because they provide cues that allow animals to anticipate the change in seasons. Factors such as calories, acute changes in ambient temperature, and social cues are often categorized as proximate variables because their effects are more direct than the effects of predictive variables. The importance of the tradeoff between predictability and opportunism in breeding decisions has been recognized, but most research has examined the importance of predictive variables in isolation from the more proximate variables that may vary from year to year.

We suggest that the interaction of several factors, when viewed in a natural context, increases the range of responses available to the individual to immediate environmental conditions so as to time the onset of puberty and the reproductive effort most effectively. There has been little effort to describe the mechanisms that underlie the integration of these variables across the life of an individual. How does the brain of a female vole integrate a stimulatory day length in April with the presence of lower-than-normal temperatures and very little new growth of grasses? How does a female montane vole integrate a winter photoperiod with the presence of green, growing grasses and an adult cohort of males? In this chapter, we demonstrate that the timing of puberty and subsequent reproduction in small mammals depends on many such interacting variables. Furthermore, we provide examples of the types of experiments that can be designed to reveal the critical variables that are used by animals in their natural habitats. Our results illustrate that erroneous or incomplete conclusions about the mechanisms that underlie reproductive strategies have been drawn because critical natural signals or cues were omitted from experimental designs.

The purpose of this chapter is to examine in greater depth how photoperiod sets a context in which other variables exert their effects. We begin by illustrating, in both adult Siberian hamsters and juvenile meadow voles, that this photoperiodic context is established by an integration of past day length exposures, that is, of photoperiod history. We describe exogenous cues that are generated by gradual, rather than abrupt changes in day length, and endogenous cues generated by the changing duration of the nightly increase in melatonin. Next, we provide evidence for the idea that photoperiod history determines the animal's interpretation of information contained in both the endogenous and exogenous signals. We go on to describe how effects of other seasonal signals, such as energy availability, the presence of stimulatory non-nutritive substances in food, and ambient temperature depend on this photoperiodic

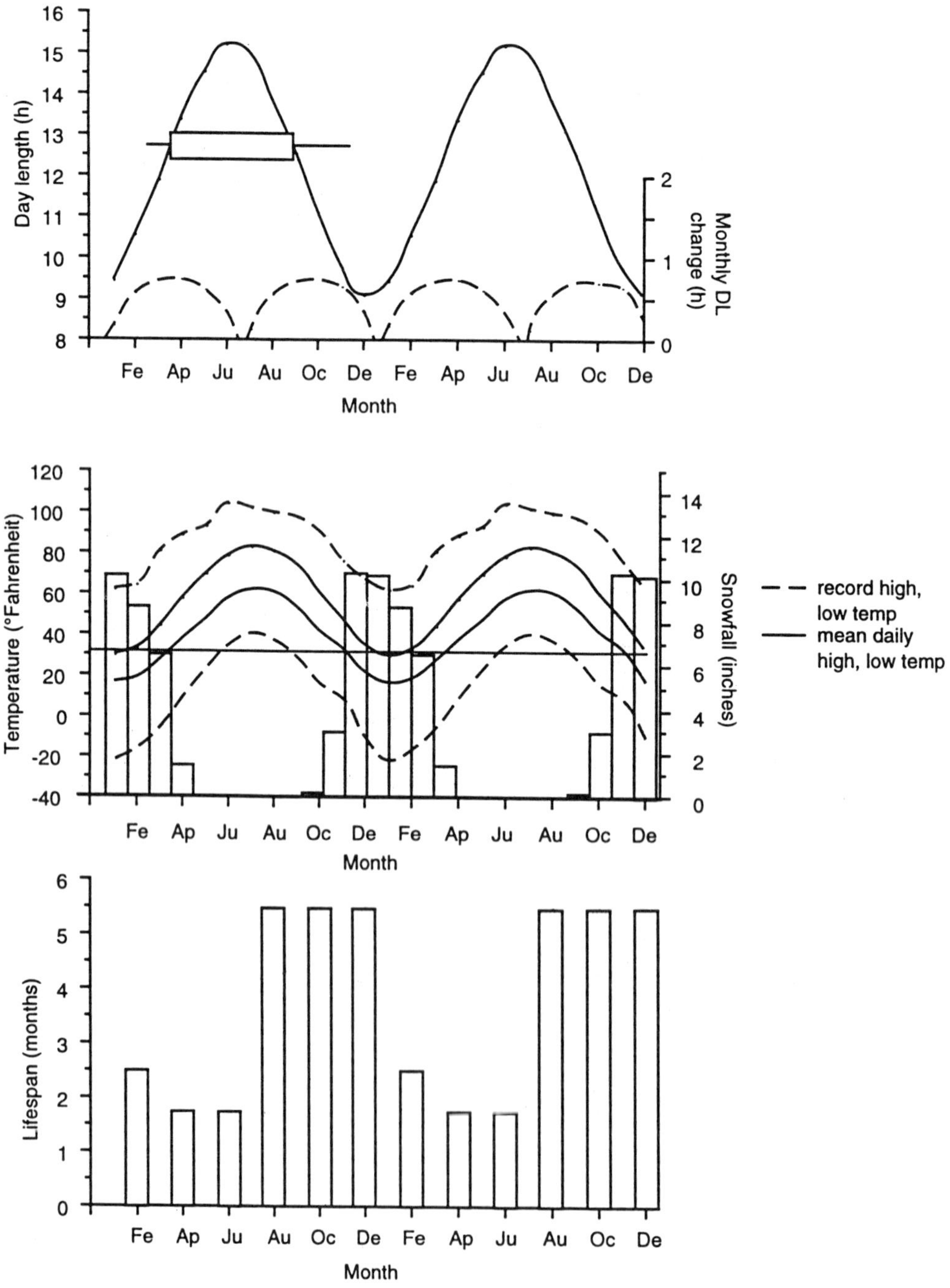

16
15
14
13
12
11
10
9
8
Day length (h)
2
1
0
Monthly DL change (h)
Fe Ap Ju Au Oc De Fe Ap Ju Au Oc De
Month
120
100
80
60
40
20
0
-20
-40
Temperature (°Fahrenheit)
14
12
10
8
6
4
2
0
Snowfall (inches)
record high, low temp
mean daily high, low temp
Fe Ap Ju Au Oc De Fe Ap Ju Au Oc De
Month
6
5
4
3
2
1
0
Lifespan (months)
Fe Ap Ju Au Oc De Fe Ap Ju Au Oc De
Month

historical context. While interactions of multiple variables may seem mechanistically complicated and thus unwieldy, their ecological significance is perhaps more straightforward and compelling; these interactions increase the range of possible responses to environmental conditions so as to time onset of puberty and reproductive effort most efficiently.

Day Length

Natural Changes Contain More Predictive Information Than Static Photoperiods

Day length is often described as the most important predictive environmental cue timing reproduction in nonequatorial regions of the world. This hierarchical ordering of seasonal cues by their presumed significance obscures the importance of the integration of photoperiod with other cues at transitional times of the year. Most often, reproductive transitions for small mammals occur in the spring and fall when day length is changing most rapidly, and when temperature, food, and water are most variable from year to year. The direction of the change in photoperiod at the equinoxes might provide additional information about how the rest of the environment might be interpreted when a static photoperiod would provide little predictive information about whether current conditions are consistent with the onset or end of the breeding season (figure 7.1A).

Early laboratory studies of several rodent species established that in the presence of ample food and water, short day lengths that mimic winter conditions inhibit reproduction, whereas long day lengths that simulate summer conditions stimulate reproduction. A rather sharp stepwise function between inhibitory and stimulatory day lengths led to the concept of the "critical day length." It was assumed that under natural conditions, animals breed until day lengths decrease below the critical day length after the summer solstice. When the daily hours of daylight fall below a critical

Figure 7.1
Example of the relationship between predictable annual environmental change, less predictable annual variations, and average life span for meadow voles born at different times of the year. (A) Double-plotted annual changes in day length and day length changes each month in southern Michigan. Note that the peak rate of day length change occurs during the equinoxes when absolute day length information provides less information about the time of year than do day lengths at the solstices. The horizontal bar represents the variability in average annual availability of fresh growing green vegetation. (B) Average high and low temperatures and extreme highs and lows are shown with average monthly snowfall. (C) Average expected life span of meadow voles born in the field at different times of the year. The longer life span of animals born during August–December, when breeding is declining, is influenced both by weather conditions, which provide significant cover under the snow in the winter, and by the reduced movement of the animals while they remain in an extended prepubertal state.

threshold, reproductive processes are inhibited. In contrast, the onset of the spring breeding season was thought to be cued by a different mechanism termed the "interval timer." After a fixed number of weeks in short day lengths in the laboratory, animals showed spontaneous regrowth of their gonads and began to show normal reproductive behavior. They appeared to become refractory, or insensitive, to the inhibitory effects of short day length (reviewed in Reiter 1980). More recently, the critical day length–refractoriness model has been extended to a more proximate level of analysis in the form of the duration hypothesis of melatonin action. The melatonin duration hypothesis posits that the number of hours of darkness in the daily light-dark cycle is positively correlated with the duration of elevated pineal melatonin secretion. Nightly melatonin signals longer than a critical duration inhibit reproductive function, whereas signals of shorter duration stimulate gonadal activity (reviewed in Goldman 1991).

This simplified view of seasonality was revised after the demonstration, first in montane voles and later in hamsters, that responses to day length depended on the history of prior photoperiodic conditions. Day length of summer solstice for montane voles would be 16 hours of light (16L). Horton (1984, 1985) reasoned that pubertal development should proceed rapidly in pups born in intermediate day lengths, for example, 14L, before the summer solstice, but should be reproductively inhibited if born into the same 14L day length after the summer solstice. The 14L day length when it occurs after the summer solstice is a cue that predicts the beginning of the season when conditions would no longer be favorable for breeding. When litters were gestated in short day lengths (e.g., 12L) to simulate the early breeding season, they responded to postnatal day lengths of 14L with rapid gonadal growth. In contrast, among litters gestated in longer day lengths (e.g., 16L) to simulate the late breeding season, exposure to 14L postnatally inhibited reproductive development. This demonstration of the context-dependence of seasonality mechanisms, a photoperiodic history effect, has since been extended to adults and juveniles of a variety of mammalian species. These results, in showing that changes in day length could be more important than absolute day length in determining photoperiodic responses, raised certain methodological problems that were left largely unaddressed until recently. If the photoperiodic context determines responses to day length, then it seems unlikely that the patterns of day length typically used in laboratory experiments would accurately simulate natural conditions.

Our understanding of photoperiodic mechanisms in mammals primarily derives from studies conducted under highly artificial conditions in which animals are exposed to day lengths or melatonin signals that are invariant for many weeks or months. Furthermore, research subjects are commonly moved from one extreme photoperiodic condition to another in a single day. Both of these conditions distort the pattern of continuously, but incrementally changing day lengths and melatonin durations experienced in nature under which photoperiodic time measurement systems have evolved. Under natural conditions, the time of year is uniquely specified by the latitude, ambient day length, and direction of change in day length (increasing or decreasing). Animals, therefore, have the potential to determine precisely the time of year by knowing the current and previous day length. Day lengths lose their temporal specificity, however, when they are unchanging, as frequently occurs in the laboratory. For example, a 6-week regimen of 12 hours of light per day does not correspond to any time of year for a Temperate Zone mammal (i.e., it cannot be mapped onto figure 7.1). If animals rely on day length as a seasonal cue because of its predictive information, it is ironic that investigators have eliminated a great deal of predictive information inherent in the natural progression of gradually changing day lengths. We sought to investigate the role of this lost temporal information in timing reproduction. In the section that follows, we demonstrate that the signals for temporal control of reproduction are generated by the gradual progression of day length. Second, we argue that unnatural patterns of photoperiod change commonly employed in the laboratory produce unnatural responses that may obscure the interaction of day length with other environmental variables in timing reproduction.

The first task was to use more natural, gradual manipulations of day length to address whether the primary signal controlling reproduction is the change in day length or absolute day length. It is conceivable that photoperiodic history effects might be artifacts of large (e.g., 2- to 4-hour), abrupt transitions in day length and might be inoperative under more natural conditions in which day lengths gradually increased or decreased to intermediate values. Alternatively, photoperiodic history effects might be augmented under naturalistic conditions in which animals experience daily, as opposed to one-time, increases or decreases in day length.

To resolve this issue, male Siberian hamsters were presented with a yearlong pattern of naturally increasing and decreasing day lengths, or simulated natural photoperiods (SNPs). These day lengths were modified so as to neither rise nor fall above or below the putative critical day length of 13L (Gorman and Zucker 1995). In the long SNP group, a long day length varied sinusoidally between 19L and 13L, so that it never fell below the putative critical day length. In the short SNP group, a shorter day length varied between 13L and 7L, so that it never rose above the critical day length. If photoperiod history effects were artifacts of abrupt transitions in day length, then it would be predicted that hamsters exposed to long or short SNPs would resemble animals held chronically in static long or short day lengths, respectively. On the other hand, if the pattern of change in day length determines photoperiodic responses, then both of these groups would be expected to exhibit normal seasonal cycles of gonadal growth, regression, and recrudescence. Results showed that animals on both short and long SNPs, as well as controls exposed to an unmodified SNP, showed cycles of gonadal development, regression, and recrudescence, as well as corresponding changes in body weight. Despite maintenance in day lengths never below the critical day length (long SNPs), these animals underwent complete gonadal regression in their shortest day lengths (13L). Likewise, animals presented with day lengths never above the critical day length (short SNPs) were reproductively stimulated during exposure to their maximum day lengths (the same 13L). These results demonstrated that the gradual pattern of change in day length could override the effects of absolute day length.

Another important result of this experiment was that the seasonal changes in gonadal condition and body weight were not identical between groups. Animals that never experienced day length below 13L spent more time in a stimulated state than animals that never saw day length greater than 13L. The reverse was true for time spent in a regressed state. Because absolute day length was the only factor distinguishing these groups, both pattern of gradual change in day length and absolute day length contribute substantially to photoperiodic responses (Gorman and Zucker 1995).

More detailed analyses of these data revealed several further contradictions to the critical day length model (Gorman 1995). For example, somatic and reproductive maturation was more rapid among animals exposed to shorter, but gradually increasing day lengths than in those kept in what were presumed to be maximally stimulatory, but unchanging, long day lengths. Additionally, as day lengths decreased in simulated late summer, the animals varied in first day length to induce gonadal regression. First day length to induce gonadal regression varied between 15.3L and 12.8L depending on the prior pattern of day length exposure. Finally, the interval between gonadal regression and spontaneous recrudescence in the following season, the presumed fixed duration of the interval timer, varied by 8 weeks between groups. These results establish that, under conditions that incorporate natural changes in day lengths, photoperiodic mechanisms are remarkably plastic in the timing of reproduction. We shall return to this issue.

It should be stressed that photoperiodic stimuli employed in these studies did not accurately simulate all natural conditions found in nature. In traditional photoperiod experiments, animals were moved once from one static photoperiod to the next. The present experiments took one step toward a more natural photoperiod by producing a more gradual change in photoperiod. This was accomplished by changing the timing of "lights on" or "lights off" each successive day. However, in nature, lights do not simply "turn on" in the morning and "turn off" at night. In our experiments, gradual dawn and dusk transitions between light and dark were omitted. Rather our studies were designed to examine the importance of one additional seasonal cue, the day-to-day change in day length, in photoperiodic regulation. The relevance of these studies carried out under more naturalistic conditions remains to be demonstrated.

The fact that gradual changes in day length carry "information" about phase of year (separate from that inherent in absolute day length) raises several important questions. What information does changing day length and, therefore, melatonin duration carry? How do responses to gradual changes in day length compare with those to abrupt changes?

Once photoperiod change was established as a mediator of seasonal responses, we sought to distinguish between two possible mechanisms. The first possibility is that photoperiodic history alters the manner in which the neuroendocrine system responds to a melatonin signal of a given duration. The alternative is that the pattern of melatonin secretion in a particular day length might differ depending on the photoperiodic history. These two possibilities were distinguished by presenting one group of hamsters with long, but gradually decreasing melatonin signals (Gorman and Zucker 1997b). These durations were chosen because they were known to produce gonadal inhibition under static conditions. Other hamsters were given short, but increasing melatonin durations that were known to stimulate gonadal growth under static conditions. If photoperiodic history effects were mediated exclusively at the level of melatonin production, then it would be predicted that the direction of change in melatonin duration should be irrelevant; only absolute melatonin duration would be expected to determine gonadal responses. Alternatively, if responses to a particular melatonin duration depend on prior melatonin patterns, then it would be predicted that the direction of the change in melatonin duration would determine whether or not the treatment is stimulatory, that is, longer-duration melatonin signals would be more stimulatory than shorter signals under naturally changing conditions.

One group of male Siberian hamsters exposed to short day lengths (SD; 10L) from birth were administered a natural pattern of nightly melatonin infusions that gradually decreased in duration from 10 hours at 6 weeks of age to 7.5 hours over the following 12 weeks. All hamsters were maintained in constant light to suppress endogenous melatonin secretion (figure 7.2A). This melatonin pattern corresponds to that presumably generated under gradually increasing day lengths in late winter or early spring. A second group of hamsters kept in long day lengths (LD; 16L) from birth was treated similarly except that melatonin infusions increased in duration from 5 hours at week 0 to 7.5 hours at week 12, mimicking signals produced naturally in late summer or early autumn. Prior to these infusions, endogenous melatonin

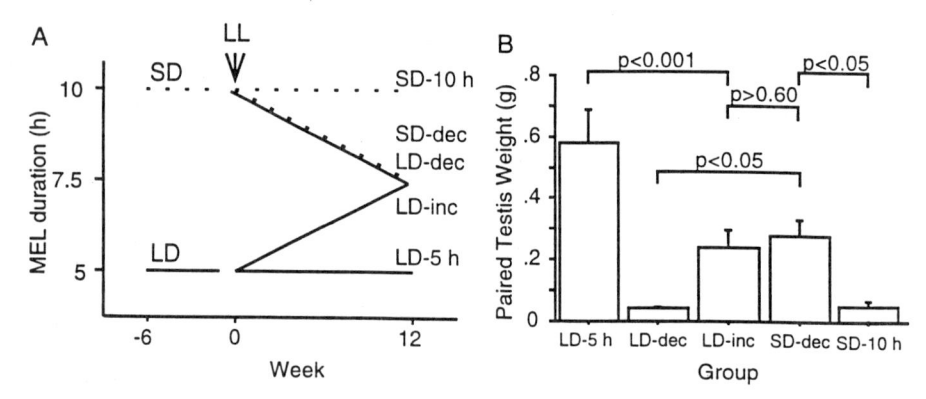

Figure 7.2
(A) Schematic representation of photoperiodic histories and melatonin (MEL) infusion patterns. Siberian hamsters were maintained in long day (LD; 16L) or short day (SD; 10L) from birth. Estimated durations of nocturnal Mel secretion were 5 and 10 hours in LD and SD, respectively. At 6 weeks of age (week 0), hamsters were catheterized for Mel infusion and transferred to constant light (LL) to suppress endogenous Mel secretion. Over the next 12 weeks, Mel (4 ng/hour) was infused as shown (hamsters with SD histories, dashed lines; LD histories; solid lines; see text for additional details). (B) Mean paired testes weight ±SEM after 12 weeks of Mel signals depicted in (A).

durations in SD and LD hamsters were estimated at 10 hours and 5 hours per night, respectively. A third group of hamsters with an LD (summer) history was given the same decreasing pattern of winter-spring melatonin duration as the SD group. This experimental manipulation introduced an abrupt change in melatonin duration (from 5 to 10 hours) that was not present in the other groups.

In the first group, the SD hamsters exposed to long but gradually decreasing melatonin signals showed significant gonadal growth compared with a control group given chronic 10-hour infusions. Stimulation continued through week 12 when melatonin duration was 7.5 hours. In the second group, the LD hamsters exposed to short but naturally increasing melatonin signals underwent significant gonadal regression within 8 weeks. Melatonin duration increased from 5 hours to 6.7 hours and continued with testicular involution through week 12 when melatonin duration was 7.5 hours. In contrast, the third group of LD hamsters that experienced an abrupt lengthening (from 5 to 10 hours) of the duration of melatonin infusions responded to this increase in melatonin duration with complete gonadal regression by the end of the 12 weeks of treatment (figure 7.2B; Gorman and Zucker 1997b).

Two findings and one inference are noteworthy. First, "long," but gradually decreasing melatonin durations stimulated gonadal development only among animals that would likely experience similar patterns in nature. The traditional critical day length hypothesis would predict that durations of this length would inhibit reproduction. To the contrary, under more natural conditions of gradually decreasing day length, gonadal growth occurred when melatonin durations were longer (8.3 hours) than the presumed critical melatonin duration (estimated to be between 7 and 8 hours in this species). Previous experiments showed that under static conditions these same durations were inhibitory. The present results replicate the photoperiodic history effects at the level of melatonin action. Thus, response to a given melatonin duration depends on the previous pattern of change in melatonin duration. This suggests that melatonin signals in natural conditions are read differently than they are under typical artificial laboratory conditions. The mechanism of such a difference is as yet unknown.

Second, the ecologically relevant "meaning" of this natural pattern, revealed by the response of SD hamsters that are exposed to these natural patterns, was obscured in the LD animals that would never have encountered such a signal under natural conditions. Instead, the mismatch between the photoperiodic history and the experimentally produced melatonin signal completely masked the capacity of hamsters to be stimulated by gradually decreasing melatonin durations. Under standard laboratory conditions, experiments routinely expose animals to photoperiodic and melatonin signals inconsistent with natural changes. This critique is analogous to the well-known distinction between physiological and pharmacological concentrations of hormone, albeit in a temporal domain. In photoperiodic systems, the concentration of melatonin over a certain threshold is of little apparent physiological significance. Rather, the amount of time that melatonin durations are elevated above threshold determines photoperiodic responses. To hamsters with an SD–long melatonin duration history, an LD–short melatonin duration signal is a highly unnatural and nonphysiological stimulus. The typical laboratory experiment presents to animals over successive days a change in melatonin duration several orders of magnitude greater than they would ever experience in nature, potentially contributing to an inflated view of the importance of day length. This critique does not deny the utility of the standard laboratory paradigm as an analytic probe with which to investigate seasonality mechanisms. However, aspects of photoperiodic mechanisms that operate under natural conditions (e.g., the stimulatory effects of long but gradually decreasing melatonin durations) may be discernible only under conditions that avoid unnatural and potentially powerful abrupt transitions in day length and melatonin duration.

Interpretation of Day Length During Development Is Influenced by Maternal Photoperiod History

The early work of Horton (1984) demonstrating that the photoperiod history experienced by young montane voles could influence their responses to intermediate day lengths was extended to show that different gestational photoperiods, mediated by the mother, could influence postnatal development (Horton 1985). When pups born to mothers in 10L were cross-fostered at birth to mothers from 16L, and litter and foster mother were moved to 14L, the pups reached puberty rapidly in accordance with their shift from prenatal (10L) to postnatal (14L) photoperiod, but not in accordance with the photoperiod change experienced by the foster mothers (16L to 14L). These findings were soon replicated and expanded in Siberian hamsters (Stetson et al. 1986) and meadow voles (Lee et al. 1987).

Reppert and colleagues had earlier reported that melatonin could cross the placenta to the fetus, so an obvious question was whether the duration of the melatonin signal in the mother determines the response to postnatal photoperiods in the young (Reppert and Klein 1978; Reppert et al. 1979). Weaver and Reppert (1986) demonstrated that the duration of melatonin in the mother during the last 3 days of pregnancy was sufficient to determine the response of neonatal Siberian hamsters to postnatal changes in day length. Lee et al. (1989) reached similar conclusions after pregnant females housed in 14L were injected with melatonin to produce a longer duration signal during late gestation and gave birth to pups whose postnatal development resembled that of SD gestated pups.

These data suggested that maternal photoperiod mediated by duration of the melatonin signal crossing the placenta provides the fetus with a prenatal photoperiod history. Offspring born into ambiguous day lengths can use this change in day length during gestation as a cue to correctly interpret the postnatal day length. Thus, animals that experience an increased day length between gestation and end of lactation undergo rapid development, while the opposite inhibits the timing of puberty. However, these data examined maternal photoperiodic influences on offspring development for births occurring in a static day length. For most small rodents the reproductive season spans a range of photoperiods, frequently from equinox to equinox, and mothers experience different photoperiod histories prior to reproducing at different times of the breeding season. Development, including the timing of puberty, is more varied during the year than the two patterns seen for animals gestated under solstice day lengths (14L and 10L). We therefore asked whether different maternal photoperiod histories were involved in determining the different seasonal patterns of development.

Among meadow voles, most reproduction occurs between March and September, but new litters also are found in the field well into October and even throughout the winter in some years (Christian 1980; Madison and McShea 1987; Mallory and Clulow 1977; Mihok 1984). Typically, litters born between March and August grow rapidly and reach puberty within 4 to 6 weeks, while those born after August grow more slowly and typically delay puberty for several months. Dams at these different times of year have experienced markedly different photoperiod histories prior to pregnancy and during gestation. While reproductive physiology and behavior differ for cohorts of field animals at different seasons (Madison and McShea 1987), without precise knowledge of the season of birth and data from early in development, it is difficult to know whether the animals are responding to ambient conditions or have different developmental patterns established prior to birth. Laboratory studies of cohorts of animals experiencing different natural photoperiod changes also demonstrate developmental timing consistent with the field cohorts (Forger and Zucker 1985), but both the prenatal and postnatal photoperiods differed between groups. Thus, the role of maternal photoperiod history on offspring development remained unclear.

We tested the hypothesis that maternal photoperiod history would influence offspring development by generating females with photoperiod histories that were similar to field animals at different seasons. Ideally we would have examined animals raised under natural conditions, breeding them at different phases of the year and examining offspring development under such conditions. However, to determine the relative importance of prenatal and postnatal maternal photoperiod history required that we have more stable photoperiods than naturally occur. We generated mothers with three different short photoperiod histories. First, female meadow voles moved to 10L (SD) 2 weeks prior to pairing with a male (SD2) mated readily, but by the time offspring were born, 3 to 4 weeks later, they had adjusted to the new photoperiod. When lactation ended they rapidly grew a winter pelage and lost weight to that typical of unmated, SD females (Lee and Zucker 1988). Females placed in SD 8 weeks prior to mating with a male (SD8), had completed the physiological changes associated with short day lengths (pelage growth, weight loss, reduced food intake), and only 25% to 30% produced litters. A third group of SD females was exposed to 26 weeks of SD, and had become photorefractory (insensitive to SD) by the time they were paired with males (SD26). This latter group was not measurably different (weight, food intake, pelage) at the time they were paired with males from females housed in 14L (LD; Lee and Zucker 1988). The physiological and reproductive conditions of these groups of females (and the LD control animals), and the developmental patterns of their pups (when it is known), are similar to those of four groups of field animals: June–July (LD), September (SD2), November–January (SD8), and January–March (SD26) (Madison et al. 1984; Madison and McShea 1987; Lee and Zucker 1988).

The offspring of nonrefractory (i.e., photosensitive) mothers (LD, SD2, SD8) developed as expected. Male pups born in 14L rapidly reached maturity and had large paired testes weights and viable sperm by 49 days of age, and females had maximally heavy uteri (later found to be consistent with graafian follicle development and high estrogen levels). Male pups in 10L, whether born to SD2 or SD8 mothers, had small testes without viable sperm, and female uterine weights were only 25% of those of pups in 14L at the same age (and not consistent with follicular development). Reproductive development of both males and females born to photorefractory dams (SD26) was intermediate to that of LD and SD pups born to photosensitive animals. Male testes were more developed than those of SD2 and SD8 animals, contained viable sperm (although less than LD males), but were approximately 50% lighter than testes from males in 14L. For SD26 females, uterine weight was significantly greater than those of females housed in 10L, but lighter than those of females housed in 14L. Additionally, the growth rate of SD26 pups was greater than that of SD2 or SD8 pups. Thus, SD26 pups reared in the same prenatal and postnatal SD photoperiod as SD2 and SD8 pups, but with mothers having different photoperiod histories, demonstrated strikingly different patterns of development.

While we found that SD26 pups developed more rapidly than SD2 or SD8 pups, development was still significantly slower than in LD pups. However, under natural conditions, photorefractory females in late winter would produce litters that were also exposed to increasing day lengths postnatally. We hypothesized that naturally increasing day lengths might normally interact with prenatal maternal photoperiod history information and hasten development of SD26 pups to parallel that of LD animals. The result in the field might be to produce a young cohort of animals born in mid- to late winter, ready to reproduce as early as the beginning of March.

To test the hypothesis that maternal photoperiod history would influence the response to postnatal changes in day length, animals born to photosensitive mothers in 14L (LD) and 10L (SD2), and pups from photorefractory 10L mothers (SD26) were placed in one of three conditions at weaning (3 weeks). SD2 and SD26 pups were either left in SD or were transferred at weaning to an SNP that increased from 10L (February 1 in southern Michigan) to 11.18L (February 28) over the next 4 weeks. LD controls remained in 14L postnatally. In static day lengths the LD, SD2, and SD26 pups developed as previously described. As expected, the SD2 pups placed in the SNP demonstrated robust increases in paired testes weight or uterine weight compared with SD2 pups that remained in 10L. Much to our surprise, the SD26 pups placed in the SNP demonstrated no increased rate of reproductive development by 49 days of age compared with 10L controls (figure 7.3). It would appear that maternal photoperiod history of photorefractoriness and increasing day lengths during late winter may provide redundant information since SD2 and SD26 pups in SNP did not differ in their reproductive development.

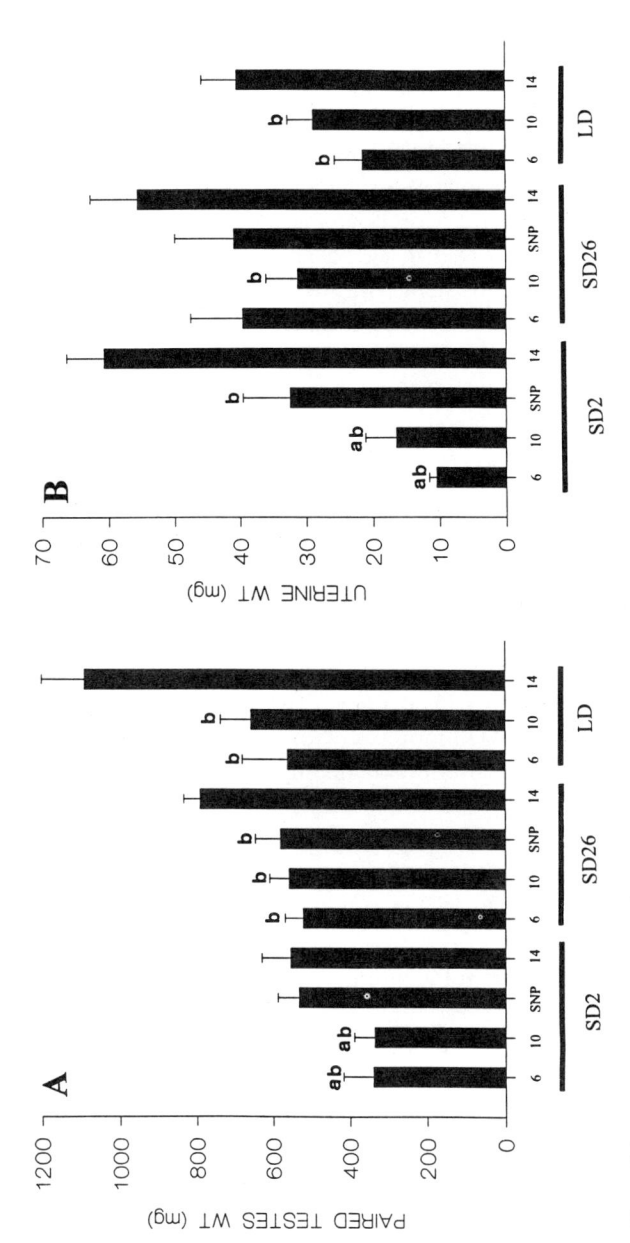

Figure 7.3
Paired testes weights (A) and uterine weights (B) for meadow voles at 7 weeks of age that were born to SD2 (mothers exposed to short days [10L] for 2 weeks prior to pairing with a male and remaining in SD thereafter), SD26 (mothers exposed to 26 weeks of SD prior to pairing with a male and remaining in SD thereafter), or long day (LD) dams (continuously exposed to 14L) and after weaning were housed in one of four conditions. Between 3 and 7 weeks of age pups either remained in the original photoperiod (SD2 and SD26: 10L; LD: 14L), were moved to a shorter photoperiod (SD2 and SD26: 6L; LD: 6L and 10L), or were exposed to a simulated natural photoperiod (SNP of February 1–March 1, SD2 and SD26 only). a, significantly less than SNP in the same maternal history condition; b, significantly less than 14L in same maternal history condition.

These data led us to consider the possibility that offspring born to photorefractory mothers would, like their mothers, be unaffected by changes in photoperiod. To test this hypothesis we placed pups born to SD2, SD26, or LD females into 6L, 10L, or 14L from birth to 7 weeks of age. The data for SD2 and SD26 pups in 10L and for LD pups in 14L are comparable to the previous experiment and the data are pooled for ease of comparison in figure 7.3. Several aspects of these data are of interest. First, SD26 pups developed faster when placed in 14L even though they were unaffected by the SNP. These data are not consistent with the hypothesis that pups of photorefractory females are insensitive to changes in photoperiod.

Second, in any given postnatal photoperiod (6L, 10L, 14L) the male SD26 pups exhibit greater reproductive development than SD2 pups, and in 6L and 10L SD26 females are also more developed than SD2 females.

Last, LD pups moved to 6L and 10L postnatally are not as delayed in development as SD2 pups treated similarly. In earlier studies, LD pups moved to SD at birth demonstrated reproductive development equivalent to SD2 pups. However, in those studies, the pups were also cross-fostered to SD mothers, while in this experiment the LD mother remained with her pups. The data suggest that photoperiod history of the mother continues to influence pup development postnatally. These results are consistent with data from an earlier study that demonstrated that meadow vole mothers affect pup development pre- and postnatally (Lee 1993). In that study, male pups cross-fostered at birth between SD2 and SD26 mothers, with both groups remaining in 10L, developed in accordance with the pattern predicted by the postnatal mother, while female pups developed like the SD26 controls if they had either a prenatal or postnatal SD26 mother. In contrast, Stetson et al. (1989) have not found any influence of the mother or photoperiod during lactation on development in Siberian hamsters, indicating species differences in the extent of maternal influence over postnatal responses to photoperiod.

If mothers are conveying information about the duration of their exposure to short day lengths (photoresponsive vs. photorefractory animals in short day lengths), how do they do so? Earlier studies demonstrated that development could be influenced by the duration of maternal melatonin (Weaver and Reppert 1986; Lee et al. 1989; Stetson et al. 1986). However, conventional wisdom indicates that the duration of the melatonin signal does not differ between photosensitive and photorefractory animals in the same photoperiod (Goldman and Elliott 1987; Watson-Whitmyre and Stetson 1987). Nevertheless, an examination of the sparse data on the pattern (duration and amplitude) of melatonin secretion in animals in the field across the year, or among animals held in constant short day lengths in the laboratory, demonstrates that most show a dampened melatonin amplitude, and sometimes a shortened duration after many weeks of short day lengths (Lerchl and Nieschlag 1992; Malpaux et al. 1988; Sheikheldon et al. 1992). The critical concentration of melatonin has not been determined for most species. Thus, it is conceivable that dampened amplitude could act as a shorter duration if the result were fewer hours in which melatonin levels were above a critical threshold during the night because less melatonin would be likely to cross the placenta to the offspring. We found that animals housed in short day lengths for 26 weeks had a dampened and shorter-duration melatonin rhythm than animals that had been in short day lengths for only 8 weeks (Lee et al., unpublished data). Thus, the mother's melatonin rhythm might encode the information about the coming end of winter. Pups are either able to discriminate these fine changes in the melatonin rhythm of the mother, or other changes associated with the development of photorefractoriness influence pup development and the interpretation of postnatal photoperiod.

If melatonin were key to the intermediate rates of development among pups of photorefractory mothers, then it would be predicted that increases in the amplitude and duration of melatonin in these mothers would result in reproductively delayed pups similar to those of SD2 mothers. Because microtine rodents eat and drink at regular intervals of 2 to 4 hours throughout the day, we elevated blood levels of maternal melatonin within physiological levels by providing the females with water containing melatonin for 14.5 to 15 hours nightly. The concentration was adjusted to the volume of water consumed by each animal so as to produce a uniform blood concentration. As predicted, SD2 controls were unaffected by the additional melatonin, and LD controls had suppressed reproductive development. Both male and female pups of photorefractory SD26 dams maintained in 10L demonstrated altered development after increased prenatal exposure to melatonin. Also as predicted, male testes size was smaller and growth rates were slower than in SD26 control males and did not differ from SD2 males. In contrast, melatonin-treated SD26 dams produced females with growth rates equivalent to SD2 females and uterine weights comparable to control SD26 females. This latter effect may be the result of continued postnatal contact with the SD26 mother no longer being treated with melatonin. As previously described, all females born to SD26 females and cross-fostered to SD2 dams (prenatal SD26) or cross-fostered from SD2 to SD26 dams (postnatal SD26) had uterine development consist with SD26 control females (Lee 1993).

These data support the hypothesis that maternal photoperiod history is communicated to fetuses by maternal melatonin crossing the placenta. However, we do not yet know whether amplitude as well as duration of the melatonin signal may be important for signaling a photorefractory state to fetuses. It is still unclear whether melatonin is the only maternal signal that provides information about season to the pups.

And finally, we have no idea what is altered in the fetuses to result in the differential development of pups born in 10L to photosensitive or photorefractory mothers. It is clear that photoperiod control of reproductive development is a far more complex process than was first envisioned. Maternal photoperiod history interacts with the postnatal photoperiod in complex ways that differ between the sexes. Integration of maternal and postnatal photoperiod information with other energetic and social information serves to orchestrate the rate of development and timing of puberty such that the offspring born at each season of the year are appropriately prepared for the environment that is about to emerge.

Winter Breeding and Photoperiod Insensitivity

As discussed by Bronson in chapter 2 and Wingfield and co-authors in chapter 4, life-history characteristics are important factors in the equation that predicts optimal seasonal breeding strategy. Empirically, some individuals in many rodent species breed into or throughout the winter, despite a clear seasonal trend of decreased reproduction for the population as a whole. A tradeoff clearly exists between the enormous benefits resulting from successful winter reproduction and the costs of a failed breeding effort. An animal, especially a female, that attempts to breed in energetically challenging conditions risks losing her reproductive investment, and also her own life and, thus, her future breeding opportunities. Because the probability of future breeding opportunities decreases with increasing age, old and young animals should differ in their propensity to attempt winter breeding. Consider, for example, a female vole born early in the breeding season. She undergoes rapid maturation and breeds in the season of her birth. In late summer, any litters she produces will delay puberty until the following breeding season. The chances of successful winter production of those young born late in the season is probably very low given the small size of the species and the energetic demands of winter, whereas they are most likely to survive the winter and breed the following season if they remain reproductively suppressed. In contrast, if the dam ceases breeding in fall, she has little chance of reproducing the following season because her longevity approaches its limits. As a result, the cost to the mature dam of a failed reproductive effort is likely to be minimal. Thus, animals might be expected to respond quite differently to environmental signals as a function of their age.

Among meadow voles, mating behavior and successful production of litters differed significantly between nulliparous females raised in LD or SD conditions (Meek and Lee 1993). All nulliparous LD females mated and 90% become pregnant, while

33% of SD females in complex environments did not mate, and of those that eventually mated only 55% became pregnant. In contrast, primiparous LD and SD females did not differ in the short latency to mate or successful litter production. Even a single previous litter was sufficient to prevent short day lengths from halting reproduction. This effect of parity, therefore, would likely extend the breeding season of the species into the fall and winter, so long as there was a cohort of primiparous females and males that had not undergone testicular regression. The latter requirement is met by approximately 20% of adult males (Dark et al. 1983; Dark and Zucker 1984).

In hamsters, failure to suppress reproduction in short day lengths appears instead to have a circadian basis, that is, photoperiod-insensitive Siberian hamsters have circadian entrainment patterns typical of animals in long day lengths regardless of photoperiod. Thus, they fail to produce the long-duration melatonin signal necessary to suppress gonadal activity (Puchalski and Lynch 1986). Artificial selection experiments demonstrated that nonresponsivness had a large genetic component, but our studies with naturalistic day lengths revealed a marked dependence on photoperiodic history. Male hamsters born into a spring solstice pattern of day lengths at 40°N latitude, at the southern limit of this species' range, were almost uniformly inhibited by decreasing day lengths the following autumn. In contrast, only 50% of hamsters maintained in a naturalistic day length of 60°N latitude, near the northern range limit, responded to short day lengths of fall and winter (Gorman and Zucker 1995). To our surprise, post hoc analyses of several previous experiments revealed no relation between nonresponsiveness and any measure of rate of change in day length (Gorman and Zucker 1997a). Rather, incidence of nonresponsivness was high among hamsters that had been exposed to very long day lengths (e.g., 16L).

We subsequently tested the hypothesis that, at a particular value above the critical day length for gonadal stimulation and maintenance, exposure to some other critical day length would induce nonresponsiveness to any short day length. We exposed SD hamsters to different stimulatory day lengths to determine if some, but not others, allowed responsiveness to short day lengths. Male juvenile Siberian hamsters housed in short day lengths (10L) from birth were all gonadally inhibited at 6 weeks of age, thereby demonstrating uniform photoresponsiveness at this age. These hamsters were then transferred to either very long day lengths (18L) or shorter long day lengths (14L) for 10 weeks. Both experimental photoperiods induced full gonadal growth. All hamsters were then challenged with short day lengths (10L) for 10 weeks; of those previously maintained in 14L, more than 90% were responsive to short day lengths as indicated by complete gonadal regression and expanded locomotor activity rhythms. In contrast, hamsters exposed to 18L were almost uniformly (greater than 90%)

nonresponsive to short day lengths; they had large gonads and compressed activity rhythms characteristic of hamsters in long day lengths, despite having been exposed to short day lengths for 10 weeks (Gorman and Zucker 1997a).

The longest summer day lengths (at the latitude of capture of the original stock of this colony) may render animals born early in the breeding season nonresponsive to future short day lengths. As with meadow voles, extended fall and winter breeding may be adaptive in these animals because they are unlikely to survive winter and breed the following spring. In contrast, animals born later in the breeding season after day length had fallen below this critical value for nonresponsiveness would not experience these photoperiods in their season of birth and hence would retain uniform responsiveness to short day lengths. This group of animals likely represents the cohort that survives winter and breeds the following spring. Photoperiodic mechanisms are sensitive to different types of "long day lengths," representing various parts of the breeding season.

Interaction of Photoperiod with Other Environmental Variables

The studies so far described have focused on the different effects of natural and artificial photoperiods and the complex interactions between different sources of photoperiod information (maternal prenatal, maternal postnatal, and ambient exposure) in the control of reproduction. A similar approach is used to examine the interaction of photoperiod with nonphotoperiodic factors such as food availability and temperature on the timing of reproduction. The general view that these nonphotoperiodic environmental factors are of minor importance relative to day length likely relates, in part, to their assessment against inappropriate photoperiod histories. However, photoperiod acts as a predictor of these environmental changes, and we would expect that the role of these factors would depend upon the photoperiod-induced physiological state of the individual.

Photoperiod and Food

Many researchers have demonstrated that sufficient calories and particular nutrients such as protein are critical for successful reproduction in rodents (reviewed by Negus and Berger 1987). In addition, some non-nutritive plant components can act to signal season of the year. In some rodents (e.g., prairie and montane voles, and house mice), 6-MBOA (6-methoxybenzoxazolinone) stimulates female uterine development and reproduction (Nelson and Shiber 1990; Nelson and Blom 1992; Berger et al. 1987; Negus and Berger 1987). This chemical only occurs in young growing plants that have tissue damage (as from biting), and therefore is a predictive signal of spring plant growth for species of microtines that inhabit areas in which mild winters induce early spring growth of plants fairly often.

Photoperiod, of course, predicts changes in ambient temperature and food availability that will alter energy balance to favor or inhibit reproduction (Bronson and Pryor 1983; Nelson et al. 1992). However, as mentioned earlier, the exact timing of the onset of good food resources varies between years. It would be adaptive to adjust the breeding season in each year to the immediate conditions—delaying when winter lasts longer than average, and beginning early when food appears earlier. One would predict, however, that sensitivity to nonphotoperiodic variables might be linked to the predictability provided by photoperiod. For example, photoperiodic species from higher latitudes would be expected to respond differently to the presence of favorable food resources in early winter than would species from mid-latitudes because the likelihood that these resources will remain present throughout pregnancy and weaning will be less assured in the higher than mid-latitudes. The appearance of green, growing shoots in early January in Michigan (as in the mild winter of 1998), for example, does not predict the onset of a successful period for reproduction as well as the same green shoots appearing in early March. The appearance of green food in January in Michigan will likely be followed by another extended cold snap in most years than will the same conditions in March (in 1998 green shoots and temperatures above freezing in early January were followed by subzero temperatures and severe weather in early March). In contrast, southern Illinois may have sufficiently mild conditions and food available throughout some winters to support reproduction.

Thus, we predicted that the gonadal development of meadow voles (from a colony developed from animals at 45°N) born to LD, SD2, or SD26 mothers would respond differently to the quality of the diet. A 6-MBOA-free control diet replaced the regular rodent chow immediately after the female gave birth and throughout the 9-week period of development. The food was calorically equivalent to the standard laboratory chow diet, but had a lower protein-carbohydrate ratio. Prairie and montane voles reared on this diet do not differ from animals raised on a standard rodent chow. Two other groups were fed the 6-MBOA-free diet supplemented daily with 4 g of sprouted wheat or alfalfa. Sprouted wheat replaced the 6-MBOA in the diet, while alfalfa is a favored food item of meadow voles which has been previously demonstrated to influence reproductive development (Martinet and Meunier 1969).

The 6-MBOA-free diet produced an unexpected suppression in gonadal development of 9-week-old LD and SD26 males and SD26 females compared with animals fed the standard laboratory chow (Meek et al. 1995). The lack of an effect in SD2 animals was not due to a basement effect, since the paired testes weights of SD26 males were significantly lower than those of SD2 males on the 6-MBOA-free diet, while SD2 females had somewhat heavier uterine weights on the 6-MBOA-free diet than on the standard Purina chow diet (figure 7.4). Using male spermatogenesis as an indicator of fertility, the 6-MBOA-free diet suppressed fertility in LD (−50%) and SD26 (−70%) animals, but not SD2 males, compared to development when fed Purina rodent chow. Adding either alfalfa or sprouted wheat to the diet produced complete recovery for LD and SD26 animals. The number of fertile SD2 males increased slightly when alfalfa was added to the diet (+10%), and significantly with the addition of sprouted wheat (+45%).

We concluded that the important change in the 6-MBOA-free diet that suppressed development in meadow voles, but did not have such an effect on montane or prairie voles, was not likely the removal of 6-MBOA but rather the decrease in protein, while calories were held constant. Alfalfa does not contain 6-MBOA, whereas sprouted wheat does, and both food supplements increased the protein content of the diet. LD and SD26 animals fed either food supplement demonstrated a robust reproductive development compared with animals fed only the 6-MBOA-free diet. Because sprouted wheat had a larger effect than alfalfa on the spermatogenesis of SD2 males, it is possible that only male meadow voles born to photosensitive mothers in short day lengths are sensitive to the reproductive enhancing cues of 6-MBOA.

In conclusion, animals born to photorefractory mothers in the field, likely between early February and May, will be exquisitely sensitive to their food resources as a final determinant of when to undergo reproductive development. A low-protein diet may lead to greater suppression of development in these animals than we have ever previously observed, while abundant green food can support a robust reproductive response. In contrast, animals born in the fall or winter (SD2) are likely unaffected by low dietary protein, and in their habitat would only rarely be exposed to sprouted wheat or fresh alfalfa. And, of course, animals born in summer conditions would likely have plenty of high-protein green foods available to support growth and reproductive development. Subsequent work in this laboratory has further demonstrated the importance of fresh green food for photorefractory females. Female young of SD26 dams reared on Purina chow have uterine weights consistent with follicular development and vaginal opening and will not mate unless they also are supplied with fresh green alfalfa (Lee and Grady, unpublished data).

Photoperiod and Temperature

A similar analysis may be fruitful in understanding the interactions of photoperiod and temperature. Temperature has largely been viewed as a modulator of photoperiodic effects, but low ambient temperatures might be a signal of great significance

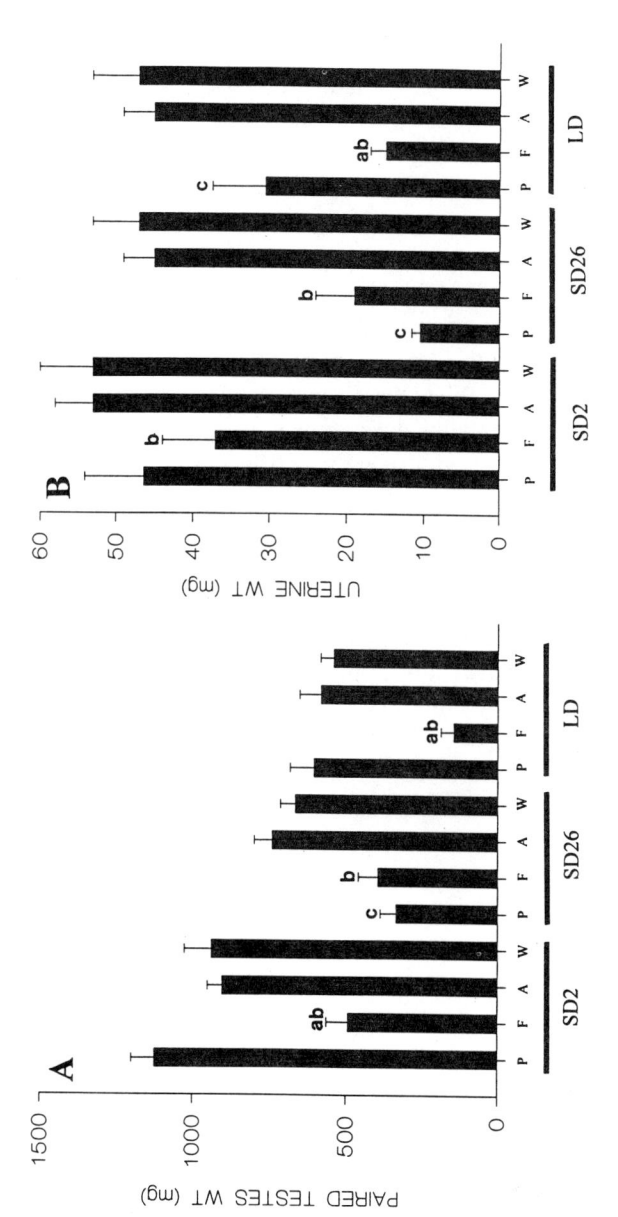

Figure 7.4
Paired testes weights (A) and uterine weights (B) for meadow voles at 9 weeks of age that were born to SD2, SD26, or LD dams (see description in figure 7.3) and were fed standard Purina chow (no. 5015; P), 6-MBOA-free chow (F), F plus fresh alfalfa (A), or F plus fresh sprouted wheat (W) throughout development. a, animals fed F diet were significantly smaller than animals in the same maternal photoperiod history condition fed P diet; b, animals fed either A or W diet were significantly more developed than animals fed F within the same maternal photoperiod history condition; c, animals fed A or W diet were significantly more developed than animals fed P within the same maternal photoperiod history condition.

in timing some physiological and behavioral transitions in particular contexts. For example, low temperatures alone do not induce gonadal regression in long day–housed hamsters or prairie voles, but they significantly accelerate gonadal regression in hamsters transferred abruptly to short day lengths (Pevet et al. 1989; Nelson et al. 1989). Does this mean that temperature is only a modulator of photoperiod effects? The role of temperature in seasonality has been assessed primarily under conditions inappropriate for observing large effects—in the context of either nonpermissive unchanging long day lengths or against a background of a supraphysiological abrupt change to short day lengths. An important role of temperature in extending or curtailing the breeding season is suggested by the data of Steinlechner et al. (1991) who tested the effects of several intermediate day lengths on gonadal growth in juvenile Siberian hamsters. In this experiment, gonadal development was completely prevented by transfers (abrupt) from 16L to 14L at 5°C but was still stimulated at 22°C. At the latter temperature, decreases to 13L were necessary to inhibit testicular growth.

A study examining the interaction of moderate (22°C) and low (8°C) ambient temperatures on the first 3 weeks of development of meadow voles reared by LD, SD2, and SD8 dams also found differences in the interaction of photoperiod history and temperature (Reeves 1994). Because the study stopped at 21 days, there was no significant reproductive development in any group. However, the growth rate (body weight and length) and development of reflex and motor patterns were influenced by an interaction of photoperiod and temperature. The growth and development of LD pups was generally delayed by low temperatures, while the development of SD pups was either unaffected or improved by low temperatures (Reeves 1994). These data are consistent with the hypothesis that different photoperiod histories prepare animals to cope with different environments.

Conclusion

Bronson has noted that "Probably seldom in the wild is a mammal's reproduction controlled at any one moment in time by a single environmental variable. Many environmental variables can alter a mammal's reproductive potential, and these factors interact with each other in complex ways" (1989, p 184). In this chapter we have illustrated the complexity of interactions that exists even in how the indirect timing agent, photoperiod, will be interpreted. The same photoperiod is interpreted differently by animals that have had different recent past photoperiod histories (increasing vs. decreasing day lengths to the current photocycle) and are at different stages of their life history. The mechanism by which photoperiod history influences differential responses in the subject, whether adult or prepubertal, involves melatonin. But how different histories result in the same pattern of melatonin producing different responses, or how the maternal melatonin pattern alters offspring responses to photoperiod, is unknown.

Similarly, animals at different phases in their life histories respond differently to many environmental factors, such as social interactions and energetic demands, that influence the onset or cessation of reproduction. In this chapter we describe how short-lived animals may continue breeding in the winter months because they no longer respond to short day lengths. In the case of Siberian hamsters, the mechanism involves an inability to lengthen the activity and melatonin duration after experiencing very short nights of summer. However, for neither the hamsters nor the voles do we understand the underlying mechanism that causes them to be "stuck" in summer breeding condition. We can argue that it is adaptive for short-lived animals to continue breeding when they have little chance of surviving until the following summer in any case, but we do not understand the proximate mechanism that causes this to occur.

Last, we provide examples of photoperiod apparently altering the reproductive response of animals to food or temperature. In these examples, the food and temperature variables are incompatible with reproduction in some photoperiods but not in others. How does photoperiod establish a physiological expectation that specific food resources or temperature conditions will exist? In these examples, it almost certainly involves changing sensitivity of the metabolic sensory system described by Schneider and Wade in chapter 3, so that food requirements and deleterious effects of cold temperatures are less evident in short day–adapted than long day–adapted animals. But, again, the exact mechanism has yet to be explicated.

Other chapters in this book examine the control of reproduction by social factors or energetic constraints. It is likely that the mechanisms that underlie the impact of these immediate variables on reproduction must be integrated with the long-term timing mechanism of photoperiod, and shorter-term timing brought about by food, water, and temperatures.

Acknowledgments

This work was supported by research grants HD-24575 from the National Institutes of Health (T.M.L.) and the Michigan Society of Fellows (M.R.G.).

References

Berger PJ, Negus NC, Rowsemitt CN (1987). Effect of 6-methoxybenzoxazolinone on sex ratio and breeding performance in *Microtus montanus*. Biol Reprod 36:255–260.

Bronson FH (1989). Mammalian Reproductive Biology. Chicago: University of Chicago Press.

Bronson FH, Heideman PD (1992). Lack of reproductive photoresponsiveness and correlative failure to respond to melatonin in a tropical rodent, the cane mouse. Biol Reprod 46:246–250.

Bronson FH, Pryor S (1983). Ambient temperature and reproductive success in rodents living at different latitudes. Biol Reprod 29:72–80.

Bronson FH, Rissman EF (1986). The biology of puberty. Biol Rev 61:157–195.

Christian JJ (1980). Regulation of annual rhythms of reproduction in temperate small rodents. *In* Testicular Development, Structure and Function (Steinberger A, Steinberger F, eds), pp 367–380. New York: Raven Press.

Dark J, Zucker I (1984). Gonadal and photoperiodic control of seasonal body weight changes in male voles. Am J Physiol 247:R84–R88.

Dark J, Zucker I, Wade GN (1983). Photoperiodic regulation of body mass, food intake, and reproduction in meadow voles. Am J Physiol 245:R334–R338.

Forger NG, Zucker I (1985). Photoperiodic regulation of reproductive development in male white-footed mice (*Peromyscus leucopus*) born at different phases of the breeding season. J Reprod Fertil 73:271–278.

Goldman BD (1991). Parameters of the circadian rhythm of pineal melatonin secretion affecting reproductive responses in Siberian hamsters. Steroids 56:218–225.

Goldman BD, Elliott JA (1987). Photoperiodism and seasonality in hamsters: role of the pineal gland. *In* Processing of Environmental Information in Vertebrates (Stetson MH, ed), pp 203–218. New York: Springer-Verlag.

Gorman MR (1995). Seasonal adaptations of Siberian hamsters: I. Accelerated gonadal and somatic development in increasing versus static long day lengths. Biol Reprod 53:110–115.

Gorman MR, Zucker I (1995). Seasonal adaptations of Siberian hamsters: II. Pattern of change in day length controls annual testicular and body weight rhythms. Biol Reprod 53:116–125.

Gorman MR, Zucker I (1997a). Environmental induction of photononresponsiveness in the Siberian hamsters, *Phodopus sungorus*. Am J Physiol 272:R887–R895.

Gorman MR, Zucker I (1997b). Pattern of change in melatonin duration determines testicular responses in Siberian hamsters, *Phodopus sungorus*. Biol Reprod 56:668–673.

Horton TH (1984). Growth and maturation in *Microtus montanus*: effects of photoperiods before and after weaning. Can J Zool 62:1741–1746.

Horton TH (1985). Cross-fostering of voles deomonstrates in utero effect of photoperiod. Biol Reprod 33:934–939.

Lee TM (1993). Development of meadow voles is influenced postnatally by maternal photoperiodic history. Am J Physiol 265:R749–R755.

Lee TM, Zucker I (1988). Vole infant development is influenced perinatally by maternal photoperiodic history. Am J Physiol 255:R831–R838.

Lee TM, Smale L, Zucker I, Dark J (1987). Influence of day length experienced by dams on postnatal development of young meadow voles (*Microtus pennsylvanicus*). J Reprod Fertil 81:337–342.

Lee TM, Spears N, Tuthill CR, Zucker I (1989). Maternal melatonin treatment influences rates of neonatal development of meadow vole pups. Biol Reprod 40:495–502.

Lerchl A, Nieschlag E (1992). Interruption of nocturnal pineal melatonin synthesis in spontaneous recrudescent Djungarian hamsters (*Phodopus sungorus*). J Pineal Res 13:36–41.

Madison DM, McShea WJ (1987). Seasonal changes in reproductive tolerance, spacing and social organization in meadow voles: a microtine model. Am Zool 27:899–908.

Madison DM, Fitzgerald RW, McShea WJ (1984). Dynamics of social nesting in overwintering meadow voles (*Microtus pennsylvanicus*): possible consequences for population cycling. Behav Ecol Sociobiol 15:9–17.

Mallory FF, Clulow FV (1977). Evidence of pregnancy failure in the wild meadow vole, *Microtus pennsylvanicus*. Can J Zool 55:1–17.

Malpaux B, Moentner SM, Wayne NL, Woodfill CJI, Karsch FJ (1988). Reproductive refractoriness of the ewe to inhibitory photoperiod is not caused by alteration of the circadian secretion of melatonin. Neuroendocrinology 48:264–270.

Martinet L, Meunier M (1969). Influence des variations saisonnières de la luzerne sur la croissance, la mortalité et l'éstablissement de la maturité sexuelle chez le campagnol des chams (*Microtus arvalis*). Ann Biol Anim Biochim Biophys 9:451–462.

Meek LR and Lee TM (1993). Female meadow voles have a preferred meeting pattern predicted by photoperiod, which influences fentility. Physiol Behav 54:1201–1210.

Meek LR, Lee TM, Gallon JF (1995). Interaction of maternal photoperiod history and food type on growth and reproductive development of laboratory meadow voles (*Microtus pennsylvanicus*). Physiol Behav 57:905–911.

Meserve PL, Yunger JA, Gutiérrez JR, Contreras LC, Milstead WB, Lang BK, Cramer KL, Herrera S, Lagos VO, Silva SI, Tabilo EL, Terrealba M-A, Jaksic FM (1995). Heterogeneous responses of small mammals to an El Niño southern oscillation event in north-central semiarid Chile and the importance of ecological scale. J Mammal 76:580–595.

Mihok S (1984). Life history profiles of boreal meadow voles (*Microtus pennsylvanicus*). *In* Winter Ecology of Small Mammals, vol 10 (Merritt, JF ed), pp 91–102. Pittsburgh: Carnegie Museum of Natural History.

Negus NC, Berger PJ (1987). Mammalian reproductive physiology: adaptive responses to changing environments. *In* Current Mammalogy, vol 1 (Genoways, HH ed), pp 149–173. New York: Plenum Publishing.

Nelson RJ (1987). Photoperiod-nonresponsive morphs: a possible variable in microtine population-density fluctuations. Am Naturalist 130:350–369.

Nelson RJ, Blom JMC (1992). 6-Methoxy-2-benzoxazolinone and photoperiod: prenatal and postnatal influences on reproductive development in prairie voles (*Microtus ochrogaster ochrogaster*). Can J Zool 71:776–789.

Nelson RJ, Shiber JR (1990). Photoperiod affects reproductive responsiveness to 6-methoxy-2-benzoxazolinone in house mice. Biol Reprod 43:586–591.

Nelson RJ, Dark J, Zucker I (1983). Influence of photoperiod, nutrition and water availability on reproduction of male California voles (*Microtus californicus*). J Reprod Fertil 69:473–477.

Nelson RJ, Frank D, Smale L, Willoughby SB (1989). Photoperiod and temperature affect reproductive and nonreproductive functions in male prairie voles (*Microtus ochrogaster*). Biol Reprod 40:481–485.

Nelson RJ, Kita M, Blom JMC, Rhyne-Grey J (1992). Photoperiod influences the critical caloric intake necessary to maintain reproduction among male deer mice (*Peromyscus maniculatus*). Biol Reprod 46:226–232.

Pévet P, Vivien-Roels B, Masson-Pévet M (1989). Low temperature in the golden hamster accelerates the gonadal atrophy induced by short photoperiod but does not affect the daily pattern of melatonin secretion. J Neural Transm Gen Sect 76:119–128.

Puchalski W, Lynch GR (1986). Evidence for differences in the circadian organization of hamsters exposed to short day photoperiod. J Comp Physiol [A] 159:7–11.

Reeves GD (1994). Influences of Photoperiod and Ambient Temperature on Maternal-Pup Interactions in the Meadow Vole (*Microtus pennsylvanicus*). PhD dissertation, University of Michigan, Ann Arbor.

Reiter RJ (1980). The pineal and its hormones in the control of reproduction in mammals. Endocr Rev 1:109–131.

Reppert SM, Klein DC (1978). Transport of maternal [^{3}H]melatonin to suckling rats and the fate of [^{3}H]melatonin in the neonatal rat. Endocrinology 102:582–588.

Reppert SM, Chez RA, Anderson A, Klein DC (1979). Maternal-fetal transfer of melatonin in a nonhuman primate. Pediatr Res 13:788–792.

Sheikheldin MA, Howland BE, Palmer WM (1992). Seasonal profiles of melatonin in adult rams. J Pineal Res 12:58–63.

Steinlechner S, Stieglitz A, Ruf T, Heldmaier G, Reiter RJ (1991). Integration of Environmental Signals by the Pineal Gland and Its Significance for Seasonality in Small Mammals. New York: Plenum Press.

Stetson MH, Elliott JA, Goldman BD (1986). Maternal transfer of photoperiodic information influences photoperiodic response of prepubertal Djungarian hamsters (*Phodopus sungorus sungorus*). Biol Reprod 34:664–669.

Stetson MH, Ray SL, Creyaufmiller N, Horton TH (1989). Maternal transfer of photoperiodic information in Siberian hamsters. II. The nature of the maternal signal, time of signal transfer, and the effect of the maternal signal on peripubertal reproductive development in the absence of photoperiodic input. Biol Reprod 40:458–465.

Watson-Whitmyre M, Stetson MH (1987). Reproductive refractoriness in hamsters: environmental and endocrine etiologies. *In* Processing of Environmental Information in Vertebrates (Stetson, MH ed), pp 219–249. New York: Springer-Verlag.

Weaver DR, Reppert SM (1986). Maternal melatonin communicates daylength to the fetus in Djungarian hamsters. Endocrinology 119:2861–2863.

8 Environmental and Social Influences on Seasonal Breeding and Immune Function

Randy J. Nelson and Sabra L. Klein

The immune system is commonly perceived as a system of specialized cells in the thymus, bone marrow, spleen, lymph, blood, and diffuse lymphatic tissues, whose function is to recognize and destroy foreign substances in the body. The mammalian reproductive system involves primarily the hypothalamic-pituitary-gonadal axis that functions to bring gametes together for fertilization and to support gestation, parturition, and lactation. For the most part, these two systems have been treated as parallel, independent systems, meeting only at the point when sexually transmitted diseases induce an immune response. However, it has become apparent in recent years that the immune system is linked to reproductive processes on several different levels. For example, in many mammals and birds, immunity is often compromised during the breeding season, and compromised immune function is a common feature of pregnancy and lactation even among nonseasonal mammals. Furthermore, immune function can be influenced by many of the same environmental, social, and intrinsic factors that affect reproductive function (Nelson and Demas 1996). Recent studies suggest that hormones traditionally considered to be "reproductive" hormones can modulate immune function, and that cytokines and other chemical regulators of immune function are intimately involved in reproductive processes such as ovulation and hypothalamic hormone secretion (Kalra and Kalra 1996; Ober and van der Ven 1997; Priddy 1997). Another important link between these two systems occurs via the behavioral aspects of reproduction, such as courtship and mate selection. Steroid-dependent secondary sex characteristics are important criteria for mate selection in a number of species. It has been hypothesized recently that females use secondary sex characteristics to assess immune function of potential mates (Able 1996; Hamilton and Zuk 1982; Zuk 1990). In summary, it is becoming increasingly apparent that the reproductive and immune systems interact and function together.

In order to understand the link between the immune and reproductive systems, it is necessary to consider each in terms of energy homeostasis. As noted in chapters 2 and 3, all organisms experience short- and long-term variability in energy requirements for heat production, foraging, general activity, and metabolic rate. Most organisms do not ingest food sufficiently often to provide a continual supply of energy and nutrients to their cells. Despite this variability in both energy supply and demand, there are cellular processes necessary for survival that do require a relatively continual supply of energy. Therefore, survival depends on the ability to achieve and maintain energy homeostasis within particular limits. Mammals consume food in excess of their immediate energy requirements and store the digested, metabolized food in the form of glycogen and fat to be utilized as needed when food is not being ingested

(Ganong 1995; Harris et al. 1979). Energy homeostasis is maintained over the course of the day by switching from the feeding (energy storage) to the fasting (energy mobilizing and oxidizing) mode of metabolism (see figures 3.1 and 3.2). In addition, a complex array of metabolic changes occur in preparation for and during seasonal cycles of climate and food availability.

Outside the tropics, food availability is generally low during the winter, while thermogenic demands are typically high. The endocrine, metabolic, morphological, and behavioral changes that enhance winter survival in mammals include a variety of cellular changes that increase thermogenic activity and capacity, increase shelter or nest building, and increase social interactions that attenuate heat loss and enhance pelage growth (Barnett 1965; Kenagy and Barnes 1984; Heldmaier et al. 1989). In addition, under energetically challenging conditions, some energetically costly activities are inhibited in an effort to conserve energy for those processes that are critical to survival. For example, reproductive activities, from courtship to weaning, are energetically demanding, particularly for female mammals, and reproduction is inhibited during the winter in Temperate Zone mammals (for review, see Bronson 1989). Reproduction is inhibited by changes in day length that may occur prior to the energetic challenges of winter in many seasonally breeding species (see chapters 4, 5, and 7). Even in the tropics, animals experience yearly fluctuations in rainfall that dramatically affect food availability, energy requirements for thermogenesis, as well as changes in reproductive capability (Bronson and Heideman 1994). As illustrated by Bronson, for species representing most mammalian orders, reproduction is consistently inhibited whenever a mammal is confronted with a negative energy balance, or when too large a portion of metabolic energy is diverted into storage leaving too little available for oxidation (see chapter 2).

In contrast to other physiological processes, immune function has been assumed to remain relatively constant across the seasons (see Sheldon and Verhulst 1996). However, recent evidence suggests that immune function actually varies substantially on a seasonal basis (Nelson and Demas 1996; Lochmiller et al. 1994). Maintaining maximal immune function is energetically expensive; the cascades of dividing immune cells, the onset and maintenance of inflammation and fever, and the production of humoral immune factors all require substantial energy (Demas et al. 1997a; Maier et al. 1994). Energy utilization involves elevation of basal metabolic rate, blood glucose, and free fatty acid levels, but ketogenesis is not accelerated. Therefore, mounting an energetically costly immune response has the potential to compromise the ability to preserve protein stores in muscle (Beutler and Cerami 1988). Additionally, mounting an immune response obviously requires resources that could otherwise be allocated to other functions (Sheldon and Verhulst 1996). Thus, it is reasonable to consider immune function in terms of energetic tradeoffs. Individuals may partition resources between immune function and other biological processes, such as reproduction, growth, or thermogenesis. Consequently, animals may maintain the highest level of immune function that is energetically possible given the constraints of processes essential to survival, growth, reproduction, thermogenesis, foraging, and other activities (Deerenberg, et al. 1997; Festa-Bianchet 1989; Richner et al. 1995). The observations that immune function fluctuates seasonally and is compromised during times of breeding, migration, and molt are consistent with this idea (John 1994; Zuk 1990).

The widely known link between immune function and "stress" can be reinterpreted in the context of an energetic perspective. "Stress" is a notoriously ethereal concept that has been used to describe any factor that increases glucocorticoid secretion, including injury, pain, infection, overcrowding, harsh ambient temperature, food deprivation, noise, restraint, and social interactions. The problem is confounded by the use of the word "stress" to include both the stressor and the stress response (Sapolsky 1992). The ecological literature illustrates that environmental factors perceived as stressors, such as low food availability, cold ambient temperatures, overcrowding, lack of shelter, or increased predator pressure recur seasonally and correlate with seasonal fluctuations in immune function among individuals and seasonal changes in population-wide disease and death rates (Lochmiller et al. 1994). Laboratory studies have established that various stressful stimuli inhibit immune function (Keller et al. 1983; Laudenslager et al. 1983), and that the effect is mediated by stress-induced increases in circulating glucocorticoid concentrations (Ader and Cohen 1993; Besedovsky and del Rey 1991; Black 1994; Claman 1972; Hauger et al. 1988; MacMurray et al. 1983; Monjan 1981). Stressors disturb energy homeostasis. Because restoring homeostasis requires more energy than maintaining homeostasis (Sapolsky 1992), exposure to a stressor increases energy demands on individuals. Sapolsky (1992) has suggested that individual differences in coping with similar stressors reflect differences in perception and processing of information about stressful situations. However, given the link between immune function and energy balance, it is also possible that individual differences in coping with stress may reflect differences in energy partitioning or energetic efficiency.

During long-term perturbations in energy homeostasis, such as during winter when temperatures are low and food is scarce, glucocorticoids are released and energy is mobilized to restore homeostatic equilibrium. There may be competition for energy among immune function, thermogenesis, and reproductive processes. Immune function is typically suppressed during the breeding season (Nelson and Demas 1996; Zuk 1990). Immune function can be suppressed by pharmacological blockade of glucose oxidation (Demas et al. 1997b), suggesting that one immunosuppressive signal is related to changes in fuel oxidation. Changes in circulating steroid hormone concentrations may initiate a switch in energy partitioning between immune and reproductive systems, because mechanisms that mediate seasonal changes in reproductive processes also appear to mediate changes in immune function (Nelson and Demas 1996). For example, immunosuppression by pharmacological blockade of metabolic fuel utilization is attenuated by housing in a short-day photoperiod (Demas et al. 1997b). This reveals the fascinating possibility that individuals of some species use photoperiod information to initiate or terminate specific seasonal responses that maintain energy homeostasis, and these mechanisms also affect both reproduction and immune function (reviewed in Bartness and Goldman 1989; Heldmaier et al. 1989; Saafela and Reiter 1994; see also chapter 7).

The mechanisms by which steroid hormones, photoperiod, and metabolic fuel partitioning interact are still unknown. A number of interesting, testable questions arise from these new findings. Do steroid hormones affect immune function directly, or do they affect immune function indirectly through their well-known effects on energy metabolism (Wade and Schneider 1992)? Short-day photoperiods can preserve immune function independent of the effects of temperature and gonadal hormones (Demas et al. 1997b). Is this a direct effect on overall metabolic efficiency or is it related to differential use of particular kinds of fuels under short-day photoperiod?

In this chapter, we evaluate the immune function of seasonally breeding species as it relates to social interactions, breeding, winter adaptations, and responsiveness to photoperiod. We propose that individuals maintain the highest level of immune function that is energetically possible within the constraints of other survival needs, including growth and reproduction, in habitats in which energy requirements and availability often fluctuate. The notion of energetic tradeoffs between immune function and other biological and behavioral processes is consistent with an adaptationist perspective. In other words, traits that increase *fitness*, defined as the ability to reproduce and rear reproductively viable offspring, are considered adaptive. Because an animal cannot maximize the functioning of all systems without imposing costs on other systems, the notion of tradeoffs has been proposed. Thus, seemingly adaptive changes in one trait may decrease functioning of other equally important traits. However, if the overall result of this energetic tradeoff is increased lifetime reproductive success, then selection for individuals that successfully partition energetic resources will persist. The observation that immune function is generally compromised during specific energetically demanding times such as winter, breeding (including pregnancy and lactation), migration, or molting is consistent with the hypothesis that immune function is energetically optimized (John 1994; Zuk 1990).

We begin with a brief overview of the immune system, including a discussion of the different types of immune response and a description and evaluation of the methods currently used to measure immune function. The remainder of the chapter is divided into environmental and social factors that influence immune function in relation to reproduction. In the environmental half of the chapter, we examine changes in immune function that occur during seasonal changes in reproduction. We also demonstrate that the predictive cues that allow animals to maximize reproductive success can also affect immune function. Next, we examine the effects of changes in ambient temperature both alone and in combination with different photoperiods on immune function. Whether we are talking about predictive cues like photoperiod, or proximate cues such as ambient temperature, the ultimate usefulness of the cues is related to timing reproductive effort with periods of sufficient energy availability. We examine the direct effects of metabolic fuel oxidation on immune function, and the interaction of photoperiodic and metabolic cues. The final half of the chapter is concerned with the notion that reproductive behavior, social interactions, and choice of mate may be influenced by the immune status of conspecifics.

Assessment of Immune Function

Components of the Immune System

Immunology is the study of the body's defense against infection and disease. The immune system can defend against infection and disease arising from external sources, such as viruses, bacteria, fungi, and parasites, as well as disease arising from internal sources such as tumors and cancers. Vertebrates possess both innate and adaptive immunity. Innate immunity is present from the time of birth. The defenses associated with innate immunity do not require prior exposure, or sensitization to a particular pathogen (i.e., any foreign substance that has entered the body, also called an antigen); therefore, this type of immunity is often referred to as nonspecific. However, many of the cells involved in innate immunity (e.g., macrophages) possess receptors for proteins common to most microbial organisms. This enables innate immunity to serve as a first line of defense against infection. Because these receptors are encoded in the germline, they are not malleable (i.e., not subject to genetic recombination) and, therefore, cannot detect structural distinctions between microbial organisms or mutations within a microbial organism. Conversely, because innate immunity uses germline-encoded proteins, its importance in the evolution of immune defense has recently been emphasized (Medzhitov et al. 1997).

Adaptive immunity is both specific and acquired. Adaptive immunity is specific because the immune cells involved possess receptors that are genetically organized to recognize and respond to a single pathogen. Adaptive immunity is acquired because the immune cells involved can only engage in an immune response based on familiarity or previous exposure to a pathogen. The importance of adaptive immunity is that the recognition and elimination of a pathogen improves each time an individual is reexposed; this improvement ultimately leads to resistance against infection and disease.

Adaptive immune responses mediated by lymphocytes are categorized as cell-mediated or humoral immunity. Cell-mediated immunity is modulated by activated T cells that can either directly kill a pathogen or release soluble factors called cytokines (i.e., chemical messengers that enable immune cells to communicate with one another) to attract other lymphocytes to the site of infection. The distinguishing characteristic of cell-mediated immunity is that it involves direct interaction between lymphocytes and the pathogen; in other words, contact with a pathogen is required for this adaptive immune response to occur. Examples of immune responses that are cell-mediated are rejection of transplanted tissue, and localized inflammation in response to a foreign substance, such as tuberculin (Borysenko 1987).

Humoral immunity is another class of adaptive immune responses that primarily involves activated B cells secreting antibodies into circulation. Antibody molecules are surface, antigen-specific receptors released from B cells in response to a particular pathogen. In contrast to T cells, B cells do not directly contact a pathogen, but rather the specific antibodies that are secreted come into contact with the pathogen. The antibodies bind to epitopes on the pathogen to form inactive complexes that may either be eliminated from circulation or serve to activate the complement cascade (a serial activation of complement immune cells) which will destroy a pathogen by boring holes in the cell membrane. Although antibodies secreted by B cells are the key participants in humoral immune responses, cytokines secreted by T cells and macrophages are very important for regulating humoral immunity because they serve to activate resting B cells.

Assessment of Immune Function

Throughout this chapter, we refer to seasonal changes in immune function as determined by one of several standard tests of immunity. To aid in the interpretation of these immune data, the methodology commonly used in these studies is reviewed below. In addition to utilizing assays of immune function, several field studies reviewed in subsequent sections report seasonal changes both in ectoparasite (e.g., ticks and mites) and blood parasite (e.g., protozoa and hematazoa) loads. Measures of parasite prevalence are not direct measures of immune function, because factors other than reduced immune function (e.g., changes in the vector, not the host) may contribute to the observed results. Nonetheless, both field and laboratory studies provide compelling evidence for seasonal fluctuations in parasite prevalence.

Immune function can be measured in the laboratory using either in vitro or in vivo techniques. There are advantages and disadvantages to each method. In vitro assays are beneficial because they can provide a constant environment devoid of the fluctuations and changes in the internal milieu that occur continuously. Additionally, in vitro assessment of immune function permits examination of the precise effects of particular neuroendocrine substances on immune responses without the confounding variable of other hormonal influences, for example, systematically exposing lymphocytes to various hormones, individually, to assess the precise contribution of each hormone to the immune response. The assessment of isolated responses is a double-edged sword. One of the pitfalls of in vitro assessments of immune function is that this technique minimizes the influences of both neural and endocrine factors, which are likely to have a profound impact on "natural" immune responses. Accordingly, the primary benefit of in vivo assessments of immune function is that immune responses are examined within the organism's natural internal milieu. Therefore, some researchers argue that in vivo assessments of immune function have the most information value regarding neuroendocrine-immune interactions (Maier and Laudenslager 1988). Immunoassays can be developed to examine either specific immune responses against a particular antigen (pathogen) or nonspecific immune responses such as the replication of cells.

Lymphocyte Proliferation Activation of lymphocytes is measured in vitro. This immune response is generally considered a good indicator of the mitotic ability of lymphocytes following exposure to a pathogen in vivo (Borysenko 1987). Following activation, lymphocytes undergo blastogenesis and begin to proliferate. The synthesis of new DNA that results during proliferation and differentiation of lymphocytes is what is measured in this assay. To stimulate or activate lymphocytes, either antigens or mitogens are used. An antigen is a foreign substance that the lymphocytes have not encountered previously. A mitogen is a protein or lectin that stimulates lymphocytes to divide. Some mitogens, like concanavalin A (ConA) from *Canavaia ensiformis* and phytohemagglutinin (PHA) from *Phaseolus vulgaris*, primarily stimulate T cells. Other mitogens, like pokeweed mitogen (PWM) from *Phytolacca americana* and lipopolysaccharide (LPS) from *Escherichia coli*, are commonly used to stimulate B cells. The lymphocytes used in this assay can be isolated from circulation or from specific immune organs, like the thymus or spleen. In terms of nonspecific measures of lymphocytes, this measure is much more reliable and informative than merely counting numbers of different types of lymphocytes.

Natural Killer Cytotoxicity The ability of natural killer (NK) cells to kill tumor cells is measured in vitro in this assay. Measurement of the cytotoxic activity of NK cells killing tumor cells does not require prior exposure to the tumor cells; therefore, this is considered a nonspecific measure of cell-mediated immunity. In this assay NK cells isolated from either blood or immune organs are incubated with tumor cells (specific for the species; e.g., mouse tumor cell lines are used to assess NK cells from mice) that have been labeled with chromium[51]. When a tumor cell is lysed by an NK cell, the tumor cell releases the chromium into the supernatant, and it is the released chromium that is measured and recorded as percent lysis. The advantage of this measure is that it can provide a general indication of the ability of NK cells to kill tumorous or virally infected cells in vivo.

Skin Tests for Delayed Hypersensitivity This is an in vivo assessment of specific cell-mediated immune responses involved in the cutaneous reaction to antigens. Delayed hypersensitivity measures the memory response of lymphocytes to a specific antigen. This response is quantified by measuring the dimensions of erythema and induration following local injection of the antigen (Borysenko 1987). This assay of immune function has direct clinical relevance. For example, delayed hypersensitivity skin testing is the basis for diagnostic tests for such agents as tuberculin. This reaction can also be used to assess resistance to various bacterial and viral infections (Dhabhar and McEwen 1996). The manifestations of this response can also be seen in allergic reactions to such agents as poison ivy, poison oak, and various cosmetics and dyes (Dhabhar and McEwen 1996).

Antibody Production The soluble factors, or antibodies, produced and released from B cells can be quantified using an enzyme-linked immunosorbent assay (ELISA). The ELISA tests for the presence of antibodies in either plasma or serum, following exposure to a novel antigen (i.e., a foreign substance that the organism has not been exposed to previously). The basic premise of the ELISA involves the binding of antibodies in circulation to a known amount of antigen. This antibody-antigen complex is then labeled with a secondary antibody followed by a nonradioactive tag. This nonradioactive enzyme changes optical density (color) when it binds to the secondary antibody that is bound to the antigen-antibody complex. It is the colorimetric change that provides quantitative information. This assay is extremely precise and can provide a direct assessment of an immune response mounted against a specific pathogen;

therefore, this measurement is considered a good indicator of specific humoral immune function. Other soluble factors, such as interleukins or interferons, can be quantified in a similar manner as antibodies, using the ELISA.

Serological Measures Using classic methods of hematology, numbers and types of immune cells, such as granulocytes, macrophages, and lymphocytes, can be assessed in blood smears. Additionally, hematocrit, hemoglobin, and red blood cell counts can be assessed using this procedure. The problem with this immune measure is that understanding how numbers of cells in circulation translates into an assessment of immune function is still unclear. Recently, immune cell subpopulations in blood and tissue have been assessed using fluorescence markers for specific immune cells and a cell sorter, called a flow cytometer. Using this technique, both relative and absolute numbers of macrophages, T cells, B cells, and NK cells can be quantified. Additionally, subpopulations of cells can be quantified using appropriate markers. For example, of T cells in circulation, the ratio of CD4+ (i.e., T helper cells) to CD8+ (i.e., T cytotoxic cells) cells can be determined using markers specific for the receptor subtypes (i.e., CD4+ or CD8+). In many immunodeficiency diseases, such as human immunodeficiency virus (HIV) infection, the ratio of T helper cells to T cytotoxic cells is altered. Accordingly, this immune measure has important clinical and basic science implications; however, the disadvantage of this assessment is that it provides no functional information regarding an individual's ability to mount an immune response (Borysenko 1987).

Environment and Immune Function

Seasonality, Photoperiodism, and Immune Function

In common with the seasonal pattern of lymphatic organ development (John 1994; Nelson and Demas 1996), elevated gonadal steroid levels associated with breeding coincide with an increased prevalence of some diseases and reduced immune function. Many studies have demonstrated a seasonal change in parasite and pathogen prevalence, and the overwhelming evidence suggests a seasonal change in the host, rather than in the parasite (John 1994). For example, house sparrows (*Passer domesticus*) infected with avian malaria (*Plasmodium relictum*) display a relapse of symptoms occurring synchronously throughout a population of infected birds and coincident with the onset of vernal breeding activities (Applegate and Beaudoin 1970). The possibility that immune function is enhanced in birds during the winter to keep parasitic activities minimal remains open (Beaudoin et al. 1971; John 1994).

Thus, it appears that *Plasmodium* infections remain latent during the winter, but parasitemia becomes evident mainly during the energetically expensive vernal migration or breeding (Beaudoin et al. 1971; Alexander and Stimson 1988). Increased prevalence or incidence of relapse of blood protozoa of *Plasmodium*, *Leucocytozoon*, *Haemoproteus*, and *Trypanosoma* species has been reported during spring and summer among many species of birds (see John 1994 for review). Suppressed immune function during the breeding season has also been reported for birds with viral infections (Vindervogel et al. 1985; Hughes et al. 1989).

Similar patterns of seasonal changes in mammalian immune function and disease prevalence have also been reported. For instance, seasonal variation in the ability of bank voles (*Clethrionomys glareous*) to infect larval ticks with Lyme disease (caused by *Borrelia burgdorferi*) has been reported (Talleklint et al. 1993). Although larval tick infestations of voles were highest in June and July, nearly 70% of *B. burgdorferi* infections occurred during August and September. Virtually no infections occurred during the winter. These data are intriguing, but whether they reflect a seasonal alteration in immune function of the host or reflect the latency to infection from year to year requires further studies.

Similarly, cotton rats (*Sigmodon hispidus*) displayed seasonal variation in both humoral and cell-mediated immunity (Lochmiller et al. 1994). Although these results strongly suggest a seasonal cycle of immune function that is consistent with photoperiodic mediation, animals captured the following February did not display a similar change in immune function in comparison to the previous year (Lochmiller et al. 1994), suggesting that factors other than photoperiod affect the timing of immune function. Although there are other equivocal results, the vast majority of studies that report seasonal fluctuations in immune function or disease prevalence indicate that immune function and disease resistance is lowest during the breeding season (Dobrowolska et al. 1974; Dobrowolska and Adamczewska-Andrzejewska 1991).

The vast majority of laboratory studies of environmental changes in immune function have manipulated ambient photoperiod. To our knowledge, evaluation of photoperiodic changes on immune function has only been conducted in small mammals, primarily rodents. Short day lengths increased splenic mass of deer mice (*Peromyscus maniculatus*) (Vriend and Lauber, 1973), and Syrian hamsters (*Mesocricetus auratus*) (Brainard et al. 1985). Splenic mass, total splenic lymphocyte numbers, and macrophage counts were significantly higher in hamsters exposed to short photoperiod as compared to animals exposed to long photoperiod (Brainard et al. 1985; Vaughan et al. 1987). However, photoperiod did not affect thymic weight or antibody production in hamsters (Brainard et al. 1988). In all of the studies that manipulated photoperiod reviewed here, activity level, food intake, and body mass have not been controlled (but see Nelson et al. 1995). Photoperiodic influences on lymphocyte number and total white blood cell (WBC) count also have been reported for deer mice (Blom et al. 1994). Animals maintained in short day lengths (8L:16D) possessed more WBCs than animals maintained in long day lengths (L16:8D); neutrophil number was unaffected by day length in adult female deer mice. In another study of adult deer mice, animals maintained in short days displayed faster healing rates than long-day mice (Nelson and Blom 1994).

In another recent study from our laboratory, splenic lymphocyte proliferation in response to ConA was assessed in male deer mice that had been maintained for 8 weeks either in long (16L:8D) or short (8L:16D) days (Demas et al. 1996). Splenic lymphocytes from short-day deer mice exhibited increased proliferation in response to ConA compared with the splenic lymphocytes from long-day animals (figure 8.1). These data establish that photoperiod influences immune function.

Young C57BL/6 mice (*Mus musculus*) were housed in a laboratory environment kept constant at $22.5 \pm 1°C$ and a 12L:12D photoperiod (Brock 1983). Peak proliferative responses of lymphocytes in response to PHA, ConA, and LPS were two to

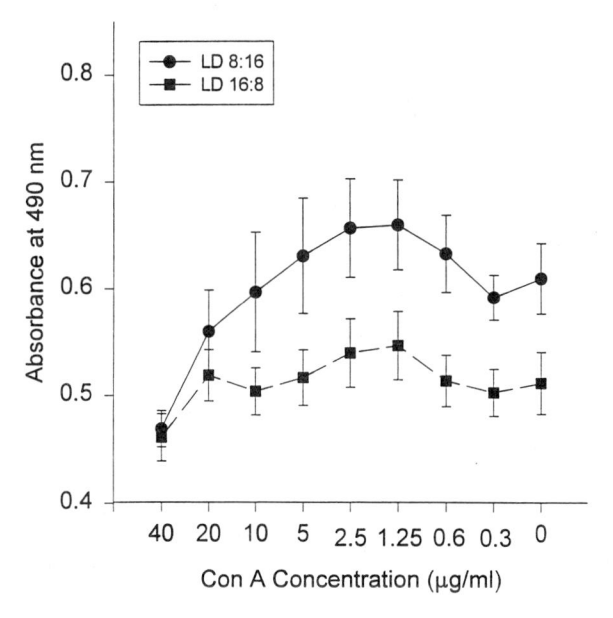

Figure 8.1
Mean ($\pm$SE) splenocyte proliferation responses to the T-cell mitogen concanavalin A (ConA) in deer mice (*Peromyscus maniculatus bairdii*). Animals housed for 8 consecutive weeks in short day lengths had higher proliferative responses than animals housed in long day lengths. (From Demas et al. 1996.)

five times higher in March and April 1978 and February and March 1977 than in either of the two previous Decembers. Summer comparisons were not reported. It should be noted that these animals, like most laboratory strains of rats, typically are reproductively nonresponsive to photoperiod (Nelson 1990). Although the reproductive systems of these animals may not respond to day length (or to an endogenous circannual rhythm), nonreproductive traits may retain responsiveness to day lengths (Goldman and Nelson 1993).

In general, short day lengths appear to bolster immune function. The most likely physiological mechanism by which photoperiod affects immune function involves changes in steroid hormone concentrations. The precise mechanisms through which photoperiod interacts with the endocrine system to exert influences on the immune system are not known. However, steroid hormones have well-known effects on energy metabolism (Wade and Schneider 1992). Thus, the effects of steroid hormones on the immune system could be mediated via their effects on metabolism. The presence of receptors both for androgens in lymphoid tissues (e.g., thymus, bone marrow, and spleen) and for estrogens in the cytosol of circulating lymphocytes and macrophages suggests that these steroid hormones play an important role in the mediation of immune function (Alexander and Stimson 1988; J.H.M. Cohen et al. 1983; Cutolo et al. 1996; Danel et al. 1983; Grossman 1985; Hall and Goldstein 1984; McCruden and Stimson 1991; Nilsson et al. 1984; Sasson and Mayer 1981; Weusten et al. 1986). Generally, androgens compromise and estrogens enhance immune function (Alexander and Stimson 1988; Billingham 1986; Grossman 1985; Olsen and Kovacs 1996; Schuurs and Verheul 1990). Concentrations of both of these steroid hormones vary seasonally. Sex steroid hormones are suppressed by short day exposure in long-day breeders (Bronson 1989). Thus, the short-day enhancement of immune function could be the result of the reduction of steroid hormone suppression of immunity, at least among males. However, gonadectomized, short day–housed deer mice display elevated immune function compared with their long day–housed counterparts, suggesting that sex steroid hormones may not mediate short-day enhancement of immune function (Demas and Nelson 1998).

Prolactin is also an immunomodulatory hormone, and secretion of prolactin is profoundly affected by day length; short photoperiods induce reductions in blood prolactin concentrations (Arkins et al. 1993; Bernton et al. 1991, 1992; Castanon et al. 1992; Goldman 1983; Goldman and Nelson 1993; Matera et al. 1992; Reber 1993). It is possible that photoperiodic effects on immune function are mediated via photoperiodic changes in blood concentrations of prolactin. The pattern of melatonin secretion is also altered by photoperiod. In most cases, melatonin enhances immune function (Caroleo et al. 1992; Kuci et al. 1988; Maestroni and Peirpaoli 1981;

Maestroni et al. 1986; Pioli et al. 1993). It remains possible that melatonin could act directly on the immune system (e.g., see Nelson and Blom 1994). However, additional studies are required to determine the specific contributions of various hormones, alone or in combination, to immune function.

The effects of photoperiod and melatonin treatment on reproductive and immune function of two subspecies of *Peromyscus maniculatus* from different latitudes of origin were assessed by Demas et al. (1996). *P.m. bairdii* (lat. 42°51′N) and *P.m. luteus* (lat. 30°37′N) were maintained individually in either long (16L:8D) or short (8L:16D) days for 8 weeks. Short-day *P.m. bairdii* displayed regressed reproductive and elevated splenocyte proliferation in response to the T-cell mitogen ConA, as compared with long-day mice. Individuals of *P.m. luteus* housed in short days displayed neither reproductive nor immunological changes in comparison with long-day conspecifics (Demas et al. 1996). In a follow-up study, individuals from these two subspecies were implanted with either empty Silastic capsules or capsules filled with melatonin. Again, melatonin treatment evoked gonadal regression and elevated splenocyte proliferation only among *P.m. bairdii* mice. Neither reproductive nor immunological function was affected by melatonin in *P.m. luteus*. These results suggest that reproductive responsiveness to melatonin mediates short-day enhancement of immune function in deer mice. Importantly, these data imply that melatonin may not possess universal immune-enhancing properties. Rather, the effectiveness of melatonin in influencing immune responses may be constrained by reproductive responsiveness to this indole-amine. However, melatonin treatment of castrated or gonadally intact *P.m. bairdii* caused elevated splenocyte proliferation regardless of gonadal state (Demas and Nelson 1998), suggesting that direct effects of melatonin or indirect effects, independent of gonadal hormones, mediate photoperiodic effects on immune function.

Temperature and Immune Function

We would predict that immune function should be compromised in animals maintained in low as compared to high temperatures. Recently, the interaction between photoperiod and temperature was examined on IgG concentrations and splenic mass in male deer mice (*P. maniculatus*) (Demas and Nelson 1996). Animals were maintained in 16L:D8 (long-day) or 8L:16D (short-day) photoperiods and either in 20° or 8°C ambient temperatures.

The results showed that cold temperatures compromised the immune response of long day–housed animals, and this effect was attenuated by housing on short-day photoperiods (figure 8.2). Short days elevated IgG concentrations relative to IgG concentrations typical of long days. Low temperatures caused a significant reduction

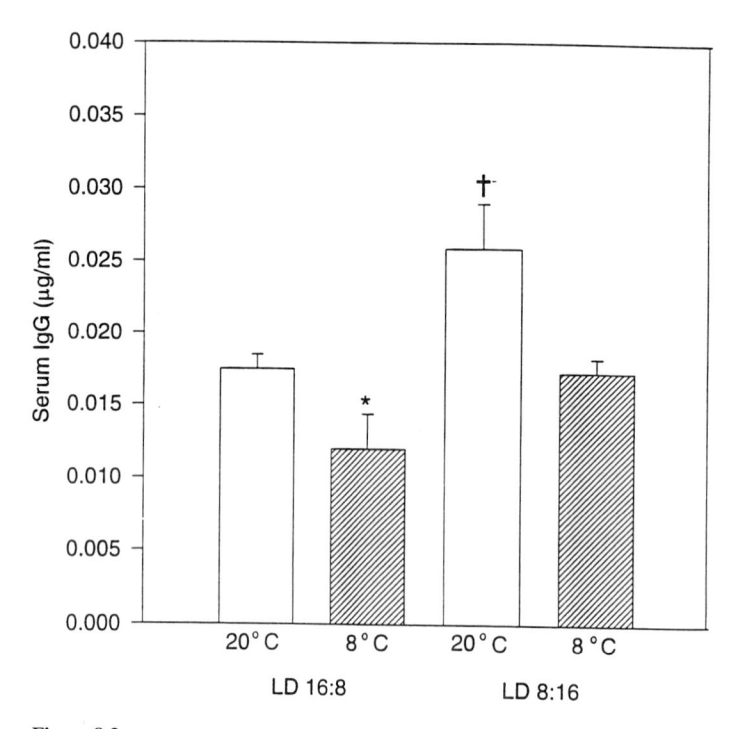

Figure 8.2
Mean (±SE) IgG concentrations from male deer mice (*Peromyscus maniculatus bairdii*) housed in long (16L:8D) or short (L8:16D) days in ambient (20°C) or low (8°C) temperatures. Bars with different symbols are statistically different. (From Demas and Nelson 1996.)

in IgG concentrations only in mice housed in long days, but not in mice housed in short days. The net effect of short-day enhancement and low temperature reduction of IgG concentrations was no appreciable difference from long-day mice kept at 20°C (Demas and Nelson 1996) (see figure 8.2). This adaptive system may help animals cope with seasonal stressors and ultimately increase reproductive fitness. In order to enhance immune function in anticipation of demanding winter conditions animals must initiate these adaptations well in advance of the demanding conditions. The most reliable environmental cue for time of year is the annual pattern of changing photoperiod.

Energetic Challenges to Immune Function

Animals must maintain a relatively constant flow of energy (i.e., energy intake and storage must be greater than energy expended) to the body, despite potentially large fluctuations in energetic availability in their environment, as was discussed in chapters 2 and 3. Mounting an immune response likely requires resources that could otherwise be allocated to other biological functions (Sheldon and Verhulst 1996). Thus, we have predicted that immune function is "optimized" so that individuals can tolerate small infections if the energetic costs of mounting an immune response outweigh the benefits (Behnke et al. 1992). In fact, few studies have assessed directly the energetic cost of mounting an antibody response, although the initiation of an immune response (i.e., inflammation, activation of cytokines, induction of fever) presumably requires substantial energy. For example, every 1°C increase in body temperature requires a 7% to 13% increase in caloric energy production, depending on the species (Maier et al. 1994). Recently, precise quantification of the energetic costs of an immune response was attained in adult house mice (*Mus musculus*) (Demas et al. 1997a). House mice were injected with a specific nonreplicating antigen, keyhole limpet hemocyanin (KLH). This substance induced an antibody response without inducing fever or making the treated animals sick (Dixon et al. 1966; Curtis et al. 1970). Both oxygen consumption and metabolic heat production increased in KLH-injected animals (figure 8.3). Mounting an immune response costs energy; presumably, mounting an immune response to a replicating antigen requires additional energy. In contrast, oxygen consumption is actually reduced in male red jungle fowl (*Gallus gallus*) infected with the nematode *Ascaridia galli*, compared with uninfected males, suggesting that overcoming infection is costly in terms of aerobic capacity (Chappell et al. 1997). Thus, a general energy deficit can increase the risk of infection and death because sufficient energy reserves may not be available to sustain immunity.

The notion that optimal immune function and reproduction are energetically incompatible was tested in captive zebra finches (*Taeniopygia guttata*), in which reproductive effort was manipulated experimentally by altering brood size. Increased brood size resulted in reduced immunoglobulin production against sheep red blood cells (SRBCs) (Deerenberg et al. 1997). Similarly, increased work load (i.e., increased hopping activity) was associated with lower antibody responses in nonbreeding zebra finches, suggesting that energetically costly activities associated with reproductive behavior are incompatible with maximal immune function (Deerenberg et al. 1997). These data provide a functional explanation for increased parasitism during the breeding season (Festa-Bianchet 1989; Richner et al. 1995).

In an attempt to examine the role of energetics in seasonal changes in immune function in deer mice, 2-deoxy-D-glucose (2DG) was used to manipulate energy availability at the input end of the energetic equation (Demas et al. 1997b). 2DG is a glucose analogue which inhibits cellular utilization of glucose, thus inducing a state

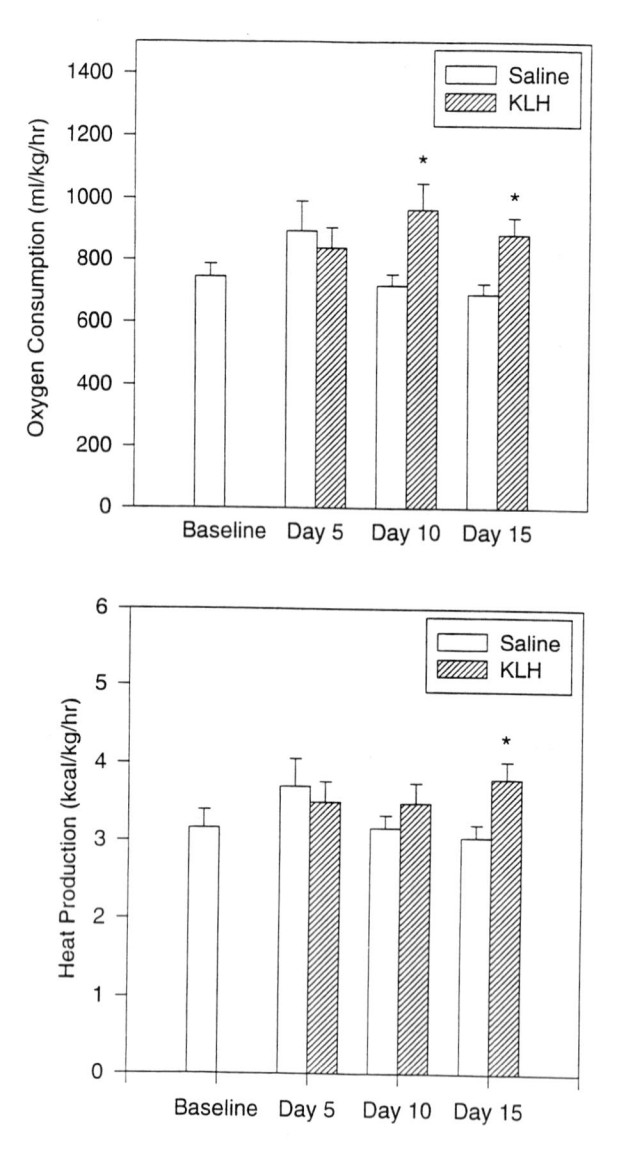

Figure 8.3
Mean (±SE) O_2 consumption and heat production in sexually mature C57BL/6J mice before and 5, 10, and 15 days following immunization with the innocuous antigen, keyhole limpet hemocyanin (KLH). Mice immunized with KLH displayed increased O_2 consumption and heat production compared with animals injected with saline. (Adapted from Demas et al. 1997a.)

of glucoprivation (Smith and Epstein 1969). 2DG competes with glucose for uptake and blocks glucose oxidation at the phosphohexosisomerase step. Treatment with 2DG results in a well-known compensatory sympathoadrenal response that involves increased secretion of epinephrine and norepinephrine, as well as glucocorticoids. Treatment with 2DG increases serum corticosterone concentrations (Lysle et al. 1988), induces anestrus in female Syrian hamsters (*Mesocricetus auratus*) (Schneider and Wade 1989), and initiates daily torpor in female Siberian hamsters (*Phodopus sungorus sungorus*) (Dark et al. 1994). 2DG-induced metabolic stress also affects immune function; 2DG administration inhibits splenic T-cell proliferation in a dose-dependent manner in laboratory strains of rats (*Rattus norvegicus*) (Lysle et al. 1988) and mice (*Mus musculus*) (Miller et al. 1994).

We used 2DG-induced glucoprivation to determine whether cues generated by housing in short-day photoperiods could buffer deer mice against energetic challenges. We found that long day–housed deer mice injected with 2DG had elevated corticosterone concentrations, as compared with long-day mice injected with saline; corticosterone concentrations were not significantly elevated in short-day mice injected with 2DG. 2DG-treated long-day mice displayed reduced splenocyte proliferation to ConA compared with saline-injected mice. Splenocyte proliferation did not differ among short-day deer mice regardless of experimental treatment; short-day animals exhibited enhanced immune function. Overall, short-day mice treated with 2DG displayed higher splenocyte proliferation than long-day mice treated with 2DG (Demas et al. 1997b) (figure 8.4). Short day–housed mice undergo a number of metabolic adjustments that might change their responsiveness to glucoprivation. For example, they may be predisposed toward the utilization of free fatty acids from lipids stored in adipose tissue. Thus it is possible that short-day animals may be predisposed to utilization of free fatty acids hydrolyzed and mobilized from adipose tissue, and thus may not respond as readily to treatment with 2DG, which affects only glucose oxidation. Conversely, short-day animals may be better able to synthesize and oxidize glucose than long-day mice. In either case, future studies must determine the precise mechanisms involved in the reduced responsiveness to 2DG among animals exposed to short photoperiods.

These data are consistent with the hypothesis that short days buffer against metabolic stress. Reduced corticosterone concentrations in animals maintained on short days or treated with melatonin are likely due to improved metabolic function (Saafela and Reiter 1994). Accordingly, improved immune function in short days represents one component of numerous winter-coping adaptations that may be mediated by melatonin.

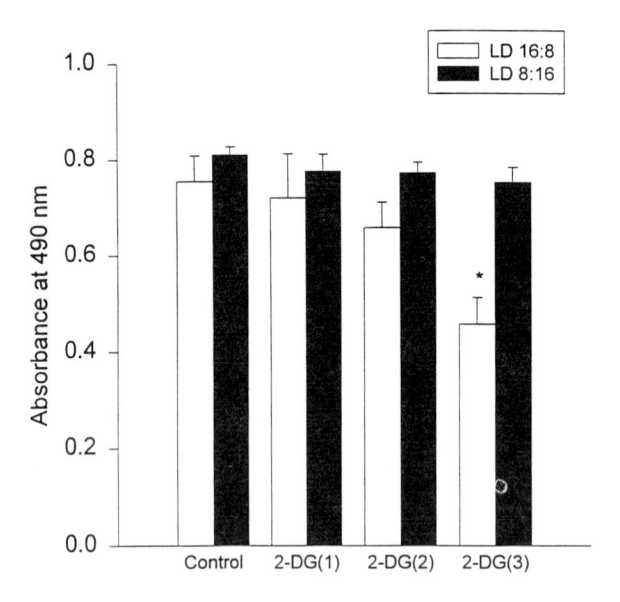

Figure 8.4
Mean (±SE) splenocyte proliferation responses to an optimal concentration (5 g/mL) of concanavalin A (ConA) in deer mice (*Peromyscus maniculatus bairdii*). Data are represented as absorbence units based on an enzymatic reaction that positively correlates with amount of cellular division. Animals receiving three consecutive injections of 2-deoxy-D-glucose (2DG) had reduced proliferative responses compared with controls. (Adapted from Demas et al. 1997b.)

Social Factors Influence Reproduction and Immune Function

Social Interactions: Differences Between the Breeding and Nonbreeding Seasons

Nontropical rodents display many distinct adaptations to cope with the challenge of winter survival (Bronson and Heideman 1994; Moffatt et al. 1993). For example, social organization often changes from highly territorial during the breeding season to highly social during the winter. Communal group formation in the winter is often an attempt to reap the thermoregulatory benefits (e.g., warmth and humidity conservation) of huddling (Ancel et al. 1997; Getz and McGuire 1997; Madison et al. 1984 McShae 1990). Animals that form communal nests with unrelated, mixed-sex conspecifics during the winter (e.g., meadow voles [*Microtus pennsylvanicus*], prairie voles [*Microtus ochrogaster*], and flying squirrels [*Glaucomys volans*]) typically cease reproductive activities, and then disperse following reproductive recrudescence (Getz and McGuire 1997; Madison et al. 1984; Layne and Raymond 1994; Wolff and Lidicker 1981).

Odor preference is an important component of seasonally changing social interactions among small rodents. For many species, chemosensory cues are used to assess the quality of potential mates and facilitate reproductive condition (e.g., estrous behavior) (Egid and Brown 1989; Potts et al. 1991; Richmond and Stehn 1976; Yamazaki et al. 1976). The extent to which odor preference varies seasonally has been studied extensively in the context of reproduction; however, seasonal variation in odor preference in relation to disease (or immunological) status has not been examined. Female meadow voles (*M. pennsylvanicus*) prefer the scent of conspecific males to that of conspecific females during the breeding season; conversely, during the nonbreeding season (i.e., winter) female meadow voles prefer the scent of conspecific females to that of conspecific males (Ferkin and Seamon 1987). Changes in day length are sufficient to cause the seasonal variation in odor preference of wild-caught female meadow voles, suggesting that photoperiod is the primary cue mediating differential social interactions during the breeding and nonbreeding seasons in this species (Ferkin and Zucker 1991). Seasonal or photoperiodic changes in social behaviors and odor preference in meadow voles are mediated by seasonal changes in circulating sex steroid hormones; that is, a preference for opposite sex conspecific odors during the breeding season is mediated by high androgen concentrations in males and high estrogen concentrations in females (Ferkin and Gorman 1992; Ferkin and Zucker 1991). Seasonal variation in circulating sex steroid hormones may also underlie changes in immune function observed between the breeding and nonbreeding seasons (Nelson and Demas 1996; Nelson et al. 1995).

Our laboratory has examined the relationship between chemosensory cues and disease prevalence during the breeding season. Female meadow voles housed in long photoperiods (i.e., 16 hours of light) can distinguish between the odors of parasitized and unparasitized males, and actually prefer the odors of unparasitized conspecific males (Klein, Gamble, and Nelson 1999) (figure 8.5). Female odor preference for unparasitized vs. parasitized males during the nonbreeding season (i.e., short photoperiods) remains to be examined. These results suggest that immunological and disease status are important criteria in mate selection. Further research is required to identify specific physiological links between immune and reproductive functions.

Social Interactions and Immune Function

Stress of Breeding Hans Selye was among the first to demonstrate that exposure to stressors evokes elevated glucocorticoid secretion, adrenal hypertrophy, and thymic involution; thus he suggested that stressors inhibit some aspects of immune function (Selye 1956). Exposure to stressors may require that stored energy be allocated to other physiological or behavioral processes, thereby reducing the amount of energy

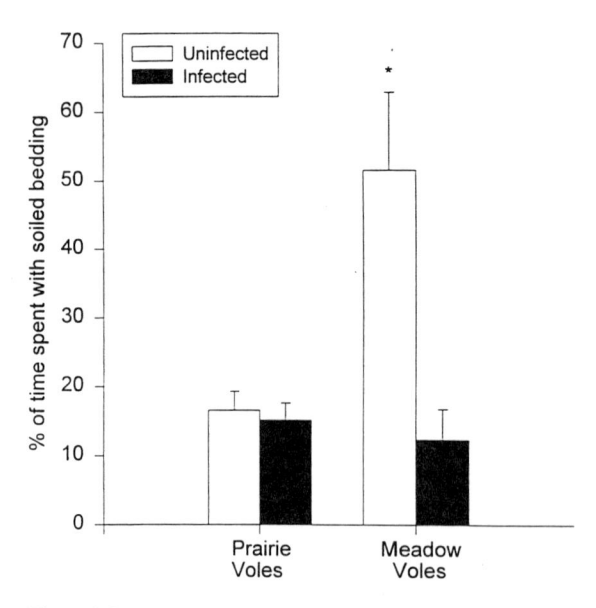

Figure 8.5
Mean (±SE) proportion of time female prairie voles (*Microtus ochrogaster*) and meadow voles (*Microtus pennsylvanicus*) spent with soiled bedding from uninfected males and males infected with *Trichinella spiralis*. Female meadow voles spent more time with the soiled bedding of uninfected as compared with infected males. (Adapted from Klein et al. 1999.)

that can be utilized for optimizing immune function. Laboratory studies of stress have established that stressors (both psychological and physical) can suppress cell-mediated and humoral immunity, as well as increase susceptibility to disease, and escalate tumor growth in rodents, nonhuman primates, and even humans (Bohus and Koolhaas 1991). However, the stressors in most animal studies—noise, shock, and restraint—are artificial and probably have not been previously encountered (Bohus and Koolhaas 1991). The extent to which the physiological sequelae of artificial stressors model natural stressors remains unspecified. Natural stressors usually include environmental and social pressures, such as temperature changes, reduced food availability, and competition for mates, associated with both the breeding and non-breeding seasons.

For males, there are several stressors associated with the breeding season, including energetically expensive courtship displays and exhaustive aggressive encounters with rivals over territories, potential mates, or other resources (e.g., food). For example, in the anuran amphibian, *Hyla versicolor*, courtship vocalizations used to attract females utilize three times as much oxygen as resting metabolism (Taigen and

Wells 1985). The development and maintenance of some physiological structures used to attract females are often energetically incompatible with maximal disease resistance (Nelson and Demas 1996; Zuk et al. 1990). For example, experimental induction of parasitic infection in male red jungle fowl (*Gallus gallus*) reduces the expression of secondary sex traits (e.g., combs and hackle feathers) used to attract mates; however, traits that are not associated with reproduction (e.g., bill size and saddle feathers) are not affected by the intensity of the infection (Zuk et al. 1990). In male reindeer (*Rangifer tarandus tarandus*), the stress of the rut season has been hypothesized to elevate glucocorticoid concentrations and reduce immunological resistance to nematode infestation (Folstad et al. 1989; Gaudernack et al. 1984; Halvorsen et al. 1985).

The stress of the mating season appears to be related to elevated mortality rates in the small marsupial, brown antechinus (*Antechinus stuartii* Macleay). In wild populations of brown antechinus, males typically engage in intense aggressive encounters with other males, as well as participate in prolonged mating bouts lasting approximately 6 to 12 hours over several successive nights during the breeding season (Marlow 1961; Woolley 1966). Physiologically, these males experience a dramatic decline in plasma corticosteroid-binding globulin (CBG) and a rise in plasma free corticosteroids during the mating season, which is generally followed by death due to involution of lymphoid organs, reduced humoral immunity, and infestation of parasites (Barker et al. 1978; Bradley et al. 1980).

Females are also exposed to stressors associated with the breeding season, including mate selection, pregnancy, and lactation. Accordingly, females, like males, experience reduced immunological functioning during the breeding season (Festa-Bianchet 1989). Cell-mediated immune responses are reduced and susceptibility to infection (viral, bacterial, parasitic, and fungal) is increased during pregnancy in both human and nonhuman animals (Priddy 1997). The prevailing hypothesis to explain compromised immune function during pregnancy is the necessity to protect the genetically incompatible fetus from rejection by the maternal immune system (Janeway and Travers 1994). Alternatively, because pregnancy is energetically costly to females (Bronson 1989), immune function may be suppressed during pregnancy in an effort to maintain a positive energy balance. Lactation is also energetically costly to females (Bronson 1989; Fairbairn 1977) and may be incompatible with maximal disease resistance. Accordingly, lactating bighorn ewes have increased parasitic infection in fecal samples as compared with nonlactating females (Festa-Bianchet 1989). However, during early lactation, compromised immune function is not correlated with serum indices of metabolic status in cows (Ropstad et al. 1989). Further studies are required to tease apart the energetic costs of breeding and immune function to determine whether these physiological systems are incompatible in other species.

Mate Choice In males, high testosterone concentrations enhance competition and aggressiveness, while also reducing both cell-mediated and humoral immune responses (Møller 1994; Wingfield, 1994; Zuk 1990, 1994). Thus, a tradeoff exists in the evolution of high testosterone concentrations. High testosterone concentrations are beneficial to males because mating success may increase; however, increased mating success may be achieved at the cost of reduced survival. Low circulating testosterone concentrations may enhance survival at the cost of reducing successful competition for mates. In essence, males with high testosterone concentrations that do not have compromised immunity are at a selective advantage as potential mates (Folstad and Karter 1992; Wedekind and Folstad 1994). The evolutionary bind for males is that they must have a way to advertise this selective advantage. Males may use the maintenance of secondary sex characters that are androgen-dependent (e.g., antlers, body size, or tail feathers) to advertise exceptional immune function (Folstad and Karter 1992). Thus, it was initially hypothesized that possession of these traits by males may inform potential mates that they can maintain androgen-dependent traits, despite the cost of high circulating testosterone concentrations on immune function (Folstad and Karter 1992). Alternatively, high-quality males may use androgen-dependent traits to advertise possession of sufficient resources for allocation to both superior immune function and secondary sex traits (Wedekind and Folstad 1994).

These hypotheses have not been empirically tested; however, adult male barn swallows (*Hirundo rustica*) with experimentally elongated tail feathers have lower immune responses against SRBC, suggesting that tail length imposes an immunological cost on males (Saino and Møller 1996). From an energetic perspective, metabolic resources, as opposed to hormone concentrations, may be the rate-limiting factor mediating the relationship between the expression of secondary sex characters and immunocompetence (Sheldon and Verhulst 1996). Thus, utilization of energy for elaborate displays or growth of epigamic characters during the breeding season may limit resources needed for maximizing immune function. Presumably, if individuals that partition more metabolic resources into traits important for reproduction gain more matings, then selection against heightened immune function during the breeding season would persist.

In terms of the evolutionary significance of these testosterone-dependent traits (i.e., male ornamentation and suppressed immune function), Hamilton and Zuk (1982) proposed that secondary sex characters signal male quality, suggesting that these ornaments are an indicator of general fitness or resistance to disease. Therefore, because there is a genetic component to disease resistance, by mating with a male that advertises "good genes" for disease resistance, a female may gain an indirect heritable advantage for her offspring (Hamilton and Zuk 1982). Alternatively, females may use the condition of secondary sex traits to gain a direct fitness advantage; that is, to avoid mating with males that carry potentially contagious diseases (Hamilton 1990; Kirkpatrick 1996; Kirkpatrick and Ryan 1991; Møller 1994). In either case, selection should favor characteristics that allow females to judge the present and past parasite load as well as the underlying immunocompetence of a male (Read 1987). Thus, male ornaments or even pheromones may function as a signal of disease resistance and female choice may serve to detect such signals to distinguish healthy males with which to mate (Kavaliers and Colwell 1995; Zuk et al. 1990).

This controversial hypothesis (Folstad and Karter 1992), appears to be applicable to polygynous species, in which males typically have high circulating testosterone concentrations, exaggerated secondary sex traits, and increased incidence of competition for mates; thus, the validity of this hypothesis for monogamous species remains ambiguous. Because reproductive roles still differ between monogamous males and females, selection for revealing traits in males may persist in the absence of any initial sexual dimorphism (Møller 1994). This is especially critical if disease resistance is an important feature of good parental care among monogamous males (Møller 1994).

The extent to which female preference for healthy males represents an indirect preference for more robust immune responses underlying disease resistance remains unexplored. Additionally, the degree to which male immunocompetence affects female choice, in relation to mating system, has not been reported. In a recent study, female preference for unmanipulated males or males that were exposed in vivo to LPS was examined in monogamous prairie voles (*M. ochrogaster*) and polygynous meadow voles (*M. pennsylvanicus*). LPS, or endotoxin, is the immunologically active component of gram-negative bacteria cell walls, usually derived from *E. coli*. Exposure to LPS results in an acute inflammatory response, characterized by the activation of immune cells and the release of cytokines into circulation. The increased release of cytokines, especially interleukin (IL)-1, results in a series of immune and endocrine responses that mimic bacterial infection. In our study, female prairie voles spent significantly less time with LPS-injected than with saline-injected male prairie voles, whereas among meadow voles, females spent an equivalent amount of time with both groups of male meadow voles (Klein and Nelson 1999) (figure 8.6). These data suggest that females can detect subtle changes in immunocompetence, but, the precise cues used by female voles to assess immune status remain unidentified. We hypothesize that females may use odor cues to distinguish healthy from immuno-

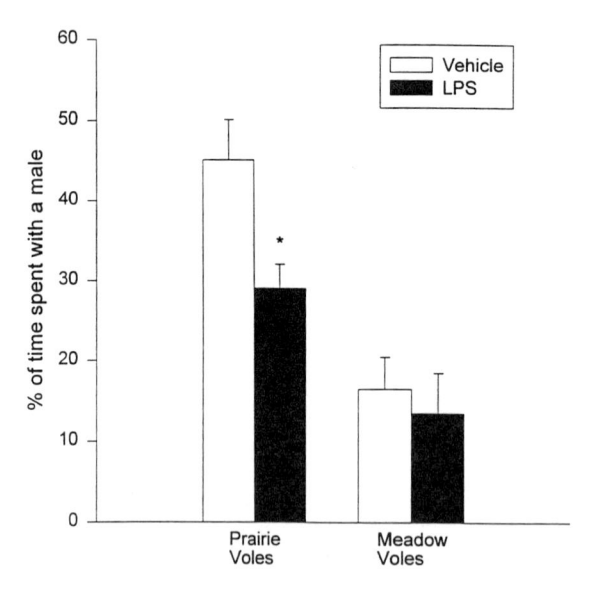

Figure 8.6
Mean (±SE) proportion of time female prairie voles (*Microtus ochrogaster*) and meadow voles (*Microtus pennsylvanicus*) spent with males injected either with vehicle or lipopolysaccharide (LPS) 3 hours before behavioral testing. Female prairie voles spent more time with vehicle injected as compared with LPS-injected conspecific males. (Adapted from Klein and Nelson 1999b.)

compromised males. Also, the extent to which females use cues associated with immunocompetence may vary seasonally (see Ferkin and Seamon 1987).

Long-Term Social Interactions Social conditions dramatically influence immune function and disease susceptibility (Barnard et al. 1994; Bohus and Koolhaas 1991; Karp et al. 1993; Rabin and Salvin 1987). Social isolation is a common laboratory method used to elevate glucocorticoids in laboratory rats, mice, and nonhuman primates. Traditionally, the stress of social isolation has been assigned to lack of contact comfort. For example, the effects of maternal separation on immune function in bonnet macaque infants is less pronounced for infants provided social interactions as compared with isolated infants (Boccia et al. 1997). Additionally, a positive relationship is observed between NK cytotoxicity and affiliative behavior for infants housed in social groups during maternal separation (Boccia et al. 1997), suggesting that social support, or possibly huddling behavior, has immunomodulatory properties. In other words, the effects of social conditions may reflect the effects of huddling behavior on energy conservation (Ancel et al. 1997).

In the laboratory, social factors, such as the number of animals housed in a cage, can have pronounced effects on immune function and disease resistance. For example, compared with mice (*Mus musculus*) housed five per cage, individually housed mice have higher cell-mediated and humoral immune responses, and increased resistance to viral, parasitic, and fungal infections (Karp et al. 1993, 1997; Plaut et al. 1969; Rabin and Salvin 1987). In these laboratory studies, immunological differences between individually and group-housed mice only become apparent 2 weeks after the onset of the respective housing conditions (Rabin and Salvin 1987); these immunological differences are not related to differences in circulating corticosterone concentrations (Rabin et al. 1987). Similarly, in red deer (*Cervus elaphus*), isolated female calves had higher proliferation of lymphocytes and antibody responses following immunization with ovalbumin compared with animals housed in groups; cortisol responses did not differ between isolated and group-housed females (Hanlon et al. 1997). Thus, these data in mice and red deer contradict the notion that isolated housing conditions elevate glucocorticoids and reduce immune function, suggesting that some effects of social interaction on immune function may be either direct (e.g., altering receptor regulation in immune cells) or via other neuroendocrine mechanisms (Hillgarth and Wingfield 1997; Rabin et al. 1994).

In contrast to laboratory studies, field studies suggest that the relationship between circulating steroid hormones and immune function may only be evident during the initial period of the housing conditions, when social relationships are first being established (Wingfield et al. 1990). Field studies in both birds (Wingfield 1994) and baboons (Alberts et al. 1992) have suggested that the relationship between hormones and aggression, as well as between hormones and the immune system, may only be apparent during times of social instability (i.e., prior to the establishment of social status). In female red deer, switching group composition every 2 weeks so that social status is not established resulted in reduced proliferation of B cells in response to KLH. Following adrenocorticotropic hormone (ACTH) injection, cortisol concentrations increase more rapidly in females exposed to group switching compared with females that remained in stable groups throughout the 6-week study (Hanlon et al. 1995). Consequently, in most studies examining the effects of differential housing conditions on immune function in rodents, steroid hormone concentrations are only measured at the termination of the housing conditions (i.e., after 14 to 30 days in their respective housing conditions), when social relationships are well established. In sum, to our knowledge there has been no attempt to bridge the field studies with laboratory data, to examine the impact of social factors on immune function in an ecological context. Additionally, the contribution of changes in social organization,

Figure 8.7

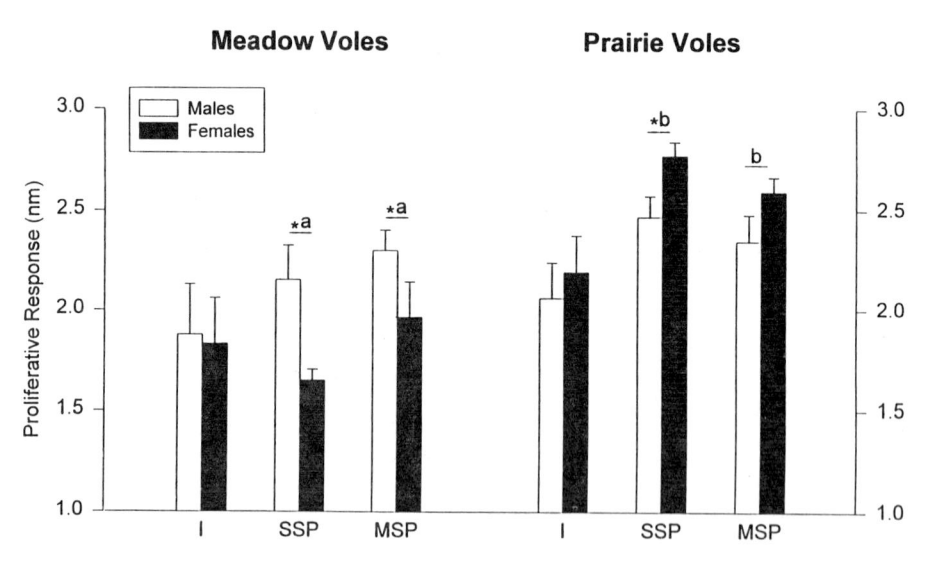

Mean (±SE) splenocyte proliferation values in response to an optimal concentration of concanavalin A (ConA) (10 g/mL) in male and female meadow (*Microtus pennsylvanicus*) and prairie voles (*Microtus ochrogaster*) that were housed individually (I), in same-sex pairs (SSP), or in mixed-sex pairs (MSP) for 28 consecutive days. For a single housing condition, an asterisk indicates significant sex differences within a species. For a single housing condition, bars with different letters (a or b) indicate significant differences between species. (Adapted from Klein et al. 1997.)

as well as the role of the social system (i.e., socially monogamous vs. polygynous), in the mediation of seasonal fluctuations of immune function has not been examined.

Recently, the effects of social interactions on cell-mediated and humoral immune function were examined in polygynous meadow voles and monogamous prairie voles housed in long photoperiods. Among individually housed voles, there were no sex or species differences in cell-mediated immunity, or more specifically the proliferative responses of splenocytes to ConA (Klein ct al. 1997). Pairing animals with either a male or female conspecific for 28 days revealed both sex and species differences in cell-mediated immune responses. Male meadow voles had higher immune responses than female conspecifics in both housing conditions and female prairie voles had higher proliferative responses than conspecific males in same-sex pairs only. Overall, prairie voles had higher proliferative responses than meadow voles (Klein et al. 1997) (figure 8.7). Conversely, the same housing regimens resulted in lower humoral immune responses (i.e., antibody production against KLH) in male meadow voles compared with conspecific females, and overall, meadow voles had higher antibody production than prairie voles (Klein and Nelson 1999a) (figure 8.8). Taken together,

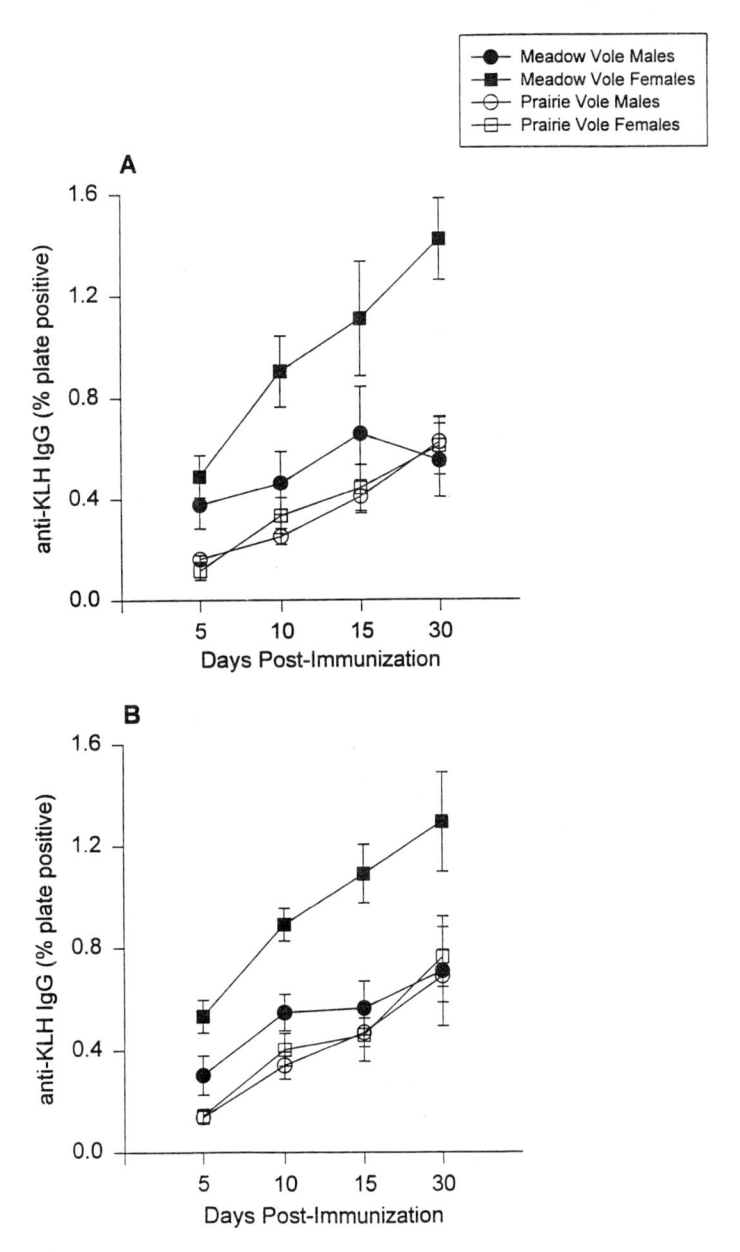

Figure 8.8
Serum anti-KLH (keyhole limpet hemocyanin) IgG concentrations (mean $\pm$SE) 5, 10, 15, and 30 days after immunization with 150 g KLH in male and female meadow (*Microtus pennsylvanicus*) and prairie voles (*Microtus ochrogaster*) that were housed in same-sex (A) and mixed-sex (B) pairs. Female meadow voles had higher antibody responses than conspecific males and, overall, meadow voles had higher responses than prairie voles. Data are presented as percent plate-positive, in which the mean of each sample was divided by the positive control sample (of the same dilution) run on the same microtiter plate. (Adapted from Klein & Nelson 1999a.)

these data suggest (1) social system may influence the effects of social interaction on immune function, (2) social isolation may mask the expression of sex and species differences in immune function typically observed in wild populations of animals (Zuk and McKean 1996), and (3) social interactions may influence the distribution of energy required for immune processes, thereby creating a tradeoff between components of the immune system (i.e., between humoral and cell-mediated immune function).

The extent to which these changes in immune function vary seasonally is currently unexamined. However, because social organization changes seasonally (Getz and McGuire 1997; McShea 1990), we predict that the effects of long-term social interaction on immune and even endocrine function may vary seasonally as well. Coincidentally, male deer mice (*Peromyscus maniculatus*) housed in short photoperiods (8L:16D) with an unrelated conspecific female are reproductively less affected by the inhibitory effects of short day length. Specifically, short-day males housed with females had larger gonads and seminal vesicles than short-day males housed in isolation (Whitsett and Lawton 1982), suggesting that social stimuli can mitigate the inhibitory effects of short day length on reproduction and may modify the effects of photoperiod on immune function as well.

Dyadic Social Interactions In general, short-term aggressive encounters in rats and mice result in reduced humoral and elevated cell-mediated immune responses in defeated animals. For example, primary antibody responses to both SRBC and KLH are lower in defeated mice and rats, respectively, compared with winners of these dyadic interactions (Beden and Brain 1982; Fleshner et al. 1989). Conversely, cell-mediated immune responses, such as proliferative responses of lymphocytes to ConA and PHA, are elevated in defeated rats (Bohus et al. 1993). However, the CD4+/CD8+ ratio is reduced to a greater extent in the spleens of defeated compared with winning rats (Bohus et al. 1993). For each immune measure the direction of the immunological change is the same for the winners of these dyadic interactions (i.e., reduced humoral and elevated cell-mediated immune responses) compared with home cage control rodents. However, the magnitude of the immunological variation is higher for losers than for winners (Bohus et al. 1991).

Genetic differences may play a fundamental role in determining the effects of dyadic interactions on immune function. For example, in mice selectively bred for high and low levels of aggression, the aggressive line of mice have reduced tumor development, increased NK cytotoxicity, and higher cell-mediated immune responses, including higher proliferative responses to ConA and elevated IL-2 and interferon (INF) production compared with mice from the low aggression line (Petitto et al. 1993, 1994).

Circulating corticosterone concentrations are not different between the two lines of mice, suggesting that the effects of genetic differences in aggression on immune function are not modified by this steroid hormone (Petitto et al. 1993). Taken together, these data suggest that genetic variation in aggression can lead to the same immunologic differences that are reported for winners and losers of dyadic interactions among nonselected strains of rats and mice (Bohus and Koolhaas 1991). Genetic variation may alter immunologic responsiveness to social interactions. The extent to which these effects of aggression on immune function occur seasonally is currently unexplored. However, aggression among males is typically reduced in rodents housed in short compared with long photoperiods (Matochik et al. 1986; Moffatt et al. 1993), which could in turn modify seasonal variation in immune function.

Population Density and Social Status Numerous laboratory and field studies suggest that high population densities can compromise reproductive and immune function, and this is primarily due to activation of the hypothalamic-pituitary-adrenal (HPA) axis (Lee and McDonald 1985; Terman 1993). Although olfactory cues can enhance the physiological responses to high population densities, laboratory studies have demonstrated that agonistic encounters are the primary mediator of elevated HPA activity and reduced reproductive and immune function, suggesting that the detrimental effects of high population densities are due to social interactions (Creigh and Terman 1988; Terman 1993). In prairie voles, population densities alter both reproductive and immune function. Reproductive organ mass is reduced by high population densities (i.e., 10.96 animals per cubic meter) compared with low population densities (i.e., 0.18 animal per cubic meter) among animals housed in long photoperiods. There is no effect of population density on reproductive organ mass among prairie voles housed in short day lengths (Nelson et al. 1996). Immunologically, prairie voles housed in high population densities had higher serum IgG concentrations than animals housed in low population densities, regardless of photoperiodic housing condition (Nelson et al. 1996). Serum corticosterone concentrations are actually lower among animals housed in high population densities than in low population densities, which may be related to the elevated immune responses in voles housed in high compared with low population densities (Nelson et al. 1996).

In contrast, the effects of population densities on immune function are not apparent among red deer calves, in which animals housed in high population densities (i.e., $1.8\,m^2$ per animal) have equivalent primary and secondary antibody responses against ovalbumin, yet have increased cortisol concentrations compared with red deer calves housed in low population densities (i.e., $4.5\,m^2$ per animal) (Hanlon et al. 1994). Coincidentally, high population densities reduce the chances of calf winter survival in wild populations of red deer (Coulson et al. 1997). Taken together, these data suggest that population density can alter immune, endocrine, and reproductive function in some mammals and the effects of population density may vary seasonally. The extent to which photoperiod interacts with population density to alter reproductive and immune function still requires further investigation.

In addition to effects of population density on immune and reproductive function, social status within a group can influence reproductive and immune function, as well as the course of infection in many mammals. Territory formation is common among many mammals when multiple males and females share the same territory. Social hierarchies form on these territories based on aggressive encounters; consequently, one male typically becomes dominant and the other males are classified as subordinates (Bohus et al. 1991). Only dominant males initiate aggressive interactions. There are multiple immunological consequences of these group dynamics. Generally, high-ranking (i.e., dominant) males have higher cell-mediated immune responses than lower-ranking males (Bohus and Koolhaas 1991). Conversely, humoral immune responses are lower in high-ranked males, and are correlated with the number of attacks initiated, suggesting that social behavior may directly modify immune responsiveness. High-ranked male mice also have slower clearance of parasitic infection than lower-ranked males (Barnard et al. 1994). Reduced clearance of infection among high-ranked males is also associated with higher testosterone and corticosterone concentrations, whereas steroid hormone concentrations are not associated with resistance to infection among lower-ranked males (Barnard et al. 1994).

Conversely, in cynomolgus monkeys, males of lower social status in both stable and unstable social environments have increased adenoviral shedding and higher cortisol concentrations than higher-ranked males (S. Cohen et al. 1997). These data suggest that differential social experiences may influence endocrine and immune function. The extent to which seasonal variation in social hierarchies influences seasonal changes in immune and endocrine function is not known. The energetic sequelae of these social interactions and their relations to immune function also remain unspecified. Consequently, if higher-ranking males have greater access to food, then higher-ranking males may have more metabolic resources to allocate to immune function.

Conclusions

Seasonal variation in endocrine, neural, and behavioral processes are well established; however, only recently has the immune system been empirically studied in this context. Immune function varies seasonally. Seasonal changes in health and disease represent a complex web of interaction between the neuroendocrine and immune systems, and ultimately represent adaptations of an organism to changing environmental conditions. The data presented in this chapter suggest that environmental cues, such as photoperiod and temperature, can alter immunologic processes and that many of these effects are due to changes in circulating hormones. Specifically, glucocorticoids and melatonin appear to be highly involved in modulating seasonal adaptations in immune function. In addition to environmental factors, social interactions can have pronounced effects on reproduction and immune function; however, the precise mechanisms mediating these complex effects remain elusive. Because social interactions vary seasonally, the relationship between environmental and social cues in mediating seasonal changes in immune function requires further examination. Finally, we propose that individuals maintain the highest level of immune function that is energetically possible within the constraints of an environment that fluctuates seasonally. During energetically demanding times, immune function is compromised, and these demands can arise from environmental pressures, such as low temperatures or reduced food availability, or from social pressures, such as mate selection or competition. In any case, animals have evolved to cope with energetic tradeoffs by reallocating energy in an attempt to maximize lifetime reproductive success.

Acknowledgments

The preparation of this review was supported by USPHS grant MH 57757 and a research grant IBN97-23420 from the National Science Foundation.

References

Able DJ (1996). The contagion indicator hypothesis for parasite-mediated sexual selection. Proc Natl Acad Sci U S A 93:2229–2233.

Ader R, Cohen N (1993). Psychoneuroimmunology: conditioning and stress. Annu Rev Psychol 44:53–85.

Alberts SC, Sapolsky RM, Altmann J (1992). Behavioral, endocrine, and immunological correlates of immigration by an aggressive male into a natural primate group. Horm Behav 26:167–178.

Alexander J, Stimson WH (1988). Sex hormones and the course of parasitic infection. Parasitol Today 4:189–193.

Ancel A, Visser H, Handrich Y, Masman D, Le Maho Y (1997). Energy saving in huddling penguins. Nature 385:304–305.

Applegate JE, Beaudoin RL (1970). Mechanism of spring relapse in avian malaria: effect of gonadotropin and corticosterone. J Wildl Dis 6:443–447.

Arkins S, Dantzer R, Kelley KW (1993). Somatolactogens, somamedins, and immunity. J Dairy Sci 76:2437–2450.

Barker IK, Beveridge I, Bradley AJ, Lee AK (1978). Observations on spontaneous stress-related mortality among males of the dasyurid marsupial *Antechinus stuartii* Macleay. Aust J Zool 26:435–447.

Barnard CJ, Behnke JM, Sewell J (1994). Social behaviour and susceptibility to infection in house mice (*Mus musculus*): effect of group size, aggressive behaviour and status-related hormonal responses prior to infection on resistance to *Babesia microti*. Parasitology 108:487–496.

Barnett, SA (1965). Adaptation of mice to cold. Biol Rev 40:5–51.

Bartness TJ, Goldman BD (1989). Mammalian pineal melatonin: a clock for all seasons. Experientia 45:939–945.

Beaudoin RL, Applegate JE, Davis DE, McLean RG (1971). A model for the ecology of avian malaria. J Wildl Dis 7:5–13.

Beden SN, Brain PF (1982). Studies on the effect of social stress on measures of disease resistance in laboratory mice. Aggressive Behav 8:126–129.

Behnke JM, Barnard CJ, Wakelin D (1992). Understanding chronic nematode infections: evolutionary considerations, current hypotheses and the way forward. Int J Parasitol 22:861–907.

Bernton EW, Bryant HU, Holaday JW (1991). Prolactin and immune function. *In* Psychoneuroimmunology (Ader R, Felten DL, Cohen N, eds), pp 403–428. San Diego: Academic Press.

Bernton E, Bryant H, Holaday J, Dave J (1992). Prolactin and prolactin secretagogues reverse immunosuppression in mice treated with cycteamine, glucocorticoids, or cyclosporin-A. Brain Behav Immun 6:394–408.

Besedovsky HO, del Rey A (1991). Feed-back interactions between immunological cells and the hypothalamus-pituitary-adrenal axis. Neth J Med 39:274–280.

Beutler, B, Cerami A (1988). The common mediator of shock, cachexia, and tumor necrosis. Adv Immunol 42:213–231.

Billingham RE (1986). Immunologic advantages and disadvantages of being a female. *In* Reproductive Immunology (Clarke DA, Croy BA, eds), pp 1–9. New York: Elsevier Science.

Black PH (1994). Central nervous system–immune system interactions: psychoneuroendocrinology of stress and its immune consequences. Antimicrob Agents Chemother 38:1–6.

Blom JMC, Gerber J, Nelson RJ (1994). Immune function in deer mice: developmental and photoperiodic effects. Am J Physiol 267:R596–R601.

Boccia ML, Scanlan JM, Laudenslager ML, Berger CL, Hijazi AS, Reite ML (1997). Juvenile friends, behavior, and immune responses to separation in bonnet macaque infants. Physiol Behav 61:191–198.

Bohus B, Koolhaas JM (1991). Psychoimmunology of social factors in rodents and other subprimate vertebrates. *In* Psychoneuroimmunology (Ader R, Felten DL, Cohen N, eds), pp 807–830. San Diego: Academic Press.

Bohus B, Koolhaas JM, De Ruiter AJH, Heijnen CJ (1991). Stress and differential alterations in immune system functions: conclusions from social stress studies in animals. Neth J Med 39:306–315.

Bohus B, Koolhaas JM, Heijnen CJ, de Boer O (1993). Immunological responses to social stress: dependence on social environment and coping abilities. Neuropsychobiology 28:95–99.

Borysenko, M (1987). Area review: psychoneuroimmunology. Ann Behav Med 9:3–10.

Bradley AJ, McDonald IR, Lee AK (1980). Stress and mortality in a small marsupial (*Antechinus stuartii*, Macleay). Gen Comp Endocrinol 40:188–200.

Brainard GC, Knobler RL, Podolin PL, Lavasa M, Lubin FD (1985). Neuroimmunology: modulation of the hamster immune system by photoperiod. Life Sci 40:1319–1326.

Brainard GC, Watson-Whitmeyer M, Knobler RL, Lubin FD (1988). Neuroendocrine regulation of immune parameters. Ann N Y Acad Sci 540:704–706.

Brock MA (1983). Seasonal rhythmicity in lymphocyte blastogenic responses of mice persist in a constant environment. J Immunol 130:2586–2588.

Bronson FH (1989). Mammalian Reproductive Biology. Chicago: University of Chicago Press.

Bronson FH, Heideman PD (1994). Seasonal regulation of reproduction in mammals. In The Physiology of Reproduction, 2nd ed, vol 2 (Knobil E, Neill JD, eds), pp 541–584. New York: Raven Press.

Caroleo MC, Frasca D, Nistico G, Doria G (1992). Melatonin as immunomodulator in immunodeficient mice. Immunopharmacology 23:81–89.

Castanon N, Dulluc J, le Moal M, Mormede P (1992). Prolactin as a link between behavioral and immune differences between the Roman rat lines. Physiol Behav 51:1235–1241.

Chappell MA, Zuk M, Johnsen TS, Kwan TH (1997). Mate choice and aerobic capacity in red junglefowl. Behaviour 134:511–529.

Claman HN (1972). Corticosteroids and lymphoid cells. N Engl J Med 287:388–397.

Cohen JHM, Danel L, Cordier G, Saez S, Revillard J (1983). Sex steroid receptors in peripheral T cells: absence of androgen receptors and restriction of estrogen receptors to OKT8-positive cells. J Immunol 131:2767–2771.

Cohen S, Line S, Manuck SB, Rabin BS, Heise ER, Kaplan JR (1997). Chronic social stress, social status, and susceptibility to upper respiratory infections in nonhuman primates. Psychosom Med 59:213–221.

Coulson T, Albon S, Guinness F, Pemberton J, Clutton-Brock T (1997). Population substructure, local-density, and calf winter survival in red deer (*Cervus elaphus*). Ecology 78:852–863.

Creigh SL, Terman CR (1988). Reproductive recovery of inhibited male prairie deermice (*Peromyscus maniculatus bairdii*) from laboratory populations by contact with females or their urine. J Mammal 69:603–607.

Curtis JE, Hersh EM, Harris JE, McBride C, Freireich EJ (1970). The human primary immune response to keyhole limpet haemocyanin: interrelationships of delayed type hypersensitivity, antibody response and *in vitro* blast formation. Clin Exp Immunol 6:473–491.

Cutolo M, Accardo S, Villaggio B, Barone A, Sulli A, Coviello DA, Carabbio C, Felli L, Miceli D, Farruggio R, Carruba G, Castagnetta L (1996). Androgen and estrogen receptors are present in primary cultures of human synovial macrophages. J Clin Endocrinol Metab 81:820–827.

Danel L, Sovweine G, Monier JC, Saez S (1983). Specific estrogen binding sites on human lymphoid cells and thymic cells. J Steroid Biochem 18:559–563.

Dark J, Johnston PG, Healy M, Zucker I (1994). Latitude of origin influences photoperiodic control of reproduction of deer mice (*Peromyscus maniculatus*). Biol Reprod 28:213–218.

Demas GE, Nelson RJ (1996). Photoperiod and temperature interact to affect immune cell parameters in adult deer mice (*Peromyscus maniculatus*). J Biol Rhythms 11:94–102.

Demas, GE, Nelson RJ (1998). Short-day enhancement of immune function is not mediated by gonadal-steroid hormones in adult deer mice (*Peromyscus maniculatus*). J Comp Physiol [B] 168:419–426.

Demas GE, Klein SL, Nelson RJ (1996). Reproductive and immune responses to photoperiod and melatonin are linked in *Peromyscus* subspecies. J Comp Physiol [A] 179:819–825.

Demas GE, Chefer V, Talan MI, Nelson RJ (1997a). Metabolic cost of mounting an antigen-stimulated immune response in adult and aged C57BL/6J mice. Am J Physiol 273:R1631–R1637.

Demas GE, DeVries AC, Nelson RJ (1997b). Effects of photoperiod and 2–deoxy-D-glucose–induced metabolic stress in female deer mice. Am J Physiol 41:R1762–R1767.

Dhabhar FS, McEwen BS (1996). Stress-induced enhancement of antigen-specific cell-mediated immunity. J Immunol 156:2608–2615.

Dixon FJ, Jacot-Guillarmod H, McConahey PJ (1966). The antibody responses of rabbits and rats to hemocyanin. J Immunol 97:350–355.

Dobrowsolska A, Adamczewska-Andrzejewska KA (1991). Seasonal and long-term changes in serum gamma globulin levels in comparing the physiology and population density of the common vole, *Microtus arvalis* Pall. J Interdisciplinary Cycle Res 22:1–9.

Dobrowolska A, Rewkiwwicz-Dziarska A, Szarska I, Gill J (1974). Seasonal changes in haematological parameters, level of serum proteins and glycoproteins, activity of the thyroid gland, supradrenals and kidneys in the common vole (*Microtus arvalis* Pall.). J. Interdisciplinary Cycle Res 5:347–354.

Egid K, Brown JL (1989). The major histocompatibility complex and female mating preference in mice. Anim Behav 38:548–549.

Fairbairn DJ (1977). Why breed early? A study of reproductive tactics in *Peromyscus*. Can J Zool 55:862–871.

Ferkin MH, Gorman MR (1992). Photoperiod and gonadal hormones influence odor preferences of the male meadow vole, *Microtus pennsylvanicus*. Physiol Behav 51:1087–1091.

Ferkin MH, Seamon JO (1987). Odor preferences and social behavior in meadow voles, *Microtus pennsylvanicus*: seasonal differences. Can J Zool 65:2931–2937.

Ferkin MH, Zucker I (1991). Seasonal control of odour preferences of meadow voles (*Microtus pennsylvanicus*) by photoperiod and ovarian hormones. J Reprod Fertil 92:433–441.

Festa-Bianchet M (1989). Individual differences, parasites, and the cost of reproduction for bighorn ewes (*Ovis canadensis*). J Anim Ecol 58:785–795.

Fleshner M, Laudenslager ML, Simons L, Maier SF (1989). Reduced serum antibodies associated with social defeat in rats. Physiol Behav 45:1183–1187.

Folstad I, Karter AJ (1992). Parasites, bright males, and the immunocompetence handicap. Am Naturalist 139:603–622.

Folstad I, Nilssen AC, Halvorsen O, Andersen J (1989). Why do male reindeer (*Rangifer t. tarandus*) have higher abundance of second and third instar larvae of *Hypoderma tarandi* than females? Oikos 55:87–92.

Ganong WF (1995). Review of Medical Physiology. Norwalk, CT: Appleton & Lange.

Gaudernack G, Halvorsen O, Skorping A, Stokkan KA (1984). Humoral immunity and output of first-stage larvae of *Elaphostrongylus rangiferi* (Nematoda, Metastrongyloidea) by infec tedreindeer, *Rangifer tarandus tarandus*. J Helminthol 58:13–18.

Getz LL, McGuire B (1997). Communal nesting in prairie voles (*Microtus ochrogaster*): formation, composition, and persistence of communal groups. Can J Zool 75:525–534.

Goldman BD (1983). The physiology of melatonin in mammals. Pineal Res Rev 1:145–188.

Goldman BD, Nelson RJ (1993). Melatonin and seasonality in mammals. *In* Melatonin: Biosynthesis (Yu HS, Reiter RJ, eds), pp 225–252. Boca Raton FL: CRC Press.

Grossman CJ (1985). Interactions between the gonadal steroids and the immune system. Science 227:257–261.

Hall NR, Goldstein AL (1984). Endocrine regulation of host immunity. *In* Immune Modulation Agents and Their Mechanisms (Fenichel RL, Chirigos MA, eds), pp 533–563. New York: Marcel Dekker.

Halvorsen O, Skorping A, Hanse K (1985). Seasonal cycles in the output of first stage larvae of the nematode *Elaphostrongylus rangiferi* from reindeer, *Rangifer tarandus tarandus*. Polar Biol 5:49–54.

Hamilton WD (1990). Mate choice near or far. Am Zool 30:341–352.

Hamilton WD, Zuk M (1982). Heritable true fitness and bright birds: a role for parasites? Science 218:384–387.

Hanlon AJ, Rhind SM, Reid HW, Burells C, Lawrence AB, Milne JA, McMillen SR (1994). Relationship between immune response, liveweight gain, behaviour and adrenal function in red deer (*Cervus elaphus*) calves derived from wild and farmed stock, maintained at two housing densities. Appl Anim Behav Sci 41:243–255.

Hanlon AJ, Rhind SM, Reid HW, Burells C, Lawrence AB (1995). Effects of repeated changes in group composition on immune response, behaviour, adrenal activity and liveweight gain in farmed red deer yearlings. Appl Anim Behav Sci 44:57–64.

Hanlon AJ, Rhind SM, Reid HW, Burells C, Lawrence AB (1997). Effects of isolation on the behaviour, live-weight gain, adrenal capacity and immune responses of weaned red deer hind calves. Anim Sci 64:541–546.

Harris RA, Mapes JP, Ochs RS, Crabb DW, Stropes L (1979). Hormonal control of hepatic lipogenesis. Adv Exp Med Biol 111:17–42.

Hauger RL, Millan MA, Lorang M, Harwood JP, Aguilera G (1988). Corticotropin releasing factor receptors and pituitary adrenal responses during immobilization stress. Endocrinology 123:396–405.

Heldmaier G, Steinlechner S, Ruf H, Wiesinger H, Klingenspor M (1989). Photoperiod and thermoregulation in vertebrates: body temperature rhythms and thermogenic acclimation. J Biol Rhythms 4:251–265.

Hillgarth N, Wingfield JC (1997). Testosterone and immunosuppression in vertebrates: implications for parasite-mediated sexual selection. *In* Parasites and Pathogens: Effects on Host Hormones and Behavior (Beckage NE, ed), pp 143–155. New York: Chapman & Hall.

Hughes CS, Gaskall RM, Jones RC, Bradbury JM, Jordon FTW (1989). Effects of certain stress factors on the re-excretion of infectious laryngotracheitis virus from latently infected carrier birds. Res Vet Sci 46:274–276.

Janeway CA, Travers P (1994). Immunobiology: The Immune System in Health and Disease. London: Current Biology.

John JL (1994). The avian spleen: a neglected organ. Q Rev Biol 69:327–351.

Kalra SP, Kalra PS (1996). Nutritional infertility: the role of the interconnected hypothalamic neuropeptide Y-galanin-opioid network. Front Neuroendocrinol 17:371–401.

Karp JD, Moynihan JA, Ader R (1993). Effects of differential housing on the primary and secondary antibody responses of male C57BL/6 and BALB/c mice. Brain Behav Immun 7:326–333.

Karp JD, Moynihan JA, Ader R (1997). Psychosocial influences on immune responses to HSV-1 infection in BALB/c mice. Brain Behav Immun 11:47–62.

Kavaliers M, Colwell DD (1995). Discrimination by female mice between the odours of parasitized and non-parasitized males. Proc Roy Soc Lond B Biol Sci 261:31–35.

Keller SE, Weiss JM, Schleifer SL, Miller NE, Stern M (1983). Stress-induced suppression of immunity in adrenalectomized rats. Science 221:1301–1304.

Kenagy GJ, Barnes BM (1984). Environmental and endogenous control of reproductive function in the Great Basin pocket mouse *Perognathus parvus*. Biol Reprod 31:637–645.

Kirkpatrick M (1996). Good genes and direct selection in the evolution of mating preference. Evolution 50:2125–2140.

Kirkpatrick M, Ryan MJ (1991). The evolution of mating preference and the paradox of the lek. Nature 350:33–38.

Klein SL, Hairston JE, DeVries AC, Nelson RJ (1997). Social environment and steroid hormones affect species and sex differences in immune function among voles. Horm Behav 32:30–39.

Klein SL, Gamble HR, Nelson RJ (1999). *Trichinella spiralis* infection alters female odor preference, but not partner preference, in voles. Behav Ecol Sociobiol, in Press.

Klein SL. Nelson RJ (1999a). Social interactions unmask sex and species differences in humoral immunity. Anim Behav 57:603–610.

Klein SL, Nelson RJ (1999b). Activation of the immune-endocrine system with lipopolysaccharide reduces affiliative behaviors in voles. Behav Neurosci, in press.

Kuci S, Becker J, Veit G, Handgretinger GR, Attanasio A, Bruchett G, Treuner J, Niethammer D, Gupta D (1988). Circadian variations in the immunomodulatory role of the pineal gland. Neuroendocrinol Lett 10:65–80.

Laudenslager ML, Ryan SM, Drugan RC, Hyson RL, Maier SF (1983). Coping and immunosuppression: inescapable but not escapable shock suppresses lymphocyte proliferation. Science 221:568–570.

Layne JN, Raymond MAV (1994). Communal nesting of southern flying squirrels in Florida. J Mammal 75:110–120.

Lee AK, McDonald IR (1985). Stress and population regulation in small mammals. Oxford Rev Reprod Biol 7:261–304.

Lochmiller RL, Vesty MR, McMurry ST (1994). Temporal variation in humoral and cell-mediated immune response in a *Sigmodon hispidus* population. Ecology 75:236–245.

Lysle DT, Cunnick JE, Wu R, Caggiula AR, Wood PG, Rabin BS (1988). 2–Deoxy-D-glucose modulation of T-lymphocyte reactivity: differential effects on lymphoid compartments. Brain Behav Immun 2:212–221.

MacMurray JP, Barker JP, Armstrong JD, Bozzetti LP, Kuhn IN (1983). Circannual changes in immune function. Life Sci 32:2363–2370.

Madison DM, FitzGerald RW, McShea WJ (1984). Dynamics of social nesting in overwintering meadow voles (*Microtus pennsylvanicus*): possible consequences for population cycling. Behav Ecol Sociobiol 15:9–17.

Maestroni GJM, Pierpaoli W (1981). Pharmacologic control of the hormonally mediated immune response. In Psychoneuroimmunology (Ader R, Cohen, N, eds), pp 405–425. New York: Academic Press.

Maestroni GJM, Conti A, Pierpaoli W (1986). Role of the pineal gland in immunity. Circadian synthesis and release of melatonin modulates the antibody response and antagonizes the immunosuppressive effect of corticosterone. J Neuroimmunol 13:19–30.

Maier SF, Laudenslager ML (1988). Inescapable shock, shock controllability, and mitogen stimulated lymphocyte proliferation. Brain Behav Immun 2:87–91.

Maier SF, Washout LR, Fleshner M (1994). Psychoneuroimmunology: the interface between behavior, brain, and immunity. Am Psychol 49:1004–1017.

Marlow BJ (1961). Reproductive behaviour of the marsupial mouse *Antechinus flavipes* (Waterhouse) (Marsupialia) and the development of the pouch young. Aust J Zool 9:203–220.

Matera L, Cesano A, Bellone G, Oberholtzer E (1992). Modulatory effect of prolactin on the resting and mitogen-induced activity of T, B, and NK lymphocytes. Brain Behav Immun 6:409–417.

Matockik JA, Miernicki M, Powers JB, Bergondy ML (1986). Short photoperiods increase ultrasonic vocalization rates among male Syrian hamsters. Physiol Behav 38:453–458.

McCruden AB, Stimson WH (1991). Sex hormones and immune function. In Psychoneuroimmunology (Ader R, Felten DL, Cohen N, eds), pp 475–493. San Diego: Academic Press.

McShea WJ (1990). Social tolerance and proximate mechanisms of dispersal among winter groups of meadow voles (*Microtus pennsylvanicus*). Anim Behav 39:346–351.

Medzhitov R, Preston-Hurlburt P, Janeway CA (1997). A human homologue of the *Drosophila* Toll protein signals activation of adaptive immunity. Nature 388:394–397.

Miller ES, Klinger JC, Akin C, Koebel DA, Sonnenfeld G. (1994). Inhibition of murine splenic T lymphocyte proliferation by 2-deoxy-D-glucose–induced metabolic stress. J Neuroimmunol 52:165–173.

Moffatt CA, DeVries AC, Nelson RJ (1993). Winter adaptations of male deer mice (*Peromyscus maniculatus*) and prairie voles (*Microtus ochrogaster*) that vary in reproductive responsiveness to photoperiod. J Biol Rhythms 8:221–232.

Møller AP (1994). Sexual Selection and the Barn Swallow. Oxford: Oxford University Press.

Monjan AA (1981). Stress and immunologic competence: Studies in animals. In Psychoneuroimmunology (Ader R, Cohen, N, eds), pp 185–228. New York: Academic Press.

Nelson RJ (1990). Photoperiodic responsiveness in laboratory house mice. Physiol Behav 48:403–408.

Nelson RJ, Blom JMC (1994). Photoperiodic effects on tumor development and immune function. J Biol Rhythms 9:233–249.

Nelson RJ, Demas GE (1996). Seasonal changes in immune function. Q Rev Biol 71:511–548.

Nelson RJ, Demas GE, Klein SL, Kriegsfeld LK (1995). The influence of season, photoperiod, and pineal melatonin on immune function. J Pineal Res 19:149–165.

Nelson RJ, Fine JB, Demas GE, Moffatt CA (1996). Photoperiod and population density interact to affect reproductive and immune function in male prairie voles. Am J Physiol 270:R571–R577.

Nilsson B, Carlsson S, Damber MK, Lindblom D, Sodergard R, Von Schoultz B (1984). Specific binding of 17-beta estradiol in the human thymus. Am J Obstet Gynecol 149:544–547.

Ober C, van der Ven K (1997). Immunogenetics of reproduction: an overview. Curr Top Microbiol Immunol 222:1–23.

Olsen NJ, Kovacs WJ (1996). Gonadal steroids and immunity. Endocr Rev 17:369–384.

Petitto JM, Lysle DT, Gariepy JL, Clubb PH, Cairns RB, Lewis MH (1993). Genetic differences in social behavior: relation to natural killer cell function and susceptibility to tumor development. Neuropsychopharmacology 8:35–43.

Petitto JM, Lysle DT, Gariepy JL, Lewis MH (1994). Association of genetic differences in social behavior and cellular immune responsiveness: effects of social experience. Brain Behav Immun 8:111–122.

Pioli C, Carleo C, Nistico G, Doria G (1993). Melatonin increases antigen presentation and amplifies specific and nonspecific signals for t-cell proliferation. Int J Immunopharmacol 15:463–468.

Plaut SM, Ader R, Friedman SB, Ritterson AL (1969). Social factors and resistance to malaria in the mouse. Effects of groups *vs* individual housing on resistance to *Plasmodium berghei* infection. Psychosom Med 31:536–552.

Potts WK, Manning CJ, Wakeland EK (1991). Mating patterns in seminatural populations of mice influenced by MHC genotype. Nature 352:619–621.

Priddy KD (1997). Immunologic adaptations during pregnancy. J Obstet Gynecol Neonatal Nurs 26:388–394.

Rabin BS, Kusnecov A, Shurin M, Zhou D, Rasnick S (1994). Mechanistic aspects of stressor-induced immune alteration. *In* Handbook of Human Stress and Immunity. (Glaser R, Kiecolt-Glaser J, eds), pp 23–51. San Diego: Academic Press.

Rabin BS, Lyte M, Hamill E (1987). The influence of mouse strain and housing on the immune response. J Neuroimmunol 17:11–16.

Rabin BS, Salvin SB (1987). Effect of differential housing and time on immune reactivity to sheep erythrocytes and Candida. Brain Behav Immun 1:267–275.

Read AF (1987). Comparative evidence supports the Hamilton and Zuk hypothesis on parasites and sexual selection. Nature 328:68–70.

Reber PM (1993). Prolactin and immunomodualtion. Am J Med 95:637–644.

Richmond M, Stehn R (1969). Olfaction and reproductive behavior in microtine rodents. In:Mammalian olfaction, reproductive processes, and behavior (Doty RL, ed), pp 197–217. New York: Academic Press.

Richner H, Christe P, Oppliger A (1995). Paternal investment affects prevalence of malaria. Proc Natl Acad Sci USA 92:1192–1194.

Ropstad E, Larsen HJ, Refsdal AO (1989). Immune function in dairy cows related to energy bala nceand metabolic status in early lactation. Acta Vet Scand 30:209–219.

Saafela S, Reiter RJ (1994). Function of melatonin in thermoregulatory processes. Life Sci 54:295–311.

Saino N, Møller AP (1996). Sexual ornamentation and immunocompetence in the barn swallow. Behav Ecol 7:227–232.

Sapolsky RM (1992). Stress, the Aging Brain, and the Mechanisms of Neuronal Death. Cambridge, MA: MIT Press.

Sasson S, Mayer M (1981). Antiglucocorticoid activity of androgens in rat thymus lymphocytes. Endocrinology 108:760–766.

Schneider, JE, Wade, GN (1989). Availability of metabolic fuels controls estrous cyclicity of Syrian hamsters. Science 244:1326–1328.

Schuurs AHWM, Verheul HAM (1990). Effects of gender and sex steroids on the immune response. J Steroid Biochem 35:157–172.

Selye H (1956). The Stress of Life. New York: McGraw-Hill.

Sheldon BC, Verhulst S (1996). Ecological immunology: Costly parasite defences and trade-offs in evolutionary ecology. TREE 11:317–321.

Smith GP, Epstein AN (1969). Increased feeding in response to decreased glucose utilization in the rat and monkey. Am J Physiol 217:1083–1087.

Taigen TL, Wells KD (1985). Energetics of vocalization by an anuran amphibian (*Hyla versicolor*). J Comp Physiol B 155:163–170.

Talleklint L, Jaenson TG, Mather TN (1993). Seasonal variation in the capacity of the bank vole to infect larval ticks (*Acari: Ixodidae*) with Lyme's disease spirochete, *Borrelia burgdorferi*. J Med Entomol 30:812–815.

Terman CR (1993). Studies of natural populations of white-footed mice: Reduction of reproduction at varying degrees. J Mammal 74:678–687.

Vaughan MK, Hubbard GB, Champney TH, Vaughan GM, Little JC, Reiter RJ (1987). Splenic hypertrophy and extrameduallry hemapoiesis induced in male Syrian hamsters by short photoperiod or melatonin injections and reversed by melatonin pellets or pinealectomy. Am J Anat 179:131–136.

Vindervogel H, Debruyne H, Pastoret PP (1985). Observation of pigeon herpesvirus 1 re-excretion during the reproduction period in conventionally reared homing pigeons. J Comp Pathol 95:105–112.

Vriend J, Lauber JK (1973). Effects of light intensity, wavelength and quanta on gonads and spleen of the deer mouse. Nature 244:37–38.

Wade GN, Schneider JE (1992). Metabolic fuels and reproduction in female mammals. Neurosci Biobehav Rev 16:235–272.

Wedekind C, Folstad I (1994). Adaptive or nonadaptive immunosuppression by sex hormones? Am Naturalist 143:936–938.

Weusten JJ, Blankenstein MA, Gmelig-Meyling FH, Schuurman HJ, Kater L, Thijssen JH (1986). Presence of oestrogen receptors in human blood mononuclear cells and thymocytes. Acta Endocrinol 112:409–414.

Whitsett JM, Lawton AD (1982). Social stimulation of reproductive development in male deer mice housed on a short-day photoperiod. J Comp Physiol Psychol 96:416–422.

Wingfield JC (1994). Hormone-behavior interactions and mating systems in male and female birds. *In* The Differences Between the Sexes (Short RV, Balaban E, eds), pp 303–330. Cambridge: Cambridge University Press.

Wingfield JC, Hegner RE, Dufty AM, Ball GF (1990). The "challenge hypothesis": Theoretical implications for patterns of testosterone secretion, mating systems and breeding strategies. Am Nat 136:829–845.

Wolff JO, Lidicker WZ (1981). Communal winter nesting and food sharing in taiga voles. Behav Ecol Sociobiol 9:237–240.

Woolley P (1966). Reproduction in *Antechinus* spp. and other dasyurid marsupials. *In* Comparative Biology of Reproduction in Mammals (Rowlands IW, ed), pp 281–294. London: Symposium of the Zoology Society.

Yamazaki K, Boyse EA, Mike V, Thaler HT, Mathieson BJ, Abbot J, Boyse J, Zayas ZA, Thomas L (1976). Control of mating preferences in mice by genes in the major histocompatibility complex. J Exp Med 144:1324–1335.

Zuk M (1990). Reproductive strategies and sex differences in disease susceptibility: An evolutionary viewpoint. Parasitol Today 6:231–233.

Zuk M (1994). Immunology and the evolution of behavior. *In* Behavioral Mechanisms in Evolutionary Ecology (Real LA, ed), pp 354–368. Chicago: University of Chicago Press.

Zuk M, McKean KA (1996). Sex differences in parasite infections: patterns and processes. Int J Parasitol 26:1009–1024.

Zuk M, Thornhill R, Ligon JD, Johnson K (1990). Parasites and mate choice in red jungle fowl. Am Zool 30:235–244.

9 Feisty Females and Meek Males: Reproductive Strategies in the Spotted Hyena

Kay E. Holekamp and Laura Smale

Reproductive processes in male and female mammals have evolved in response to different sets of selection pressures (Darwin 1871). The fundamental sexual dimorphism of differential gamete size that characterizes all anisogamous organisms is uniquely enhanced in mammals by internal fertilization, gestation, and lactation, and these processes generate enormous differences between males and females regarding their potential rates of reproduction (Fisher 1930; Bateman 1948; Trivers 1972). Consequently, male mammals can father young at much higher rates than females can bear and rear them. In most mammals, females generally dedicate a huge proportion of their total reproductive effort to nurturing a relatively small number of young. By contrast, males generally invest most of their reproductive effort in finding and winning mates, and in copulating with as many of these as possible. Reproductive success (RS) among female mammals is limited by access to resources necessary for successful rearing of young, whereas RS of the male is usually limited by the number of fertile females with whom he can copulate.

One important consequence of these sex differences is that the RS of male mammals tends to be far more variable than does the RS of females. Among northern elephant seals (*Mirouanga angustirostris*), for example, only 9% of males breed at all during their lifetimes, yet a single breeding male may inseminate over 200 females (LeBoeuf 1974; LeBoeuf and Reiter 1988). Only 4% of adult males sire 85% of the pups born each year (LeBoeuf and Peterson 1969). Whereas the mean ($\pm$SD) number of offspring surviving to weaning age is 3.0 ± 13.7 for male *Mirouanga*, it is only 0.8 ± 1.7 for female conspecifics (LeBoeuf and Reiter 1988). Note that there is a fourfold sex difference in these means, but an eightfold sex difference in variance. Differences of this magnitude between males and females with respect to variance in RS are common among mammals (e.g., see Clutton-Brock 1988).

Because male RS is so often highly variable, a great deal of research has focused on the factors influencing it, particularly on those variables affecting outcomes of male-male competition and female mate choice (Darwin 1871; Andersson 1994). This work has shed a great deal of light on the selection pressures promoting the evolution of sex differences in morphology and behavior (e.g., see Short and Balaban 1994). Meanwhile, variance in female RS has received little attention. Recent studies have revealed that, although variance in RS is usually greater among male than female mammals, female RS can vary considerably, particularly among members of mammalian species in which a single pair is responsible for all reproduction. Examples of "singular breeders" include naked mole rats (*Heterocephalus glaber*), dwarf mongooses (*Helogale parvula*), various canids, and many callithrichid primates (reviewed in Solomon and French 1997). Lifetime female RS among singular breeders tends to exhibit a bimodal distribution, with a few females producing many offspring and most females producing none (Sherman et al. 1995). It is interesting to note that variance in female RS can also be extremely high among "plural breeders." These are group-living species in which all adult females reproduce, and include most gregarious ungulates, elephants, ground squirrels, pinnipeds, Old World primates, and several species of carnivores (e.g., see Jarman 1974; Estes 1974; Murie and Michener 1984; Clutton-Brock 1988; Holekamp et al. 1996). Lifetime female RS among plural breeders tends to occur in a unimodal distribution, yet the intrasexual reproductive skew in these species can nevertheless be considerable (Reeve and Ratnieks 1993; Keller and Vargo 1993). Among plural breeders, variance in female RS is especially pronounced when conspecific females differ in their ability to obtain resources critical for successful reproduction (e.g., see Barton et al. 1996; Roberts 1996).

In this chapter, we examine contextual influences on reproduction in one plural breeder that exhibits extreme variance in female RS, the spotted hyena (*Crocuta crocuta*). This species offers opportunities for useful comparison with both singular and plural breeders, not only at ecological and evolutionary levels but also at the level of the physiological mechanisms mediating variation in reproductive success among individuals. All female spotted hyenas breed, but female RS is strongly dependent on access to food. Food availability in this species often varies seasonally, and access to food also varies as a function of outcomes of agonistic competitive interactions among females. In fact, feeding competition is so intense among spotted hyenas that it appears to have had a profound influence on the evolution of unusual reproductive strategies in both sexes. Ecological constraints often impose limits on the extent to which sexual selection can operate (Emlen and Oring 1977), and constraints imposed by severely limited access to food among spotted hyenas have apparently favored the evolution of traits, including agonistic behavior, in which a variety of sex differences are reversed from the norms seen in other polygynous mammals. The reversal of aggressive and submissive roles in spotted hyenas is one of the most extreme in the Class Mammalia.

What types of selection pressures have led to this dramatic reversal of aggressive and submissive roles? What might be the adaptive advantage of being a brash, aggressive female or of being a meek, submissive male? How does RS in this species compare with that of other plural breeders? Is there a relationship between role-reversed agonistic behaviors and RS in this species? What are the contextual factors that link agonistic interactions to reproductive success? These questions are only now beginning to be asked with regard to mammalian females (e.g., see Kappeler 1993).

In this chapter we synthesize a variety of ecological considerations with some of the reproductive data from our field studies of free-living populations of hyenas in their natural habitat in Kenya.

Our goal here is to examine hyena reproduction in the ecological and social contexts in which breeding actually occurs in nature. Although we have accumulated more information about contextual influences on reproduction in females than males, we have identified some of the important variables influencing reproductive performance in both sexes, and we review that information here. After introducing our subject animals and summarizing the methods we use to study them, we consider contextual influences on hyena reproduction, first in females and then in males. Finally we conclude by evaluating the notion that, because reproductive success varies so greatly among female *Crocuta*, and because female RS in this species depends so critically on outcomes of agonistic interactions, natural selection has favored large, aggressive, choosy females, which in turn, via the action of sexual selection, has promoted the evolution of male characteristics that differ markedly from those found in other plural breeders. Specifically, it appears that the dual pressures of natural and sexual selection confronted by spotted hyenas over the course of their evolutionary history have respectively favored feisty females and meek males.

Spotted Hyenas

Spotted hyenas are large, gregarious carnivores that occur throughout sub-Saharan Africa, from approximately 25° south of the equator to approximately 17° North latitude. *Crocuta* are generalists that exploit a wide variety of food resources, including carrion and invertebrates, but 80% to 90% of their diet consists of live ungulate prey that they hunt themselves (Kruuk 1972; Cooper et al. in press; Holekamp et al. 1997). *Crocuta* can thus live at extremely high densities, and the size of their social groups, called clans, can exceed 100 individuals in the prey-rich savannas of eastern Africa. However, in the drier regions of southern Africa where food is scarce, they may live in groups containing as few as four individuals (Mills 1990). *Crocuta* live in fission-fusion societies in which all clan members recognize each other, defend a common territory against members of neighboring clans, and rear their cubs at a single communal den (Kruuk 1972; Henschel and Skinner 1991). Although individual hyenas spend much of their time alone or in small groups, particularly when foraging, they join together during territorial defense, interactions with lions and other competitors, and at ungulate kills. The carcasses of large ungulates represent extremely rich, rare, and ephemeral food patches that occur unpredictably in space and time, so clan members compete intensively over them. Every *Crocuta* clan is structured by a rigid linear dominance hierarchy reflecting predictable patterns of outcomes in dyadic agonistic interactions (Kruuk 1972; Tilson and Hamilton 1984; Frank 1986b). An individual's position in the clan's hierarchy determines its priority of access to food during competitions with other clan members at kills (Kruuk 1972; Tilson and Hamilton 1984; Frank 1986b; Mills 1990).

Each *Crocuta* clan contains one to several matrilines of natal females and their offspring, as well as one to several adult immigrant males. All males disperse voluntarily from their natal clans at 24 to 60 months of age, but females typically spend their entire lives in their natal clans (Frank 1986a; Henschel and Skinner 1987; Smale et al. 1997). Within a clan, all adult females are socially dominant to adult males not born in the clan (Smale et al. 1993,1997). Before cubs reach reproductive maturity, they attain ranks in the clan's dominance hierarchy immediately below those of their mothers (Holekamp and Smale 1993; Smale et al. 1993). Individuals of both sexes retain their maternal ranks as long as they remain in the natal group. However, when males leave their natal clans, they behave submissively to all new hyenas encountered, and this is the point during ontogenetic development at which females come to dominate males (Smale et al. 1993). Intrasexual social rank among immigrant males is highly correlated with immigrants' tenure in their new clans, such that those arriving first dominate those arriving later (Smale et al. 1997).

Female spotted hyenas are exposed early in development to an unusual array of androgenic hormones, and are heavily masculinized with respect to many aspects of their morphology and behavior (Lindeque and Skinner 1982; Glickman et al. 1987,1992; Licht et al. 1992). For example, the female *Crocuta* is more aggressive than the male in a variety of contexts, particularly those associated with feeding and with the acquisition or maintenance of social rank (Frank et al. 1989; Smale et al. 1993,1995). Females are also approximately 10% larger than males, and their external genitalia are heavily masculinized, such that the clitoris is modified to form a fully erectile pseudopenis and the vaginal labia form structures closely resembling the male's scrotum (Matthews 1939).

The mating system in *Crocuta* is polygynous. Male hyenas pass through puberty late in their second year, and have viable sperm in their testes by 24 months of age (Matthews 1939). Females in the wild may conceive their first litters as early as 25 months of age, but rarely do so before 36 months (Holekamp et al. 1996). The gestation period in the spotted hyena is 110 days (Schneider 1926; Kruuk 1972). In most *Crocuta* populations studied to date, births occur throughout the year, but in many of these populations patterns of births include distinct seasonal peaks or troughs (Smithers 1966; Kruuk 1972; Cooper 1993; Lindeque and Skinner 1982; Frank

1986a; Mills 1990). Litters born to wild hyenas usually contain one or two cubs, although triplets occur occasionally among captive hyenas (Frank et al. 1991). Life span among free-living hyenas may exceed 18 years (Frank et al. 1995).

Methods

Since June 1988, we have been studying a single *Crocuta* clan inhabiting the Talek area of the Masai Mara National Reserve, Kenya, situated approximately 2° south of the equator. This is an area of open, rolling grassland grazed year-round by several ungulate species. Each year these resident antelope are joined for 3 to 4 months by large migratory herds of wildebeest (*Connochaetes taurinus*) and zebra (*Equus burchelli*) from the southern part of the Serengeti ecosystem. Thus prey abundance fluctuates seasonally in the Talek area.

Only 5% of the diet of Talek hyenas is composed of carrion, and the remainder of the diet consists of vertebrate prey these hyenas hunt themselves (Cooper et al. 1998). During most months of the year, approximately 85% of the diet consists of Thompson's gazelles (*Gazella thompsoni*) and topi (*Damascilus korrigum*) (Holekamp et al. 1997), which are antelope weighing approximately 20 and 140 kg, respectively (Kingdon 1982; Estes 1991). However, during months when the migratory ungulates are present in Talek, the resident hyenas mainly hunt wildebeest and zebra (Holekamp et al. 1997; Cooper et al. 1998), which weigh about 200 and 400 kg, respectively (Kingdon 1982; Estes 1991). Talek hyenas usually hunt alone or with one other hunter when capturing gazelles, topi, and wildebeest, but zebra hunts occur in groups of about nine hyenas (Holekamp et al. 1997). In contrast to the hunting behavior of big cats like lions and cheetahs, hyenas rarely use stealth as a hunting tactic. Instead hyenas are cursors which capture prey by tenacious pursuit over distances of up to several kilometers. Only about one of every three or four hunts results in successful prey capture (Holekamp et al. 1997). Thus prey capture is extremely expensive for hyenas in terms of both time and energy invested in hunting. Once a small ungulate has been brought down, its carcass can usually be defended by a single individual, who can thus monopolize access to food. Larger carcasses are usually fed upon by multiple hyenas until individual limbs or girdles are torn away. Each of these body parts can then be monopolized by a single hyena.

The Talek clan usually contains sixty-five to seventy hyenas, including twenty to twenty-three breeding females, nine to twelve adult immigrant males, and twenty to thirty-five juveniles and subadults. This clan defends a group territory of roughly 60 km^2. Observers have been present in the Talek clan's territory approximately 25 days per month since June 1988, monitoring demography and behavior of resident hyenas on a daily basis. We recognize individual hyenas by their unique spot patterns, and we distinguish males from females on the basis of penile morphology (Frank et al. 1990). Mother-offspring relations are established on the basis of regular nursing associations. We estimate birth dates of litters to within 1 week, based on the appearance and behavior of cubs when they are first seen above ground.

We observe Talek hyenas 3 to 6 hours each day from vehicles parked 1 to 30 m from our subjects, using binoculars or night vision goggles. We use critical incident sampling data (Altmann 1974), representing outcomes of several thousand dyadic agonistic interactions, to assess dominance relationships among adult clan members, as described by Holekamp and Smale (1990) and Smale et al. (1993). By convention, the highest-ranking (alpha) individual is assigned a rank position of 1. We also record all occurrences of courtship behavior (as described by Kruuk 1972 and Mills 1990), mounting, and copulation as critical incidents. We estimate prey availability within the Talek clan's home range at 2-week intervals by counting all ungulates within 100 m of two 4-km transect lines (Holekamp et al. 1993), and we record rainfall daily. We measure temporal variation in intensity of feeding competition among Talek hyenas by counting the number of hyenas present at every kill, and also by calculating the number of aggressive interactions observed per hour per hyena feeding at each kill (Holekamp et al. 1993).

Since 1988, we have monitored several different indicators of female reproductive performance in the Talek clan (Holekamp et al. 1996). We establish female age at first parturition by continually monitoring the condition of the posterior surface of the female's phallus, which tears at her first parturition, leaving a large patch of pink scar tissue (Frank and Glickman 1994). We determine litter size and composition based on observations made when cubs first appear above ground. Intervals between successive litters are calculated as the number of months between consecutive parturitions when at least one member of the first litter survives to weaning. Weaning conflicts and cessation of nursing indicate when cubs are weaned (Holekamp et al. 1996).

Since 1990, all Talek hyenas have been immobilized one to four times in order to collect blood samples and morphological measures. Anesthetic is delivered in a lightweight plastic syringe from a CO_2 rifle. Each animal is darted between 0700 and 0900 hours while sleeping alone on open plain. While anesthetized, each hyena is weighed and measured, whole blood is collected from femoral or jugular veins, and plasma samples are frozen for subsequent analysis of circulating hormone concentrations via radioimmunoassay (RIA). In collaboration with Cheryl Sisk, we have recently developed and validated an assay for measurement of plasma luteinizing hormone (LH) in *Crocuta*, using a primary antibody to rat LH. To evaluate rank-related differences in the physiological mediation of female reproduction, since 1995 we have followed immobilization of all adult Talek females with gonadotropin-releasing hormone (GnRH) challenge. That is, at 5-minute intervals for 165 minutes after immobilzation, we collect blood samples from which plasma is drawn for LH RIA. Baseline LH levels are monitored for the first 45 minutes, and then a bolus of synthetic luteinizing hormone–releasing hormone (LHRH) (1 µg/kg body mass; Sigma Chemical Co., St. Louis, MO) is injected intravenously, followed by 120 additional minutes of repeated blood sampling. Control animals are injected with saline solution after 45 minutes. We then simply examine changes in plasma LH as a function of time relative to the LHRH or saline injection.

Ecological Influences on Female Reproduction

Births occur in every month of the year in the Talek study population (Frank 1986a). However, the number of births varies significantly by month, with a distinct annual birth trough occurring from February through May (figure 9.1). Ecological variables that might mediate this seasonal pattern of reproduction include day length, rainfall, and prey abundance. Because this study site is situated immediately below the equa-

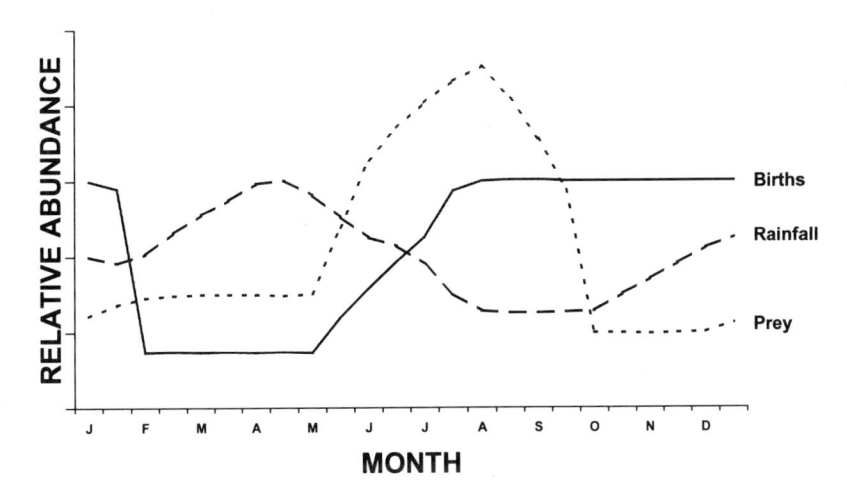

Figure 9.1
Schematic diagram depicting seasonal patterns of births, rainfall, and prey abundance in the home range of the Talek hyenas, based on data collected between June 1988 and June 1998.

tor, day length varies by only a few minutes throughout the year. Thus, although we cannot rule out day length as a possible proximate cue, it seems unlikely that reproductive seasonality in spotted hyenas is cued by changing day length as it is in other carnivore species at higher latitudes (e.g., see Dunn and Chapman 1983; Bissonnette 1932). Furthermore, patterns of reproductive seasonality differ dramatically between two other equatorial *Crocuta* populations separated by less than 1° of latitude (Kruuk 1972), a finding inconsistent with predictions of the day length hypothesis.

Both rainfall and prey abundance vary seasonally in Talek, but their annual patterns are not isomorphic (see figure 9.1). In fact, the least rain falls when prey are most abundant. Flooding of hyena dens occurs occasionally during heavy rains, suggesting that *Crocuta* might time births to avoid bearing young during the spring rains. However, we can reject this hypothesis because numbers of births, conceptions, and cubs dying each month are all unrelated to the amount of rainfall during those months (Holekamp et al. 1999).

We have used our biweekly ungulate counts to evaluate the hypothesis that availability of food functions as a proximal cue for reproductive seasonality in Talek hyenas. Prey abundance peaks each year with the presence of the migratory herds from June through September. The annual period of lowest prey abundance occurs during the months following the departure of the migration each fall, and prey densities only rise to moderate levels after rain restimulates growth of local vegetation (see figure 9.1). We have found a strong positive correlation between prey abundance and the number of litters conceived each month (figure 9.2). The annual trough in hyena births occurs approximately one gestation period length after the phase of the annual cycle during which prey animals are least abundant in our study area. This pattern suggests that seasonal changes in energy or nutrient availability are responsible for seasonal variation in conception rates among Talek hyenas. This conclusion is also supported by results from studies of *Crocuta* elsewhere in Africa. Throughout the range of this species it appears that reproductive seasonality occurs when prey abundance fluctuates seasonally, and peaks and troughs in conception rates occur in conjunction with peaks and troughs, respectively, in local food abundance (Kruuk 1972; Cooper 1993; Lindeque and Skinner 1982; Mills 1990). Thus seasonal reproduction in the spotted hyena appears to be a direct consequence of seasonal changes in availability of energy and nutrients. That energy is likely to be a more important determinant of reproductive performance than specific nutrients is suggested by data collected from a variety of other mammalian species in both the wild and in captivity (see chapters 2 and 3). However, for information about the reproductive effects of non-nutritive substances in herbivorous animals, see chapter 7.

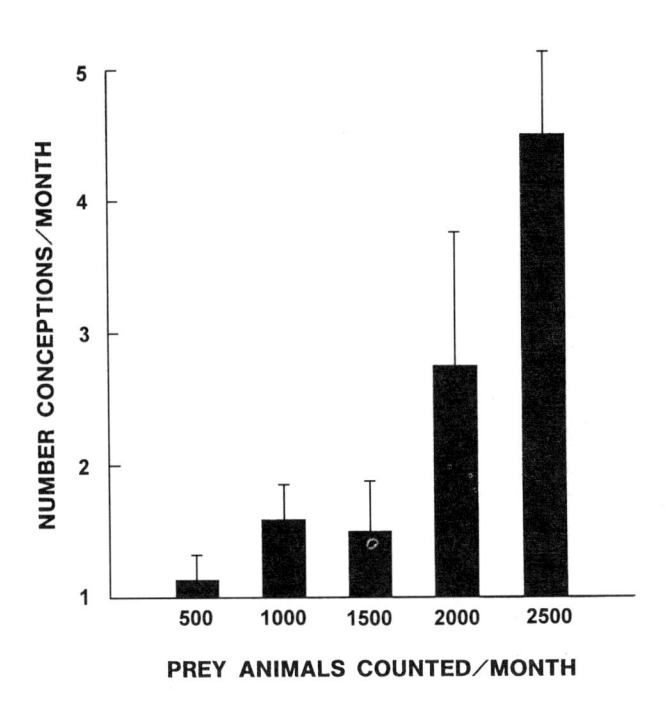

Figure 9.2
Monthly conception rates observed among Talek females as these vary with numbers of prey animals counted in the Talek home range each month.

Social Influences on Female Reproduction

The social rank of any individual clan member determines its ability to feed at kills in the presence of conspecifics, and thus profoundly influences its energy intake. Although Talek hyenas most frequently hunt alone or with one other animal, they tend to feed in groups with an average size of eleven individuals (Holekamp et al. 1997). Several hyenas scattered over a large area often rush simultaneously to the site of a kill, and the resulting group of hungry animals can completely devour a 200-kg antelope in as little as 13 minutes. When feeding in this social context, a hyena's dominance status is the primary determinant of how much food it will be able to ingest, because high-ranking animals can aggressively displace lower-ranking ones from the carcass (figure 9.3; Frank 1986b). Thus competitive ability varies with social rank in this species, and fight outcomes determine the amount of food available to individual females. Here we consider how this might influence patterns of variation in the number of offspring females can produce.

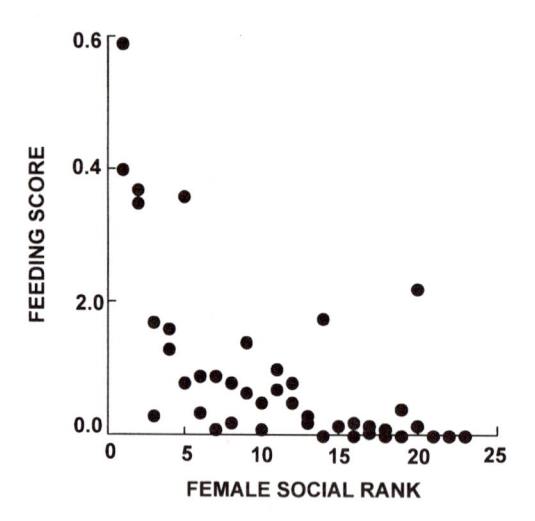

Figure 9.3
Feeding scores of adult female hyenas in the Talek clan as a function of their social rank. An individual's feeding score was the proportion of all 5-minute intervals at all kills during which that animal was observed feeding. (Modified from Frank 1986.)

Variance in production of viable offspring by female mammals can be influenced by variation in the length of the reproductive life span or in the rate of offspring production during the reproductive years. The reproductive life span of a female may be lengthened by earlier onset, later offset, or both. Her reproductive rate can be increased through changes in litter size, or by reduction in the length of intervals between successive litters. Overall female RS also depends importantly on rates of offspring mortality. Improvement in one or more of these variables should result in superior reproductive success (Clutton-Brock et al. 1982, 1988). Here we review the effects of social rank on these variables in female *Crocuta*.

Reproductive Life Span

Although age at death does not vary with social rank among female hyenas, social rank has a major impact on the age at which reproduction begins (Frank et al. 1995; Holekamp et al. 1996). Daughters of high-ranking females start bearing offspring at significantly younger ages than do their lower-ranking counterparts (figure 9.4), and variance in age of first parturition among lower-ranking females is over three times greater than is variance in this measure among females of the alpha matriline (alpha matriline, N = 8: $X \pm SD = 31.9 \pm 3.2$ months vs. lower-ranking females, N = 17: $X = 48.5 \pm 11.0$ months). When they conceive their first litters, the lowest-ranking

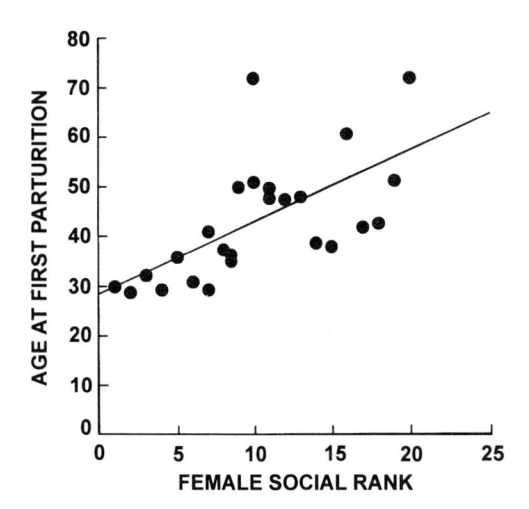

Figure 9.4
Age at first parturition as a function of female social rank in Talek hyenas.

females are 60 to 70 months old, whereas the highest-ranking females are only 26 months old, a difference of roughly 36 months. In captivity, where food is effectively unlimited, age at first parturition can be as low as 21 months (Frank et al. 1995), suggesting that rank effects on onset of reproduction in the wild are mediated by differential access to food, and hence differential energy availability.

The neuroendocrine mechanisms that mediate the effects of social rank on female reproduction have now been explored in a variety of species (e.g., see Creel et al. 1992,1995). Among singular breeders like the naked mole rat and common marmoset (*Callithrix jacchus*), baseline levels of plasma LH are lower in nonbreeding than in breeding females, as are LH responses to injections of GnRH (Faulkes and Abbott 1997; Abbott et al. 1990). Among reproductively suppressed marmosets, those that are lowest ranking exhibit a smaller LH response to GnRH challenge than do their higher-ranking counterparts. In fact, the third-ranking female exhibits a smaller LH response than does the second-ranking female, who exhibits a smaller LH response than does the alpha female. We have begun to examine the neuroendocrine mechanisms by which social rank influences female reproduction in the spotted hyena by measuring reproductive hormones before and after administration of GnRH in the wild. Data collected to date suggest that the regulation of LH may be quite different in low-ranking female hyenas than it is in nonbreeding adult female marmosets and naked mole rats. Specifically, low-ranking adult female hyenas that have not yet reproduced appear to exhibit baseline LH titers and an LH response to GnRH challenge comparable to those seen in fully mature multiparous females (figure 9.5). Maximum LH response to GnRH injection peaks at only 3 or 4 ng/mL in most females darted earlier during the lactation interval than was the female shown in figure 9.5B, who was in her 16th month of lactation, and control females injected with saline show no increase whatsoever in plasma LH. Although these data are preliminary, they suggest that the delayed rate of reproductive development seen in low-ranking female hyenas is not caused by a failure of the pituitary gland to secrete adequate baseline LH or to respond to GnRH.

The Rate of Reproduction

Litter Size Litter size among Talek hyenas does not vary with maternal rank (Holekamp et al. 1996). Mean litter size in this population is 1.43 ± 0.13 (Holekamp et al. 1996). Production of triplet litters is not uncommon among captive females maintained with food ad lib. (e.g., see Frank et al. 1991). However, during 11 years of careful observation of wild hyenas, we have only seen one triplet litter, born to a high-ranking female.

Duration of the Interbirth Interval Dominant female hyenas experience shorter interbirth intervals than do subordinate females (Holekamp et al. 1996). This effect is due primarily to rank-related variation in interlitter interval length following birth of a singleton litter (figure 9.6A). Raising a twin litter may place heavy energetic demands on all females, regardless of their social rank (Holekamp et al. 1996). To determine which components of the interbirth interval are influenced by social rank, we assume that the gestation period is constant at 110 days, and evaluate variation in lengths of only the lactation interval and the interval between weaning of one litter and conception of the next. Cub ages at weaning increase dramatically with decreasing social rank (figure 9.6B). The highest-ranking females are able to wean their litters when cubs are only 7 to 8 months old, whereas low-ranking females may require as long as 21 months. The lowest-ranking females can thus take up to 2.6 times longer to wean their young than do the highest-ranking females. High-ranking females are also able to concurrently support pregnancy and lactation more frequently than lower-ranking females, and the length of the mother's recovery period after weaning her litter increases with decreasing maternal rank (figure 9.6C).

Annual Rate of Offspring Production The annual rate of cub production by Talek females decreases significantly with social rank (figure 9.7A). The highest-ranking females are able to produce an average of approximately 2.5 offspring per year, while some of the lowest-ranking females do so at rates as low as approximately 0.5 offspring per year (see figure 9.7A), a fivefold difference. The annual rate of production

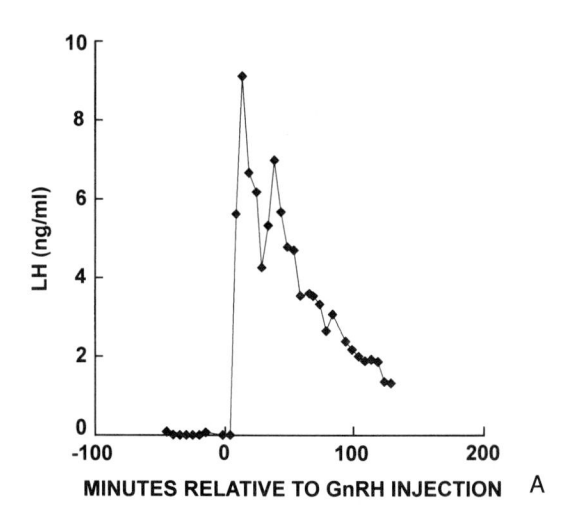

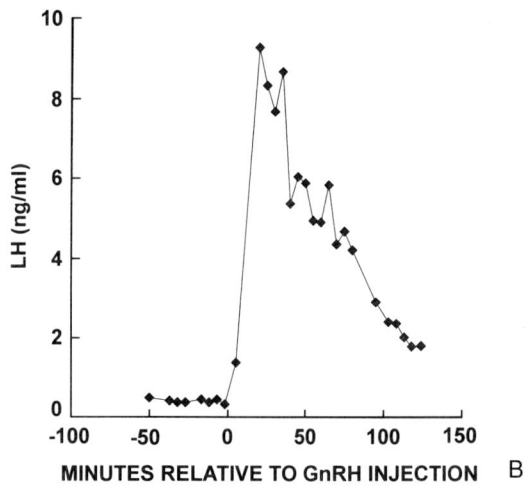

Figure 9.5
Change in plasma luteinizing hormone (LH) levels in response to intravenous injection of a bolus of gonadotropin-releasing hormone (GnRH) for (A) a 2.5-year-old nulliparous female who did not conceive her first litter until 15 months after this GnRH challenge, and (B) a 9.5-year-old multiparous female who conceived her next litter 4 months after GnRH challenge. At the time of sampling, both females were middle-ranking in the overall female dominance hierarchy.

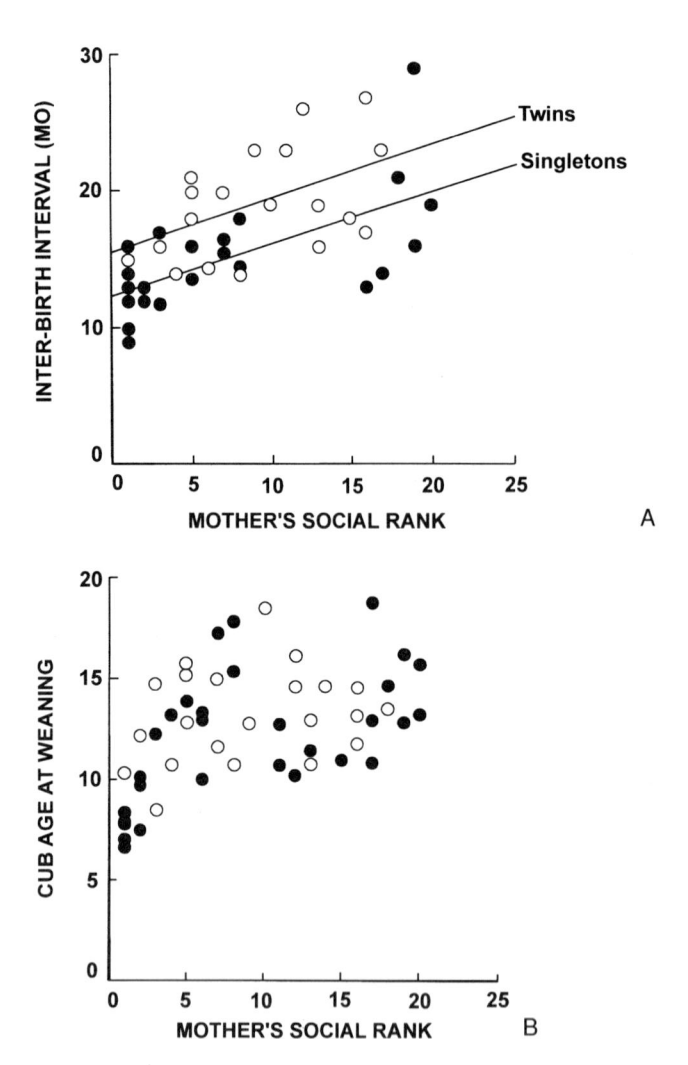

Figure 9.6
The relationship between female social rank and (A) interbirth intervals following the birth of singleton and twin litters, (B) ages at which members of singleton and twin litters were weaned, and (C) duration of the "recovery" period between weaning of one litter and conception of the next when the first litter survives to weaning. Singleton litters are represented by solid circles and twin litters by empty circles. (From Holekamp, Smale, Szykman 1996.)

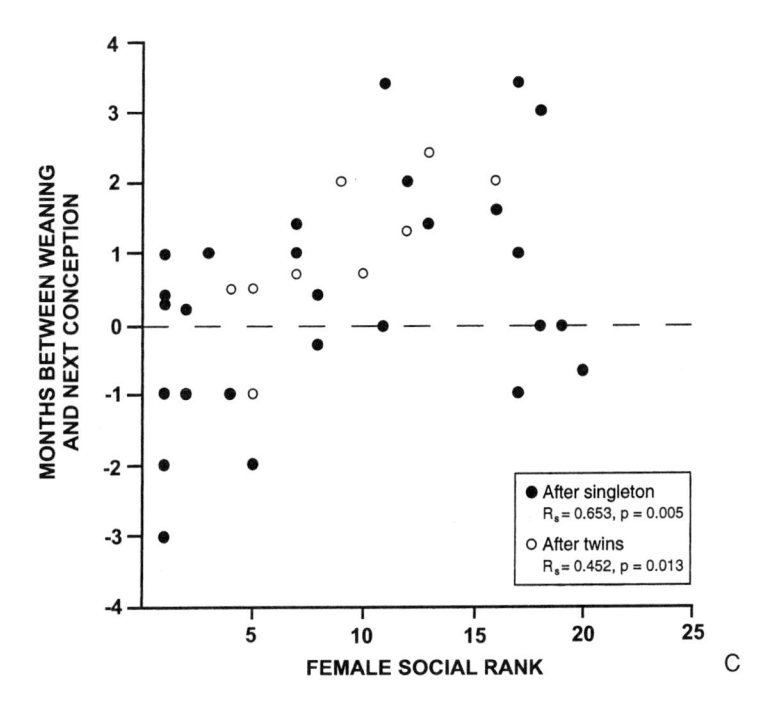

Figure 9.6 (continued)

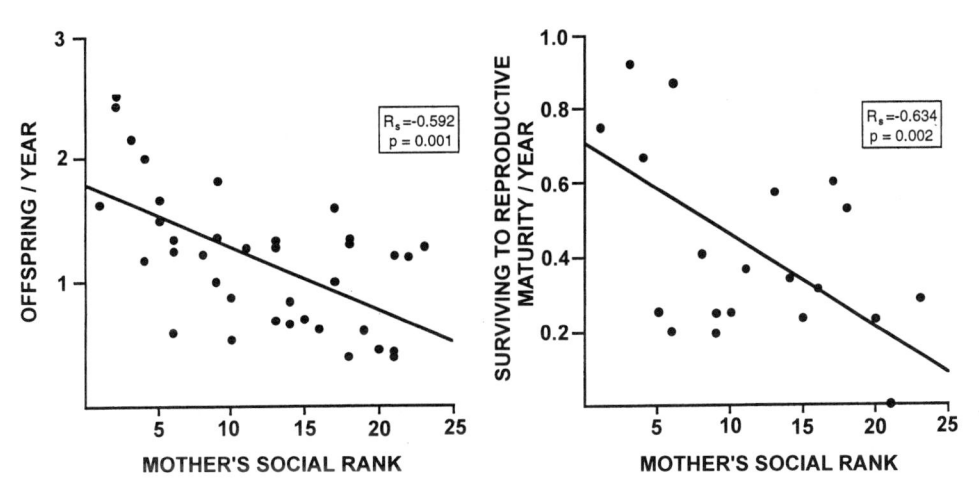

Figure 9.7
The relationship between female social rank and (A) the total number of offspring produced per year, and (B) the number of offspring surviving to reproductive maturity each year. (From Holekamp, Smale, Szykman 1996.)

of cubs surviving to reproductive maturity also varies significantly with maternal rank (figure 9.7B). Thus, high-ranking females bear young at higher rates than do low-ranking females, and their offspring enjoy better survivorship than do offspring of low-ranking females (Holekamp et al. 1996).

Interaction Effects of Food and Rank on Reproductive Performance

We use our biweekly prey counts (described above under Methods) in the Talek area to evaluate the effect of prey availability on the relationship between social rank and reproductive performance in *Crocuta*. Specifically, for high- and low-ranking females rearing singleton litters we compare weaning ages of cubs during periods when prey is relatively scarce with those recorded during periods when game is relatively plentiful. Females in the alpha matriline always wean their singleton cubs at younger ages than do lower-ranking mothers, regardless of prey abundance in the clan's home range (figure 9.8). However, whereas weaning ages of high-ranking cubs do not vary

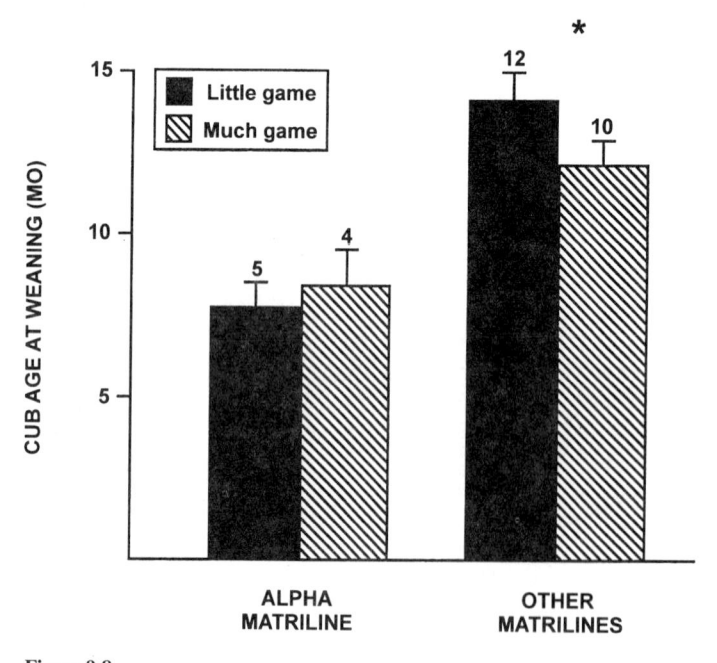

Figure 9.8
Relationship between prey abundance and duration of lactation for singleton cubs in the alpha (n = 4 females) and lower-ranking (n = 9 females) matrilines. Numbers of singleton litters born are depicted above each bar, and the asterisk indicates $P < .02$, using Student's t-test to compare mean values. (From Holekamp, Smale, Szykman 1996.)

between high and low game conditions, cubs of low-ranking mothers are weaned at significantly younger ages when game is abundant than when it is scarce. This suggests that, within the range of prey abundance observed in the Talek area, the reproductive performance of high-ranking females may be largely unaffected by fluctuations in the food supply. However, enhanced food availability can significantly shorten lactation intervals among lower-ranking females. High rank thus effectively appears to buffer some aspects of female reproductive performance from temporal variation in energy availability.

Rank and Female Reproductive Success: *Crocuta* vs. Other Plural Breeders

In our study population of spotted hyenas, most measures of female reproductive performance are characterized by a great deal of variance, and much of this variance is related to intrasexual social rank (Holekamp et al. 1996). Here we attempt to compare the effects of social rank on RS in spotted hyenas with those observed in other polygynous mammals. We restrict our discussion to those gregarious mammals with polygynous mating systems like that of the spotted hyena, in which multiple females normally breed. Plural breeders for which some data have been collected germane to the relationship between rank and RS among females include elephant seals, various ungulates, and several species of Old World primates.

In some polygynous mammals female body size, social rank, age, and RS are interrelated (e.g., red deer [*Cervus elaphus*], Clutton-Brock et al. 1984; Hanuman langurs [*Presbytis entellus*], Borries et al. 1991). For example, female elephant seals over 6 years old are larger than, and dominant to, females 3 to 5 years old, and they are twice as successful at rearing pups to weaning than are younger females (Reiter et al. 1981). The effects of age and rank in this species are largely due to variation in the time and place at which females give birth. Among female seals within a given age category, social rank does not appear to influence reproduction (Reiter et al. 1981). Among adult female *Crocuta*, body mass does not vary significantly with social rank (Holekamp et al. 1996), and neither age nor body size are good predictors of outcomes of agonistic interactions (e.g., see Smale et al. 1993).

An influence of female social rank on rate of reproductive maturation has been documented in various ungulates and Old World primates. High-ranking female gazelles and red deer hinds conceive at younger ages than do low-ranking females (Alados and Escos 1992; Clutton-Brock et al. 1984), but these rank effects are small and may also be confounded by effects of age and body size (e.g., see Thouless 1990). Among savanna baboons (*Papio cynocephalus*), social rank influences rates of reproductive maturation in some populations (Packer et al. 1995; Altmann et al. 1988). For example, in Amboseli, where food is patchily distributed and seasonally limited, conception first occurs at 2300 days of age among the lowest-ranking female baboons and at 1800 days among highest-ranking females (Altmann et al. 1988). Thus these low-ranking female baboons conceive at ages 1.28 times greater than ages at which high-ranking females first conceive. However, at Gilgil, where food availability is consistently greater than in Amboseli, no effect of social rank on age at first conception is observed in the same baboon species (Smuts and Nicolson 1989). Thus the relationship between rank and female RS among these baboons appears to depend on habitat characteristics and on the potential for resource monopolization by dominant individuals (Bercovitch and Strum 1993; Barton et al. 1996).

Ellis (1995) reviewed 17 studies of macaques (*Macaca* spp.) in which investigators examined the relationship between female social rank and age at first reproduction. Eleven of these studies found high-ranking females conceiving or giving birth at significantly younger ages than low-ranking females, five studies found no significant differences, and one concluded that low-ranking females started breeding before high-ranking individuals (Ellis 1995). Thus the relationship between female social rank and age at first breeding may vary with habitat type and the potential for resource monopolization in macaques as it does in baboons.

Among chimpanzees (*Pan troglodytes*), the age at which females show their first full sexual swelling is correlated with social rank, and ranges from 9 years for daughters of high-ranking females to 13 years for daughters of low-ranking females (Pusey et al. 1997). Thus, the lowest-ranking female chimpanzees reach puberty when they are 1.44 times older than are the highest-ranking females when they start cycling. In each of the species mentioned here, female rank has a significant influence on the rate of female reproductive maturation, but the magnitude of this effect is considerably smaller than the 2.5-fold difference observed in *Crocuta*, in which the highest-ranking females start breeding roughly 36 months before their lowest-ranking peers (Holekamp et al. 1996). Assuming an overall life span of 12 years for female hyenas of all ranks (Frank et al. 1995), the earlier onset of breeding among high-ranking female hyenas effectively increases the length of their reproductive life span by approximately 25%.

Litter size does not vary with social rank among spotted hyenas, nor have effects of dominance status on litter size been reported for other gregarious mammals. However, length of intervals between litters does vary with female social status in some species. The influence of female status on interbirth interval length has been examined in four macaque populations. In two of these, high-ranking females have significantly shorter interbirth intervals than do low-ranking females (Drickamer 1974; Dittus 1986). However, in the other two macaque populations, interbirth interval length does not vary with female social rank (Silk et al. 1981; Gouzoules et al. 1982).

The influence of social rank on interbirth interval length has been examined in three different East African populations of savanna baboons. No influence of rank on conception rates is observed among Amboseli baboons (Altmann et al. 1988). At Gombe the duration of postpartum amenorrhea is significantly correlated with female rank (Packer et al. 1995), and at Gilgil low-ranking females have average interbirth intervals that are 20% longer than those recorded for high-ranking females (Smuts and Nicolson 1989). When we use the same calculations as those used by Smuts and Nicolson (1989), we find that low-ranking female hyenas have average interlitter intervals that are 33% longer than those of high-ranking females. The relationship we observe between maternal rank and interlitter interval in hyenas thus appears to be considerably stronger than that recorded within any baboon population to date. This conclusion is consistent with the notion that the effect of intrasexual rank on RS is unusually large among female spotted hyenas. However, the generality of this finding is difficult to assess in the absence of comparable data from other polygynous species.

Among some plural-breeding mammals the annual rate of offspring production is related to female social rank, but in other species no such relationship is observed. Among red deer, female rank does not influence the overall reproductive rate as indicated by the proportion of years in which a hind produces a calf (Clutton-Brock et al. 1984). In this cervid species high social rank appears to enhance female reproductive success primarily through increased production of large and relatively successful sons (Clutton-Brock et al. 1984). Similarly, the overall conception rate is unrelated to dominance rank among savanna baboons (Altmann et al. 1988). Among Gelada baboons (*Theropithecus gelada*), the difference observed between the highest- and lowest-ranking females with respect to total number of offspring produced in a lifetime is two (Dunbar 1980; Dunbar and Dunbar 1977). By contrast, assuming equal life spans of 12 years, the highest-ranking spotted hyena is likely to produce at least seven more cubs in her lifetime (0.9 cub surviving per year × 9 breeding years) than is the lowest-ranking female (0.2 cub surviving per year × 6 breeding years). Among chimpanzees, the annual rate of production of offspring surviving to weaning age (5 years) decreases with female rank (Pusey et al. 1997). In *Crocuta*, both annual rates of offspring production and numbers of cubs surviving to reproductive maturity decrease dramatically with female rank (Holekamp et al. 1996).

In comparison with females of other gregarious plural-breeding species, female spotted hyenas exhibit markedly greater variance in RS, and the influence of social rank on specific components of female RS is far more profound in *Crocuta* than in these other mammals. Indeed, on the basis of data collected from the Talek population before 1992, Frank et al. (1995) calculated that enhanced reproductive success among members of the alpha matriline would allow that matriline to comprise the entire adult female component of the clan after only 15 generations, or approximately 45 years. The rate at which this is happening has actually increased since that estimate was generated. In 1988, when we began studying this population, there were twenty-two breeding females in the clan, as there are today. The alpha and beta females in 1988 were a mother-daughter pair. Today, nine of the twenty-two adult females present in the clan are descendants of those two females, and the alpha matriline has thus increased its representation within the population of breeding females from 9% to 41%. Meanwhile, thirteen of the original twenty-two adult females have failed to leave any descendants at all in the Talek clan. If the current two top-ranked females are as successful during the next 10 years as the top two have been during the past decade, then the entire adult female population will consist of descendants of the alpha matriline, and all other matrilines will be extinct. That is, within 20 years, two of twenty-two adult females will have left twenty-two female descendants, while the other twenty females will no longer be represented in the clan. Thus, high dominance rank confers an exceptionally large selective advantage to female *Crocuta*. The mechanism by which this selective advantage is conferred is most likely the frequent rank-related aggressive interactions occurring during feeding competition, which determine a female's access to food (Frank 1986b). Any genes that might contribute to acquisition or maintenance of high social rank should thus be very heavily favored in female hyenas by natural selection.

The strong relationship between social rank and female reproductive success in the spotted hyena may have played a critical role in the evolution of sexual monomorphism and the peculiar patterns of sexual differentiation that set this species apart from other mammals (Frank 1997). That is, because the female descendants of a single high-ranking hyena can come to constitute an entire clan within only a few generations, effective population size (N_e, the number of individuals who actually contribute to reproduction; Wright 1931; Crow and Kimura 1970) is in fact far smaller than the number of adults currently present in the clan. The smaller the effective population size, the more rapidly adaptive mutations can become fixed in a population. If only a few mutations were necessary for the occurrence of female "masculinization" during the evolutionary history of the spotted hyena, then these could have quickly become fixed within a clan, and spread to other clans via dispersing individuals.

Contextual Influences on Male Reproduction

Whereas reproductive performance of female *Crocuta* is determined by ecological and social constraints on energy availability, this is unlikely to be an important determinant of male RS. Since male hyenas have no dependent offspring, they need only secure enough energy for self-maintenance and performance of courtship behavior. The hypothesis that differential access to food influences variation in male RS is inconsistent with field data showing that all male *Crocuta* voluntarily emigrate from their natal clans after puberty (Smale et al. 1997). Consider the implications of dispersal for food access among male hyenas. As long as the male remains in his natal clan, he maintains his maternal rank and enjoys virtually the same priority of access to carcasses as does his mother. Even sons of the lowest-ranking females in a clan can displace all immigrant males from food. Alpha sons can displace all clan mambers except their mothers and younger siblings (Holekamp and Smale 1993). However, when they disperse, even alpha sons experience the worst possible priority of food access in their new clans. Not only are all immigrant males ranked below all natal animals but social rank among immigrant males only increases with tenure in the new clan, so recent arrivals have the lowest priority of access to food in the entire clan (Smale et al. 1997). Thus males incur substantial energetic costs by dispersing, and it appears that dispersal represents a critical "currency exchange" for males: they swap superior energy access for superior access to females who might accept them as mates.

Male hyenas may remain in their natal clan up to 38 months after puberty (Smale et al. 1997). As a result, there are two classes of reproductively mature males in every clan: postpubertal natal males older than 24 months, and immigrant males that have arrived from other clans. These two groups of adult males behave quite differently in their intra- and intersexual social interactions, and they also differ markedly with respect to their circulating hormone levels (Holekamp and Smale 1998). Hourly rates at which males exhibit aggressive behavior are over six times higher among natal than immigrant males. Moreover, in all agonistic interactions between adult natal and immigrant males, the former inevitably beat the latter. Not only can natal males displace immigrants from food, but they can also displace immigrants from proximity to adult females (Holekamp and Smale 1998).

When Talek females are courted simultaneously by adult natal and immigrant males, they behave differently toward the latter than the former. Whereas females often approach and sniff immigrant males, they ignore natal males. In fact, whereas females appear to treat all immigrants as potentially serious suitors, females treat natal males as they would treat any pesky youngster of either sex. Although anecdotal, these sorts of observations suggest that females are more strongly sexually attracted to immigrants than to adult natal males. Considering that small N_e is associated with high levels of inbreeding and genetic homozygosity, and that these pose certain dangers to sexually reproducing organisms (e.g., see Crow and Kimura 1970), it would not be surprising to find that female hyenas prefer immigrant males as mates.

Although many adult natal males make sexual overtures to resident adult females, they do so at hourly rates roughly four times lower than those observed for adult immigrant males (Holekamp and Smale 1998). Thus the sexual ardor exhibited by adult males toward females from their natal clan appears to be less intense than that shown by immigrants. Although adult natal males have been seen mounting resident females, we have never observed them participating in complete copulations. This finding presumably reflects not only lower levels of sexual motivation among adult natal than among immigrant males but also female preferences for immigrants over natal males as mates. Natal male hyenas are dominant to immigrants and can displace them from proximity to adult females, but female *Crocuta* need never mate with natal males if they prefer not to. Adult females are larger than males, and socially dominant to them. Furthermore, the female hyena's bizarre genitalia require that the male insert his erect penis into her flaccid pseudopenis during copulation. Thus copulation in this species is difficult even when both parties are willing, and copulation simply does not occur unless the female chooses to. If resident females prefer immigrants over adult natal males, then dispersal behavior should increase the probability that a male will be selected as a sex partner. This could explain why male dispersal behavior has been favored by natural selection despite the fact that dispersal impairs the male's ability to compete effectively for food at kills.

Interestingly, although body and testis size measures obtained from adult natal and immigrant males do not differ, plasma testosterone levels are significantly lower among adult natal than among immigrant males (Holekamp and Smale 1998). Based on tooth wear data, we can restrict our comparison of natal and immigrant males exclusively to those falling within a single narrow age range, yet the behavioral and endocrine differences between them still persist. Thus a male's dispersal status alone can account for the striking variation we have observed among adult males with respect to their hormone levels and behavior. Perhaps persistent residence of a postpubertal male in his natal clan suppresses plasma testosterone levels, and this inhibition is released at dispersal. Suppressed testosterone levels among postpubertal males still residing in the natal group are also observed in other species of gregarious mammals (e.g., callithricid primates; French and Schaffner 1995). It is possible that the low testosterone levels observed among adult natal *Crocuta* males might account for their low levels of sexual motivation.

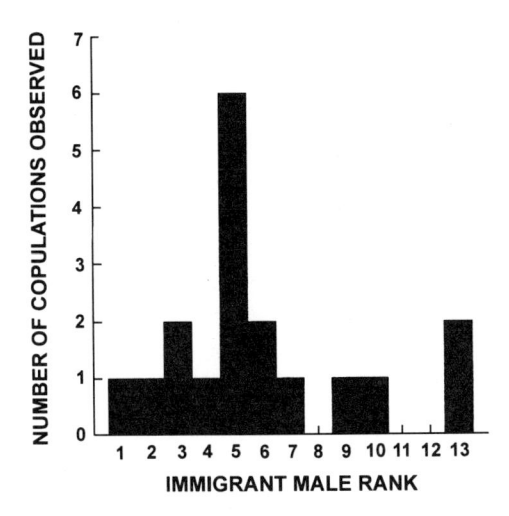

Figure 9.9
Numbers of matings observed in the Talek clan from May 1988 through May 1998 involving immigrant males of various social ranks.

As yet we know little about variance in male RS in *Crocuta*. Although molecular genetic analyses of paternity among Talek hyenas are still in progress, we have now observed eighteen copulations among members of this population. These matings have involved immigrant males ranked first to thirteenth in the male hierarchy (figure 9.9). Although these mating data accrue frustratingly slowly, they nevertheless effectively illustrate two important points. First, the alpha male is clearly not monopolizing copulations with Talek females, since low-ranking males have been observed to mate relatively often. Second, the males in the top half of the male hierarchy are likelier to mate than are those in the bottom half. Thus, although relative rank among males can accurately predict ability to feed at kills with other immigrants, male rank is only loosely associated with male RS, as indicated by observed copulations. We emphasize, however, that the number of observed copulations may not be correlated in this species with the number of offspring sired. Whereas outcomes of agonistic interactions are critical determinants of intrasexual rank among female hyenas, intrasexual rank among immigrant males is determined by their order of arrival in the clan, and new immigrants initiate social interactions with submissive behavior that can be quite extreme (Smale et al. 1997). Thus it appears that sexual selection has favored traits in male *Crocuta* that make them unique among plural-breeding mammals in which males usually exhibit larger body size, more powerful weapons, and greater pugnacity than do females.

Conclusion

The spotted hyena expresses sexually dimorphic patterns of intrasexual competition that are reversed from mammalian norms. In this species resource competition among females occurs at uniquely high intensity, and intrasexual competition might in fact be more intense among females than it is among males. As is the case in many animal species, RS among female *Crocuta* is limited by food access while male RS is limited by access to receptive females. However, due to the distribution and defensibility of the resources on which they feed, female-female competition over food is so intense in *Crocuta* that it appears to have selected for a suite of unusual traits in both males and females, resulting in feisty, "masculinized" females and meek males.

Reproductive performance among female spotted hyenas varies with energy availability. Energy availability in turn varies with seasonal changes in prey abundance, but is most profoundly influenced by female social rank in the frenzied feeding situations characteristic of this species. Agonistic interactions occur at high rates when hyenas compete over food. Natural selection has apparently therefore favored aggressiveness and large body size in females, features normally present in male mammals.

Although multiple factors undoubtedly influence male reproductive performance in *Crocuta*, the only variable that has been identified to date as an unambiguous determinant of male RS is dispersal status: immigrant males copulate but natal males do not (Holekamp and Smale 1998). A male's ability to immigrate into a new clan does not appear to be contingent upon his fighting ability or aggressiveness. In fact a prospective immigrant may meet less resistance from resident animals if he initiates contact with them by behaving submissively. Serious fights among immigrant males are rarely observed (Frank 1986b,1997) and, in contrast to most male mammals, male *Crocuta* are surprisingly meek. Nevertheless, the lowly social status of immigrant males appears to exert little influence on their ability either to nourish themselves or to mate.

Elucidating the evolution of any social system must begin with an understanding of the determinants of female behavior, since the behavior of males is largely adapted to that of females (Wrangham and Rubenstein 1986; Altmann 1990). The behavior of female spotted hyenas is largely determined by the distribution, defensibility, and richness of the carcasses on which they feed. These features of the food resource base exploited by female *Crocuta* have resulted in extremely intense direct competition, the outcome of which depends on aggressive behavior. Females vary greatly in their competitive ability under these circumstances, with dominant females enjoying far greater success. When females vary greatly in their competitive ability, we should expect them to employ malelike reproductive strategies (Berglund et al. 1993). It appears that this is precisely what has occurred during the evolution of the spotted hyena. Once selection produced large, aggressive females, aggressive competition for mates among males may have declined in importance as an operative mode of sexual selection in this species. Large females able to dominate males are effectively unconstrained in their choice of mates from among the adult males present in the clan. If females choose immigrants as mates, then the key to male success is gaining admittance to a new clan. This in turn, albeit difficult due to limited opportunities for dispersal, might be most easily accomplished if a prospective immigrant can indicate to resident clan members that he will pose no significant threat to them. Perhaps meek males find it easier to immigrate than do feisty ones.

Important questions remain unanswered here, such as why this type of social system has evolved in *Crocuta* in particular, and why the suite of behavioral and morplological traits characteristic of this species has only evolved once. Energy is at a premium at the top of the food chain. Feeding competition is rarely so intense at lower trophic levels as it is among top carnivores, and spotted hyenas are the only extant carnivore in which such large numbers of females compete for rich and defensible food patches on a daily basis. Of all the other plural-breeding mammals considered in this chapter in which multiple unrelated females compete for resources within their social unit, chimpanzees exploit a resource base most similar to that utilized by *Crocuta* with respect to richness, distribution, and defensibility (Wrangham 1979; Wrangham and Smuts 1980; Goodall 1986). Female chimpanzees enjoying the greatest RS are those that can successfully defend a rich food patch, such as a tree full of ripe fruit. It is interesting that, of the polygynous species considered here, the chimpanzee shows the strongest positive correlation between female social rank and RS, and the strength of this relationship is second only to that observed in *Crocuta*. Even so, the variance in female RS is far less among female chimpanzees than among female hyenas. Female *Crocuta* appears to be entirely unique in their frequent reliance on aggression to access the energy necessary for successful reproduction.

Acknowledgments

We thank the office of the president of Kenya for permission to conduct this research. We also thank the Kenya Wildlife Service, the Narok County Council, and the senior warden of the Masai Mara National Reserve for their cooperation. We thank the following individuals for their excellent assistance in the field: S. M. Cooper, C. I. Katona, N. E. Berry, K. Weibel, M. Durham, J. Friedman, G. Ording, T. H. Harty,

P. Garrett, E. E. Boydston, M. Szykman, and A. Engh. This work was supported by NSF grants BNS8706939, BNS9021461, IBN9309805, and IBN9606939, and by fellowships to K.E.H. from the David and Lucille Packard Foundation and from the Searle Scholars Program/Chicago Community Trust.

References

Abbott DH, George LM, Barrett J, Hodges JK, O'Bryne KT, Sheffield JW, Sutherland IA, Chambers GR, Lunn SF, Ruiz do Elvira M-C. (1990). Social control of ovulation in marmoset monkeys: a neuroendocrine basis for the study of infertility. *In* Socioendocrinology of Primate Reproduction. (Ziegler TE, Barcovitch FB, eds), pp 135–158. New York: Wiley-Liss.

Alados CL, Escos, JM (1992). The determinants of social status and the effect of female rank on reproductive success in Dam and Cuvier's gazelles. Ethol Ecol Evolution 4:151–164.

Altmann J (1974). Observational study of behavior: sampling methods. Behaviour 49:227–267.

Altmann J (1990). Primate males go where the females are. Anim Behav 39:193–195.

Altmann J, Hausfater G, Altmann SA (1988). Determinants of reproductive success in savannah baboons, *Papio cynocephalus*. *In* Reproductive Success. (Clutton-Brock, TH ed), pp 403–418. Chicago: University of Chicago Press.

Andersson M (1994). Sexual Selection. Princeton, NJ: Princeton University Press.

Barton RA, Byrne RW, Whiten A (1996). Ecology, feeding competition and social structure in baboons. Behav Ecol Sociobiol 38:321–329.

Bateman AJ (1948). Intra-sexual selection in *Drosophila*. Heredity 2:349–368.

Bercovitch FB, Strum SC (1993). Dominance rank, resource availability, and reproductive maturation in female savanna baboons. *Behav Ecol Sociobiol* 33:313–318.

Berglund A, Magnhagen C, Bisazza A, Konig B, Huntingford F (1993). The adaptive bases of female sexual behavior: reports from a workshop. Behav Ecol 4:184–187.

Bissonnette TH (1932). Modifications of mammalian sexual cycles: reactions of ferret (*Putoris vulgaris*) of both sexes to electric light added in November and December. Proc R Soc Lond B Biol Sci 110:332–336.

Borries C, Sommer V, Srivastava A (1991). Dominance, age, and reproductive success in free-ranging female Hanuman langurs. Int J Primatol 12:231–257.

Clutton-Brock TH (1988). Reproductive Success. Chicago: University of Chicago Press.

Clutton-Brock TH, Guiness FE, Albon SD (1982). Red deer: behavior and ecology of two sexes. Chicago: University of Chicago Press.

Clutton-Brock TH, Albon SD, Guiness FE (1984). Maternal dominance, breeding success and birth sex ratios in red deer. Nature 308:358–360.

Clutton-Brock TH, Albon SD, Guiness FE (1988). Reproductive success in male and female red deer. *In* Reproductive Success. (Clutton-Brock, TH, ed), pp 325–343. Chicago: University of Chicago Press.

Cooper SM (1993). Denning behaviour of spotted hyaenas (*Crocuta crocuta*) in Botswana. Afr J Ecol 31:178–180.

Cooper SM, Holekamp KE, Smale L (1998). A seasonal feast: long-term analysis of foraging behaviour in the spotted hyaena *Crocuta crocuta* (Erxleben). Afr J Ecol, in press.

Creel S, Creel NM, Wildt DE, Monfort SL (1992). Behavioral and endocrine mechanisms of reproductive suppression in Serengeti dwarf mongooses. Anim Behav 43:231–245.

Creel S, Creel NM, Mills GL, Monfort SL (1995). Rank and reproduction in cooperatively breeding African wild dogs: behavioral and endocrine correlates. Behav Ecol 8:298–306.

Crow JF, Kimura M (1970). An Introduction to Population Genetics Theory. New York: Harper & Row.

Darwin C. (1871). The Descent of Man and Selection in Relation to Sex. London: John Murray.

Dittus WPJ (1986). Sex differences in fitness following a group take-over among Toque macaques: testing models of social evolution. Behav Ecol Sociobiol 19:257–266.

Drickamer LC (1974). A ten-year summary of reproductive data for free-ranging *Macaca mulatta*. Folia Primatol 21:61–80.

Dunbar RIM (1980). Determinants and evolutionary consequences of dominance among female gelada baboons. Behav Ecol Sociobiol 7:253–265.

Dunbar RIM, Dunbar EP (1977). Dominance and reproductive success among female gelada baboons. Nature 266:351–352.

Dunn JP, Chapman JA (1983). Reproduction, physiological responses, age structure and food habits of racoons in Maryland, USA. Z Säugetiere, 48:161–175.

Ellis L (1995). Dominance and reproductive success among nonhuman animals: a cross-species comparison. Ethol Sociobiol 16:257–333.

Emlen ST, Oring LW (1977). Ecology, sexual selection and the evolution of mating systems. Science 197:215–223.

Estes, RD (1974). Social organization of the African Bovidae. *In* The Behaviour of Ungulates and Its Relationship to Management. (Geist V, Walther F, eds), pp 166–205. Morges, Switzerland: IUCN.

Estes RD (1991). The Behaviour Guide to African Mammals. Berkeley: University of California Press.

Faulkes CG, Abbott DH (1997). The physiology of a reproductive dictatorship: regulation of male and female reproduction by a single breeding female in colonies of naked mole-rats. *In* Cooperative Breeding in Mammals. (Solomon NG, French JA, eds), pp 302–334. Cambridge: Cambridge University Press.

Fisher RA (1930). The Genetical Theory of Natural Selection. Oxford: Macdonald.

Frank LG (1986a). Social organization of the spotted hyaena (*Crocuta crocuta*) I. Demography. Anim Behav 35:1500–1509.

Frank LG (1986b). Social organization of the spotted hyaena (*Crocuta crocuta*) II. Dominance and reproduction. Anim Behav 35:1510–1527.

Frank LG (1997). Evolution of genital masculinization: why do female hyaenas have such a large "penis?" Trends Ecol Evolution 12:58–62.

Frank LG, Glickman SE (1994). Giving birth through a penile clitoris: parturition and dystocia in the spotted hyaena, *Crocuta crocuta*. J Zool Lond 234:659–665.

Frank LG, Glickman SE, Zabel CJ (1989). Ontogeny of female dominance in the spotted hyena: perspectives from nature and captivity. *In* The Biology of Large African Mammals in Their Environment. Zoological Society of London Symposium No. 61 (Jewell PA, Maloiy GMO eds), pp 127–146. Oxford: Clarendon Press.

Frank LG, Glickman SE, Licht P (1991). Fatal sibling aggression, precocial development, and androgens in the neonatal spotted hyaenas. Science 252:702–704.

Frank LG, Glickman SE, Powch I (1990). Sexual dimorphism in the spotted hyaena. J Zool Lond 221:308–313.

Frank LG, Holekamp KE, Smale L (1995). Dominance, demography, and reproductive success of female spotted hyaenas. *In* Serengeti II: Dynamics, Management and Conservation of an Ecosystem. (Sinclair ARE, Arcese P, eds), pp 364–384. Chicago: University of Chicago Press.

French JA, Schaffner CM (1995). Social and developmental influences on urinary testosterone levels in male black tufted-ear marmosets (*Callithrix kuhli*). Am J Primatol 36:123.

Glickman SE, Frank LGF, Davidson JM, Smith ER, Siiteri PK (1987). Androstenedione may organize or activate sex-reversed traits in female spotted hyenas. Proc Natl Acad Sci U S A 84:3444–3447.

Glickman SE, Frank LG, Pavgi S, Licht P (1992). Hormonal correlaes of "masculinization" in female spotted hyenas (*Crocuta crocuta*). I. Infancy to sexual maturity. J Reprod Fertil 95:451–462.

Goodall J (1986). The Chimpanzees of Gombe. Cambridge, MA: Harvard University Press.

Gouzoules H, Gouzoules S, Fedigan L (1982). Behavioural dominance and reproductive success in female Japanese monkeys (*Macaca fuscata*) Anim Behav 30:1138–1150.

Henschel JR, Skinner JD (1987). Social relationships and dispersal patterns in a clan of spotted hyaenas, *Crocuta crocuta*, in the Kruger National Park. S Afr Tydskr Dierk 22:18–23.

Henschel JR, Skinner JD (1991). Territrial behavior by a clan of spotted hyaenas, *Crocuta crocuta*. Ethology 88:223–235.

Holekamp KE, Smale L (1990). Provisioning and Food Sharing by lactating spotted hyenas, *Crocuta crocuta*. Ethology 86:191–202.

Holekamp KE, Smale L (1993). Ontogeny of dominance in free-living spotted hyaenas: juvenile rank relations with other immature individuals. Anim Behav 46:451–466.

Holekamp KE, Smale L (1998). Dispersal status influences hormones and behavior in the male spotted hyena. Horm Behav 33:205–216.

Holekamp KE, Ogutu JO, Dublin HT, Frank LG, Smale L (1993). Fission of a spotted hyena clan: consequences of prolonged female absenteeism and causes of female emigration Ethology 93:85–299.

Holekamp KE, Smale L, Szykman M (1996). Rank and reproduction in the female spotted hyena. J Reprod Fertil 108:229–237.

Holekamp KE, Smale L, Cooper S (1997). Hunting rates and hunting success in the spotted hyenas. J Zool Lond 242:1–15.

Holekamp KE, Szykman M, Boydston EE, Smale L (1999). Association of seasonal reproductive patterns with changing food availability in an equatorial carnivore. J Reprod Fertil, in press.

Jarman PJ (1974). The social organization of antelope in relation to their ecology. Behaviour 48:215–267.

Kappeler, PM (1993). Female dominance in primates and other mammals. Perspect Ethol 10:143–158.

Keller L, Vargo EL (1993). Reproductive structure and reproductive roles in colonies of eusocial insects. *In* Queen Number and Sociality in Insects (Keller L, ed), pp 16–44. Oxford: Oxford University Press.

Kingdon J (1982). East African Mammals. Chicago: University of Chicago Press.

Kruuk H (1972). The Spotted Hyaena: A Study of Predation and Social Behavior. Chicago: University of Chicago Press.

LeBoeuf BJ (1974). Male-male competition and reproductive success in elephant seals. Am Zool 14:163–176.

LeBoeuf BJ, Peterson RS (1969). Social status and mating activity in elephant seals. Science 163:91–93.

LeBoeuf BJ, Reiter J (1988). Lifetime reproductive success in northern elephant seals. *In* Reproductive Success (Clutton-Brock TH ed), pp 403–418. Chicago: University of Chicago Press.

Licht P, Frank LG, Pavgi S, Yalcinkaya TM, Siiteri PK, Glickman SE. (1992). Hormonal correlates of "masculinization" in female spotted hyaenas (*Crocuta crocuta*). 2. Maternal and fetal steroids. J Reprod Fertil 95:463–474.

Lindeque M, Skinner JD (1982). A seasonal breeding in the spotted hyaena *Crocuta crocuta* in southern Africa. Afr J Ecol 20:271–278.

Matthews LH (1939). Reproduction in the spotted hyaena, *Crocuta crocuta* (Erxleben). Philos Trans R Soc Lond B Biol Sci 230:1–78.

Mills MGL (1990). Kalahari Hyaenas: The Behavioural Ecology of Two Species. London: Unwin Hyman.

Murie JO, Michener GR, eds. (1984). The Biology of Ground-Dwelling Squirrels. Omaha: University of Nebraska Press.

Packer C, Collins DA, Sindimwo A, Goodall L (1995). Reproductive constraints on aggressive competition in female baboons. Nature 373:60–63.

Pusey A, Williams J, Goodall J (1997). The influence of dominance rank on the reproductive success of female chimpanzees. Science 277:828–831.

Reeve HK, Ratnieks FLW (1993). Queen-queen conflicts in polygynous societies: mutual tolerance and reproductive skew. *In* Queen Number and Sociality in Insects. (Keller L, ed), pp 45–85. Oxford: Oxford University Press.

Reiter J, Pankin KJ, LeBoeuf BJ (1981). Female competition and reproductive success in northern elephant seals. Anim Behav 29:670–687.

Roberts SC (1996). The evolution of hornedness in ruminants. Behaviour 133:399–442.

Schneider KM (1926). Über Hyänenzucht. Pelztierzucht 2:1–14.

Sherman PW, Lacey EA, Reeve HK, Keller L (1995). The eusociality continuum. Behav Ecol 6:102–108.

Short RV, Balaban E (1994). The Differences Between the Sexes. Cambridge: Cambridge University Press.

Silk JB, Clark-Wheatkey CB, Rodman PS, Samuels A (1981). Differential reproductive success and facultative adjustment of sex ratios among captive female Bonnet Macaques (*Macaca radiata*). Anim Behav 29:1106–1120.

Smale L, Frank LG, Holekamp KE (1993). Ontogeny of dominance in free-living spotted hyaenas: Juvenile rank relations with adults. Anim Behav 46:467–477.

Smale L, Holekamp KE, Weldele M, Frank LG, Glickman SE (1995). Competition and cooperation between littermates in the spotted hyaena, *Crocuta crocuta*. Anim Behav 50:671–682.

Smale L, Nunes S, Holekamp KE (1997). Sexually dimorphic dispersal in mammals: patterns, causes, and consequences. *In* Advances in the Study of Behavior. (Slater PJB, Rosenblatt JS, Milinski M, Snowdon CT, eds), pp 181–250. San Diego: Academic Press.

Smithers RHN (1966). The Mammals of Rhodesia, Zambia and Malawi. London: Collins.

Smuts B, Nicolson N (1989). Reproduction in wild female olive baboons. Am J Primatol 19:229–246.

Solomon NG, French JA, eds. (1997). Cooperative Breeding in Mammals. Cambridge: Cambridge University Press.

Thouless CR (1990). Feeding competition between grazing red deer hinds. Anim Behav 40:105–111.

Tilson RT, Hamilton WJ (1984). Social dominance and feeding patterns of spotted hyaenas Anim Behav 32:715–724.

Trivers RL (1972). Parental investment and sexual selection. *In* Sexual Selection and the Descent of Man. (Campbell B, ed), (pp 136–179). Chicago: Aldine.

Wrangham RW (1979). Sex differences in chimpanzee dispersion. *In* The Great Apes. (Hamburg DA, McCown ER, eds), pp 481–490. Menlo Park, CA: Benjamin Cummings.

Wrangham RW, Smuts B (1980). Sex differences in the behavioral ecology of chimpanzees in the Gombe National Park, Tanzania. J Reprod Fertil Suppl 28:13–31.

Wrangham RW, Rubenstein DI (1986). Social evolution in birds and mammals. *In* Ecological Aspects of Social Evolution (Rubenstein DI, Wrangham, WR, eds), pp 452–470. Princeton, NJ: Princeton University Press.

Wright S (1931). Evolution in Mendelian Populations. *Genetics* 6:111–178.

IIA SOCIAL CONTEXT AND REPRODUCTIVE BEHAVIOR

The first part of this book addressed how the physical environment, whether it be energetics, seasonality, or weather, affects reproductive behavior. Chapters 4 through 9 also integrated environmental factors with cues from conspecifics, which might be considered as social factors. This section is focused more uniformly on how social context affects reproductive behavior. For example, the following chapters do not involve day length, energy availability, and ambient temperature. However, it is a false dichotomy to distinguish environmental from social influences, for together they constitute the context in which reproductive behavior occurs. Their separation should be seen more as a convenient heuristic tool to emphasize the special place that the social environment holds in regard to behavior.

Historically, much of what we think we understand about basic mechanisms of behavior has come from studying animals and humans divorced from their species-typical social environment. In many cases, a single individual in a tightly controlled environment has been used to develop basic behavioral principles. Yet even these minimal environments constitute a social environment, albeit one where the subject must deal with separation from social interactions and social support while performing whatever behavior is under study. Typically, social influences refer to how the presence of one or more conspecifics affects behavior. In its extreme form the social context controls the behavior, as in notions of social construction of human sexual behavior or gender. More typically, in nonhumans the interest is in how interactions between members of a social group modify or modulate the behavior of individuals. In this sense, social influences are similar to social constructionism in that the pattern of behavior displayed results from the social context as well as the state of the individual.

If studies of animals in natural or seminatural contexts had been the principle method of studying behavior, we might not be focusing on social context at all. Behavior would automatically be seen as reflecting the social context and not as something generated by the individual whose expression is modified by social context. Because of the initial success of controlled laboratory studies with their reliance upon minimalist environments, it has been necessary to rediscover the social context of behavior. In some cases, this has meant re-creating a physical and social environment in which behavioral patterns can be seen in their natural relationship to the historical contexts that probably shaped their evolution. In other cases, this has meant systematically manipulating the complexity of the social environment, for example, the relative number of males and females or mothers and offspring, to determine the relative contribution of different social components to the expression of behavior. In other cases, the social variable has been the degree of social experience and its influence on behavior.

The first two chapters specifically investigate the impact of social context on sexual behavior in nonhuman primates. Chapter 10, by Kim Wallen, addresses how the social context modulates the extent to which gonadal hormones affect rhesus monkey sexual behavior. This chapter illustrates that the variability in responsiveness to hormones can only be understood when the social context of the behavior is taken into consideration. Furthermore, this chapter illustrates that sexual behavior in a complex environment involves risk, and that overcoming reticence to engage in sexual behavior under socially risky conditions may have contributed to the evolution of a system of sexual motivation driven by gonadal hormones that couples sexual behavior to fertility.

Chapter 11, by Jeffrey French and Colleen Schaffner, looks at primates that differ markedly from rhesus monkeys in that they are not promiscuous and are considered to be monogamous. As these authors demonstrate in their inventive studies, monogamy may be in the eye of the beholder. Monkeys that show every evidence of being bonded to a mate, when given a choice between that mate and a familiar nonmate, will enthusiastically interact sexually with the nonmate if the mate is out of sight. This striking effect of social context may reflect basic motivational mechanisms or may reflect the existence of two behavioral strategies, one to maintain a social relationship and the other to find a new or replacement mate. The lesson of such studies is that manipulating social context reveals an underlying complexity that would be unrecognized in a single context.

Chapter 12, by Martha McClintock, addresses possible mechanisms that might transmit social influences on behavior. Our theories of how social influences are transmitted are weak and typically stop at the claim that social context affects a behavioral endpoint. The exact mechanism by which that influence occurs is left to future studies to determine, studies which are rarely carried to conclusion. However, it is the unstated assumption of many of these studies that the mechanism has something to do with perceptions of and cognitions about the social context that influence behavior. McClintock's chapter addresses whether pheromones emanating from individuals in the social context modulate the physiology and behavior of others. She presents a novel model, and evidence to support it, which suggests a complex pheromonal synchronization of ovarian cycles, and therefore the timing of fertility between women. Synchrony in ovarian cycles has been an enduring topic since McClintock's early evidence of its occurrence among college women. The model presented here shows that synchrony is a special case of a more general pattern of interfemale regulation of ovarian cycles. Her extensive review of the comparative literature shows pheromonal modulation of sexual behavior in a variety of species, with clear evidence of modulation in humans. However, the nature of this modulation in human beings is in some ways unlike the types of modulation seen in other species.

10 Risky Business: Social Context and Hormonal Modulation of Primate Sexual Desire

Kim Wallen

Sexual passion can only exist outside normal life
—Benoîte Groult (1992)

Sexual intercourse is dangerous, posing social and physical challenges to the health and well-being of the partners. It is physically risky, the penetration of one body by another providing opportunities for physical injury and infection. It is energetically expensive, directing resources necessary for survival to producing other individuals. It is psychologically distracting, diverting attention from avoiding predation or finding food and shelter to seeking, courting, and consorting with a mate. Although many other behaviors, such as eating, involve similar dangers, they are essential to survival, justifying the tradeoff between risk and reward. In contrast, sexual behavior provides no immediate survival benefit, only risk and, possibly, pleasure. However, because reproductive success in internally fertilizing species requires sexual intercourse, its dangers cannot be avoided by simply not engaging in the behavior. Thus the physical and behavioral mechanisms assuring the occurrence of sexual intercourse must have been under heavy selective pressure, for any mechanism that increases the probability of reproducing directly affects reproductive success. Although many factors influence reproductive output, mechanisms limiting sexual behavior to the brief period of female fertility maximize reproductive benefit and minimize the risks of sexual behavior. It is not surprising, therefore, that most internally fertilizing species show a tight coupling between female fertility and sexual behavior.

However, evolution has produced more than one solution coordinating sexual behavior with fertility and these vary in how tightly coupled sexual behavior is to female fertility. In some species, such as insects, hormonal or pheromonal mechanisms release a stereotyped pattern of behavior when the female is fertile, leaving little to choice on the part of the mating pair (Izard 1983). In other species, hormones regulate the physical capacity to mate, either through hormonally regulated vaginal closure, as in the guinea pig (Stockard and Papanicolaou 1919), or through hormonally regulated female spinal reflexes required for intromission, as in rats (Diakow 1974; Pfaff et al. 1978). In these species intercourse is physically possible only when the female is fertile (Wallen 1990, 1995). Other species are less strictly hormonally regulated and this chapter focuses on the system employed by primates in whom the hormones that produce fertility do not regulate the physical capacity to mate, but increase the likelihood of sexual behavior occurring by modulating sexual motivation. This loose coupling between sexual behavior and female fertility is an adaptation accommodating sexual activity in a complex social environment where sex is socially disruptive and risky. Such loose coupling allows a much wider range of expression of sexual behavior, both in form and timing during the ovarian cycle, than in species with tight fertility-behavior coupling. This variability in the timing of sexual behavior has produced debate about the role of reproductive hormones in primate sexual behavior.

The chapter begins with a description of the dangers inherent in sexual intercourse and a brief discussion of the mechanisms that have evolved to ensure that this dangerous activity occurs. This section presents the distinction between the physical capacity to engage in sex and the psychological desire to engage in sex, and introduces the idea that physical capacity for sex has been emancipated from the control of gonadal hormones. This section is followed by a historical view of human and nonhuman primate sexuality and the influence of hormones on the expression of sexuality. For example, the first study to focus on the role of hormones affecting female sexual motivation was based on the description of rhesus monkey sexual behavior in field settings. A description of the early days of primate behavioral endocrinology is followed by studies from the modern era which emphasized research from the laboratories of Richard Michael, who emphasized male control, and W. C. Young, who emphasized female control of sexual interactions. Early studies in the field led to conclusions different from those of many of the laboratory experiments. The section closes with the return of field behavioral endocrinology, which was carried out under the controlled conditions of large outdoor populations. These new studies of rhesus monkeys in a more natural context ultimately led to a reconciliation of the differences in the data that results from studies of groups vs. studies of male-female pairs.

The next section describes the evidence that social context modulates hormonal influences on female sexual motivation, followed by similar evidence from studies of women. This is followed by a discussion of risk as a modulator of sexual behavior in humans. This section suggests that pregnancy avoidance affects the type of sexual behavior displayed by women at different points in their cycle and further develops the thesis that perceived risk affects the extent to which hormonally modulated female sexual motivation affects human sexual behavior. In the next section, the principles derived from studies of females are applied to male human and nonhuman primates and found to apply equally well to both sexes. The last section presents some final considerations and addresses the relative sex drive of males and females, why female control of sexuality seems limited in American society, and discusses the role of sexuality in social cohesion. While this chapter emphasizes risk as an important factor in the evolution of hormonal modulatory systems, it also reflects the theme of this book, that knowing the context of behavior is crucial to understanding its regulation.

Early Views of Hormones and Female Primate Sexual Behavior

Full expression of sexual behavior in female primates, including humans, requires ovarian steroid hormones; however, these hormones are not necessary for the occurrence of sexual behavior (Wallen 1995). This apparent contradiction has fueled a controversy for more than 60 years about the role of ovarian hormones in regulating female primate sexual behavior. The controversy stems primarily from a view, derived from nonprimate mammals, that hormonal effects on sexual behavior should be regulatory and not permissive. Early views of human and nonhuman primate sexual behavior often emphasized the flexibility of human sexual behavior as opposed to the fixed, and biologically determined, nature of nonhuman primate sexual behavior. For example, the anthropologist Malinowski (1927) contrasted the sexual behavior of humans with that of apes, describing the control and rigidity of nonhuman primate sexual behavior as follows:

Among apes the courtship begins with a change in the female organism, determined by physiological factors and automatically releasing the sexual response in the male. The male then proceeds to court according to the selective type of wooing which prevails in a given species.... All the factors which define animal behaviour at this stage are common to all individuals of the species. They work with such uniformity that for each animal species one set of data and only one has to be given by the zoologist ... within the species the variations, whether individual or otherwise are so small and irrelevant that the zoologist ignores them and is fully justified in doing so. (p. 194)

In contrast, human sexual behavior was described thus:

In the first place we see that in man there is no season of rut, which means that man is ready to make love at any time and woman to respond to him—a condition which, as we all know does not simplify human intercourse. There is nothing in man which acts with the same sharp determination as does the onset of ovulation in any mammalian female. (p. 195)

Malinowski, however, saw sexual behavior in both humans and animals as socially disruptive and described social and cultural controls that regulate human sexual behavior and prevent it from interfering with social order. In contrast, he saw the circumscribed and physiologically controlled period of nonhuman primate sexual behavior as serving the same function:

Considering the great danger from outside enemies and the disruptive forces within, which are associated with courtship, the elimination of the sex interest from normal times and its concentration on a definite short period is of great importance for the survival of animal species. (p. 198)

Sexual behavior is dangerous and must be limited either culturally, as in human society, or through strict biological mechanisms, as in nonhuman primates. The view that humans have been essentially freed from strong biological determination of sexual behavior while nonhuman primates and other mammals limit their sexual behavior to brief circumscribed periods is still popular today. However, contemporaries had pointed out that nonhuman primate sexual behavior was more variable and less stereotyped than Malinowski and others believed.

Heape (1900), in his discussion of mammalian estrus, by which he meant a period of intense sexual activity, recognized that in monkeys "estrus" was not necessarily linked to ovulation (Heape 1900, cited in Nadler 1994). Thirty years later Zuckerman (1930) described the basic conundrum of primate sexuality, that sexual behavior occurred at any time but was also much more likely to occur when females were near ovulation. As he put it,

The matings of lower mammals are confined to short periods circumscribed by the activity of the follicular hormone. The matings of the primate are diffused over the entire cycle, paralleling the continued action of the follicular hormone, but varying in frequency according to the varying degrees of activity of the hormone. (Zuckerman 1930, cited in Miller 1931)

Zuckerman was correct in his description of the timing of mating behavior in primates, but incorrect in attributing its occurrence throughout the cycle to the action of follicular hormone, by which he meant estrogen, across the cycle. Miller (1931), in his comparative analysis of nonhuman primate and human sexual behavior, identified the key adaptation in primates that produced this loose coupling between the female's hormonal state and sexual behavior. Miller recognized that in primates, unlike most other mammals, the physical capacity to engage in sexual intercourse was uncoupled from hormonal control. Although Miller recognized this hormonal emancipation, he still argued for humans having a unique adaptation in that

... in man alone of all mammals is the male known to be able to force his sexual will on the unconsenting or unconscious female, a peculiarity that seems to arise from human ingenuity combined with human pelvic adjustments to the upright posture ... (p. 406)

Although Miller saw the physical capacity for rape as uniquely human, which it is not, his insight that physical ability to mate had become uncoupled from fertility is essential to understanding how gonadal hormones affect primate sexual behavior. Zuckerman's description of both the highly variable nature of nonhuman primate sexual behavior, as well as its increased probability when the female was fertile, is the other piece of the puzzle, though it took almost 60 years before these two notions were linked together (Wallen 1990). In the intervening years many investigators attempted to discover the relationship between changes in female hormones and nonhuman primate behavior. This work has almost exclusively used rhesus monkeys and thus they are the focus of this chapter. However, it is believed that the basic principles described apply to a wide range of simian primates and only the paucity of detailed information about other primate species prevents critical assessment of the commonality of these principles.

Early Studies of Primate Behavioral Endocrinology

Josephine Ball and Carl Hartman undertook the first controlled attempt to discover the relationship between female ovarian function and sexual behavior in rhesus monkeys (Ball and Hartman 1935). They observed the behavior of single male-female pairs separated from all other social context, a technique they developed and that dominated nonhuman primate studies of sexual behavior for the next 50 years. They reported that sexual behavior occurred at all times in the female's cycle, and in many cases, but not all, was more frequent near midcycle when the female was likely to be ovulating (Ball and Hartman 1935). They also pointed out that the female's interest in sex varied with her cycle, stating that female

... sexual excitability typically increases just before ovulation and falls thereafter, even though the drop is not so complete as to mean a consistent refusal to mate. (p. 117)

This study was quite remarkable in that ovulation was verified by manual palpation, a level of precision that was not to be seen again until hormonal assays were developed in the 1970s. This was the first, and for almost 40 years the only study that accurately related behavioral change to ovulation and provided the first controlled evidence that, at least in some monkeys, sexual behavior varied predictably with the menstrual cycle. However, Ball and Hartman's study also clearly demonstrated that whatever endocrine events occurred during the ovarian cycle did not strictly limit sexual behavior to a single brief period; some sexual activity occurred at all times in the female's cycle. The authors' finding of an apparent cyclicity in female excitability was the first to focus on behavioral change in the female independent of male behavior. Furthermore, the finding that sexual behavior never ceased completely during the female's ovarian cycle characterized the findings of primate behavioral endocrinology, using pairs of animals for the next 50 years (Wallen 1989, 1990). In contrast, studies of intact social groups of monkeys suggested a more tightly coupled relationship between hormones and behavior.

Field behavioral endocrinology was essentially nonexistent when Carpenter studied the sexual behavior of rhesus monkeys on Cayo Santiago in the 1940s. Tracking female cycles in semi–free-ranging rhesus monkeys, Carpenter reported a striking periodicity midway between menstruations in both copulation with males and in female sexual solicitations (Carpenter 1942a,b). Unlike Ball and Hartman's studies of rhesus monkey pairs, Carpenter reported that female rhesus monkeys mated and interacted with males intensely only for a few days during their menstrual cycle. Particularly striking was his evidence that females intensely followed males during this midcycle period and initiated sexual activity through a variety of sexual solicitations (Carpenter 1942a). Twenty years later, other field researchers studying free-ranging populations of rhesus monkeys reported a similar limited period of mating (Altmann 1962; Conaway and Koford 1964; Southwick et al. 1965). However, whether it was because Carpenter studied nonlaboratory populations of monkeys, concerns about field estimates of menstrual cycles, the inability to validate ovulation, or because it took other field workers 20 years to corroborate his findings, Carpenter's study initially had little impact on subsequent research and was rarely cited by researchers when primate behavioral endocrinology experienced a resurgence in the 1960s.

Primate Behavioral Endocrinology: The Modern Era

Although a small number of studies of nonhuman primate behavioral endocrinology were done in the late 1940s and 1950s, it was the convergence of two trends in the 1960s that resulted in the resurgence of primate behavioral endocrinology. The first was the detailed understanding of nonprimate behavioral endocrinology that had come from studies of rodents, particularly the guinea pig in William C. Young's laboratory (Young 1961) and the rat in Frank Beach's laboratory (Beach 1942; 1975; 1981; Beach and Levinson 1950). The second was an increased interest in nonhuman primates, possibly stemming from the successful use of rhesus monkeys in developing a polio vaccine in the late 1950s, which ultimately resulted in the creation of a group of regional primate research centers by the National Institutes of Health in the early 1960s. This period saw increased field studies of rhesus monkeys and very active laboratory investigations of rhesus monkey sexual behavior.

Two laboratories, W. C. Young's at the Oregon Regional Primate Research Center, later under the direction of Robert W. Goy after Young's death in 1966, and Richard P. Michael's in London, dominated primate behavioral endocrinology for almost 20 years. Both laboratories employed controlled studies of male-female pairs, but developed quite different philosophical approaches to their research. The hundreds of studies published during this period varied widely, but two principles stand out. Michael's laboratory, with his colleagues Robert Bonsall, Barry Everitt, Joe

Herbert, E.B. Keverne, and Doris Zumpe, emphasized male control of mating and focused on how ovarian hormones influenced female attractiveness to males (Michael and Herbert 1963; Michael and Welegella 1968; Herbert and Trimble 1967; Trimble and Herbert, 1968; Michael and Zumpe 1970).

Specifically, these investigators emphasized hormonally induced vaginal olfactory cues they claimed released male copulatory behavior, and either minimized or were unable to detect the role females played in behaviorally regulating sexual interactions (Michael and Keverne 1968; Keverne 1976; Michael et al. 1982; Michael and Bonsall 1977a,b). The role of vaginal olfactory cues, however, was controversial and has been largely discredited. The stimulation of sexual behavior via vaginal olfactory cues reported by Michael and colleagues could not be replicated in a different laboratory (Goldfoot et al. 1976). In addition, it was found that anosmic male rhesus monkeys showed cyclic variation in copulatory behavior even though they could not detect odor cues (Goldfoot et al. 1978). Finally, Goldfoot (1981) produced similar behavioral changes to those reported by Michael's laboratory using nonbiological odors and a testing paradigm comparable to Michael's. Although Michael argues that the failure of other laboratories to detect reliable effects of vaginal olfactory cues stems from procedural differences (Michael and Zumpe 1993), his latest review of rhesus monkey sexual behavior acknowledges that ovarian hormones affect female sexual motivation and no longer emphasizes vaginal pheromones as primary regulators of monkey sexual behavior (Michael and Zumpe 1993).

Michael's laboratory presented the first controlled data since Ball and Hartmann (1935) on the occurrence of ejaculation in relation to the female's ovarian cycle and demonstrated the pattern Zuckerman described 40 years earlier of some mating throughout the cycle, with a midcycle elevation (Michael and Herbert 1963; Michael and Welegalla 1968; Michael and Zumpe 1970). However, the authors found no evidence that these changes in mating across the cycle were related to changes in female behavior, and argued instead that they reflected changes in female attractiveness to the male which increased and decreased his sexual initiation. Only when females were required to perform an operant to gain access to a male and it was found that females performed this more rapidly at midcycle than at other times in the cycle, did Michael's group acknowledge that female sexual motivation might also vary with the female's hormonal condition (Keverne 1976; Bonsall et al. 1978).

In contrast, Young and his colleagues, Robert W. Goy, Charles P. Phoenix, and John Resko, focused on how the female's hormonal state affected her sexual initiation (Goy and Resko 1972). This emphasis on the female may have resulted from the extremely pronounced effect of ovarian hormones on female guinea pig behavior which Young's laboratory had spent 25 years investigating. At that time, there was an increased interest in female sexuality in America, possibly reflecting the effects of Masters and Johnson's landmark studies of human sexual response (Masters and Johnson 1965, 1966) which gave male and female sexual response equal attention. Alternatively, or in addition, it may have been related to the rise of the American women's movement during the 1970s, which resulted in both a greater number of women entering all fields of biology and psychology and a greater general awareness of women's sexuality.

Prior to this time, the field of behavioral endocrinology, which had seen many contributions from women, was dominated by three patriarchs, Frank Beach, Daniel Lehrman, and W. C. Young, and their mostly male academic progeny. Finally, after decades of emphasizing the role of male rodents in sexual interactions, Frank Beach published a landmark article in 1976 coining the term "proceptivity" to describe the active solicitation of sexual activity (Beach 1976). Though this term could ostensibly be applied equally to males and females, his description focused exclusively on females and it has subsequently been applied principally to females. It will be left to others to determine whether this focus on female sexuality was related to the rise of the women's movement or to the dramatic increase in female graduate students in behavioral endocrinology, or to both. The fact remains that interest in the sexuality of female primates increased in the 1970s along with interest in how hormones affected female sexual behavior. Studies from Young's, and later from Goy's, laboratory provided evidence of ovarian influences on female sexual initiation when male control of the sexual interaction was restricted (Czaja and Beilert 1975; Pomerantz and Goy 1983). However, when the occurrence of ejaculation was studied in pairs across the female's ovarian cycle, the pattern reported by Michael's laboratory, of some mating each day with a midcycle elevation, was also found by Goy's laboratory (Goy 1979). This was not always the case. Johnson and Phoenix (1978), also from Young's laboratory, failed to find any evidence of cyclic variation in sexual behavior of rhesus monkeys pairs.

After lying fallow for almost 40 years behavioral endocrinology moved back into complex monkey social groups in the 1980s. The creation of the Primate Center Program dramatically increased opportunities to study nonhuman primate physiology and behavior. At the Yerkes, Regional Primate Research Center, Irwin Bernstein, Thomas P. Gordon, and Robert Rose had successfully investigated the behavioral endocrinology of male rhesus monkeys using techniques they developed for sampling hormones in group-living monkeys (Rose et al. 1975; Gordon et al. 1976, 1978). Gordon applied these techniques to group-living females and in 1981 published a landmark study showing that mating behavior in outdoor-housed mixed-sex groups of rhesus monkeys was limited to a few months of the year and to a few days within each female's ovarian cycle (Gordon 1981). Unlike Carpenter's study, Gordon's animals were visible at all times and blood samples, which could now be assayed for estradiol and progesterone, were collected in addition to observing menstruation. Thus, for the first time unequivocal evidence was obtained that rhesus monkey sexual behavior was strongly influenced by the female hormonal state, with long periods during the cycle with no sexual activity.

Subsequent work suggested that the female's behavior varied with her ovarian cycle (Cochran 1979), a view that was verified when daily behavioral and hormonal samples were taken on group-living monkeys (Wallen et al. 1984). These studies found, in contrast to those of isolated pairs of animals, that female sexual initiation increased sharply with increases in estradiol and that sexual activity was limited to a small number of days within the female's 28-day ovarian cycle. This pattern was found whether the multiple female group had a single male (Wallen et al. 1984) or multiple males (Wilson et al. 1982). Clearly, this pattern differed markedly from that seen in the behavior of pairs of monkeys, but why?

Social Modulation of Hormonal Influences on Female Sexual Behavior

In retrospect it seems obvious that rhesus monkey sexual behavior would be less tightly coupled to female hormonal state in pair than in group tests, but this perspective developed slowly. Studies of rhesus monkey pair tests so dominated nonhuman primate behavioral endocrinology that results from other contexts, which often contradicted pair-test data, had little impact. However, surveying the last 70 years of study it is clear that the elements needed to explain these divergent findings have been present from the very beginning and were encapsulated in the early recognition that sexual behavior in primates occurs throughout the female's cycle and is more likely to occur when the female is fertile. In addition, the exact relationship between the female's cycle and the occurrence of sexual behavior is strongly influenced by social context.

This principle is illustrated in figure 10.1 which shows the occurrence of ejaculation, or the percentage of females receiving ejaculations, in relation to the female's ovarian cycle under three different social contexts. Single pairs of monkeys tested for 12 minutes in a small area show the pattern, first described by Zuckerman, of some mating every day with a midcycle elevation (Goy 1979). In comparison, when a single male is tested with a group of females in a large area for 30 minutes, all mating occurs in the 8-day period around the midcycle estradiol peak, even though more time is available for mating (Wallen et al. 1984). When multiple males and multiple

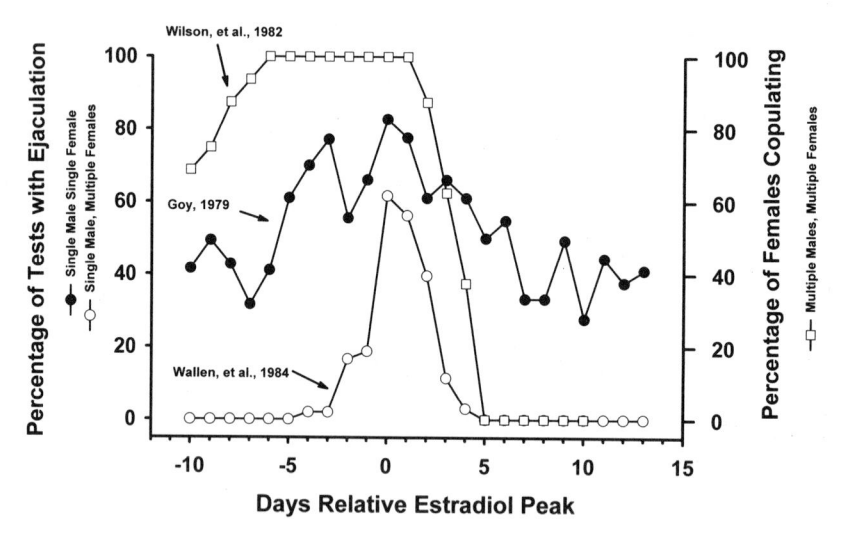

Figure 10.1
Relationship between female ovarian cycle day, aligned relative to the midcycle estradiol peak, under three testing conditions that differ in social context, testing area, and observation time. Tests of single-male, single-female pairs display the most continuous sexual behavior across the female's cycle even though the least amount of time was available for mating to occur. (Data from Goy 1979; Wilson, et al. 1982; and Wallen et al. 1984.)

females are observed for 3 hours in a large area, more mating is seen in the follicular phase, while mating ceases completely during the luteal phase, just as it does when only a single male is present and despite the fact that more than six times as much time was available for mating in this testing situation. Though these studies differ in many ways the conclusion that social context affects the degree of coupling between the female's cycle and sexual behavior is inescapable. Why does social context have this effect? The first part of the answer is simply "because it can."

Social modulation this striking would not be found if one studied female guinea pigs instead of female monkeys. Unlike female monkeys, female guinea pigs are physically capable of mating for only 2 to 3 days of their 14- to 16-day cycle because a hormonally controlled membrane closes their vagina except around the time when they are fertile (Stockard and Papanicoulaou 1919). Similarly, female rats without any hormonal stimulation would be unable to display the lordosis posture necessary for male intromission (Diakow 1974; Pfaff et al. 1978). Unlike these mammalian females, female primates, with the exception of some prosimian primates (Hrdy and Whitten 1987), are always capable of engaging in sex, with or without hormonal stimulation. However, this primate capacity to engage in sex at any time does not explain why sexual behavior is more tightly coupled to female hormonal state under some social conditions and less tightly coupled under others.

Explaining this aspect requires the notion that the primary psychological function of gonadal hormones is to influence sexual motivational systems in primates (Wallen 1990, 1995). When sexual activity is physically possible at any time, sexual motivation will influence sexual behavior only when circumstances require high sexual motivation for the behavior to occur. Evidence that gonadal hormones modulate female sexual motivation comes from studies that varied the effort needed to seek a sexual partner or to engage in sex. As previously mentioned, requiring a female to perform an operant to gain access to a male partner revealed a previously hidden midcycle increase in performance with the female accessing the male more slowly at other times in the cycle (Keverne 1976). Similarly, the sexual activity of a male-female pair was more strongly affected by the female's hormonal state, being high at midcycle and low during the luteal phase, when the pair was tested in an area 100 times larger than that typically used for pair tests (Wallen 1982). Thus, simply increasing the physical effort necessary for sex or providing more behavioral alternatives increased the effect that the female's hormonal state had on the occurrence of sexual behavior.

Hormonally modulated female sexual motivation only becomes important under conditions where sex requires more effort or behavioral choices must be made. However, even when increased effort is required, larger physical areas must be traversed, or an operant performed, sexual activity does not cease completely in all pairs during the nonovulatory portion of the cycle. In contrast, females in social groups of monkeys uniformly do not mate early in the follicular phase or during the luteal phase of the cycle (Carpenter 1942b; Gordon 1981; Wilson et al. 1982; Wallen et al. 1984). A study comparing sexual behavior in a pair test with sexual behavior with the same male and a group of familiar females found little difference during the periovulatory portion of the cycle, but sexual behavior during the luteal phase occurred only in the pair test and was completely absent when multiple females were present (Wallen and Winston 1984). This difference could not reflect differences in physical effort between the two types of tests as both took place in the same $625\,m^2$ arena. However, the group tests introduced a social complexity, interactions between familiar females, not present in the pair tests. It was this added social factor that accounted for the greater influence of female hormonal state on sexual behavior in the group tests. Some description of the social context rhesus monkeys live in is necessary to develop this point.

Rhesus monkeys are a female-bonded society (Wrangham 1980) in which relations between matriarchs and their families form the core of the social structure (Sade

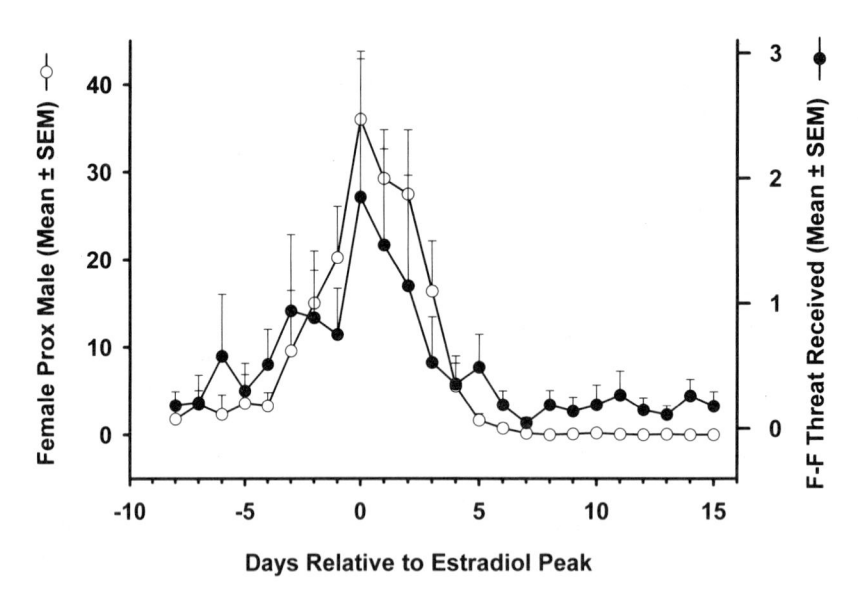

Figure 10.2
The occurrence of female initiation of proximity (Prox) and threatening between females in relation to the female's ovarian cycle day, aligned relative to the midcycle estradiol peak in a social group of a single male and multiple females. As females interact more intensely with the group male, they are threatened more by other group females (F-F). (Adapted from Wallen and Tannenbaum 1997.)

1965; Missakian 1972; Gouzoules and Gouzoules 1987). Males are transitory members of this social structure, leaving their natal group and emigrating to a new rhesus monkey group during the breeding season with whom they typically live for 5 years or less (Koford 1966; Lindburg 1969; Drickamer and Vessay 1973). In this social environment males serve an important, but transitory social role and the crucial social interactions are between females and these are not all affiliative. Antagonistic relations between females vary with their ovarian cycle in group-living rhesus monkeys, with peak occurrence during the periovulatory portion of the menstrual cycle (Mallow 1981; Walker et al. 1983; Wallen and Tannenbaum 1997). As shown in figure 10.2, data from a single-male, multiple-female group reveal increased female threatening of other females at the same time as increased female initiation of proximity with the group male occurred. Not only does sexual activity in a social group require greater physical effort than in a pair test, but it also entails social risk, at least for lower-ranking females. This social risk makes the sexual motivational state of the female a critical regulator of sexual behavior in a social group because under such conditions females must be intensely interested in interacting with males to risk neg-

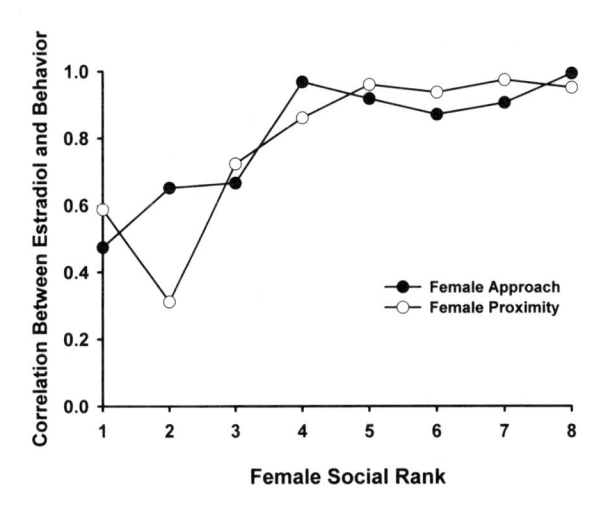

Hormones and Sexual Motivation

Figure 10.3
The relationship between social rank and the magnitude of the correlation between serum estradiol on the 8 days prior to, and including, the estradiol peak in relation to the daily occurrence of female initiation of approach and proximity (sitting within 20 cm) to the group male in a single-male, multiple-female group. Correlations for the two highest-ranking females were not statistically significant, whereas those of the five lowest ranking females were and all approached 1.0. These data support the notion that low-ranking females are more dependent upon hormonal influences than are high-ranking females.

ative social interactions with other females that their attention to a male will produce. Furthermore, when many adult males are in the group, female interest in one male may elicit negative social interactions from other group males (Smuts and Smuts 1993) creating another social impediment to expressing female sexual interest.

Further support for the idea that hormonally modulated sexual motivation is an important modulator of female sexual behavior in a group context is found in analyzing the relationship between a female's social rank and the extent to which her cyclic variation in estradiol predicts her sexual initiation behavior. For high-ranking females, sex entails little social risk, whereas low-ranking females are potentially exposed to a higher social risk. Thus if gonadal hormones modulate female sexual motivation, then within a social group the hormonal state of a high-ranking female should be a poor predictor of her behavior, since she is not strongly dependent upon her sexual motivation to engage in sex. In contrast, female hormonal state should be an excellent predictor of a low-ranking female's sexual behavior since she is much more strongly dependent upon her sexual motivation.

Figure 10.3 illustrates for eight females who mated with males in a social group the relationship between female social rank and the magnitude of the correlation between daily changes in estradiol and daily changes in behavior, approaching, and proximity initiation for the 8 days prior to the estradiol peak. For the first- and second-ranked females, the correlations for both behaviors were not significant and only significant at $P = .05$ for the third-ranking female. In contrast, for the rest of the group females, all correlations were statistically significant and greater than .8 with many approaching 1.0. Since estradiol secretion did not vary with female social rank, these data most likely reflect an increased dependence of low-ranking females on estradiol-induced increased sexual motivation. Further evidence that high-ranking females are less dependent upon hormones to mate is seen in the finding that they mated on more days of their ovarian cycle and started mating earlier in the follicular phase, when they would have been exposed to less estradiol, than did low-ranking females (Wallen 1990). Taken together, these findings support the idea that hormonal modulation of female sexual motivation is necessary for sexual behavior to occur in a socially complex setting where both physical effort and social risks affect how easily mating occurs.

Detecting the female's degree of sexual motivation is more difficult in pair tests because the effort required in selecting a partner and the social risks of sexual activity have been markedly reduced. In addition, the economic realities of indoor nonhuman primate research resulted in using small test cages (2 m × 3 m floor areas) that accidentally duplicated the proximity cues that female rhesus monkeys use to convey their sexual interest in males (Wallen, et al. 1984; Wallen 1989). Thus in pair tests not only are physical effort and social risk low but the small physical space causes females to emit behavioral cues which, under free-ranging conditions, occur only when females are highly motivated to interact sexually with males.

It could be argued that the cyclic changes in female behavior do not reflect motivational state, but are a response to unidentified male-generated cues. The strongest evidence against this view comes from studies where female's hormonal state and sexual behavior varied with little or no response by the male. One approach has been to treat ovariectomized group-living female rhesus monkeys during the nonbreeding season when males are sexually nonresponsive. The two studies using this method found evidence that female sexual motivation increases with estradiol treatment, though the specific manner in which this was expressed varied between the two studies. Pope et al. (1987) reported that estradiol-treated ovariectomized females showed increased female-female mounting during the nonbreeding season and increased heterosexual copulation during the breeding season. Thus estradiol induced increased sexual activity that was expressed with groups females when the males were sexually quiescent. Zehr et al. (1998), studying these same females 12 years later when they were more sexually experienced, found estradiol increased female approach,

Hormones and Sexual Motivation

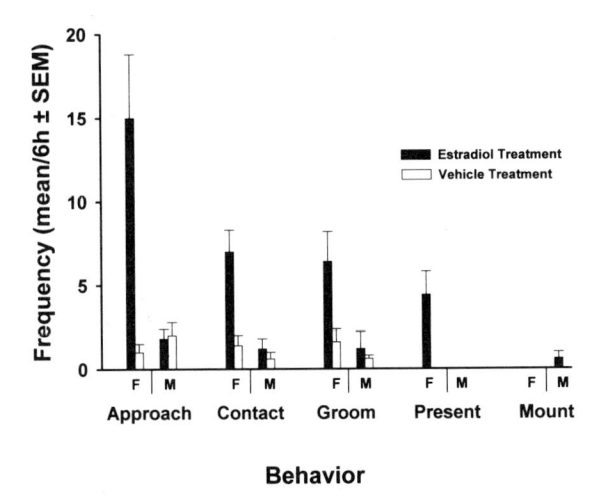

Figure 10.4
The occurrence of selected social and sexual behaviors by group-living ovariectomized females or group males during female estradiol or vehicle treatment. F on the x-axis indicates that behavior was initiated by one of the females; M indicates a male initiated the behavior. Only female-initiated behavior during estradiol treatment differed significantly from the vehicle condition. (Adapted from Zehr et al. 1998.)

contact, grooming, and presentations to group males, even though the males displayed no significant change in sexual interactions with the females. Figure 10.4 summarizes these findings and supports the notion that estradiol directly affects female sexual motivation and that the female's behavior is not dependent upon the male's sexual response.

Is Sexual Desire in Women Related to Ovarian Hormones?

Whether the ovarian cycle of women influences their sexual interest has been debated since the 1930s. Tinklepaugh (1933), in a review of data from more than two thousand women, concluded that two periods of increased sexual desire occurred during the menstrual cycle: one just prior to menstruation and a second immediately after menses, corresponding to the fertile period. Tinklepaugh raised the possibility that the premenstrual period of increased female sexual desire might reflect knowledge that this time has the lowest risk of pregnancy during the cycle, suggesting that female sexual interest is affected by perception of risk of pregnancy. However, Tinklepaugh's review had little impact on views of female sexuality. In subsequent research, if female sexual desire was considered at all, it was typically thought to be unaffected by ovarian function, a view that was bolstered by the claim that ovariectomy had no detectable impact on female sexuality (Filler and Dresner 1944; Waxenberg et al. 1959).

Paralleling the interest in female nonhuman primate sexuality, increased investigation of women's sexuality started in the late 1960s. Reports appeared showing cyclic variation in human sexual intercourse (Udry and Morris 1968) and female sexual initiation (Adams et al. 1978). However, the exact relationship of sexual behavior to the menstrual cycle was unclear (Udry and Morris 1977) and the occurrence of intercourse was affected by cultural conventions such as the weekend (Palmer et al. 1982) and by psychological factors such as fear of pregnancy (Tsui et al. 1991). In addition, human studies typically investigated the occurrence of intercourse which confounded male sexual initiation with the influence of the female (Wallen and Lovejoy 1993). However, studies accumulated showing a midcycle increase in female sexual activity in lesbian couples where male influences were eliminated (Matteo and Rissman 1984) and in newlyweds, whose sexual activity would be expected to be heightened (Hedricks et al. 1987).

A more general view developed that the ovarian cycle influenced women's sexual activity, not just male's initiation (Wallen and Lovejoy 1993; Hedricks 1994). For example, when female sexual desire was explicitly investigated by asking women to report the first day they felt increased sexual desire, it peaked at midcycle (Stanislaw and Rice 1988). The pattern of increased sexual desire reported by the more than four thousand women in Stanislaw and Rice's study strikingly paralleled the changes in sexual initiation displayed by female rhesus monkeys across the menstrual cycle, as shown in figure 10.5. Similarly, unlike Filler and Dresner's original study, which only reported the effect of ovariectomy on the occurrence of intercourse, subsequent studies that explicitly investigated female sexual desire found it almost completely eliminated by ovariectomy (Dennerstein and Burrows 1977; Sherwin and Gelfand 1987; Sherwin et al. 1985), just as pharmacological suppression of ovarian function eliminated female sexual initiation in female rhesus monkeys (Wallen et al. 1986). While direct comparisons between humans and nonhuman primates may seem presumptuous, it appears that sexual motivation in women is as influenced by ovarian function as is that of female rhesus monkeys.

Does Risk Affect the Expression of Sexual Desire in Women as It Does in Female Rhesus Monkeys?

While no study has specifically investigated this question, two intriguing pieces of evidence suggest that increased risk affects the way women's sexual behavior is

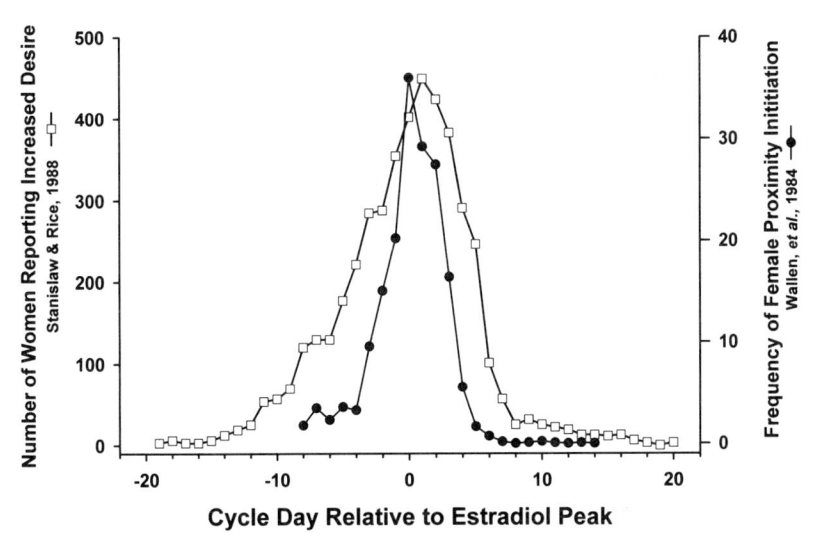

Figure 10.5
The number of women reporting an increase in sexual desire in relation to ovarian cycle day (Stanislaw and Rice 1988). The similarity of this curve to changes in female-initiated proximity to a male by group-living female rhesus monkeys (Wallen et al. 1984) is striking. Cycles are aligned by day of peak estradiol (rhesus monkeys) or putative peak estradiol day derived from changes in basal body temperature (humans).

expressed in relation to their ovarian cycle. The first evidence comes from a study of female sexual initiation and how it is affected by hormonal birth control pills (Adams et al. 1978). In addition to investigating the effect of hormonal contraceptives on female sexual initiation, which the pill suppressed, this study compared the relationship between type of contraceptive used and cyclic variation in female sexual initiation and autosexual behavior in women with regular sexual partners. The women in the study using nonhormonal contraceptives are considered here. About half of these used highly reliable contraceptives that did not intrude on sexual interactions, such as surgical sterilization or an intrauterine device (nonintrusive contraceptive users). The rest used unreliable contraceptives that intruded on sexual interactions, such as condoms or diaphragms (intrusive contraceptive users).

The type of nonhormonal contraceptive used affected the pattern of female sexual behavior shown across the cycle. Nonintrusive contraceptive users showed an increase in sexual initiation around reverse cycle day 14, near presumed ovulation (figure 10.6A). In contrast, the midcycle peak in female sexual initiation was muted in women using the unreliable intrusive contraceptives. This difference might reflect self-selection such that women who are sexually less active, or less interested in sex,

Adapted from Adams, Gold, and Burt, 1978

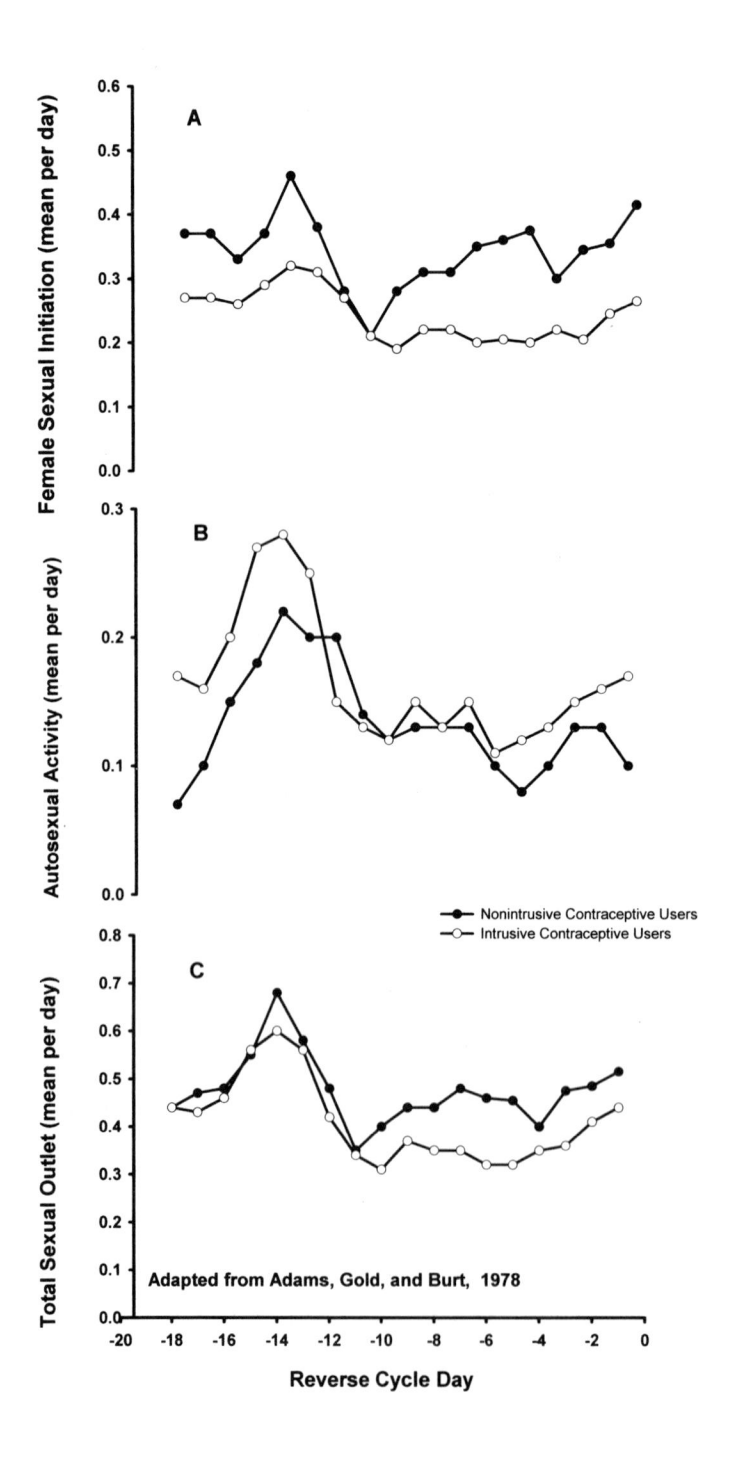

use less permanent forms of contraception. However, when the autosexual activity of these two groups of women are compared (figure 10.6B), both groups of women show a midcycle elevation, with the women using the unreliable intrusive contraceptives showing a higher peak than the nonintrusive contraceptive users. Combining both sexual initiation and autosexual activity to measure total daily sexual outlet (figure 10.6C) shows no apparent differences between the two groups of women, with both showing a midcycle elevation in sexual activity followed by a luteal decrease. Thus these two groups of contraceptive users differ not in their overall level of sexual activity, but in how it is distributed between heterosexual sexual initiation and masturbation.

While different interpretations of these findings are possible, it seems likely that contraceptive-using women are aware of when they can most easily become pregnant in their cycle (Small 1996) and of the effectiveness of the contraceptive method they use. Thus one interpretation is that women using unreliable contraceptives perceive sexual initiation at midcycle as risky and inhibit their heterosexual activity and substitute a higher level of autosexual activity than women using highly reliable contraceptives. In this case perception of the risk of pregnancy causes the midcycle increase in sexual desire to be expressed as masturbation rather than initiation of sexual intercourse. A second example suggests that under some conditions socially risky sexual activity is more tightly coupled to female hormonal state than is less risky sex.

Bellis and Baker (1990) used a cross-sectional method to obtain information about the sexual activities of 2708 English women who had a primary male sexual partner. Subjects were asked to report whether their last copulation was with their primary sexual partner or with an extrapair partner and to provide enough information so that the menstrual cycle length and the cycle day of the last copulation could be calculated. In addition, the 162 women who claimed that their most recent copulation had been with an extrapair partner also indicated when they last had intercourse with their primary partner. Bellis and Baker were interested in obtaining evidence that women manipulated sperm competition between males, but their data are compatible with a quite different interpretation.

Figure 10.6
The occurrence of (A) female sexual initiation, (B) autosexual activity (B), and (C) total sexual outlet (the combination of sexual initiation and autosexual activity) in women using nonhormonal contraception. Nonintrusive contraceptives were permanent or semipermanent forms of contraception, such as surgical sterilization or intrauterine devices that did not intrude on the sexual interaction. Intrusive contraceptives were those that need to be utilized close to the time of intercourse, such as condoms, vaginal foam, or diaphragms. The distribution of female sexual activity varied according to the type of contraceptive used and ovarian cycle. (Data from Adams et al. 1978.)

Bellis and Baker (1990) divided the women's cycles into three phases roughly corresponding to the follicular, periovulatory, and luteal portions of the cycle. They found that sexual intercourse with the women's primary partner varied across their cycle, being highest luteally and lowest during the follicular phase. Sexual intercourse outside of the primary relationship also varied with the cycle but had a periovulatory peak and was lower both follicularly and luteally. Extrapair sexual intercourse is a socially risky sexual activity and it more closely followed the female's cycle than did the less socially risky sexual intercourse with the woman's partner. Furthermore, the higher luteal phase sexual intercourse with the partner is consistent with pregnancy avoidance affecting the occurrence of intercourse within an established pair. The tighter coupling between sexual intercourse and the woman's cycle was even more pronounced when the distribution of sexual intercourse for the 50 women in the study who had sexual intercourse with an extrapair male and their primary partner within 5 days of each other was considered.

As shown in figure 10.7 these so-called double-matings (Bellis and Baker 1990) were not randomly distributed across the female's cycle, but peaked on the presumed day of maximal fertility (Barrett and Marshall 1969). While Bellis and Baker interpret these data as evidence that these women were promoting sperm competition, as sperm remain viable for at least 5 days in the female reproductive tract (Barrett and

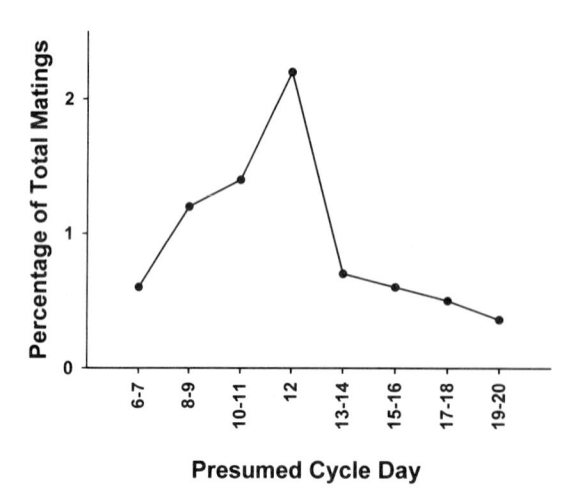

Figure 10.7
Distribution of "double-matings," in which women had sexual intercourse with their primary partner and an extrapair partner within 5 days of each other, in relation to the female's cycle as a percentage of total copulations. Day 12 is the day of maximal fertility (Barret and Marshall 1969). (Data from Bellis and Baker 1990.)

Marshall 1969), they are also completely consistent with the notion that this risky form of sex is more likely to occur when the women's sexual motivation is highest, resulting in a tighter coupling of the behavior to her ovarian cycle. In this view extrapair intercourse occurs more at midcycle as a result of heightened female sexual interest and is therefore more opportunistic with little consideration of social consequences. Further support for this interpretation comes from the finding that for all intercourse, a greater proportion with the primary partner used contraception than was used during extrapair copulations (Bellis and Baker 1990). Similarly, for double-matings, a significantly greater proportion of extrapair sexual intercourse during the most fertile period did not use contraception in comparison with intercourse with the primary partner (26% for extrapair intercourse vs. 14% for intercourse with the primary partner; Bellis and Baker 1990). Thus, not only were double-matings more likely to occur when the women was maximally fertile but they were also less likely to use contraception during the extrapair mating, suggesting less consideration of the consequences of sexual intercourse.

While many caveats apply to these studies, for example, the estimates of cycle phase are less precise than those obtained in nonhuman primate studies, these results are intriguing because they do not simply suggest that sexual behavior varies with the female cycle, but that specific types of sexual behavior will be most strongly affected by ovarian influences.

Investigations of human sexuality have focused primarily on the sexual behavior of established couples, a condition that seems unlikely to have shaped the evolution of hormonally modulated female sexual desire. Sexual intercourse within an established couple is more likely to be influenced by nonhormonal factors such as the day of the week, how hard the workday has been, whether one is on vacation, or gets a job promotion (Blumenstein and Schwartz 1983) than it is by the blood levels of ovarian hormones. Humans, like all primates, do not require specific hormonal conditions to engage in sex, providing great latitude in the conditions under which sex occurs and allowing sex to be used for many social purposes in addition to reproduction. Instead of regulating sexual activity in established couples, it seems more likely that the system of hormonally modulated sexual desire evolved to solve the problem of seeking and engaging a sexual partner when one is not routinely available. This system provides the motivation to take the social and physical risks necessary to find a mate. In addition to motivational effects, ovarian hormones alter women's perceptions affecting their sensitivity to reproductively salient cues (Griffith and Walker 1975; Krug et al. 1994). Although it has not been specifically investigated, one suspects that social gregariousness and risk-taking would be found to be highest in women at midcycle. The view that female sexual desire is strongly coupled to fertility and increases social risk-taking has important social implications.

American society tries to manage adolescent sexuality, particularly that of young women, by prohibitions against becoming sexually active. Currently, the abstinence pledge, in which young men and women pledge to remain virgins until marriage, is popular among teenagers, particularly teenage girls, with more than 2.75 million teenagers pledging since 1993 (Bearman and Bruckner 1998). If adhered to, abstinence pledges delay the time of transition to first intercourse (Bearman and Bruckner 1998). However, there is little consideration of the conditions under which the pledge is likely to be broken. Young women who have pledged abstinence are unlikely to be taking hormonal contraceptives, which mute female sexual initiation (Adams et al. 1978) and thus these young women experience the full complement of hormonal changes during their ovarian cycle, probably with little discussion of how their ovarian cycle might affect their sexual feelings.

The model presented here, unfortunately, suggests that abstinence pledges are most likely to break down when teenage girls are most fertile, since at that point in their cycle their heightened sexual motivation could cause them to abandon their societally imposed abstinence and follow their sexual urges without consideration of the consequences. Since these young women are unlikely to use contraception this has potential public health implications in terms of both teenage pregnancy and disease transmission. While it is probably wishful thinking in today's anti-sex climate, a greater impact on teenage pregnancy could probably be achieved by combining a recognition that hormones influence female sexual desire with frank sexual education designed to increase awareness of, and psychological tools for, managing sexual desire. Already it has been found that abstinence pledges lose their effectiveness when more than 40% of the members of a high school participate (Bearman and Bruckner 1998). Thus a system of simple prohibition is likely to break down for both social and biological reasons.

Do the Principles Described in Females Apply to Males?

In contrast to females, where there was debate about whether gonadal hormones played any role at all, it has always been assumed that male sexuality was under testicular control. It was common, though erroneous, knowledge that testicular function was necessary in males for penile erections. Kinsey's data on the sexual behavior of males presented striking evidence that the onset of puberty was associated in males with an almost immediate increase in many aspects of male sexuality. For example, approximately 10% of males had experienced orgasm by age 12, but 100% had by 18 years of age. In contrast, while 10% of 12-year-old girls had experienced orgasm, by age 22 only 60% had and the increase was steady and gradual, with no sudden and abrupt change around puberty (Kinsey et al. 1953). Thus the notion was both popular and consistent with published information, that male sexuality was turned on by testicular activity and that males could not be sexually active without testicles. Support for this notion came from studies of rodents in which castration completely eliminated male sexual behavior (Beach 1942; Beach and Levinson 1950). It was assumed that the same would hold true for nonhuman primate males as it did for mice and men.

The first full-scale studies of castration in rhesus monkeys provided strikingly different results from similar studies in rodents. Studied under controlled laboratory conditions, in single male-female pairs, castration produced a gradual and steady decline, with some males continuing to achieve intromission and show ejaculatory reflexes years after castration, their capacity to actually produce seminal emissions having disappeared soon after the removal of their testicles, (Phoenix et al. 1973; Michael and Wilson 1974; Phoenix 1978). Similar results were reported for castration in the stumptail macaque, a species closely related to the rhesus monkey (Schenk and Slob 1986). Thus, neither the capacity for erection nor for intercourse itself appeared to be under testicular control. However, sexual behavior did decline following castration, suggesting that testicular hormones modulated male sexual motivation.

Similar evidence for the independence of sexual activity in men came from a retrospective study of 39 men castrated in Europe for sex crimes (Heim 1981). These men, after they were released from prison, were asked to estimate their frequency of sexual intercourse and masturbation before and after castration. Castration significantly reduced both types of sexual activity, but reduced masturbation significantly more than it did intercourse. Prior to castration 25 (64%) of the 39 subjects had sexual intercourse once per month or more often and 34 (87%) masturbated with the same frequency. Four to 7 years after castration, 14% continued to have intercourse one or more times per month and only 3% continued to masturbate at that frequency ($\chi^2 = 4.1; P = .04$).

This study demonstrates that castration does not eliminate male sexual activity, but markedly reduces male sexual motivation. Sexual intercourse reflects both the sexual motivation of the male and the desires of his partner, whereas masturbation results from internal sexual desire; thus the greater reduction in masturbation than in intercourse probably reflects the decrease in male sexual motivation following castration. This study also suggests that the physical capacity to get an erection is not under testicular control, a view that was confirmed in more recent studies which found that hypogonadal men (males with endogenous castrate levels of testosterone) achieved erections in response to sexually explicit films as rapidly as males with normal testosterone levels (Kwan et al. 1983; Bancroft and Wu, 1983; Carani et al. 1992). These same males rarely showed spontaneous erections but were perfectly capable of erections in response to erotic stimuli. Thus, as in the case of female primates, the physical ability to engage in sex in males is not under gonadal hormonal control. Do males show a similar social modulation of the importance of gonadal hormones in modulating sexual behavior as that seen in females?

Social Modulation of Hormonal Effects on Male Sexual Behavior

Both studies of the effect of castration on male rhesus monkeys using pair tests reported a gradual decline in sexual behavior following castration (Phoenix et al. 1973; Michael and Wilson 1974) which lasted for more than 6 years (Phoenix 1978). Evidence that this slow decline reflected the relatively lower importance of male sexual motivation in pair tests came from comparing the effect of castration in pair tests with the effect of suppressing male testicular function with a gonadatropin-releasing hormone (GnRH) antagonist (Wallen et al. 1991). Figure 10.8 illustrates the more

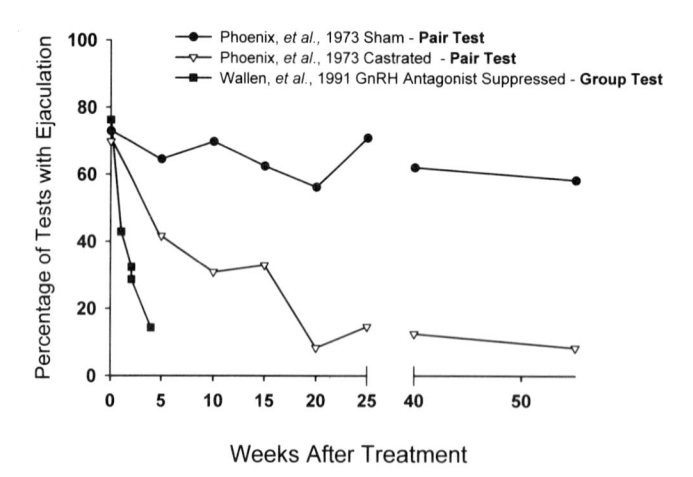

Figure 10.8
Comparison of the effect of castration in male rhesus monkeys, tested in male-female pairs, and the effect of testicular suppression in seven group-living male rhesus monkeys on the percentage of observation periods with ejaculation. Pair tests were 10 minutes in duration and group tests were 120 minutes. The decline in ejaculation following removal of testicular function was more rapid in the group setting than in the pair test. GnRH, gonadotropin-releasing hormone. (Pair-test data from Phoenix et al. 1973; group-test data from Wallen et al. 1991.)

rapid decline in ejaculatory behavior that occurred in males with suppressed testicular function who were tested in a multimale, multifemale social group in comparison to the effects of castration in pair tests. This more rapid decline in sexual behavior in the group setting occurred in spite of the fact that males had 120 minutes to interact sexually, whereas only 10 minutes were available in the pair tests, a difference which biases against finding the more rapid decline in group-tests presented here.

Additional evidence of social modulation of the effect of suppressing testosterone comes from comparing an earlier study that used a GnRH agonist to suppress male testicular function (Davis-DaSilva and Wallen 1989), but tested males in a single-male, multifemale group rather than the multimale, multifemale groups used in the GnRH antagonist study. Testosterone suppression affected male sexual behavior in the single-male condition, but the effect was more gradual and not as complete as that seen in the GnRH antagonist study. Two males tested in both studies continued to ejaculate after 4 weeks of testosterone suppression in the single-male condition, but stopped mating after 1 week of testosterone suppression in the multimale condition. Thus, as with female rhesus monkeys, the opportunity for intrasexual competition affected the importance of hormonal state in maintaining sexual behavior. Also, as in females, there was evidence that male social rank influenced how extensively an individual male's sexual behavior was affected by testosterone suppression.

Figure 10.9 illustrates the magnitude of the overall correlation between male testosterone level and the occurrence of ejaculation prior to receiving GnRH antagonist treatment during the first 4 weeks post GnRH antagonist treatment when testosterone levels were uniformly suppressed, and during the last 4 weeks of the study when testosterone secretion was returning (Wallen et al. 1991). Two of the seven sexually active males in the group were only 4 years old and experiencing their first breeding season, whereas the other five males averaged 12 years of age and had extensive sexual experience. Both of the sexually inexperienced males, who were natal males and the offspring of high-ranking females, occupied the top two positions in the male hierarchy, yet both stopped mating within the first week of testosterone suppression. Thus, as shown in figure 10.9, there is no overall correlation between male testosterone level and the frequency of ejaculation. However, there is a significant correlation when only the five sexually experienced males are considered (see figure 10.9). Prior to GnRH antagonist treatment, male testosterone level did not predict ejaculation frequency, neither in all males nor in the sexually experienced males, though the lack of significance in the latter case clearly stems from the small number of sexually experienced males. Following GnRH antagonist treatment, male testosterone level

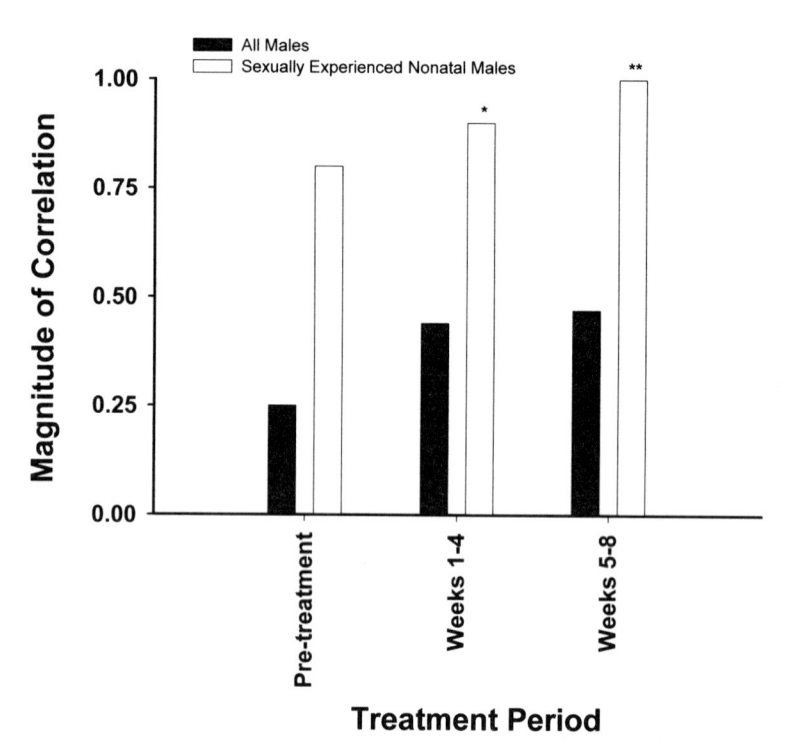

Figure 10.9
Magnitude of the overall correlation between male testosterone level and the weekly frequency of ejaculation prior to receiving gonadotropin-releasing hormone (GnRH) antagonist treatment during the first 4 weeks post GnRH antagonist treatment when testosterone levels were uniformly suppressed, and during the last 4 weeks of the study when testosterone secretion was returning, for all males (black bars) and for the five older, sexually experienced group males (open bars). *$P = .037$; ** = $P < .001$.

significantly predicted male ejaculation frequency, both when testosterone was suppressed and when testosterone secretion was returning (see figure 10.9).

Figure 10.10 shows the magnitude of individual correlations during all 12 weeks of the study between testosterone and weekly ejaculation frequency. Within both the sexually inexperienced and experienced males the magnitude of the correlation were higher for males with lower social rank. Thus, the third-ranking male (the highest-ranking of the sexually experienced males) showed almost no correlation between testosterone and his behavior. This was primarily because GnRH antagonist treatment profoundly suppressed his testosterone level, but this had no detectable effect on his copulatory behavior. In contrast, the lowest-ranked sexually experienced male had an almost perfect correlation ($r = .93$) between his testosterone level and his

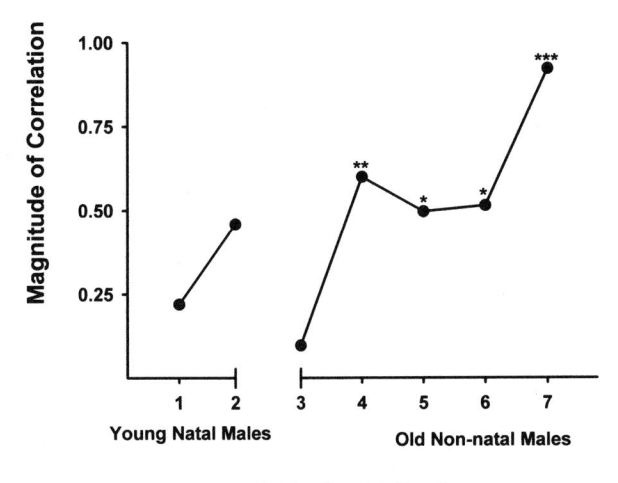

Hormones and Sexual Motivation

Figure 10.10
Individual correlations between testosterone level before, during, and after gonadotropin-releasing hormone antagonist–induced testicular suppression and weekly ejaculatory frequency in relation to male social rank. The two highest-ranking males were 4 years old and sexually inexperienced, but were high-ranking as a result of being natal males born to high-ranking mothers. The five lower-ranking males, averaging more than 12 years of age and sexually experienced, were recent immigrants into the social group. Within each subgroup of males, lower-ranking males have higher correlations between testosterone and behavior, suggesting a greater dependence upon hormonal stimulation for the occurrence of sexual activity. $^*P < .10$; $^{**}P = .04$; $^{***}P < .001$.

ejaculations. When his testosterone level was high he mated, and when it was low he did not. As was argued for female rhesus monkeys, these findings support the notion that hormonal influences on sexual motivation are more critical in competitive social situations and either less critical or not necessary at all under noncompetitive conditions.

Although male sexual motivation does not undergo monthly cycles and is relatively constant with full testicular function, it appears that male sexual motivation serves a similar function in males as it does in females. For both sexes, sexual motivation is a critical modulator of sexual behavior only under specific social conditions, and sexual behavior can occur without any apparent hormonal input at all. The dramatic effects of social context, particularly the opportunity for intrasexual competition, in both male and female rhesus monkeys suggests that this system of hormonally modulated sexual motivation evolved as an adaptation to the problems of sexual activity in a complex social environment.

Final Considerations

In recent years the relative sex drives of males and females has been debated, in part in the discussion of gender equality (Oliver and Hyde 1993), and partly in the discussion of evolution (Symons 1980; Buss 1989). The evidence presented in this chapter suggests resolving this debate by reframing the issue. It is not whether men and woman have equal sex drives, but whether they have the same pattern of sex drive? The primary difference between males and females across mammalian species is that the male sex drive is more or less continual, whereas the female sex drive is discontinuous and, in most cases, cyclic. When females are sexually motivated they are as intensely, or perhaps even more intensely, interested in sexual activity as are males. However, this heightened interest in sex occurs less frequently for females than it does for males. Data on human sexual behavior, ranging from the frequency of masturbation (Oliver and Hyde 1993; Leitenberg et al. 1993) to the number of sexual partners (Oliver and Hyde 1993) support the notion that sexual interest in females is less continuous and demanding.

The intermittent nature of female sexual motivation may contribute to the higher incidence of low sexual desire reported in women (Leif 1977; Segraves 1988). Compared with a male standard of relatively constant sexual interest, women would appear to have lower sexual desire. Recognition that many women's sexual interest varies across their menstrual cycle and that hormonal preparations that suppress ovarian function will affect female sexual desire might result in less sexual dissatisfaction in relationships and lead to different therapeutic assessments and treatments.

In addition, comparisons of male and female sexuality are clouded by the fact that female sexual arousability, as in males (Kwan et al. 1983; Carani et al. 1992), appears to be little influenced by her hormonal state (Schreiner-Engel et al. 1981; Slob et al. 1991, 1996), but cycle phase may affect her initial response when she is observed in a laboratory (Slob et al. 1991, 1996). Women with low sexual motivation, who are less likely to initiate sexual activity, are still sexually arousable and respond sexually to the initiation of their partner. Thus the occurrence of sexual intercourse often reflects the woman's capacity to be sexually aroused and not her underlying degree of sexual motivation. Distinguishing sexual motivation from sexual arousal will lead to more sensible discussions of male and female sexuality.

Why isn't female control of sexual activity as prominent in human society as it is in rhesus monkeys? The same underlying hormonally modulated motivational system appears in both, yet human sexual activity, for the most part, appears more male-controlled than that of rhesus monkeys; a woman's pattern of sexual activity is more likely to reflect her partner's sexual motivation than her own. One likely possibility is that most human societies have minimized systems of female social control of sexual interactions, obscuring the cyclic nature of female sexual motivation with the result that sexual interactions usually occur for the benefit of males (Smuts 1995; Hrdy 1997). It may appear to some that marriage and the nuclear family put males and females on an equal footing, but the data suggest that these are factors that put females at a sexual disadvantage. Despite a cyclic fluctuation in desire and motivation, women show a relatively constant pattern of sexual activity.

In rhesus monkeys, a strongly female-bonded society (Wrangham 1980) combined with a minority of males allows females to regulate sexual intercourse according to their desires. Under these circumstances mating is regulated by females and occurs over a relatively limited number of days. In this regard it is interesting that the conditions wherein rhesus monkey males seem to control mating occurs in pair tests in small areas (Michael and Zumpe 1970; Bonsall et al. 1978; Michael and Bonsall 1979; Michael et al. 1982). Similarly, male sexual aggression is nonexistent in multifemale, multimale groups where females outnumber males by as much as nine to one, whereas sexual aggression is common on Cayo Santiago where the numbers of adult males and females are more balanced (Carpenter 1942a,b; Smuts and Smuts 1993). Thus, it appears that the relative role of females in regulating sexual behavior is sensitive to social conditions and that many human societies may have minimized female control of sexual activity.

It remains to be discovered whether this is related to the relatively constant sexual interest of males, which is best served by continual female sexual availability, or is purely an economic or political phenomenon (Hrdy 1997). The fact remains that aspects of popular culture, such as assertions of a superior sex drive in men or of an equal sex drive in men and women at all times, the institution of marriage, and the use of hormonal contraceptives by females all serve to minimize or obscure the cyclic nature of female sexual motivation, and thus may reduce female opportunity to control sexual interactions that is evident in other cyclically ovulating species. A view of gender differences in sexuality that incorporates both biological predispositions and social context (Baldwin and Baldwin 1997) may result in greater equity in sexual relations. The fact will remain, however, that only women can become pregnant and thus sexual activity for women will always be more risky for women than for men. Whether this inherent inequity must determine the character of sexual relations remains to be seen.

Zuckerman argued that sexual behavior was the cement that bonded primate society (Zuckerman 1932/1981). Some interpreted this as requiring the continual occurrence of sexual behavior and criticized Zuckerman's view because many primates are seasonal breeders and sex is unavailable for much of the year (Lancaster and Lee 1965;

van Horn 1980), a fact Zuckerman was well aware of from his studies of baboons. Evidence has now accumulated, at least in rhesus monkeys, that mating affects social affiliations long after mating has ceased (Wallen and Tannenbaum 1997; Tannenbaum 1997; Tannenbaum and Wallen 1997). A brief mating bout can permanently alter patterns of affiliation between a male and a female (Tannenbaum 1997; Wallen and Tannenbaum 1997). Thus, while sexual behavior in a social group may be socially disruptive and risky, it has long-term benefits in terms of individual reproductive success and enduring social cohesion. The hormonal mechanisms linking increased sexual motivation with female fertility ensure that the complicated mix of risks and rewards produced by sexual activity will occur in complex primate societies. As novelists and poets have long recognized, sexual passion is one of the most dangerous, yet rewarding, of human emotions. Because of its disruptive nature society limits it in such a way that it occurs outside of "normal" times.

Acknowledgments

Former and current graduate students Maryann Davis-DaSilva, John Eisler, Jennifer Lovejoy, Pamela Tannenbaum, and Julia Zehr are thanked for their support and encouragement during the course of the studies described here. Without their assistance these data would never have been collected. Colleagues Tom Gordon, Dario Maestripieri, David Mann, and Mark Wilson are thanked for the many discussions that helped formulate the ideas presented here. The assistance of Timothy Wallen in preparing the references is gratefully acknowledged. This chapter is dedicated to my wife Daiga A. Dunis who has been unstinting in her support over many years. Research described in this chapter was supported by National Science Foundation grants BNS 81-17627, BNS 84-07295, BNS 89-19888, and by NIH grant RR-00165 from the Center for Research Resources to the Yerkes Regional Primate Research Center. The Yerkes Regional Primate Research Center is fully accredited by the American Association for Accreditation of Laboratory Animal Care. Preparation of this chapter was supported by NIH Research Scientist Development Award K02-MH01062.

References

Altmann SA (1962). A field study of the sociobiology of rhesus monkeys (*Macaca mulatta*) Ann NY Acad Sci 102:338–435.

Adams D, Gold AR, Burt AD (1978). Rise in female-initiated sexual activity at ovulation and its suppression by oral contraceptives. N Engl J Med 299:1145–1150.

Baldwin JD, Baldwin JI (1997). Gender differences in sexual interest. Arch Sex Behav 26:181–210.

Ball J, Hartman CG (1935). Sexual excitability as related to the menstrual cycle in the monkey. Am J Obstet Gynecol 29:117–199.

Bancroft J, Wu FCW (1983). Changes in erectile responsiveness during androgen replacement therapy. Arch Sex Behav 12:59–68.

Barrett JC, Marshall J (1969). The risk of conception on different days of the menstrual cycle. Popul Stud 23:455–461.

Beach FA (1942). Analysis of the factors involved in the arousal, maintenance and manifestation of sexual excitement in male animals. Psychosom Med 4:173–198.

Beach FA (1975). Behavioral endocrinology: an emerging discipline. Am Sci 63:178–187.

Beach FA (1976). Sexual attractivity, proceptivity, and receptivity in female mammals. Horm Behav 7:105–138.

Beach FA (1981). Historical origins of modern research on hormones and behavior. Horm Behav 15:325–376.

Beach FA, Levinson G (1950). Effects of androgen on the glans penis and mating behavior of castrated rats. J Exp Zool Lond 114:159–171.

Bearman PS, Bruckner H (1998). The structure of commitment: social context and the transition to first intercourse among American adolescents. *In* Abstracts of the 24th Meeting of the International Academy of Sex Research, Sirmione, Italy, June 6, 1998.

Bellis MA, Baker RR (1990). Do females promote sperm competition? Data for humans. Anim Behav 40:997–999.

Bonsall RW, Zumpe D, Michael RP (1978). Menstrual cycle influences on operant behavior of female rhesus monkeys. J Comp Physiol Psychol 92:846–855.

Blumenstein P, Schwartz P (1983). American Couples: Money, Work, and Sex, New York: Morrow.

Buss DM (1989). Sex differences in human mate preferences—evolutionary hypotheses tested in 37 cultures. Behav Brain Sci 12:1–14.

Carani C, Bancroft J, Granata A, Del Rio G, Marrama P (1992). Testosterone and erectile function, nocturnal penile tumescence and rigidity and erectile response to visual erotic stimuli in hypogonadal and eugonadal men. Psychoneuroendocrinology 17:647–654.

Carpenter CR (1942a). Sexual behavior of free-ranging rhesus monkeys (*Macaca mulatta*). I. Specimens, procedures, and behavioral characteristics of estrus. J Comp Psychol 33:113–142.

Carpenter CR (1942b). Sexual behavior of free-ranging rhesus monkeys (*Macaca mulatta*). II. Periodicity of estrus, homosexual, autoeroticism and nonconformist behavior. J Comp Psychol 33:143–162.

Cochran CG (1979). Proceptive patterns of behavior throughout the menstrual cycle in female rhesus monkeys. Behav Neurol Biol 27:342–353.

Conaway CH, Koford CB (1964). Estrous cycles and mating behavior in a free-ranging band of rhesus monkeys. J Mammal 45:577–588.

Czaja JF, Bielert CF (1975). Female rhesus sexual behavior and distance to a male partner: relation to stage of the menstrual cycle. Arch Sex Behav 4:583–597.

Davis-DaSilva M, Wallen K (1989). Suppression of male rhesus testicular function and sexual behavior by a gonadotropin-releasing-hormone agonist. Physiol Behav 54:263–268.

Dennerstein L, Burrows GD (1977). Sexual response following hysterectomy and oophorectomy. Obstet Gynecol 49:92–96.

Diakow C (1974). Motion picture analysis of rat mating behavior. J Comp Physiol Psychol 88:704–712.

Drickamer LC, Vessey SH (1973). Group changes in free-ranging male rhesus monkeys. Primates 14:359–368.

Filler W, Dresner N (1944). The results of surgical castration in women under forty. Am J Obstet Gynecol 47:122–124.

Goldfoot DA (1981). Olfaction, sexual behavior, and the pheromone hypothesis in rhesus monkeys: a critique. Am Zooloqist 21:153–164.

Goldfoot DA, Kravetz MA, Goy RW, Freeman SK (1976). Lack of effect of vaginal lavages and aliphatic acids on ejaculatory responses in rhesus monkeys: behavioral and chemical analyses. Horm Behav 7:1–27.

Goldfoot DA, Essock-Vitale SM, Asa CS, Thornton JE, Leshner AI (1978). Anosmia in male rhesus does not alter copulatory activity with cycling females. Science 199:1095–1096.

Gordon TP (1981). Reproductive behavior in rhesus monkeys: social and endocrine variables. Am Zooloqist 21:185–195.

Gordon TP, Rose RM, Bernstein IS (1976). Seasonal rhythm in plasma testosterone levels in the rhesus monkey (*Macaca mulatta*): a three year study. Horm Behav 7:229–243.

Gordon TP, Bernstein IS, Rose RM (1978). Social and seasonal influences on testosterone secretion in the male rhesus monkey. Physiol Behav 21:623–627.

Goy RW (1979). Sexual compatibility in rhesus monkeys: predicting sexual behavior of oppositely sexed pairs of adults. Ciba Found Symp 62:227–255.

Goy RW, Resko JA (1972). Gonadal hormones and behavior of normal and pseudohermaphroditic nonhuman female primates. Recent Prog Horm Res 28:707–733.

Gouzoules S, Gouzoules H (1987). Kinship. In Primate Societies (Smuts BB, Cheney DL, Seyfarth RM, Wrangham RW, Struhsaker TT, eds), pp 299–305. Chicago: University of Chicago Press.

Griffith M, Walker CE (1975). Menstrual cycle phases and personality variables as related to response to erotic stimuli. Arch Sex Behav 4:599–603.

Groult B (1992). Salt on Our Skin. London: Hamish Hamilton.

Heape W (1900). The "sexual season" on mammals and the relation of the "pro-estrum" to menstruation. Q J Microsc Sci 44:1–70.

Hedricks CA (1994). Female sexual activity across the human menstrual cycle. Annu Rev Sex Res 5:122–172.

Hedricks C, Piccinino LJ, Udry JR, Chimbira THK (1987). Peak coital rate coincides with onset of luteinizing hormone surge. Fertil Steril 48:234–238.

Heim N (1981). Sexual behavior of castrated sex offenders. Arch Sex Behav 10:11–19.

Herbert J, Trimble MR (1967). Effect of oestradiol and testosterone on the sexual receptivity and attractiveness of the female rhesus monkey. Nature 216:165–166.

Hrdy SB (1997). Raising Darwin's consciousness—female sexuality and the prehominid origins of patriarchy. Hum Nat 8:1–49.

Hrdy SB, Whitten PL (1987). Patterning of sexual activity. In Primate Societies (Smuts BB, Cheney DL, Seyfarth RM, Wrangham RW, Struhsaker TT, eds), pp 370–384. Chicago: University of Chicago Press.

Izard MK (1983). Pheromones and reproduction in domestic animals. In Pheromones and Reproduction in Mammals (Vandenberg JG, ed). New York: Academic Press.

Johnson DF, Phoenix CH (1978). Sexual behavior and hormone levels during the menstrual cycles of rhesus monkeys. Horm Behav 11:160–174.

Keverne EB (1976). Sexual receptivity and attractiveness in female rhesus monkeys. Adv Stud Behav 7:155–200.

Kinsey AC, Pomeroy WB, Martin CE, Gebhard PH (1953). Sexual Behavior in the Human Female. Philadelphia: WB Saunders.

Koford CB (1966). Population changes in rhesus monkeys: Cayo Santiago, 1960–64. Tulane Stud Zool 13:1–7.

Krug R, Pietrowski R, Fehm HL, and Born J (1994). Selective influence of the menstrual cycle on perception of stimuli with reproductive significance. Psychosom Med 56:410–417.

Kwan M, Greenleaf WJ, Mann J, Crapo L, Davidson JM (1983). The nature of androgen action on male sexuality: a combined laboratory–self-report study on hypogonadal men. J Clin Endocrinol Metab 57:557–562.

Lancaster JB, Lee RB (1965). The annual reproductive cycle in monkeys and apes. *In* Primate Behavior. DeVore I, ed. (pp 486–513). New York: Holt, Rhinehart & Winston.

Leif HI (1977). Inhibited sexual desire. Med Aspects Hum Sexuality 7:94–95.

Leitenberg H, Detzer MJ, and Srebnik D (1993). Gender differences in masturbation and the relation of masturbation experience in preadolescence and/or early adolescence to sexual behavior and sexual adjustment in young adulthood. Arch Sex Behav 22:87–98.

Lindburg DG (1969). Rhesus monkeys: mating season mobility of adult males. Science 166:1176–1178.

Malinowski B (1927). Sex and Repression in Savage Society. Chicago: University of Chicago Press.

Mallow GK (1981). The relationship between aggressive behavior and menstrual cycle stage in female rhesus monkeys (*Macaca mulatta*). Horm Behav 15:259–269.

Masters WH, Johnson VE (1965). The sexual response cycle of the human female, 1. Gross anatomic considerations. *In* Sex Research: New Developments (J Money, Ed), pp 53–89. New York: Holt Rhinehart and Winston.

Masters WH, Johnson VE (1966). Human Sexual Response. Boston: Little, Brown.

Matteo S, Rissman EF (1984). Increased sexual activity during the midcycle portion of the human menstrual cycle. Horm Behav 18:249–255.

Michael RP, RW Bonsall (1977a). Chemical signals and primate behavior. *In* Chemical Signals in Vertebrates. (Mozell MM, ed), pp 251–271, New York: Plenum.

Michael RP, Bonsall RW (1977b). Periovulatory synchronization of behaviour in male and female rhesus monkeys. Nature (Lond) 265:463–465.

Michael RP, Bonsall RW (1979). Chemical signals and primate behaviour. *In* Endocrine Control of Sexual Behavior. (Beyer C, ed), pp 251–271. New York: Raven Press.

Michael RP, Herbert J (1963). Menstrual cycle influences grooming behavior and sexual activity in rhesus monkeys. Science 140:500–501.

Michael RP, Keverne EB (1968). Pheromones and the communication of sexual status in primates. Nature (Lond) 218:746–749.

Michael RP, Welegalla J (1968). Ovarian hormones and the sexual behaviour of the female rhesus monkey (*Macaca mulatta*) under laboratory conditions. J Endocrinol 41:407–429.

Michael RP, Wilson M (1974). Effects of castration and hormone replacement in fully adult male rhesus monkeys (*Macaca mulatta*). Endocrinology 95:150–159.

Michael RP, Zumpe D (1970). Rhythmic changes in the copulatory frequency of rhesus monkeys (*Macaca mulatta*) in relation to the menstrual cycle and a comparison with the human cycle. J Reprod Fertil 21:199–201.

Michael RP, Zumpe D (1993). A review of hormonal factors influencing the sexual and aggressive behavior of macaques. Am J Primatol 30:213–241.

Michael RP, Zumpe D, Bonsall RW (1982). Behavior of rhesus during artificial menstrual cycles J Comp Physiol Psychol 96:875–885.

Miller GS (1931). The primate basis for human sexual behavior. Q Rev Biol 6:379–410.

Missakian EA (1972). Geneological and cross-geneological dominance relations in a group of free-ranging rhesus monkeys (*Macaca mulatta*) on Cayo Santiago. Primates 13:169–180.

Nadler RD (1994). Walter Heape and the issue of estrus in primates. Am J Primatol 33:83–87.

Oliver MB, Hyde JS (1993). Gender differences in sexuality: a meta-analysis. Psychol Bull 114:29–51.

Palmer JD, Udry JR, Morris NM (1982). Diurnal and weekly, but no lunar rhythms in human copulation. Hum Biol 54:111–121.

Pfaff, DW, Diakow C, Montgomery M, Jenkins FA (1978). X-ray cinematographic analysis of lordosis in female rats. J Comp Physiol Psychol 92:937–941.

Phoenix CH (1978). Steroids and sexual behavior in castrated male rhesus monkeys. Horm Behav 10:1–9.

Phoenix CH, Slob AK, Goy RW (1973). Effects of castration and replacement therapy on sexual behavior of adult male rhesus monkeys. J Comp Physiol Psychol 84:472–481.

Pomerantz SM, Goy RW (1983). Proceptive behavior of female rhesus monkeys during tests with tethered males. Horm Behav 17:237–248.

Pope NS, Wilson ME, Gordon TP (1987). The effect of season on the induction of sexual behavior by estradiol in female rhesus monkeys. Biol Reprod 36:1047–1054.

Rose RM, Bernstein IS, Gordon TP (1975). Consequences of social conflict on plasma testosterone levels in rhesus monkeys. Psychosom Med 37:50–61.

Sade DS (1965). Some aspects of parent-offspring and sibling relationships in a group of rhesus monkeys, with a discussion of grooming. Am J Phys Anthropol 23(1):1–18.

Schenk PE, Slob AK (1986). Castration, sex steroids, and heterosexual behavior in adult laboratory-housed stumptailed macaques (*Macaca arctoides*). Horm Behav 20:336–353.

Schreiner-Engel P, Schiavi RC, Smith H, White D (1981). Sexual arousability and the menstrual cycle. Psychosom Med 43:199–214.

Segraves RT (1988). Hormones and libido. *In* Sexual Desire Disorders (Lieblum R, Rosen RC, eds), pp 271–311. New York: Guilford Press.

Sherwin BB, Gelfland MM (1987). The role of androgens in the maintenance of sexual functioning in oophorectomized women. Psychosom Med 49:397–409.

Sherwin BB, Gelfland MM, Brender W (1985). Androgen enhances sexual motivation in females: a prospective, crossover study of sex steroid administration in surgical menopause. Psychosom Med 47:339–351.

Slob AK, Ernste M, van der Werff ten Bosch JJ (1991). Menstrual cycle phase and sexual arousability in women. Arch Sex Behav 20:567–577.

Slob AK, Bax CM, Hop WCJ, Rowland DL, van der Werff ten Bosch JJ (1996). Sexual arousability and the menstrual cycle. Psychoneuroendocrinology 21:554–558.

Small MF (1996). "Revealed" ovulation in humans? J Hum Evolution 30:483–488.

Smuts BB (1995). The evolutionary origins of patriarchy. Hum Nat 6:1–32.

Smuts BB, Smuts RW (1993). Male aggression and sexual coercion of females in nonhuman primates and other mammals: evidence and theoretical implications. Adv Stud Behav 22:1–63.

Southwick CH, Beg MA, Siddiqi MR (1965). Rhesus monkeys in North India. *In* Primate Behavior. de Vore I, ed. (pp 111–159). New York: Holt, Rinehart & Winston.

Stanislaw H, Rice FJ (1988). Correlation between sexual desire and menstrual cycle characteristics. Arch Sex Behav 17:499–508.

Stockard CR, Papanicolaou GN (1919). The vaginal closure membrane, copulation, and the vaginal plug in the guinea-pig, with further consideration of the oestrous rhythm. Biol Bull 37:222–245.

Symons D (1980). Précis of The evolution of human sexuality. Behav Brain Sci 3:171–214.

Tannenbaum PL (1997). Biobehavioral Mechanisms of Male Rhesus Monkey Social Integration. Unpublished dissertation, Emory University.

Tannenbaum PL, Wallen K (1997). Sexually initiated affiliation facilitates rhesus monkey integration. Ann N Y Acad Sci 807:578–582.

Tinklepaugh OL (1933). The nature of periods of sex desire in women and their relation to ovulation. Am J Obstet Gynecol 26:2–12.

Tsui AO, de Silva SV, Marinshaw R (1991). Pregnancy avoidance and coital behavior. Demography 28:101–117.

Udry JR, Morris NM (1968). Distribution of coitus in the menstrual cycle. Nature 220:593–596.

Udry JR, Morris NM (1977). The distribution of events in the human menstrual cycle. J Reprod Fertil 51:419–425.

Van Horn RN (1980). Seasonal reproduction patterns in primates. Prog Reprod Biol 5:181–221.

Walker ML, Wilson ME, Gordon TP (1983). Female rhesus monkey aggression during the menstrual cycle. Anim Behav 31:1047–1054.

Wallen K (1982). Influence of female hormonal state on rhesus sexual behavior varies with space for social interaction. Science 217:375–376.

Wallen K (1989). Nonfertile mating in rhesus monkeys: sexual aggression of miscommunication? Primate Rep 23:23–34.

Wallen K (1990). Desire and ability: hormones and the regulation of female sexual behavior. Neurosci Biobehav Rev 14:233–241.

Wallen K (1995). The evolution of female sexual desire. In Sexual Nature Sexual Culture (Abramson PR, Pinkerton SD, eds), pp 57–79. Chicago: University of Chicago Press.

Wallen K, Lovejoy J (1993). Sexual behavior: endocrine function and therapy. In Hormonally Induced Changes in Mind and Brain (Schulkin J, ed), pp 71–97. New York: Academic Press.

Wallen K, Tannenbaum PT (1997). Hormonal modulation of sexual behavior and affiliation in rhesus monkeys. Ann N Y Acad Sci 807:185–202.

Wallen K, Winston LA (1984). Social complexity and hormonal influences on sexual behavior in rhesus monkeys (*Macaca mulatta*). Physiol Behav 32:143–162.

Wallen K, Winston LA, Gaventa S, Davis-DaSilva M, Collins DC (1984). Periovulatory changes in female sexual behavior and patterns of ovarian steroid secretion in group-living rhesus monkeys. Horm Behav 18:431–450.

Wallen K, Mann DR, Davis-DaSilva M, Gaventa S, Lovejoy J, Collins DC (1986). Chronic gonadotropin-releasing hormone agonist treatment suppresses ovulation and sexual behavior in group-living female rhesus monkeys. Physiol Behav 36:369–375.

Wallen K, Eisler JA, Tannenbaum PL, Nagell KM, Mann DR (1991). Antide (NAL-LYS GnRH antagonist) suppression of pituitary-testicular function and sexual behavior in group-living rhesus monkeys. Physiol Behav 50:429–435.

Waxenberg SE, Drellich MG, Sutherland AM (1959). The role of hormones in human behavior. I. Changes in female sexuality after adrenalectomy. J Clin Endocrinol 19:193–202.

Wilson ME, Gordon TP, Collins DC (1982). Variation in ovarian steroids associated with the annual mating period in female rhesus (*Macaca mulatta*). Biol Reprod 27(3):530–539.

Wrangham RW (1980). An ecological model of female-bonded primate groups. Behaviour 75:262–300.

Young WC (1961). The hormones and mating behavior. In Sex and Internal Secretions, vol 2 (Young WC, ed), pp 1173–1239. Baltimore: Williams & Wilkins.

Zehr JL, Maestripieri D, Wallen K (1998). Estradiol increases female sexual initiation independent of male responsiveness in rhesus monkeys. Horm Behav 33:95–103.

Zuckerman S (1930). The menstrual cycle of the primates. Part I. General nature and homology. Proc Zool Soc Lond Pt 3:691–754.

Zuckerman S (1932/1981). The Social Life of Monkeys and Apes. Boston: Routledge & Kegan Paul.

11 Contextual Influences on Sociosexual Behavior in Monogamous Primates

Jeffrey A. French and Colleen M. Schaffner

Monogamy, as a context in which reproductive behavior is expressed, has attracted a disproportionate amount of attention from behavioral biologists in the past century (Dewsbury 1988). The origins of this interest are twofold. From a scientific perspective, the existence of a monogamous mating strategy in organisms that display differential investment in gametes or offspring care flies in the face of predictions derived from modern theories of the evolution of reproduction (e.g., see Trivers 1972). Second, from the perspective of modern human culture, there are profound moral, ethical, and religious traditions that draw our attention to monogamous social and reproductive relationships. Regardless of the source of interest, monogamy has attracted considerable attention from behavioral biologists interested in both the evolutionary origins and functions of monogamy (e.g., see Kleiman 1977; Wittenberger and Tilson 1980; van Schaik and Dunbar 1990) and in the proximate features that regulate the development and expression of monogamy (Anzenberger 1992; Carter and Roberts 1997).

Recent elegant work has addressed the neurobiological substrates of some key features of monogamy in rodents, such as partner fidelity, social and sexual exclusiveness, and parental behavior. This work has suggested a strong link between species-specific neurobiology and the display of these traits associated with monogamy (Carter et al. 1995; Insel and Shapiro 1992). Although cautionary remarks accompany the interpretations of these data, the authors' stated implication is that monogamous tendencies are strongly determined by underlying neurobiological factors. Thus, "monogamous behaviors, including pair bonding, are influenced by hormones throughout life" (Carter et al. 1997, p. 270) and the reference to the oxytocin receptor gene as the " 'candidate gene' for monogamy" (Insel et al. 1997, p. 303). In fairness, Insel et al. (1997) admit that the evidence to date on the neurobiological substrates of monogamy is correlational at best, yet a strong deterministic theme is implicit in the work of these authors.

Although monogamy tends to be the modal breeding system in prairie voles (*Microtus ochrogaster*), a growing body of more recent evidence suggests that there is considerable variability in the expression of monogamy and its associated traits in this species (e.g., see Roberts and Carter 1997; Dharmadhikari et al. 1997). Thus, these traits can vary among (or even within) individuals, in spite of similar neurobiological systems. The sources of this variability in the expression of monogamy-like traits are likely to be diverse, and probably include ecological, individual, populational, seasonal, and strain-related differences. An additional candidate that certainly plays an important role in the expression of a monogamous profile is *social context*, as defined by intra- and intergroup social relationships, social structure, and group demography.

In addition to rodents, monogamy is widespread throughout several other mammalian orders, including carnivores and primates. Given the complexity of social structure and relationships in nonhuman primates (Smuts et al. 1986), this taxonomic group is appropriate for study of contextual influences on the expression of monogamy and its associated traits. The overall goal of this chapter, then, is to highlight the multiple roles that social context plays in regulating reproductive behavior, broadly defined to include breeding group formation, courtship, copulation, and mate guarding, in monogamous primates.

During the course of this discussion, we first explore the multiple definitions of monogamy to illustrate the complexity and sophistication of the question, What is monogamy? Second, we provide a broad taxonomic review of primate species that are reported to display monogamous sociosexual relationships. This section highlights variation in the ways in which monogamy is made manifest in species across the primate order, and especially focuses on the fairly widespread occurrence of "departures from monogamy" in most species that have been well studied in the wild. In the final section, the focus is on the contextual variables that promote monogamous sociosexual relationships in the taxon that has been studied in greatest detail in an experimental sense, the callitrichid primates (marmosets and tamarins). The section contrasts the relative roles of features intrinsic to the pair (e.g., "emotional attachments," partner preferences) and features extrinsic to the pair (e.g., intolerance and aggression toward same-sex strangers) in the maintenance of monogamous social systems. The section also highlights the complex interplay between social context, endocrine factors, and the expression of reproductive behavior.

The general thesis of this chapter is threefold: (1) monogamous sociosexual relationships occur with some regularity among certain species of nonhuman primates, but there is considerable intraspecific variability in this trait; (2) monogamy appears to be maintained more by features *extrinsic* to the heterosexual pair than by *intrinsic* features of the individuals involved in the pair; and (3) social context appears to be a more critical determinant than endocrine factors in shaping the occurrence and patterning of sexual behavior within these monogamous social units.

Definitions of Monogamy

In spite of the fairly wide agreement about the nature of monogamy in human societies, there has been considerable debate regarding the features to consider as one attempts to classify a species, or a population, or a pair, or an individual member of a pair, as "monogamous." Some authors offer fairly restrictive definitions of monogamy, based solely on the ways in which gametes combine in any given breeding season (e.g., see Gowaty 1981). Others are more liberal in their definitions, basing the definition upon patterns of social or sexual affiliation (e.g., see Wittenberger and Tilson 1980). For readers interested in a thoughtful historical survey of these issues, we refer them to the excellent discussions by Wickler and Seibt (1983) and especially Dewsbury (1988).

The difficulties in defining a complex higher-order phenomenon such as a mating system should come as no surprise. Monogamy is a trait that is inferred from the behavior of a collection of individuals, each of whom can be viewed as making a series of reproductive decisions, played out in sociosexual interactions, in order to maximize reproductive output. It is not therefore a fixed characteristic of a species, as is a morphological trait like feathers or fur, but rather an emergent phenomenon that arises out of a specific pattern of decisions by individuals. Much of the difficulty in defining monogamy, as many have recognized (e.g., see Dewsbury 1988; Anzenberger 1992) is that competing definitions that deal with different levels of analysis are often pitted against one another, and hence cannot be directly compared. A synthesis of these levels of analysis is presented in table 11.1, which describes the level at which a social system can be defined, the criteria on which the classification is made at that level, and an example of how the definition might be applied to a monogamous system. Recognizing these different levels reduces the conceptual blocks that may arise as an offshoot of defining the concept of monogamy.

In spite of the diversity of definitions, Dewsbury (1988) has identified three elements or dimensions that seem to be common to most orientations toward monogamy, especially in mammals. First, monogamy is characterized by more or less

Table 11.1
Levels of definition of mating systems, and application of definition to monogamy

Level of Analysis	Definition	Criterion as Applied to Monogamy
Demographic	Number of males and females of breeding age per social group	Single breeding adult male and female residing in group
Social	Pattern of affiliative interactions and relationships between males and females	Male and female have close temporal and spatial proximity, "pair-bonded," joint territory defense, and offspring care
Sexual	Distribution of copulatory behavior between individuals within each sex	Single male and female have exclusive sexual relationship
Genetic	Representation of paternity and maternity in offspring	All offspring in group are produced by a single male and female

exclusivity of mating relationships. That some would argue about the relative merits of exclusivity of copulation vs. exclusivity of paternity simply reflects the fact that the first refers to monogamy at a *sexual* level, while the second refers to monogamy at a *genetic* level. Second, monogamy is characterized by a *selective association* between males and females, including joint nesting, joint travel, and joint defense of resources and territories. Finally, and especially in mammals, monogamy tends to be associated with *joint parental care.* As will be seen in the next section, which surveys monogamy in extant nonhuman primates, these three features serve as the key evidence for classifying primates as monogamous.

Taxonomic Survey of Monogamy in Primates

Monogamous social systems have been reported in six families of primates, including prosimians, Old and New World primates, and lesser apes. For each species, a description of the nature of the observational evidence that supports the claim to monogamy will be presented, along with any recent data that may call the categorization into question. No attempt is made to be exhaustive, but only to cite the most relevant papers that support the notion of monogamy. The source for taxonomic information on the distribution of monogamy is derived from the most recent exhaustive review of primate social organization (appendix A1 in Smuts et al. 1987). Table 11.2 provides a brief overview of the information contained in this section. The most noteworthy aspect of Table 11.2 is the absence of information on monogamy among

Table 11.2
Primate species in which monogamy has been reported as prevalent mating system

Species Name	Common Name	Nature of Evidence
Indri indri	Indri	Demographic
Mizra coquereli	Coquerel's dwarf lemur	Weak demographic
Aotus trivirgatus	Owl (night) monkey	Demographic, social
Pithecia spp.	Saki monkeys	Demographic, social
Callicebus spp.	Titi monkeys	Demographic, social
Cebuella pygmaea	Pygmy marmosets	Demographic, social
Callithrix spp.	Marmosets	Demographic, social, sexual
Saguinus spp.	Tamarins	Demographic, social, sexual
Leontopithecus spp.	Lion tamarins	Demographic, social, sexual
Cercopithecus neglectus	De Brazza's guenon	Demographic
Presbytis potenziani	Mentawai langur	Demographic
Hylobates spp.	Gibbons and siamang	Demographic, social, sexual

primates, as assessed at the genetic level. In spite of the passage of a decade since the development of molecular genetic markers for paternity exclusion-inclusion analysis (see review in Martin et al. 1992), there has up to now been no strong test of genetic monogamy in any species of nonhuman primate. The impact of such testing in avian species has been dramatic, revealing the widespread existence of extrapair fertilizations in some species (Burke et al. 1989; Westneat and Sherman 1993) and, perhaps more surprising, the *lack* of extrapair fertilizations in others (Piper et al. 1997). Studies are in progress for some species and hopefully the next decade will yield answers to the fundamental question about the occurrence and distribution of genetic monogamy among primates.

Prosimians

Indridae Early field studies on the indri (*Indri indri*) revealed that this species lives in small groups of two to five individuals, with an adult pair and offspring composing each group (Pollock 1975; Klopfer and Boskoff 1982). Groups live at low densities and have frequent vocal contact with neighboring groups. Sexual behavior is limited to the adult male and female in the group, and close intergroup encounters are rare, but vocally mediated intergroup interactions occur with regularity. During a field study which included 2500 hours of direct observation on three groups of indris, no stranger was ever seen inside the identified territorial boundaries of another group. However, a sampling of individuals on the basis of vocal characteristics suggested that some individuals, especially males, may cross territorial boundaries and freely range across territories (Pollock 1975).

Cheirogaleidae Among the nocturnal malagasy prosimians, Coquerel's dwarf lemur (*Mirza* (*Microcebus*) *coquereli*) has been classified as exhibiting monogamy, although in a slightly different form from all other purportedly monogamous primates. Original field studies (reviewed in Pages 1980) revealed that males and females independently defend territories against same-sex conspecifics, and particular male-female "pairs" reside in partially overlapping home ranges. Given patterns of distribution of males and females over time within the home ranges, Pages suggested that privileged social contacts may exist between the pair, resulting in "loose" pair bonds.

A more recent field study of a large number of marked individuals reveals a different mating system than that originally envisaged by Pages. Kappeler (1997) provides a host of inferential evidence that suggests that the social system of *M. coquereli* is more likely to be characterized by polygyny than by monogamy. Among the evidence gathered at Kappeler's field site is the following: (1) during the short mating season, testis size in males increased fivefold; (2) during the mating period, male territories quadrupled in size, from less than 5 hectares to greater than 17 hectares, while at the same time female territory size remained constant at 2 to 4 hectares; and (3) the incidence of males with physical injuries (most likely from fighting) rose dramatically during the breeding season, and was much higher than the incidence in females. Captive studies of this species also suggest a higher incidence of male-male competition during the breeding season (Stanger et al. 1995). Together, the evidence that male testis size increases during the breeding season, that males expand their territories, and that males incur more injuries from fighting suggests high male-male competition. When these features are combined with an extremely short period of female estrus, it may be more accurate to describe the mating system in dwarf lemurs as scramble competition polygyny.

New World Primates

Cebidae Among the cebid primates, monogamy has been claimed for three genera: the owl monkeys (*Aotus*), titi monkeys (*Callicebus*), and saki monkeys (*Pithecia*). The nocturnal *Aotus* is clearly recognized as a monogamous form, living in small groups of three to five with a single adult male and female, along with juveniles and dependent infants (Wright 1978; Robinson et al. 1987; Kinzey 1997). No variation in social structure has been reported in field studies, nor have extrapair copulations been noted. Captive studies reveal a high investment in parental care by males (Dixson and Fleming 1981). Further, fights between parents and offspring are noted when group size exceeds five individuals, suggesting that emigration of subadult offspring prevent agonistic interactions that may affect the integrity of the social group (Dixson 1982).

Titi monkeys of the genus *Callicebus* are also demographically "monogamous," with all published data on stable groups suggesting that groups contain only a single adult male and female (Easley and Kinzey 1986; Robinson et al. 1987; Müller 1996; Kinzey 1997). Adult males are extensively involved in parental care (Mendoza and Mason 1986; Robinson et al. 1987; Hoffman et al. 1995). There is one significant report of a departure from sexual monogamy in a wild population of *Callicebus moloch* (Mason 1966). During the course of intergroup interactions, Mason witnessed several instances of temporary "liaisons" between a resident female and a male from the neighboring group. These interactions typically lasted only minutes, but in some cases persisted for several hours. Not surprisingly, these interactions tended to occur when the resident female was also sexually receptive to her pairmate. These observations are of particular interest for this genus, since recent reports indicate the existence of floating solitary individuals in and among existing groups of titi monkeys (Robinson et al. 1987; Müller 1996), and also since titis are known to develop strong behavioral attachments to pairmates in captive contexts (see below). In any event, Mason's observations suggest that mating exclusivity may be compromised in at least some groups under some occasions.

Saki monkeys of the genus *Pithecia* have historically been considered to be monogamous; however, this classification was based solely on demographic scans from field studies of extremely short duration (e.g., see Buchanan et al. 1981). Observations of groups in captivity suggest that demographically monogamous groups are stable, while groups that depart from monogamy are not stable. Savage et al. (1993) created both polygynous groups (two females plus one male) and polyandrous groups (two males and one female) in captivity. The multimale group was stable for only 12 days, and was terminated because of male-male aggression. Interestingly, the female solicited and copulated with both males prior to the group dissolution, and the mating interactions themselves did not appear to be a proximate cue for the initiation of fighting between the males. The polygynous group was less stable than the polyandrous group, and one of the females and the male evicted the second adult female in an apparent joint action. More recent field data suggest that wild groups may depart from a classic monogamous demography, including groups with multiple adult males and females, and observations of matings between a single male and multiple females in a group (see review in Kinzey 1997).

Callitrichidae Marmosets and tamarins have long been considered a model taxon for monogamous monkeys (Kleiman 1977; Snowdon 1990), primarily on the basis of results from brief field studies and on extensive captive experiments and zoo breeding. More recent field studies have revealed that callitrichid primates (including marmosets of the genera *Cebuella* and *Callithrix*, tamarins of the genera *Saguinus* and *Leontopithecus*, and Goeldi's monkey, *Callimico*) exhibit departures from both demographic monogamy, with groups containing, on average, more than one adult male and adult female (Goldizen 1987), and sexual monogamy, with multiple males and females engaging in sexual activity within a single group (reviewed in French 1997). We consider the callitrichid primates in greater detail below.

Old World Primates

Cercopithecidae Two species of Old World primates have been reported to have monogamous social structures: the Mentawai langur (*Presbytis potenziani*) and De Brazza's guenon (*Cercopithecus neglectus*). Tilson and Tenaza (1976) first reported on the Mentawai langur, and inferred a monogamous social structure for this species based on a variety of features. First, strong demographic evidence suggested monogamy, since the majority of groups sampled consisted of a single adult male and female along with younger animals—presumably offspring. Second, there is much less sexual dimorphism in the Mentawi langur than in other Old World primates, a trait that is correlated with monogamy (Kleiman 1977). Finally, the existence of a vocal "duet" between adult males and females during intergroup encounters provided behavioral evidence of a bond between a particular male and female.

In the De Brazza's monkey, evidence for monogamy is again based primarily upon demographic data. Gautier-Hion and Gautier (1978) studied a population in Gabon and reported that the modal group pattern among six groups of monkeys was a single male and female with one or two dependents. On occasion, several groups would merge for short periods of time into a larger aggregation of seven to nine individuals. Because of the thick vegetation at the field site, and the monkeys' cryptic nature, more detailed observations on social interactions during these multiple group aggregations were not reported. More recent field evidence from populations of *C. neglectus* in Kenya (Wahome et al. 1993) suggest that social grouping in this species is variable, since group structure in this study population was more typical for cercopithecine monkeys (i.e., a single resident male, three to four adult females, and multiple infants). Thus, while some populations may tend toward monogamy, there is certainly considerable variability among populations in this trait.

Hylobatidae

Next to the callitrichid primates, the most often cited example of primate monogamy is the gibbons of the monogeneric (*Hylobates*) family Hylobatidae. As with other forest-dwelling primates of the Old and New World, the classification of gibbons as monogamous is based on short-term field studies and primarily on demographic (single adult male and female per group) and morphological (high degree of monomorphism between the sexes) data (e.g., see Gittins and Raemaekers 1980; Tilson 1979, 1981; Chivers 1984; Mitani 1984, 1987). Indeed, as recently as a decade ago, Leighton (1987) characterized gibbons as *"invariably monogamous"* (emphasis added, p. 137). As with callitrichid primates (see above), however, the evidence gathered since that time has led to a new appreciation of variability in gibbon mating structure and individual mating strategies.

The nature of the data suggesting that gibbons are not strictly monogamous comes in several forms. First, group structure in gibbons is not limited to demographic monogamy, and some species may exhibit demographic and reproductive polygyny (e.g., *Hylobates concolor*: Haimoff et al. 1987; Bleisch and Chen 1991; *Hylobates pileatus*: Srikosamatara and Brockelman 1987). Second, longer-term field studies of well-habituated and well-identified animals have led to the discovery of previously unsuspected and observed instances of pair instability and mate desertion or switching. For example, of eleven "pair-bonded" males and females under study by Palombit (1994a), two were terminated by the death of one of the partners and almost half (five pairs) were permanently terminated by the desertion of the current mate. In other cases, pairs were disbanded for shorter periods of time, and the deserting partner returned to the original mate (Palombit 1996). Third, there are three cited cases of sexual polygyny in otherwise stable groups of gibbons (Srikosamatara and Brockelman 1987; Bleisch and Chen 1991; Ahsan 1995). In all three cases, males consorted and copulated with multiple adult females that were resident in the social group. Finally, there have been two reports of extrapair copulations (EPCs) in two species of gibbons (*Hylobates lar* and *Hylobates syndactylus*) *within the context of an existing monogamous pairing* (Palombit 1994b; Reichard 1995). Both studies report EPCs between a resident female and subadult or adult males from neighboring groups during intergroup encounters at territory boundaries. Some of the EPCs occurred during the suspected period of conception (when backdated from the time of parturition) so it is possible that the nonpair male could have sired the offspring. It is clear from these recent studies that adult gibbons do not mate for life, the genetic structure of groups is likely to vary from a genetic "nuclear family," and male and female gibbons engage in considerable mating activity outside of the pair relationship.

Contextual Influences on Sociosexual Behavior in Callitrichid Primates

As described above in the taxonomic review of primate monogamy, marmosets and tamarins have many of the correlated characters classically associated with monogamy (long-term spatial association, sociosexual pair bonds, and biparental care), but also reside in groups that range demographically from polyandrous to monogamous to polygynous. Further, given high rates of dispersal and immigration (Goldizen and Terborgh 1989), groups are likely to be composed of unrelated as well as related individuals. Thus, there is likely to be considerable variation under natural conditions in the social contexts in which reproductive behavior is expressed, and this section reviews a variety of experimental studies that address the influence of this variability on behavior among partners.

Context Influences on Partner Preferences

A hallmark of vole monogamy is the tendency for individuals in a long-term pair relationship to display a persistent preference for their partner, and antagonism or aversion toward unfamiliar animals of the opposite sex (e.g., see Williams et al.

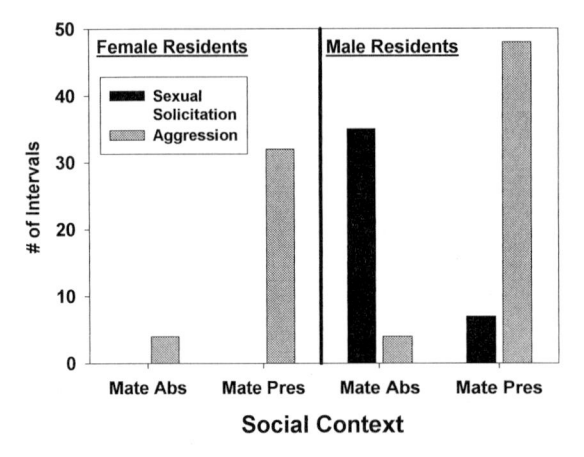

Figure 11.1
Responses of female (left panel) and male (right panel) common marmosets to the presentation of strangers of the opposite sex. Animals were tested with the stranger either in the presence of their own pair mate (Mate Pres) or in the absence of the pair mate (Mate Abs). Dark bars represent the mean number of observational intervals in which sexual solicitations were directed toward the opposite-sex stranger, and shaded bars represent the number of observation intervals in which agonistic behavior was directed toward the stranger. (Modified from Evans 1983.)

1992). Under normal conditions in undisturbed family groups of callithrichid primates, there is also clear evidence of a special affiliative relationship between adult pairmates, as measured by time in proximity, food-sharing, and patterns of affiliative social behavior (see, e.g., Schaffner et al. 1995). However, there is clear and convincing evidence from studies on four species that demonstrates just how dependent partner preference is upon the social context in which it is tested.

Two studies have addressed partner preferences (and interactions with unfamiliar individuals) in common marmosets (*Callithrix jacchus*). Evans (1983) allowed members of a long-term pair to interact for 15 minutes with an unfamiliar individual of the opposite sex in either the presence or the absence of their pairmate. Males and females differed in their responses to the strangers, but the qualitative and quantitative aspects of their interactions were dramatically affected by social context (figure 11.1). In the presence of the heterosexual partner, both males and females showed high levels of agonistic behavior toward the stranger. When tested in the absence of their mates, females showed substantially reduced aggression and increased amicable interactions with the unfamiliar males. Males likewise showed reduced levels of agonistic behavior, increased amiable interactions with the female, and spent over one third of the test period in sexual solicitation patterns (tongue-flicking and open mouth displays; Schaffner et al. 1995) when tested in the absence of the mate.

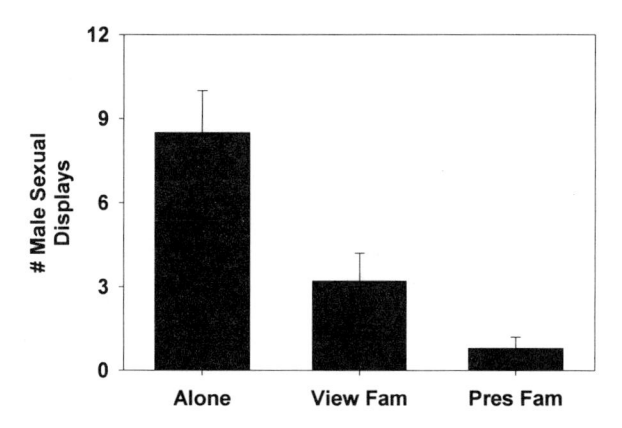

Figure 11.2
The number of sexual displays ("genital control") displayed by male common marmosets during an encounter with an unfamiliar breeding female. Testing occurred either while the pair was alone in the cage (Alone) while the pair could view the female's home group (but not vice versa; View Fam), or while in the physical presence of the female's family group (Pres Fam). (Modified from Anzenberger 1985.)

In a slightly different design, Anzenberger (1985) arranged encounters between all possible heterosexual combinations of unfamiliar breeding males and females (and also adult-aged but subdominant male and female offspring; see Abbott 1993) under three distinct social contexts: in the first, animals were tested with an unfamiliar partner alone (alone); in the second context, the familiar partners were visible to the interacting strangers (but not vice versa) through a one-way screen (view family); and finally, familiar partners were visible to the unfamiliar pair through a wire mesh screen (presence family). Both alpha and subdominant males displayed sexual behavior during encounters with unfamiliar females, and in all cases rates of behavior were highest in the alone condition and lowest in the presence family condition (figure 11.2). Huddling behavior, an index of affiliation in heterosexual partners, was also high when unfamiliar males and females were tested in the alone condition and decreased dramatically in presence family testing. Thus, in both studies with common marmosets, the presence of a long-term pairmate is associated with low sociosexual interactions with unfamiliar individuals. However, providing marmosets the opportunity to interact with opposite-sex strangers in the absence of the pairmate dramatically changes patterns of interactions, and there is no evidence of selective affiliation or fidelity to the partner.

Several studies have tested partner preference in callitrichid primates utilizing a simultaneous-choice paradigm, where tamarins can make a simultaneous choice between a familiar vs. unfamiliar partner. Epple (1990) provided saddleback tamarins

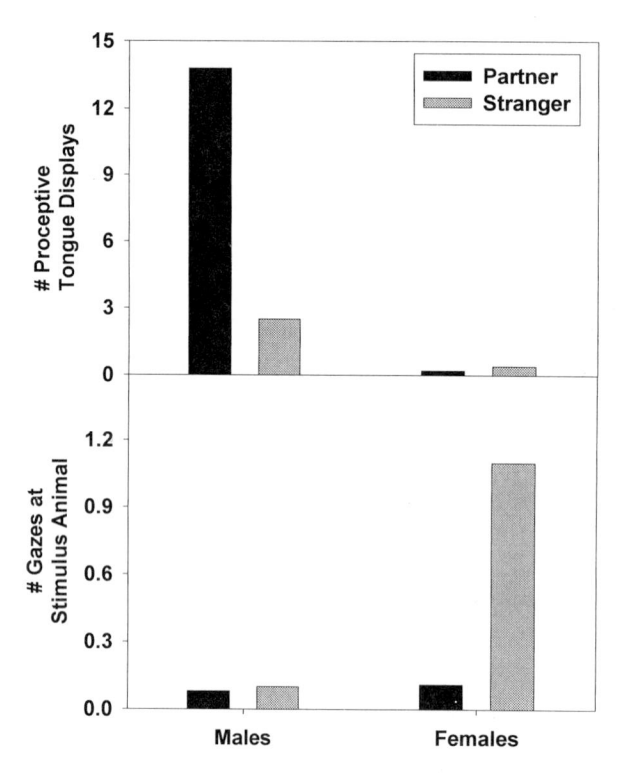

Figure 11.3
Mean rates of proceptive tongue displays (top panel) and gazes (bottom panel) directed toward long-term pairmates (Partner) or toward unfamiliar animals of the opposite sex (Stranger) in saddleback tamarins. (Modified from Epple, 1990.)

(*Saguinus fuscicollis*) with a choice between their pairmate vs. a stranger in a modified Y-maze. Evidence for male social fidelity was obtained, as males spent significantly more time in close proximity to their mates, approached them more frequently, and sniffed them more often than strange females. Males displayed sexual solicitation patterns toward both stimulus females, but directed significantly more tongue flicks toward their partner than toward the strange female (figure 11.3). Females, however, exhibited no significant preferences for their mates relative to a strange male, and in fact, directed more proceptive "gazes" at strange males than toward their mates. Thus, in the saddleback tamarin there are important sex differences in the manifestation of partner preferences, with males exhibiting more pronounced preferences than females for the long-term mate.

The impact of social context on the expression of partner preference in lion tamarins was revealed in a study in which males and females were given the opportunity to interact with a strange tamarin of the opposite sex in three conditions (Inglett et al. 1990). In the first, tamarins could interact with the stranger while in visual contact with the pairmate (in mate's view). In the second condition, interactions with the stranger could occur, but a visual barrier prevented the partner from viewing the interactions (out of mate's view). In the final condition, lion tamarins could interact with a stranger of the opposite sex in the absence of the pairmate (alone). In the in mate's view condition, both males and females exhibited significant preferences for their mates, and showed low levels of affiliative interactions with the stranger (figure 11.4). However, interactions with strangers of the opposite sex were clearly influenced by context: lion tamarins of both sexes approached strangers more, spent more time in close proximity to strangers, and sniffed them more frequently when the pairmate was either out of sight or out of the testing situation (see figure 11.4). Once again, preference for strangers of the opposite sex in pair-bonded tamarins is highly dependent upon the context in which it is tested.

In the final study in this series, the strength of partner attachment was tested in red-bellied tamarins (*Saguinus labiatus*) by giving them a choice between spending time with either their heterosexual partner, an unfamiliar partner of the opposite or same sex, or an empty cage (Buchanan-Smith and Jordan 1992). Both males and females displayed preference for their partner when the alternative choice was an empty cage of an individual of the same sex as the subject. In contrast, however, when the alternative choices were partner or opposite-sex stranger, tamarins spent equivalent amounts of time with each stimulus animal.

Together, the experiments reviewed in this section do not provide convincing evidence for strong and selective partner preference in "pair-bonded" marmosets and tamarins, at least when assessed under certain contextual conditions (i.e., the absence of the partner). The absence of selective preferences may facilitate the formation of new sociosexual relationships upon the death or disappearance of a long-term partner. In addition, even within an existing pair relationship, weak partner preferences (or, alternatively, attraction to opposite-sex strangers) may allow marmosets and tamarins to engage in opportunistic matings with individuals other than the pairmate. The ability to pursue a mixed reproductive strategy (monogamy with opportunistic matings outside the pair) has obvious implications for male reproductive success, and may benefit females through increased genetic diversity in offspring or selection for good genes. The data from laboratory partner preference tests are consistent with recent observations on copulations outside the pair in wild callitrichids, particularly during encounters between neighboring groups (for review, see French

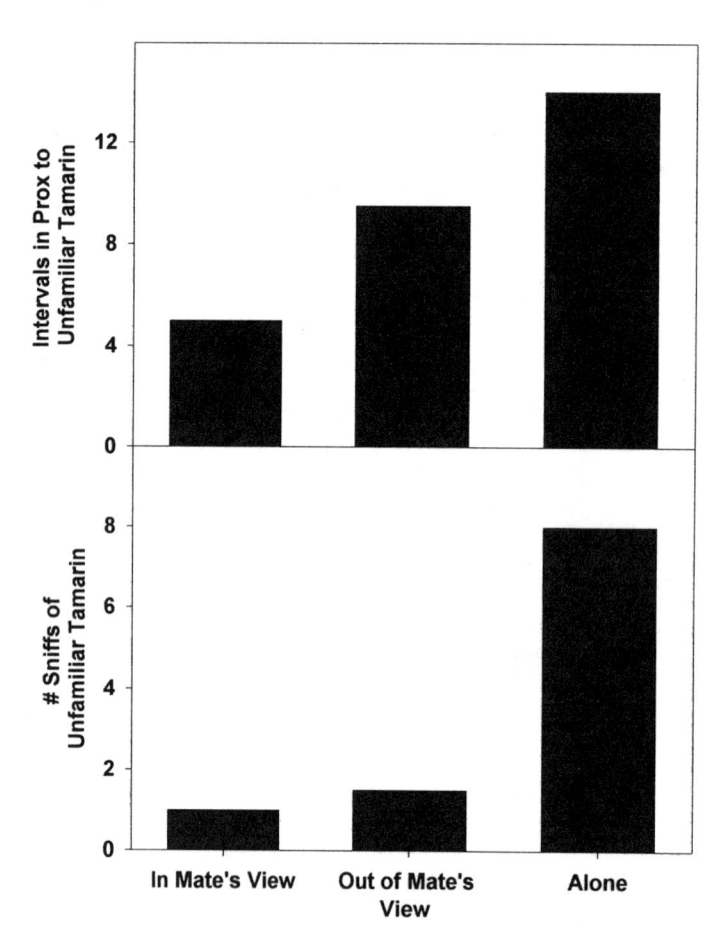

Figure 11.4
Mean number of intervals in proximity to a stranger of the opposite sex (top panel) and mean number of olfactory investigations of the stranger in lion tamarins. Responses to opposite-sex strangers occurred during visual access of the pairmate (In Mate's View), while the pairmate's view was obstructed (Out of Mate's View), and while tested alone (Alone) with the stranger. (Modified from Inglett et al. 1990.)

1997). Therefore, selective preference for long-term partners in callitrichid primates is demonstrable under the appropriate testing conditions, but it is equally clear that this trait is considerably more plastic in marmosets and tamarins than in prairie voles.

Intolerance for Same-Sex Conspecifics

The process of pair-bonding in prairie voles is associated with two dramatic changes in behavioral states (for review, see Carter and Roberts 1997). First, as described above for callitrichid primates, social preference for a particular partner develops. Second, and concurrent with the development of preference patterns for the mate, is the development of intolerance for same-sex individuals, expressed in high levels of aggression toward intruders or opponents in staged encounters. This behavioral change, particularly prominent in males, may be important as an extrinsic factor that, when combined with partner preferences, helps to maintain social and sexual exclusivity in the monogamous voles.

A host of studies with callitrichid primates have utilized the "intruder" or "stranger" paradigm to test intolerance for nonpair conspecifics. Most of the studies have contrasted the responses of both male and female residents to the presentation of either male or female intruders, and monitored the responses of each member of the resident pair separately. The results of these studies are summarized in table 11.3. While the results differ slightly among species, there are several consistent themes that run through the patterns of results. First, as in the prairie vole, there is a consistent pattern of intrasexual aggression toward strangers in callitrichid primates. With few exceptions, same-sex aggression is consistently higher than between-sex

Table 11.3
Species differences in intolerance and aggression toward same-sex strangers in intruder encounters

Species	Same-Sex Aggression?	More Aggressive Sex	Intersexual Aggression?	References
Saddleback tamarin (*Saguinus fuscicollis*)	Yes	M ≈ F	Low	Epple 1981; Epple & Alveario 1985
Cotton-top tamarin (*Saguinus oedipus*)	Yes	M > F	No	French & Snowdon 1981; Harrison & Tardif 1988
Common marmoset (*Callithrix jacchus*)	Yes	M ≈ F	Low	Epple 1970; Sutcliffe & Poole 1984; Anzenberger 1985
Golden lion tamarin (*Leontopithecus rosalia*)	Yes	F > M	No	French & Inglett 1989
Tufted-ear marmoset (*Callithrix kuhli*)	Yes	F > M*	Low	French et al. 1995; Schaffner & French 1997

*Sex difference in aggression only apparent in large groups (breeders plus offspring).

aggression. Only in cases where there are extremely high levels of agonistic interactions in within-sex encounters is the expression of between-sex aggression apparent during intruder encounters. Second, there are consistent species differences in the sex that displays higher agonistic intolerance of intruders. In cotton-top tamarins, males show higher levels of aggression toward intruders than do females. In saddleback tamarins and common marmosets, no pronounced sex differences are observed during intruder encounters. In contrast, however, female tufted-ear marmosets and golden lion tamarins tend to express higher levels of aggressive behavior toward intruders than resident males do. These taxonomic variants in sex-based aggressiveness are correlated with species differences in the mechanisms of reproductive control of subordinates. In cotton-top and saddleback tamarins, subordinate females are prevented from breeding by pronounced physiological suppression of ovulatory function (French et al. 1984; Epple and Katz 1984; Ziegler et al. 1987). In marmosets and lion tamarins, subordinate females can express ovulatory function, and fertility regulation appears to be based on behavioral, rather than physiological, mechanisms (French and Inglett 1989; French et al. 1989; Saltzman et al. 1997a,b; Smith et al. 1997).

The response of callitrichid primates to intruders suggests a common tendency toward intolerance of same-sex conspecifics. This intolerance of same-sex conspecifics may be as important a determinant of social and sexual exclusivity for a pair as the intrinsic sociosexual attachment that develops between members of the pair. Thus, factors extrinsic to the pair, and independent of the socioemotional attachment between partners in the pair, may determine the nature of the reproductive unit in callitrichid primates.

Contextual Influences on Reproductive Behavior

This section reviews evidence from marmosets and tamarins that suggests that context is a critical determinant of the timing, frequency, and intensity of sociosexual interactions. We cover the following variables that have been shown to affect the expression of affiliative and sexual interactions: the length of time the pair has resided together, the demography of the group in which the interactions occur, the relative social status of the participants, and, finally, the interaction between ovarian status and contextual factors.

Duration of the Pair Bond Among the most prominent features that influence the patterning of sexual behavior in callitrichid primates is the length of time the pair has been together. In all species studied to date, rates of sexual interactions are higher in newly paired animals than in pairs that have resided together for months or years

(Kleiman 1977; Evans and Poole 1984; Savage et al. 1988; Ruiz 1990). Our laboratory recently examined, in some detail, the time course of changes in social and sexual behavior in black tufted-ear marmosets (*Callithrix kuhli*) as the relationship between the male and female developed over the first 80 days of cohabitation (Schaffner et al. 1995). We monitored interactions in six first-time pairmates, and plotted the developmental course of the social and sexual relationship. We discovered that these two key features of the relationship followed dramatically different trajectories (figure 11.5). Affiliative behavioral patterns, such as grooming, proximity, and intrapair contact calls, developed gradually over the course of the first 80 days—in other words, the social dimension of the relationship emerged only with the passage of time. In contrast, the sexual dimension of the relationship between the pairmates was not dependent upon a close social relationship, since copulatory activity was high

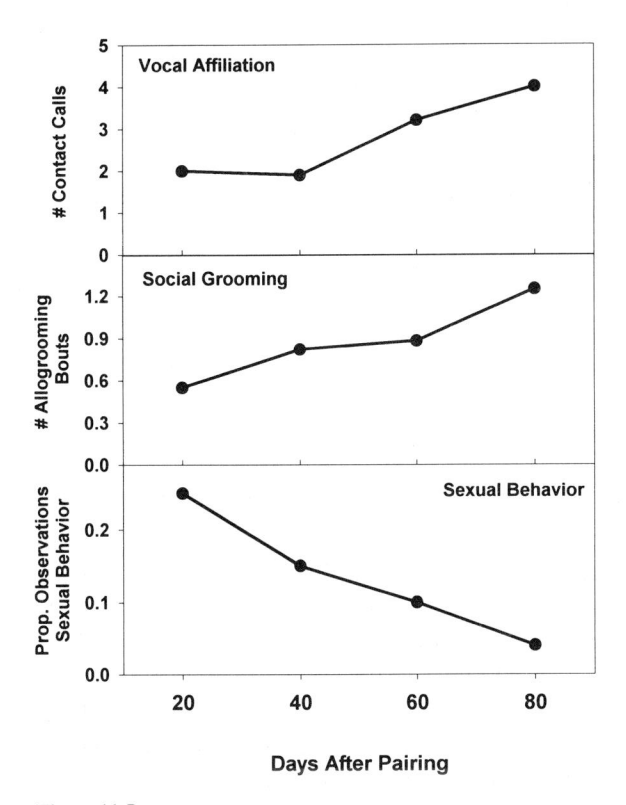

Figure 11.5
Development of social and sexual aspects of heterosexual relationship in tufted-ear marmosets during the first 80 days of pairing. (Modified from Schaffner et al. 1995.)

upon initial pairing and decreased steadily throughout the 80-day period (regardless of the female's reproductive status).

We believe that these data are telling with regard to the impact of pair duration on the sociosexual relationship. The social and sexual aspects are clearly regulated by different processes, with the social (and possibly emotional) aspects of the relationship emerging only after an opportunity for repeated interactions with the pairmate. The sexual aspects of the relationship are independent of these processes, emerging in full form shortly after pairing. Kleiman (1977) has reminded students of monogamy, "mating exclusivity has genetic consequences, whereas an emotional bond does not," a prescient commentary that anticipated the pattern of results we report here.

Group Demography We know, as a result of a host of field studies conducted in the last decade, that reproductive behavior in callitrichid social groups is expressed in a variety of demographic contexts. Two studies have experimentally manipulated group demography to explore the changes this variable produces in the patterning of reproductive behavior. Because of the difficulty of housing multiple females together in captivity, both studies have examined the impact of multiple males on patterns of reproductive behavior. Kleiman (1978) studied lion tamarins in two male–one female trios, and noted rates of sexual and social behavior. In the seven trios in which neither male was related to the female, there was a clear differentiation between males in both social and sexual access to the female. In some trios, both males received sexual access to the female, but one male had higher rates of sexual interactions (0.51 ± 0.16 mount per 30 minutes for the sexually active male vs. 0.03 ± 0.02 mount per 30 minutes for the sexually inactive male). The sexually active male tended to have greater social access to the female as well, spending more time huddling, grooming, and sniffing the female than the sexually inactive male.

In a more recent study on group demography, Schaffner (1996) explicitly contrasted the development of sociosexual relationships in heterosexual pairs with the pattern of formation observed in two male—one female polyandrous trios. Eight trios were created, of which five were eventually stable. The three unstable trios lasted only 2 hours, 2 days, and 4 days, respectively, before high levels of male—male aggression necessitated the removal of one or both males from the group. All three unstable groups were composed of two unrelated males (who had nonetheless been cohabiting with each other in the absence of a female for weeks to months without incident). The five successful polyandrous trios consisted of related males (brothers or fathers and sons). These data certainly suggest that kinship between males is critical with regard to the origins of polyandrous groups in wild populations of marmosets and tamarins (Goldizen 1987).

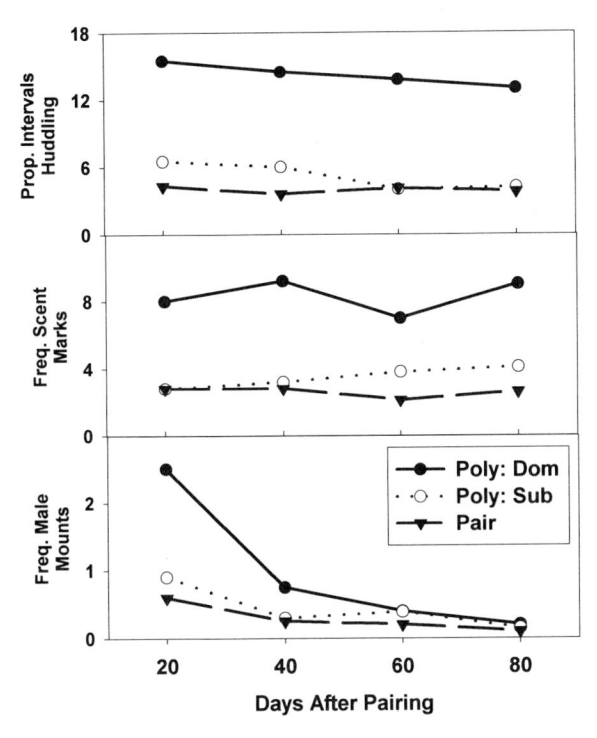

Figure 11.6
Development of sociosexual relationships in tufted-ear marmoset males while housed in a traditional heterosexual pair (Pair) or as one of two males in a two male–one female polyandrous trio. Poly: Dom, male with highest levels of sexual access to the female; Poly: Sub, male in trio with lowest levels of sexual access to the female. (Modified from Schaffner 1996.)

Contrasting the patterns of male-female interactions in pair males vs. males who must compete for access to females reveals a fairly dramatic impact of group demography upon social and sexual behavior. As shown in figure 11.6, males who interact with females in the presence of a potential reproductive competitor spend less time huddling with their pairmates than males in pairs, have higher rates of scent marking (at least the dominant male), and tend to engage in higher rates of sexual activity.

Social Status The influence of subordinate social status on the suppression of ovarian and testicular function in callitrichid primates is well established (for review, see French 1997; Abbott 1993), and it is not surprising to see a correspondingly low rate of sexual behavior in these suppressed individuals. However, there is growing recognition among workers in the field that not all subordinates experience endocrine insufficiency (e.g., see Smith et al. 1997; Saltzman et al. 1997a) yet these individuals continue to experience low rates of sexual behavior. Saltzman et al. (1997b) have recently suggested that profound behavioral inhibitions against subordinate sexuality are common in callitrichid social groups, and contribute to low levels of sexual behavior in animals other than the alpha pair. This hypothesis seems reasonable, since removal of the adult-aged subordinate individuals from their natal group and providing them with unrelated partners of the opposite sex (Hubrecht 1989; Abbott 1986; Inglett 1993) or removing one of the breeding adults (French et al. 1984; Heistermann et al. 1989; Saltzman et al. 1997b) can lead to immediate activation of sexual behavior.

Interaction of Social Context and Ovarian Status Although there have been claims that primate sexual behavior is liberated from the influence of gonadal hormones, there is considerable evidence that the quality and quantity of sexual behavior are profoundly influenced by endocrine variables, especially the ovarian hormones estrogen and progesterone, as discussed by Wallen in chapter 10. However, it is equally clear that physical and especially social contexts can substantially modify the way in which ovarian hormones influence the patterning of reproductive behavior across the ovarian cycle (Wallen 1982; Michael and Zumpe 1988; Wallen and Winston 1984; Wallen et al. 1984). Given the intraspecific variability in group structure and demography among callitrichids, then, this group should be an ideal candidate for elucidating contextual—endocrine interactions.

Early studies addressing the influence of the ovarian cycle on sociosexual dynamics generally reported a lack of association between ovarian status (assessed directly through excretion of urinary steroids or indirectly by backdating from parturition) and sexual behavior (Hampton et al. 1966; Rothe 1975; French 1982; Brand and Martin 1983; Stribley et al. 1987). There has been one clear demonstration of endocrine influences on male-female interactions. In the mid-1980s Kendrick and Dixson (1984) systematically evaluated the influence of spontaneous and experimentally induced variation in ovarian steroid and interaction patterns in male and female tamarins brought together for pair tests three times per week for 30 minutes. During normal ovarian cycles, female proceptive and receptive tongue flicking, stares, and male mount and ejaculations were higher during the periovulatory period than during the follicular or luteal phase. The proportion of male mounts that led to successful intromission was lower in the luteal phase of the cycle than during other phases (Kendrick and Dixson 1983). Ovariectomy did not eliminate female proceptive behavior completely (Kendrick and Dixson 1984), but treatment of ovariectomized females with estradiol increased female proceptive behavior and reduced mount refusals (Kendrick and Dixson 1985; Dixson 1986) and treatment with progesterone reduced proceptive behavior and increased mount refusals (Kendrick and Dixson 1985). That social context plays an important role in shaping the way in which ovarian steroids influence behavior is shown by a later study involving long-term pairmates. The link between ovarian status and mating behavior was much less pronounced under these conditions than had been noted in the pair-test paradigm, with both proceptive behavior and copulations distributed more evenly throughout the follicular, periovulatory, and luteal phases of the cycle (Dixson and Lunn 1987).

Recent studies on the roles of social context and ovarian status on sociosexual behavior in lion tamarins also point out the equipotent roles of exogenous and endogenous regulation of sexual behavior in this species. To explore the role of social context, Inglett (1993) systematically varied the extent to which the pair was allowed to develop a long-term social relationship. Males and females were allowed access to each other for 30 minutes three times per week in three conditions. In the first, females were paired with a random succession of vasectomized male partners for a series of 30 pair tests, then returned to their original isosexual housing (random pairing). In the second condition, females were paired with the identical vasectomized partner for all thirty tests (constant pairing; although, as in the first condition, the males and females lived in separate, isosexual housing conditions). In the final condition, females resided on one half of a cage which contained an unrelated and vasectomized adult male on the other side of a wire mesh partition (cohabitation). The partners could smell, see, touch, groom, and huddle near each other (essentially all social interactions with the exception of copulation were possible). Like the lion tamarins in the first condition, pairs were given 30 minutes of access to each other for a total of thirty trials. In all cases, female reproductive status was assessed by monitoring the excretion of urinary steroids, and cycle phases were divided into follicular, periovulatory, and luteal on the basis of this cycle (French and Stribley 1985; French et al. in press).

Inglett's (1993) results are shown in figure 11.7, and they reveal that the most influential factor regulating the expression of sociosexual behavior was the quality of the relationship between males and females. Affiliative behavioral patterns, including huddling, grooming, and time in proximity, increased in the cohabitation condition, relative to random and constant pairing. Likewise, attempted mounts, mounts, and copulations also increased in the cohabitation phase of the study. Ovarian status also was a potent influence on sexual behavior, with more mounting by males during the follicular and periovulatory phase than in the luteal phase. Although partner familiarity affected overall levels of sexual behavior, increasing social familiarity did not alter the overall pattern of ovarian influence on behavior. Thus, in both common

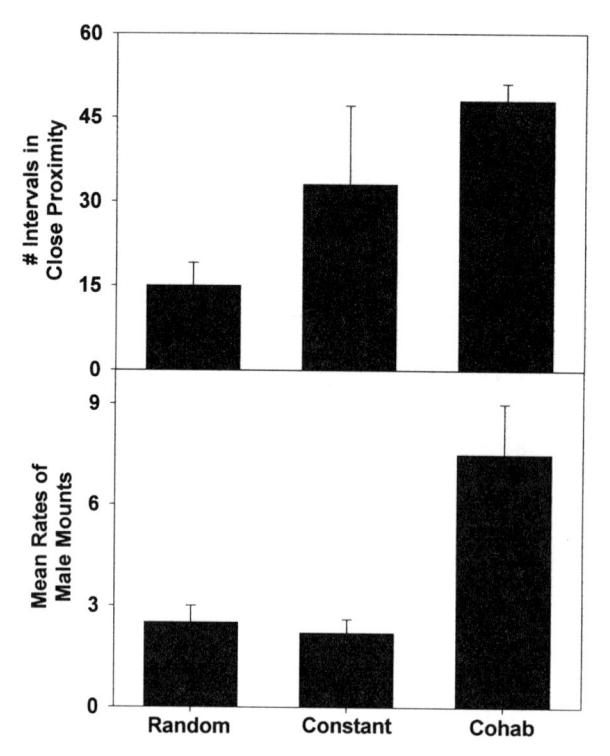

Figure 11.7
Proximity (top panel) and male mounting behavior (bottom panel) in heterosexual pairs of lion tamarins. Social familiarity with the partner was manipulated by testing randomly selected partners (Random), by pairing the same noncohabiting partners throughout the testing period (Constant), and by testing cohabiting males and females (Cohab) over thirty tests. (Modified from Inglett 1993.)

marmosets and lion tamarins, social context can exert profound regulatory influences on sexual behavior, in ways that are independent of, or orthogonal to, the effects of ovarian steroids.

Recent work on the socioendocrinology of ovarian cycles in long-term pairs has revealed that there are subtle behavioral (Converse et al. 1995) and olfactory (Ziegler et al. 1993) means by which male and female callitrichids may coordinate sexual interactions with the periovulatory phase of the ovarian cycle. However, given the fact that sexual behavior is expressed throughout all phases of reproduction, including luteal and pregnancy stages, combined with the fact that the temporal patterning of sexual behavior is so easily modified by social factors, it seems safe to conclude that context will continue to be recognized as perhaps the most important determinant of sociosexual behavior in callitrichids.

Conclusions

Our tour through the behavioral biology of monogamy in primates has revealed some important differences from the pattern that emerges from the study of arvicoline rodents, especially the prairie vole. A preliminary conclusion could be phrased simply: marmosets and tamarins are not prairie voles. Three features would characterize a more formal summary of the status of monogamy in primates. First, the evidence for monogamy throughout certain taxonomic groups in primates based primarily on demographic data has been weakened for many of these species, as reputable reports of departures from sexual monogamy become more prevalent in the literature. It is likely that more species will fall from the pedestal of monogamy as information on the distribution of paternity within social groups becomes available from molecular genetic analyses. Second, it is equally clear that there is considerable intraspecific variation in the likelihood of expressing monogamous social and mating relationships. The source(s) of this variability, whether ecological, demographic, or dispositional in origin, remains a central question for the future. Third, patterns of partner preference and the patterning of sexual behavior, especially among the callitrichid primates, is exquisitely sensitive to social contexts. Having stated that nonhuman primates are not voles, it is ironic to find that recent work on variation in monogamy and its associated traits in Kansas-derived prairie voles suggests that this species may be more like marmosets and tamarins than previously suspected (Roberts and Carter 1997; Dharmadhikari et al. 1997). Thus, in nonhuman primates, as well as in voles, social context may be the proximate trigger for determining which social and mating strategy, of multiple reproductive strategies available, is selected by individuals living within groups. Appreciating the significance of intraspecific variability and the sensitivity of monogamy to context, both on a functional and mechanistic level, may be the Rosetta stone to understanding the mysteries posed by mammalian monogamy.

Acknowledgments

We acknowledge the contributions of Jeff Fite, Jennifer Hunter, and Dan Jorgensen in locating the diverse and sometimes arcane literature on monogamy (or the lack thereof) in primates. We thank Kim Wallen and Jill Schneider for their invitation to participate in this effort, and for encouragement and feedback on the ideas presented in this chapter. Tessa E. Smith also provided useful comments on previous versions of the manuscript. The preparation of this chapter, and the work on black tufted-ear marmosets and lion tamarins described herein, was supported by funds from the National Science Foundation awarded to J.A.F (CRB 90-00094, IBN 92-09528, OSR 92-55225, and IBN 97-23842). J.A.F thanks his long-term pairmate, Mary, and the products of their sociosexual bond, Aaron and Anna, for moral support during the preparation of this work.

References

Abbott DH (1986). Social suppression of reproduction in subordinate marmoset monkeys (*Callithrix jacchus jacchus*). *In* A Primatologia no Brasil (de Mello MT, ed), pp 1–16. Brasilia: Sociedade Brasileira de Primatologia.

Abbott DH (1993). Social conflict and reproductive suppression in marmoset and tamarin monkeys. *In* Primate Social Conflict (Mason WA, Mendoza SP, eds), pp 331–372. Albany: State University of New York Press.

Ahsan F (1995). Fighting between two females for a male in the hoolock gibbon. Int J Primatol 16:731–737.

Anzenberger G (1985). How stranger encounters of common marmosets (*Callithrix jacchus jacchus*) are influenced by family members: the quality of behavior. Folia Primatol 45:204–224.

Anzenberger G (1992). Monogamous social systems and paternity in primates. *In* Paternity in Primates: Genetic Tests and Theories (Martin RD, Dixson AF, Wickings EJ, eds), pp 203–224. Basel: Karger.

Bleisch W, Chen N (1991). Ecology and behavior of wild black-crested gibbons (*Hylobates concolor*) in China with a reconsideration of evidence for polygyny. Primates 32:539–548.

Brand HM, Martin RD (1983). The relationship between female urinary estrogen excretion and mating behavior in cotton-topped tamarins, *Saguinus oedipus oedipus*. Int J Primatol 4:275–290.

Buchanan, HM, Mittermeier RA, Roosmalen MGM (1981). The saki monkeys, genus *Pithecia*. *In* Ecology and Behavior of Neotropical Primates, vol 1 (Coimbra-Filho AF, Mittermeier RA, eds). Rio de Janeiro: Academia Brasileira de Ciencias.

Buchanan-Smith HM, Jordan TR (1992). An experimental investigation of the pair bond in the callitrichid monkey, *Saguinus labiatus*. Int J Primatol 13:51–72.

Burke T, Davies NB, Bruford MW, Hatchwell BJ (1989). Parental care and mating behaviour of polyandrous dunnocks (*Prunella modularis*) related to paternity by DNA fingerprinting. Nature 338:249–251.

Carter CS, Roberts RL (1997). The psychobiological basis of cooperative breeding in rodents. *In* Cooperative Breeding in Mammals (Solomon NG, French JA, eds), pp 231–266. Cambridge: Cambridge University Press.

Carter CS, DeVries AC, Getz LL (1995). Physiological substrates of monogamy: the prairie vole model. Neurosci Biobehav Rev 19:303–314.

Carter CS, DeVries AC, Taymans SE, Roberts RL, Williams JR, Getz LL (1997). Peptides, steroids, and pair bonding. *In* The Integrative Neurobiology of Affiliation (Carter CS, Ledenhendler II, Kirkpatrick B, eds), pp 260–272. New York: New York Academy of Sciences.

Chivers DJ (1984). Feeding and ranging in gibbons: a summary. *In* The Lesser Apes: Evolutionary and Behavioral Biology (Preuschoft H, Chivers DJ, Brockelman W, Creel N, eds). Edinburgh: Edinburgh University Press.

Converse, LJ; Carlson, AA; Ziegler, TE & Snowdon, CT (1995). Communication of ovulatory state to mates by female pygmy marmosets, *Cebuella pygmaea*. Anim Behav 49:615–621.

Dewsbury DA (1988). The comparative psychology of monogamy. *In* Comparative Perspectives in Modern Psychology (Leger DW, ed), pp 1–50. Lincoln: University of Nebraska Press.

Dharmadhikari A, Lee YS, Roberts RL, Carter CS (1997). Exploratory behavior correlates with social organization and is responsive to peptide injections in prairie voles. *In* The Integrative Neurobiology of Affiliation (Carter CS, Ledenhendler II, Kirkpatrick B, eds), pp 610–612. New York: New York Academy of Sciences.

Dixson AF (1982). Some observations on the reproductive physiology and behaviour of the owl monkey. Int Zool Yearbook 22:115–119.

Dixson AF (1986). Proceptive displays of the female common marmoset (*Callithrix jacchus*): effects of ovariectomy and oestradiol 17b. Physiol Behav 36:971–973.

Dixson AF, Fleming D (1981). Parental behaviour and infant development in owl monkeys (*Aotus trivirgatus griseimembra*). J Zool Lond 194:25–39.

Dixson AF, Lunn SF (1987). Post-partum changes in hormones and sexual behaviour in captive groups of marmosets (*Callithrix jacchus*). Physiol Behav 41:577–583.

Easley SP, Kinzey WG (1986). Territorial shift in the yellow-handed titi monkey (*Callicebus torquatus*). Am J Primatol 11:307–318.

Epple G (1970). Quantitative studies on scent marking in the marmoset *Callithrix jacchus*. Folia Primatol 13:48–62.

Epple G (1981). Effect of pair-bonding with adults on the ontogenetic manifestation of aggressive behavior in a primate, *Saguinus fuscicollis*. Behav Ecol Sociobiol 8:117–123.

Epple G (1990). Sex differences in partner preferences in mated pairs of saddle-back tamarins. Behav Ecol Sociobiol 27:455–459.

Epple G, Alveario MC (1985). Social facilitation of agonistic responses to strangers in pairs of saddleback tamarins (*Saguinus fuscicollis*). Am J Primatol 9:207–218.

Epple G, Katz Y (1984). Social influences on estrogen excretion and ovarian cyclicity in saddle-back tamarins (*Saguinus fuscicollis*). Am J Primatol 6:215–227.

Evans S (1983). The pair bond of the common marmoset, *Callithrix jacchus jacchus*: an experimental investigation. Anim Behav 31:651–658.

Evans S, Poole TB (1984). Long-term changes and maintenance of the pair-bond in common marmosets, *Callithrix jacchus jacchus*. Folia Primatol 42:33–41.

French JA (1982). The Role of Scent Marking in Social and Sexual Communication in the Tamarin, *Saguinus oedipus*. Ph.D. thesis, University of Wisconsin, Madison.

French JA (1997). Regulation of singular breeding in callitrichid primates. *In* Cooperative Breeding in Mammals. Solomon NG, French JA, eds. (pp 34–75). Cambridge: Cambridge University Press.

French JA, Inglett BJ (1989). Female-female aggression and male indifference in response to unfamiliar intruders in lion tamarins. Anim Behav 37:487–497.

French JA, Snowdon CT (1981). Sexual dimorphism in responses to unfamiliar intruders in the tamarin, *Saguinus oedipus*. Anim Behav 29:822–829.

French JA, Stribley JA (1985). Patterns of urinary oestrogen excretion in female golden lion tamarins (*Leontopithecus rosalia*). J Reprod Fert 75:537–546.

French JA, Abbott DH, Snowdon CT (1984). The effect of social environment on estrogen excretion, scent marking, and sociosexual behavior in tamarins (*Saguinus oedipus*). Am J Primatol 6:155–167.

French JA, Inglett BJ, Dethlefs TM (1989). The reproductive status of nonbreeding group members in captive golden lion tamarin social groups. Am J Primatol 18:73–86.

French JA, Schaffner CM, Shepherd RE, Miller ME (1995). Familiarity with intruders modulates agonism toward outgroup conspecifics in Wied's black tufted-ear marmoset (*Callithrix kuhli*: Primates, Callitrichidae). Ethology 99:24–38.

French JA, Smith TE, Vleeschouwer K, Elsacker L, Heistermann M, Monfort SL, Ribeiro E, Pissinatti A, Bales K (in press). Reproductive endocrinology and behavior in female lion tamarins (*Leontopithecus*). *In*

Biology and Conservation of Lion Tamarins: Twenty-Five Years of Research (Kleiman DG, Rylands AB, Santos IB, eds). Washington DC: Smithsonian Press.

Gautier-Hion A, Gautier JP (1978). Le singe de Brazza: un stratégie originale. Z Tierpsychol 26:84–104.

Gittins SP, Raemaekers JJ (1980). Siamang, lar and agile gibbons. *In* Malayan Forest Primates (Chivers DJ, ed), pp 63–106. New York: Plenum Press.

Goldizen AW (1987). Tamarins and marmosets: communal care of offspring. *In* Primate Societies (Smuts BB, Cheney DL, Seyfarth RM, Wrangham RW, Struhsaker TT, eds), pp 34–43. Chicago: University of Chicago Press.

Goldizen AW, Terborgh J (1989). Demography and dispersal patterns of a tamarin population: possible causes of delayed breeding. Am Naturalist 134:208–224.

Gowaty PA (1981). An extension of the Orians-Verner-Willson model to account for mating systems besides polygyny. Am Naturalist 118:851–859.

Haimoff EH, Yang X-J, He S-J, Chen N (1987). Preliminary observations of wild black-crested gibbons (*Hylobates concolor concolor*) in Yunan province, People's Republic of China. Primates 28:319–335.

Hampton JK, Hampton SH, Landwehr BT (1966). Observations on a successful breeding colony of the marmoset, *Oedipomidas oedipus*. Folia Primatol 4:265–287.

Harrison ML, Tardif SD (1988). Kin preference in marmosets and tamarins: *Saguinus oedipus* and *Callithrix jacchus* (Callitrichidae, Primates). Am J Phys Anthropol 77:377–384.

Heistermann M, Kleis E, Pröve E, Wolters, H-J (1989). Fertility status, dominance, and scent marking behavior of family-housed female cotton-top tamarins (*Saguinus oedipus*) in the absence of their mothers. Am J Primatol 18:177–189.

Hoffman KA, Mendoza SP, Hennessy MB, Mason WA (1995). Responses of infant titi monkeys, *Callicebus moloch*, to removal of one or both parents: evidence for paternal attachment. Dev Psychobiol 28:399–407.

Hubrecht RC (1989). The fertility of daughters in common marmoset (*Callithrix jacchus*) family groups. Primates 30:423–432.

Inglett BJ (1993). The Role of Social bonds and the Female Reproductive Cycle on the Regulation of Social and Sexual Interactions in the Golden Lion Tamarin (*Leontopithecus rosalia rosalia*). Ph D thesis, University of Nebraska.

Inglett BJ, French JA, Dethlefs TM (1990). Patterns of social preference across different social contexts in golden lion tamarins (*Leontopithecus rosalia*). J Comp Psychol 104:131–139.

Insel TR, Shapiro LE (1992). Oxytocin receptor distribution reflects social organization in monogamous and polygamous voles. Proc Nat Acad Sci U S A 89:5981–5985.

Insel TR, Young L, Wang Z (1997). Molecular aspects of monogamy. *In* The Integrative Neurobiology of Affiliation (Carter CS, Ledenhendler II, Kirkpatrick B, eds), pp 302–316. New York: New York Academy of Sciences.

Kappeler PM (1997). Intrasexual selection in *Mirza coquereli*: evidence for scramble competition polygyny in a solitary primate. Behav Ecol Sociobiol 45:115–127.

Kendrick KM, Dixson AF (1983). The effect of the ovarian cycle on the sexual behaviour of the common marmoset (*Callithrix jacchus*). Physiol Behav 30:735–742.

Kendrick KM, Dixson AF (1984). Ovariectomy does not abolish proceptive behaviour cyclicity in the common marmoset (*Callithrix jacchus*). J Endocrinol 101:155–162.

Kendrick KM, Dixson AF (1985). Effects of oestradiol 17β, progesterone and testosterone upon proceptivity and receptivity in ovariectomized common marmosets (*Callithrix jacchus*). Physiol Behav 34:123–128.

Kinzey WG (1997). Synopsis of New World primates (16 genera). *In* New World Primates: Ecology, Evolution, and Behavior (Kinzey WG, ed), pp 169–305. New York: Aldine de Gruyter.

Kleiman DG (1977). Monogamy in mammals. Q Rev Biol 52:39–69.

Kleiman DG (1978). The development of pair preferences in the lion tamarins (*Leontopithecus rosalia*): male competition or female choice? *In* Biology and Behaviour of Marmosets (Rothe H, Wolters H-J, Hearn JP, eds), pp 181–188. Göttingen: Eigenverlag-H. Rothe.

Klopfer PH, Boskoff KJ (1982). Maternal behavior in prosimians. *In* The Study of Prosimian Behavior (Doyle GA, Martin RD, eds). New York: Academic Press.

Leighton DR (1987). Gibbons: territoriality and monogamy. *In* Primate Societies (Smuts BB, Cheney DL, Seyfarth RM, Wrangham RW, Struhsaker TT, eds), pp 135–145. Chicago: University of Chicago Press.

Kleiman DG (1977). Monogamy in mammals. Q Rev Biol 52:39–69.

Martin RD, Dixson AF, Wickings EJ (1992). Paternity in Primates: Genetic Tests and Theories. Basel: Karger.

Mason WA (1966). Social organization of the South American monkey, *Callicebus moloch*: a preliminary report. Tulane Stud Zool 13:23–28.

Mendoza SP, Mason WA (1986). Contrasting responses to intruders and to involuntary separation by monogamous and polygynous New World monkeys. Physiol Behav 38:795–801.

Michael RP, Zumpe D (1984). Interactions of social, spatial and hormonal factors on the behavior of rhesus monkeys (*Macaca mulatta*). Primates 25:462–474.

Michael RP, Zumpe D (1988). Determinant of behavior rhythmicity during artificial menstrual cycles in rhesus monkeys (*Macaca mulatta*). Am J Primatol 15:157–170.

Mitani JC (1984). The behavioral regulation of monogamy in gibbons (*Hylobates muelleri*). Behav Ecol Sociobiol 15:225–229.

Mitani JC (1987). Territoriality and monogamy among agile gibbons (*Hylobates agilis*). Behav Ecol Sociobiol 20:265–269.

Müller H-K (1996). Emigration of a masked titi monkey (*Callicebus personatus*) from an established group, and the foundation of a new group. Neotrop Primates 4:19–21.

Pages E (1980). Ethoecology of *Microcebus coquereli* during the dry season. *In* Nocturnal Malagasy Primates (Charles-Dominique P, Cooper HM, Hladik A, Hladik CM, Pages E, Pariente GF, Petter-Rosseaux A, Schilling A, eds), pp 97–116. New York, Academic Press.

Palombit RA (1994a). Dynamic pair bonds in hylobatids: implications regarding monogamous social systems. Behaviour 128:65–101.

Palombit RA (1994b). Extra-pair copulations in a monogamous ape. Anim Behav 47:721–723.

Palombit RA (1996). Pair bonds in monogamous apes: a comparison of the siamang *Hylobates syndac*tylus and the white-handed gibbon *Hylobates lar*. Behaviour 133:321–356.

Piper WH, Evers DC, Meyer MW, Tischler KB, Kaplan JD, Fleischer RC (1997). Genetic monogamy in the common loon (*Gavia immer*). Behav Ecol Sociobiol 41:25–31.

Pollock JI (1975). Field observations on *Indri indri*: a preliminary report. *In* Lemur Biology (Tattersall I, Sussman RW, eds), pp 287–311. New York: Plenum Press.

Reichard U (1995). Extra-pair copulations in a monogamous gibbon (*Hylobates lar*). Ethology 100:99–112.

Roberts RL, Carter CS (1997). Intraspecific variation and the presence of a father can influence the expression of monogamous and communal traits in prairie voles. *In* The Integrative Neurobiology of Affiliation (Carter CS, Ledenhendler II, Kirkpatrick B, eds), pp 559–562. New York: New York Academy of Sciences.

Robinson JG, Wright PC, Kinzey WG (1987). Monogamous cebids and their relatives: intergroup calls and spacing. In: Primate Societies (Smuts BB, Cheney DL, Seyfarth RM, Wrangham RW, Struhsaker TT, eds), pp 44–53. Chicago: University of Chicago Press.

Rothe H (1975). Some aspects of sexuality and reproduction in groups of captive marmosets (*Callithrix jacchus*). Z Tierpsychol 37:255–273.

Ruiz JC (1990). Comparison of affiliative behavior between old and recently established pairs of golden lion tamarins, *Leontopithecus rosalia*. Primates 31:197–204.

Saltzman W, Schultz-Darken NJ, Abbott DH (1997a). Familial influences on ovulatory function in common marmosets (*Callithrix jacchus*). Am J Primatol 41:159–177.

Saltzman W, Severin JM, Schultz-Darken NJ, Abbott DH (1997b). Behavioral and social correlates of escape from suppression of ovulation in female common marmosets housed with the natal family. Am J Primatol 41:1–21.

Savage A, Ziegler TE, Snowdon CT (1988). Sociosexual development, pair bond formation, and mechanisms of fertility suppression in female cotton-top tamarins (*Saguinus oedipus oedipus*). Am J Primatol 14:345–359.

Savage A, Shideler S, Lasley B (1993). A preliminary report on the behavior and reproductive patterns of captive white-faced sakis (*Pithecia pithecia*). *In* 1993 Regional Studbook of the White-faced Saki (Vecchio T, Miller A, eds). Providence, RI: Roger Williams Park Zoo.

Schaffner CM (1996). Social and Endocrine Factors in the Establishment and Maintenance of Sociosexual Relationships in Wied's Black Tufted-Ear Marmosets (*Callithrix kuhli*). Ph D thesis, University of Nebraska.

Schaffner CM, French JA (1997). Group size and aggression: "recruitment incentives" in a cooperatively breeding primate. Anim Behav 54:171–180.

Schaffner CM, Shepherd RE, Santos CV, French JA (1995). Development of heterosexual relationships in Wied's black tufted-ear marmosets (*Callithrix kuhli*). Am J Primatol 36:185–200.

Schaik CP van, Dunbar RIM (1990). The evolution of monogamy in large primates: a new hypothesis and some crucial tests. Behaviour 115:30–62.

Smith TE, Schaffner CM, French JA (1997). Social modulation of reproductive function in female black tufted-ear marmosets (*Callithrix kuhli*). Horm Behav 31:159–168.

Smuts BB, Cheney DL, Seyfarth RM, Wrangham RW, Struhsaker TT (1986). Primate Societies. Chicago: University of Chicago Press.

Snowdon CT (1990). Mechanisms maintaining monogamy in monkeys. *In* Contemporary Issues in Comparative Psychology (Dewsbury DA, ed), pp 225–251. Sunderland, MA: Sinauer Press.

Srikosamatara S, Brockelman WY (1987). Polygyny in a group of pileated gibbons via a familial route. Int J Primatol 8:389–393.

Stanger K, Coffman B. Izard MK (1995). Reproduction in Coquerel's dwarf lemur (*Mizra coquereli*). Am J Primatol 36:223–237.

Stribley JA, French JA, Inglett BJ (1987). Mating patterns in the golden lion tamarin (*Leontopithecus rosalia*): continuous receptivity and concealed estrus. Folia Primatol 49:137–150.

Sutcliffe AG, Poole TB (1984). An experimental analysis of social interactions in the common marmoset (*Callithrix jacchus jacchus*). Int J Primatol 5:591–607.

Tilson RL (1979). Behavior of hoolock gibbons (*Hylobates hoolock*) during different seasons in Assam, India. J Bombay Nat Hist Soc 76:1–16.

Tilson RL (1981). Family formation strategies of Kloss's gibbons. Folia Primatol 35:259–287.

Tilson RL, Tenaza RR (1976). Monogamy and duetting in an Old World monkey. Nature 263:320–321.

Trivers RL (1972). Parental investment and sexual selection. *In* Sexual Selection and the Descent of Man, 1871–1971 (Campbell BG, ed), pp 136–179. Chicago: Aldine.

Wahome JM, Rowell TE, Tsingalia HM (1993). The natural history of de Brazza's monkey in Kenya. Int J Primatol 14:445–466.

Wallen K (1982). Influence of female hormonal state on rhesus sexual behavior varies with space for social interaction. Science 217:375–377.

Wallen K, Winston LA (1984). Social complexity and hormonal influences on sexual behavior in rhesus monkeys (*Macaca mulatta*). Physiol Behav 32:629–637.

Wallen K, Winston LA, Gaventa S, Davis-DaSilva M, Collins DC (1984). Periovulatory changes in female sexual behavior and patterns of ovarian steroid secretion in group-living rhesus monkeys. Horm Behav 18:431–450.

Westneat DF, Sherman PW (1993). Parentage and the evolution of parental behavior. Behav Ecol 4:66–77.

Wickler W, Seibt U (1983). Monogamy: an ambiguous concept. *In* Mate Choice (Bateson P, ed), pp 33–50. Cambridge: Cambridge University Press.

Williams JR, Catania KC, Carter CS (1992). Development of partner preferences in female prairie voles (*Microtus ochrogaster*): the role of social and sexual experience. Horm Behav 26:339–349.

Wittenberger JF, Tilson RL (1980). The evolution of monogamy: hypotheses and evidence. Annu Rev Ecol Syst 11:197–232.

Wright PC (1978). Home range, activity pattern, and agonistic encounters of a group of night monkeys (*Aotus trivirgatus*) in Peru. Folia Primatol 29:43–55.

Ziegler TE, Bridson WE, Snowdon CT, Eman S (1987). The endocrinology of puberty and reproductive functioning in female cotton-top tamarins (*Saguinus oedipus*) under varying social conditions. Biol Reprod 37:618–627.

Ziegler TE, Epple G, Snowdon CT, Porter TA, Belcher AM and Küderling I (1993). Detection of chemical signals of ovulation in the cotton-top tamarin, *Saguinus oedipus*. Anim Behav 45:313–322.

12 Human Pheromones: Primers, Releasers, Signalers, or Modulators?

Martha K. McClintock

Pheromones are airborne chemical signals produced by one member of a species and received by another, triggering neuroendocrine responses underlying behavior, fertility, or development (Karlson and Luscher 1959; Karlson and Butenandt 1959). Although their existence is well established in many invertebrate and vertebrate species, research on human pheromones is in an unusual state. The lay public and certain fragrance companies presume that human pheromones exist and the media often state this presumption as fact. Thus, we found ourselves in the unusual position of presenting definitive evidence for the existence of human pheromones (Stern and McClintock 1998) to a lay audience that already believed in their existence, but to a scientific audience that was quite skeptical. These long-anticipated data, together with earlier work, point to a hitherto unstudied sensory and brain mechanism, with many potential effects on human physiology, psychological state, and social behavior —effects that may range from fertility and sexuality to aggression and fear.

Overview

Environmental Information via Chemosignals

In all phyla, chemosignals provide environmental information that guides behavior and metabolic activity. Single cells, whether they are organisms themselves or part of a multicellular organism, move in response to specific chemical gradients in their environment and change their metabolic activity based on chemical information. Neurons alter their internal metabolism and migrate in response to neurotrophic growth factors. Primordial germ cells migrate to the developing gonads and once arrived, respond to the local chemical environment with differential gene expression that makes them either spermatids or ova.

Simple organisms, such as bacteria, utilize chemosignals for orientation in their environment (Shapiro 1998). Among vertebrates, it is well-known that carnivores can follow an olfactory trail to track prey (Macdonald 1985) and that salmon follow a concentration gradient of a specific chemical profile to return to their birthplace to spawn. Similarly, yellow baboons use odors while foraging for survival on the savannas of Amboseli in Kenya (Altmann 1998). Less well-known is information that the behavior of birds can be guided by chemosignals. For example, European starlings discriminate among plant volatiles (Clark and Mason 1987) and Antarctic seabirds use olfaction in foraging (Nevitt and Reid 1998). In addition, navigation by homing pigeons is affected by the olfactory system (Wallraff 1986).

Pheromones Across Phyla

Individual responses to the environment can be fine-tuned and enhanced by social information shared by members of the same species. Such social chemosignals are termed *pheromones* and have their effects by altering the recipient's neuroendocrine system and thereby its behavior or development. However, not all chemosignals are pheromones. Some human odors are detectable not only by other people but also by other species, such as dogs—yet these chemosignals are not pheromones because they are not used for species-specific social communication, nor do they act by privileged neuroendocrine or molecular pathways.

Pheromones are ubiquitous among phyla. Bacteria and yeast have pheromone systems that mediate sexual reproduction and coordinate their social behavior (Loumaye et al. 1982; Shapiro 1998). Insect pheromones serve as sexual attractants operating across many miles and coordinate development from juvenile to pupal stage of development with resource availability (Agosta 1992). All mammalian orders, with the possible exception of Cetacea (including dolphins, killer whales, and porpoises), have anatomical evidence of a vomeronasal organ (VNO), specialized for environmental chemoreception. This organ is also found in fish, amphibians, and reptiles (Wysocki 1979).

In mammals, pheromones enhance response to environmental signals and sharpen the peak of seasonal breeding (McClintock 1981a). They also play an essential role in a variety of social interactions. Lemurs identify territorial boundaries and dominance relationships with chemosignals produced by the antebrachial organ and brachial gland (Jolly 1966) and solitary female gerbils attract males across arid desert territory so that they can mate (Randall 1993); ungulates respond to the compounds in the urine of competitors in their territory with an elaborate flehman response (Moehlman 1985).

Universal functional principles underlie pheromonal systems found in all phyla. The widespread occurrence of such systems, even among bacterial taxa, suggests that they are essential to an organism living successfully in its natural context, coordinating intracellular processes with its physical environment through social interactions. Bacterial cells communicate and have decision-making capabilities that produce autoaggregation and formation of complex colonies (Shapiro 1998). In some bacterial species "quorum-sensing signal molecules" have been identified as N-acylhomoserine lactones (AHLs). Making the bridge to multicellular taxa, unicellular regulatory circuits are proving to be homologous to the molecular basis of intercellular communication in myxobacteria and streptomycetes (Shapiro 1998). It is in this phylogenetic context that we consider what might be the structure and function of human social chemosignals and pheromones.

The Microsmaty Fallacy

It is generally assumed that humans are microsmatic, that is, that our olfactory structures border on vestigial and play a minimal, if not completely nonessential, role in governing our behavior and physiological states. Broca first divided mammals into microsmatic and macrosmatic taxa (Broca 1888). His classification, however, was based solely on the size of olfactory systems relative to neuroanatomical structures. Broca, and subsequently others (Stephan et al. 1970; Bauchot 1981), classified taxa both by comparison across species and among the various neurological systems of a single species. It was, and is, erroneously assumed that this neuroanatomical distinction underlay a comparable distinction in terms of behavior (reviewed in Schaal and Porter 1991).

Man and primates generally are microsmatic, and the sense of smell evidently plays a relatively minor role in their behavior. (Herrick 1924 after Schaal and Porter 1991)

Although this tentative conclusion might be true when considering the enormous behavioral repertoire of humans, it is not necessarily true within a specific social context or behavioral domain. Moreover, arguing strictly in neuroanatomical terms, Keverne (1983a) states:

The fact that olfactory information is coded as a pattern which requires a higher order of recognition means that animals with a greater neural backup have the potential for a more sophisticated pattern recognition. Since the olfactory bulb seems to be acting mainly as a filter, and decoding of the olfactory message a more central event, animals with greater neocortical support systems have the greatest ability to make use of this system.

Nonetheless, perhaps because our particular culture enables us to verbalize our experience in terms of sights and sounds more easily than in terms of odors, our understanding of mammalian olfaction, compared to vision and audition, is in its infancy. The study of human chemosensory systems could benefit from the view that chemosensory systems are highly efficient and parsimonious, refined over a long evolutionary history, in contrast to visual systems, which have evolved the ability to detect constancy of form and motion only relatively recently.

Popular Culture, Science, and Myths

Perfumers have long used animal pheromones as essential ingredients in perfumes (Engen 1991). These compounds are used because they are stable and effective fixatives (Berliner et al. 1991). Moreover, humans find some animal pheromones to have a pleasant odor at high concentrations (Gower and Ruparelia 1993).

The relationship between odors and sexuality was underscored by Freud, under the influence of his charismatic friend, Wilhelm Fliess (Engen 1991). Fliess noted that the nasal mucosa swells during sexual excitement and hypothesized that there were "nasal reflex neuroses" similar to the neurasthenias described by Freud. He hypothesized that these neuroses arose from vasomotor disturbances caused by sexual dysfunction and asserted that they could be cured by sniffing cocaine.

Although Freud later abandoned this hypothesis, the association between the nose and sexuality continues to resonate in popular culture. In particular, that humans enjoy perfumes made of animal sex-attractant pheromones has proved to be an irresistable association. The popular press and the fragrance industry have capitalized on popular interest, weaving small fibers of scientific data into a mythical tapestry of marketing claims that perfumes "turn women from "no, no, no" to "yes, yes!, YES!." Jovan, a fragrance purported to contain pheromones, is marketed as "scientifically designed to attract."

Many of these marketing claims play on human desires and insecurities. The 1996 annual report of the Erox Corporation (1995) contains testimonials for Realm, a perfume they claim contains human pheromones:

Wearing your product, I felt very comfortable talking with every woman in the little grocery I shop at. But more unusual is, they too, seemed eager to talk to me.

Testimonials are also used to market Athena 10X, a perfume additive claimed to contain human pheromones:

It makes the men crazy. My friends say: "Abby, what is it about you? The men just love you." And I am not going to share it with them. (Athena 1998)

Do Humans Communicate with Pheromones?

This fundamental question is the topic of this chapter. Do chemosignals, released by other people in our social environment, regulate any aspect of our physiology or behavior or psychological state? In what contexts might this occur and with what effects on our behavior? The goal of the chapter is first to review the extant data, and then to integrate that information to craft a comprehensive answer.

An accurate answer requires evaluating an implicit assumption held not only by marketing campaigns and the popular press but by scientists as well. Because pheromones are so ubiquitous phylogenetically, it is assumed that pheromonal communication is a highly conserved trait, functioning in mammals and humans in the same way it does in "ancestral" species. Perhaps, humans have sex attractants as elegant in their simplicity as those used by insects. Female silkworm moths emit bombykol, a molecule that compels males to follow aerial molecular concentration gradients to find her and mate. In any number of contexts, a single molecule reliably triggers specific male behavior (Schneider 1974).

The ubiquity of pheromones also led scientists to expect that a comparative approach will reveal universal insights. This creates the illusion that it is possible to study olfactory and pheromonal systems by cutting away the more recently evolved neocortex and focusing on disembodied olfactory systems (figure 12.1). In contrast, many are leery of even considering the hypothesis that pheromones might affect human behavior precisely because humans have evolved a complex neocortex. Certainly, mammalian pheromonal systems differ dramatically from those of other orders, in part because mammals have evolved additional mechanisms to serve the endocrine and behavioral functions mediated by pheromones in more ancestral orders (McClintock 1998b).

In this chapter we demonstrate the existence of human pheromones and illustrate how computer simulations and animal models allow us to design experiments to test predictions about their effects on ovarian function. We review in detail the evidence that humans have pheromones that meet the functional criteria for the three traditional types of pheromones: primers, releasers, and signalers. We then hypothesize that there is likely to be a fourth type of pheromone, modulators, which are more likely to regulate human behavior than are the traditional types, which were conceptualized in terms of insect systems.

Given that humans have priming pheromones, we consider how a pheromonal system would work in humans: receptor systems, modes of release, potential chemical structures, including their production, activation, and vagility. We call into question the assumption that pheromones work as isolated systems. Because pheromones integrate the convergence of environmental, social, and developmental information, we argue that pheromone systems are likely to be context-dependent (McClintock 1998b) certainly in humans and in other mammals as well. Thus, we conclude that if humans use pheromones to regulate their behavior, then such pheromones should operate only in specific contexts and should affect behavior only by modulating physiological and psychological states as part of the integration of other essential sensory, emotional, motivational, and cognitive processes.

Types of Pheromones

Traditionally, there are three types of animal pheromones distinguished by the time course and nature of their effects. We address in the following whether humans

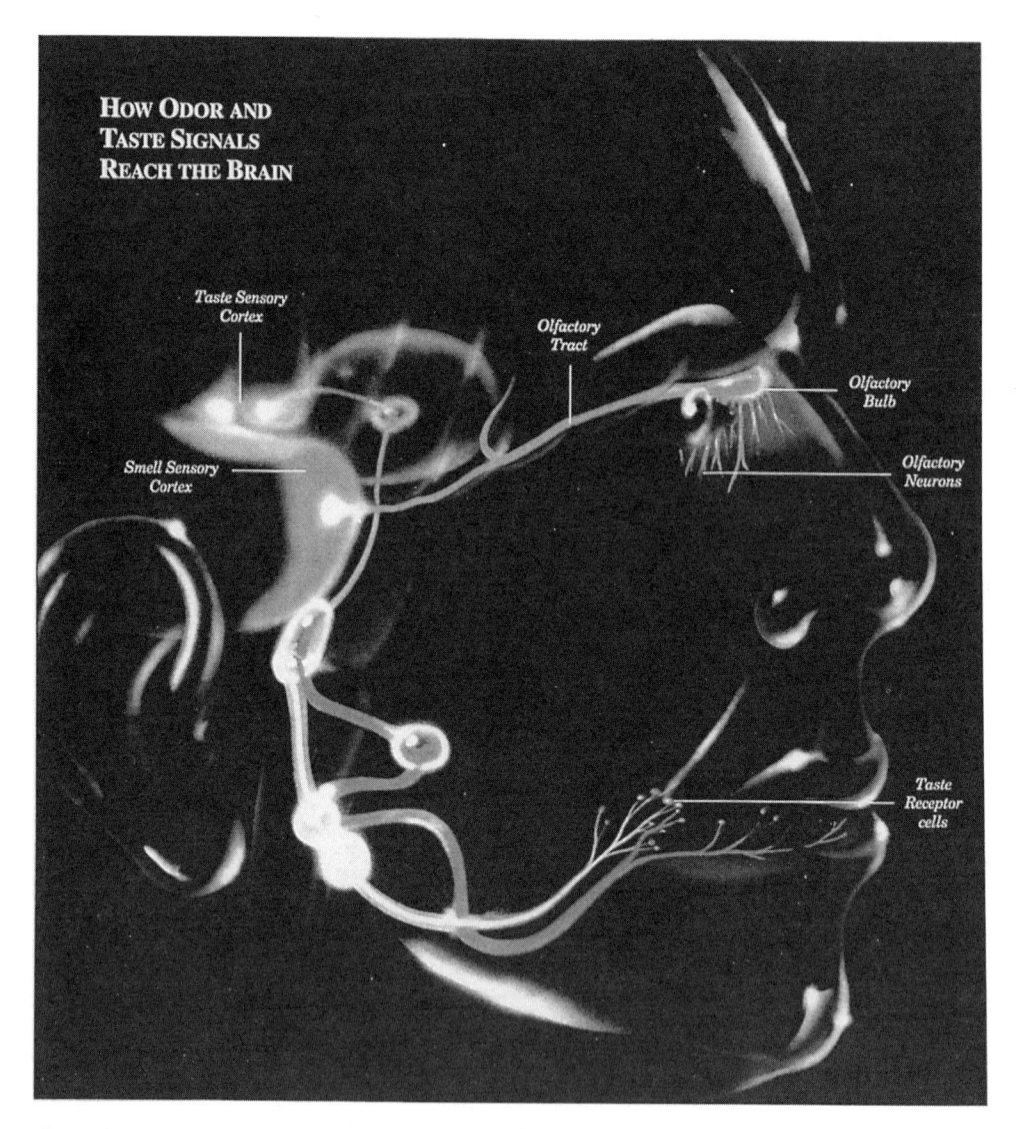

Figure 12.1
A schematic of the human olfactory system depicting it as a circuit operating in relative isolation from the rest of the brain, without reciprocal interactions with information from cortical structures. (Scott T. Barrows. © National Geographic Society)

have pheromones that meet the criteria for any of these previously recognized types and whether there are additional types of pheromones, perhaps unique to humans.

Priming pheromones produce changes in the endocrine system and operate over days or weeks (Albone 1984). For example, a pregnant female mouse exposed to specific compounds in the urine of a strange male experiences diminished pulsatile release of prolactin causing an abortion within a few days (Novotny et al. 1997); female mice living together experience suppression of estrus and spontaneous ovarian cycles (Van der Lee and Boot 1956; McClintock 1984a).

Releasing pheromones operate much more quickly, triggering within seconds to minutes changes in the nervous system that release stereotyped behaviors. A few molecules of androstenol in the breath of a boar immobilizes estrous sows in a mating stance that allows the boar to mount (Gower 1972).

Signaling pheromones operate in an intermediate time course and promote less stereotyped responses. Female hamsters typically live alone and, when not in heat, will aggressively oust male intruders. A female in heat, however, marks the perimeter of her territory with her copious vaginal secretions to signal the male that she is aggressive no longer, and indeed sexually receptive (Johnston 1977; Fiber and Swann 1996).

Modulating pheromones is a new term we propose here—a fourth type of pheromone that changes stimulus sensitivity, salience, and sensorimotor integration. Their effects are less dramatic than releasing pheromones, but more specific than priming or signaling pheromones. Recognition of this class of modulatory effects may help resolve some of the debate over the definition of what constitutes a pheromonal system in mammals and perhaps humans (Beauchamp et al. 1976; Katz and Shorey 1979; Martin 1980; Sachs 1997b).

Regulation of Physiology by Priming Pheromones

The vast majority of research on priming pheromones has focused on regulation of gonadal function. In mammals, pheromones enhance reproductive success by coordinating the fertility of males and females; ovarian function is regulated by male pheromones and female pheromones regulate testicular function. Pheromonal interactions among females also regulate the ovary and increase reproductive success, in this case by increasing the chance that a female will undertake the high costs of pregnancy and lactation in a supportive physical and social environment (McClintock 1984b; Mennella et al. 1990; Blumberg et al. 1991).

Pheromonal Regulation of Ovarian Function

Interactions Among Females

Pheromonal Mediation of Ovarian Synchrony Women who live together and interact over the course of several months can, under some conditions, experience synchronization of their menstrual cycles (McClintock 1971; Weller and Weller 1993a; see review in Graham 1993; Weller and Weller 1997). There has been much research into the circumstances in which menstrual synchrony occurs. The key environmental conditions are that women spend significant amounts of time in close proximity, have highly structured activities and social schedules, and eat similar diets. Under these special conditions, the effects of these common factors, including other salient modulators of the menstrual cycle, do not override the social interactions that increase synchrony within a social group. There has also been a great deal of confusion about the best methods of detecting and quantifying menstrual synchrony in small groups of women (McClintock 1998; Wilson 1987; Weller and Weller 1993a). Some of the confusion stems from failing to recognize that menstrual synchrony is a dynamic process, as well as from an inadequate understanding of coupled oscillating systems and the best methods to measure them (McClintock 1998).

When female rats live in groups under controlled laboratory conditions and are afforded the opportunity to interact over several ovarian cycles, their cycles also become synchronized (McClintock and Adler 1978). In rats, we demonstrated conclusively that airborne chemosignals are sufficient to enable ovarian synchrony using an apparatus that prevented rats from communicating with each other except via a shared recirculated air supply (McClintock and Adler 1978; Schank et al. 1995). We used this apparatus to demonstrate that pheromones in rats could modulate not only cycle length but also disrupt cycles and produce a state of acyclicity (McClintock and Adler 1978; McClintock 1983a,b 1998).

Two studies have attempted to test the hypothesis that chemosignals mediate menstrual synchrony in humans. Russell et al. (1980) and Preti et al. (1986) provided evidence that priming pheromones had the potential to mediate menstrual synchrony. These studies were criticized, however, for their small sample size, methodology, and statistical issues (Doty 1981; Wilson 1987; Weller and Weller 1993a; Stern 1992).

A Computer Simulation Our detailed understanding of ovarian synchrony in rats enabled us to create a computer model and simulate experiments impossible to do in the real world (Schank and McClintock 1992). Our model transformed our conception of the problem (McClintock 1998) and formed a strong foundation for studying the broader issue of pheromonal regulation of the ovarian cycle in women. In particular, the model focused our attention on the pheromonal mechanisms that could produce divergent effects of pheromonal communication within a group of females, not just on the special case of synchrony. In particular we verified that ovarian synchrony could theoretically result from two different pheromones produced during distinct phases of the rat's ovarian cycle: one advances the next ovulation and the other delays it, shortening or lengthening the ovarian cycle, respectively (McClintock 1984a).

We began by conceptualizing ovarian synchrony as the result of the mutual entrainment of coupled oscillators; the ovarian cycle is an oscillating system and the two ovarian pheromones exchanged within a social group "couple" their ovarian cycles, enabling mutual entrainment and synchrony (McClintock 1983a, 1984a). Indeed, in coupled oscillating systems as diverse as neural networks and firefly flashing, synchrony is achieved within a group through the interaction of two signals with opposite effects on the timing of the oscillating system (McClintock 1983a; Schank and McClintock 1990, 1992).

We developed a formal computer model of ovarian synchrony that would simulate the phenomenon and confirm, with analysis by synthesis, the role of these two different pheromones in generating synchrony (Schank and McClintock 1990, 1992). Our model did far more than meet our original goal, in part because it was designed to model the known dynamics of the particular hypothalamic mechanisms causing synchrony, as well as the function or interaction pattern of the system of interacting cycles (Schank and McClintock 1993). The results of these simulations forced us to redefine the phenomenon itself: the fundamental process in spontaneous cycles is changes in the timing of ovulation within an individual, which produce not only synchrony within the group but also significant asynchrony, as well as significant stasis, that is, an absence of change over time or phase-locking within the group. It is noteworthy that this extremely stable state is a balance of synchronous and asynchronous groups; there are two subgroups, synchronized within each group but asynchronous with each other (i.e., 180 degrees out of phase).

The model changed the way we measure synchrony; we are now able to determine the random level of synchrony as well as the shape of its distribution ranging from significant asynchrony at one end to synchrony at the other. Finally, the simulations demonstrated that social regulation of ovulation in a small group is an example of chaos in an oscillating system; the final state of the group is determined by the phase conditions when they first begin to interact (Schank and McClintock 1992; Glass and Mackey 1988) (figure 12.2).

Most important, the model made a host of biologically reasonable and testable predictions; many were not intuitively obvious or derivable without the computer

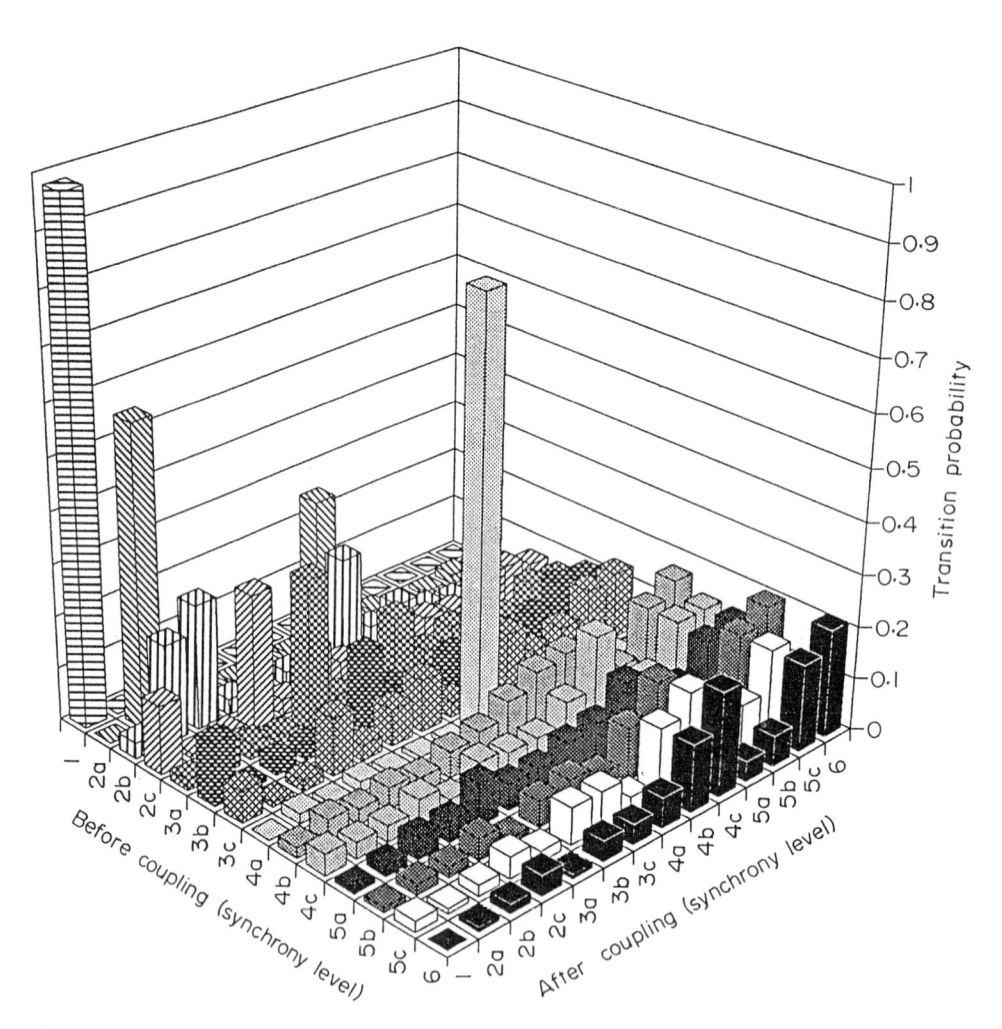

Figure 12.2
A, In a computer simulation experiment, pheromonal interactions within the group produced three outcomes, not just synchrony (levels 4–6; level 1 is asynchronous and level 6 is perfect synchrony). The interactions also produced asynchrony (level 1) and stasis (level 4a); this extremely stable state occurs when there are two subgroups synchronized within each group but asynchronous with each other (i.e., 180 degrees out of phase). The outcome, or final level, of synchrony achieved by the group was predicted by the initial level of synchrony when the group began to interact. In each graph, the x-axis represents the initial level of synchrony of the group at the beginning of their pheromonal interactions (i.e., coupling of the oscillators), the y-axis represents the final level achieved after pheromonal interaction, and the z-axis represents the number of groups that experienced the particular combination of initial and final states. When simulations were run with individuals that had different sensitivities to pheromones (A), perfect synchrony (level 6) was relatively rare. Without these individual differences in pheromone sensitivity (B), the resultant level of synchrony was usually perfect (level 6). Since perfect synchrony is rarely seen among groups of rats,

Human Pheromones

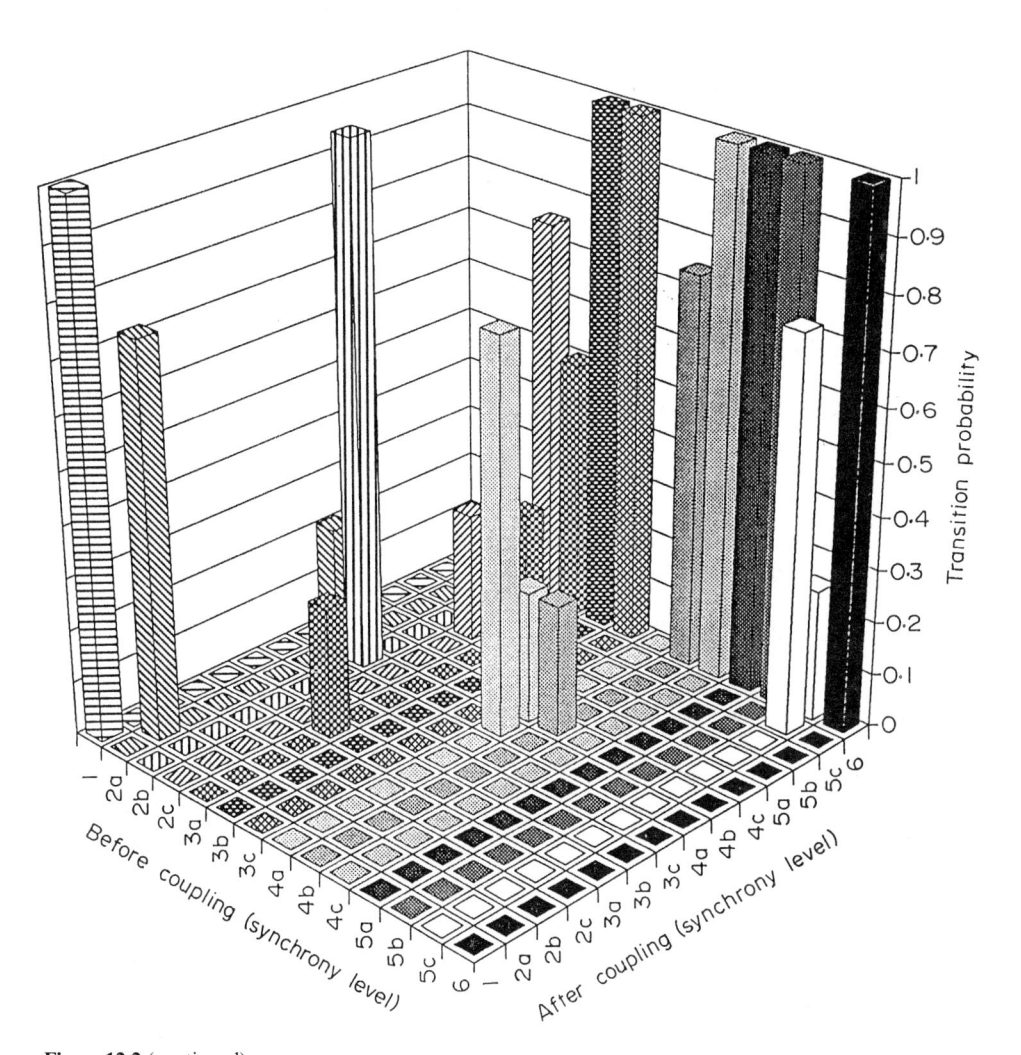

Figure 12.2 (continued)
these results strongly suggest that individual differences in sensitivity are to be expected. Note also that even when groups typically achieved perfect synchrony, statis and asynchrony were still likely outcomes of the pheromonal interactions.

model. Particularly important was the discovery that menstrual synchrony reflects a too narrow focus. Menstrual cycle coordination is a multifaceted dynamic process with multiple outcomes at the level of the social group. Therefore, an attempt to identify the role of human pheromones would be most tractable not at the social level, with its diverse outcomes, but at the individual level and the ovarian cycle, where the fundamental interaction is simply changing the time of ovulation to lengthen or shorten the cycle.

Regulation of Menstrual Cycle Length

Group Patterns or Individual Cycle Lengths?

Our next step was to design a study to test our hypothesis that humans have priming pheromones that regulate ovarian function. Our computer simulations informed us that we should focus on the two pheromonal mechanisms that change cycle length, rather than on the group level of analysis and the myriad patterns of ovarian entrainment. At the level of individual cycle lengths, the outcomes of pheromonal treatments are similar across different social environments, and so we were able to design a tightly controlled study with the power to detect the effects of pheromonal signals on menstrual cycle length and ovarian function.

Where to Collect Putative Pheromones?

As in other species, human pheromones might be produced by apocrine glands (active only during reproductive maturity), eccrine glands (which produce sweat which contains compounds found also in saliva and urine), exfoliated epithelial cells, hair, or bacterial action. (Nixon et al. 1988). We collected compounds from the axillae (armpits) because they contain all five of these potential sources and because the two studies on this subject used axillary compounds (Russell et al. 1980; Preti and Huggins 1975; Zeng et al., submitted). The nine donors bathed daily, without perfumed products, and then wore thin 4 in. × 4 in. cotton pads in their axillae for at least 8 hours. Each pad was cut into four sections for distribution to different recipients, treated with four drops of 70% isopropyl alcohol (Stern and McClintock 1998), and then frozen immediately at −80°C in a glass vial.

How to Collect Pure Samples?

We first needed to ensure that we were collecting pure samples of putative pheromones from hormonally distinct phases of the cycle. In the model described above, the parameter values representing the different effects of the two ovarian-dependent pheromones were based directly on our empirical data from rats (figure 12.3). During the preovulatory phase of the cycle, near the end of follicular development, a phase-advance pheromone is produced which lengthens the cycle. Immediately afterward, the signal switches to a phase-delay pheromone which shortens the cycle. This switchover was linked to the preovulatory surge of luteinizing hormone (LH). After ovulation and the rupture of the follicle, we no longer detected compounds with any effect on the length of the cycle. At this time progesterone is relatively higher and estradiol and the gonadotropin are low.

Thus, the model told us that pinpointing the time of the LH surge was essential to collecting pure samples of preovulatory and ovulatory pheromones, for the surge tightly predicts the transition from one type of pheromone to another. If compounds are collected during a period that overlaps both phases, our model tells us that their mixed effects would cancel each other, leading to the erroneous conclusion that the compounds had no effect on the recipient.

Donors also provided daily evening urine samples, which we assayed for LH to detect the onset of the LH surge that triggers ovulation (Stern and McClintock 1995). This singular hormonal event unambiguously demarcates the follicular from the ovulatory phases of the cycle. Along with data on vaginal secretions, menses, basal body temperature, and a rise in progesterone glucuronide (PG) in the postovulatory luteal phase, this surge was used to classify each pad as containing compounds produced during the follicular phase (2 to 4 days before the onset of the LH surge) or the ovulatory phase (the day of the LH surge and the 2 subsequent days). To ensure a similar stimulus for all recipients regardless of individual differences among donors, all nine donors contributed equally to the follicular and ovulatory compounds received by each subject.

Odorless Compounds

Pheromones do not require conscious detection to produce their effects. Indeed, in the mammalian species in which the neural pathways mediating pheromonal responses have been identified, there is no involvement of the main olfactory system and olfactory cortices which process conscious detection of odors (at least as inferred from human neurology patients!). Instead, the sensory inputs project from the accessory olfactory system directly to the amygdala and hypothalamus. This is not to argue that pheromones do not have odors at high concentrations, but to point out that a detectable odor is not necessary for their effects.

Therefore, it is of central theoretical importance that all participants in our study were blind to the experiment's hypothesis and the source of the compounds. The

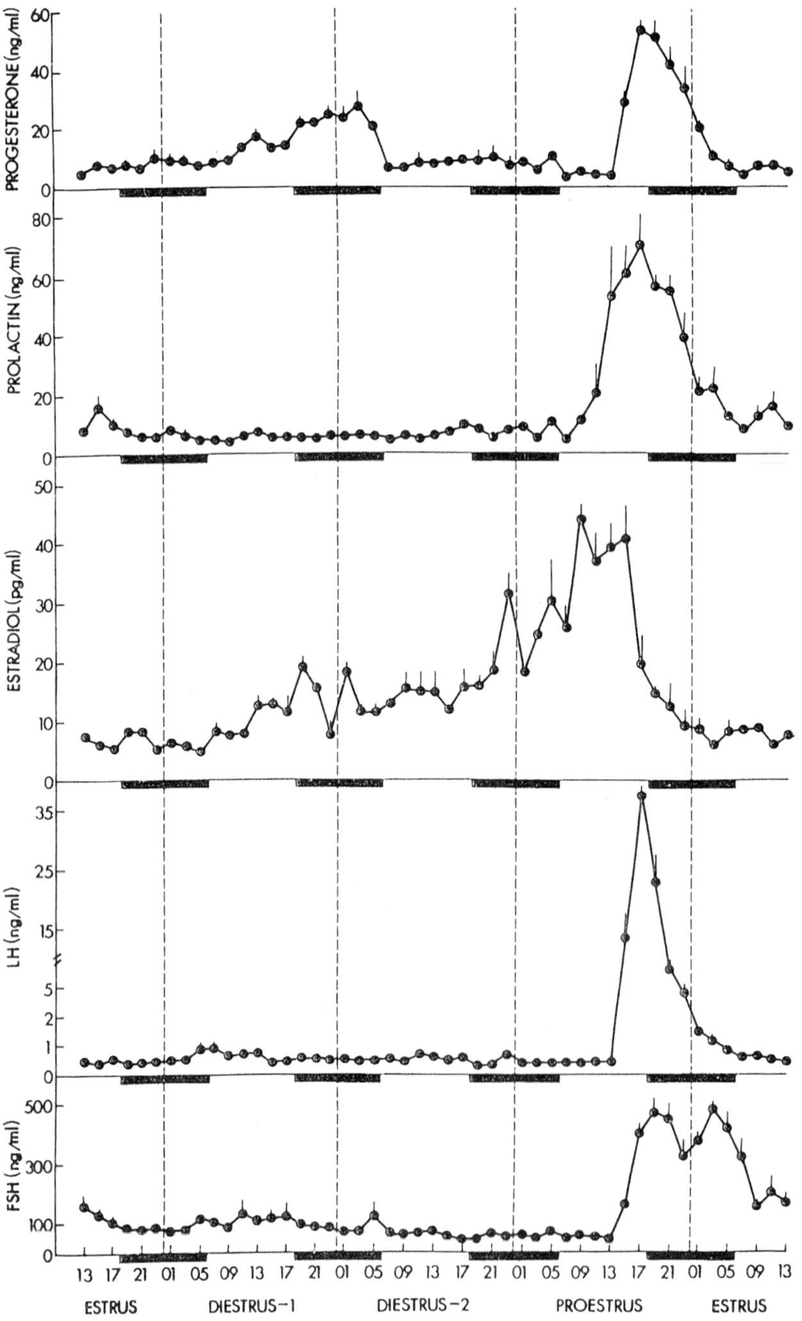
PROGESTERONE (ng/ml)
PROLACTIN (ng/ml)
ESTRADIOL (pg/ml)
LH (ng/ml)
FSH (ng/ml)
13 17 21 01 05 09 13 17 21 01 05 09 13 17 21 01 05 09 13 17 21 01 05 09 13
ESTRUS
DIESTRUS-1
DIESTRUS-2
PROESTRUS
ESTRUS

study was presented as focused primarily on the development of noninvasive methods of detecting ovulation and only secondarily on sensitivity to the odor of small amounts of "natural essences" (consent was obtained for a list of thirty compounds). The women in our study did not consciously detect the compounds as odors. In fact, many believed themselves to be in a control group that was not presented with an essence or odorant. It is the lack of conscious detection that fulfills an essential criterion for calling these compounds pheromones.

How Should the Compounds be Presented?

We asked the women subjects to come into the laboratory daily and wipe a thawed pad just under their noses, across the part of the upper lip called the frenulum (Stern and McClintock 1998). After applying the compounds, recipients were free to go about their normal activities but were asked not to wash their face for the next 6 hours. Our working hypothesis was that the women could then breathe in the compounds as long as they lingered on the upper lip. The women reported that the only odor they were conscious of was the alcohol and the smell of the pad while they were wiping their upper lip.

In rats, we had used a similar method to successfully change ovarian cycle length by continuously presenting pheromones produced by females that were in a particular reproductive state. This mode of continuous pheromone presentation was realistic for naturally occurring conditions such as pregnancy, lactation, and reproductive senescence. It was, however, an artificial mode for pheromones normally produced in pulses, and only briefly during the follicular and ovulatory phases of the cycle; obviously, female rats are never naturally in a state of continuous follicular development or continuous ovulation. Nonetheless, continuous presentation of the entraining stimulus is a classic technique used to determine effects on periodicity of an oscillator. For example, continuous light of different intensities shortens or lengthens free-running activity rhythms.

How to Detect a Response

In our original study of menstrual synchrony, the median change in onset of the cycle was approximately 2 days. Because this is substantially less than the individual vari-

Figure 12.3
Hormone levels during the ovarian cycle of (A) rats and (B) humans. The luteinizing hormone (LH) surge demarcates the follicular phase from the ovulatory phase of the cycle. C, The computer model of ovarian synchrony was built with a parameter representing the production of phase-advance pheromones during the follicular phase and phase-delay pheromones during the ovulatory phase.

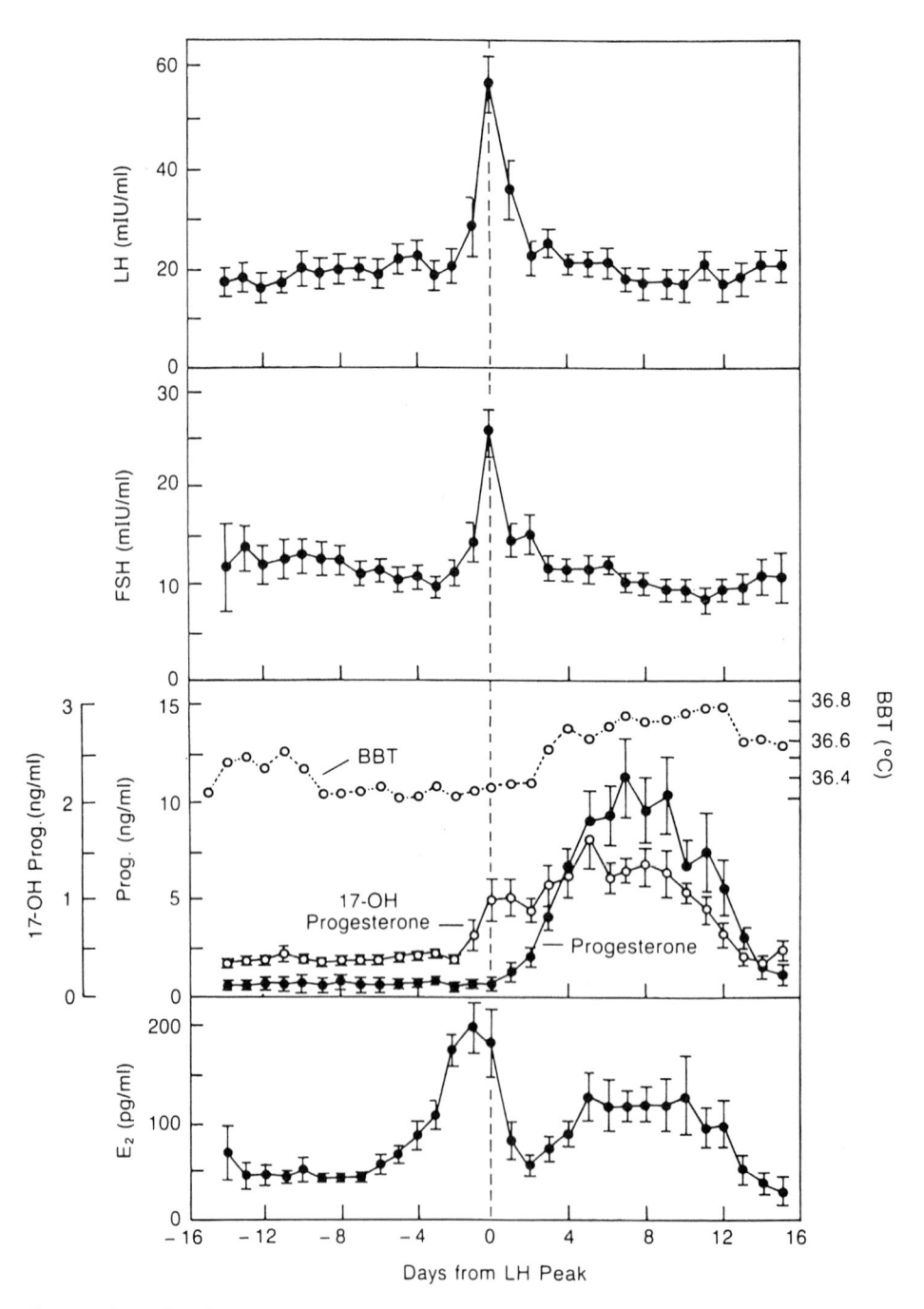

Figure 12.3 (continued)

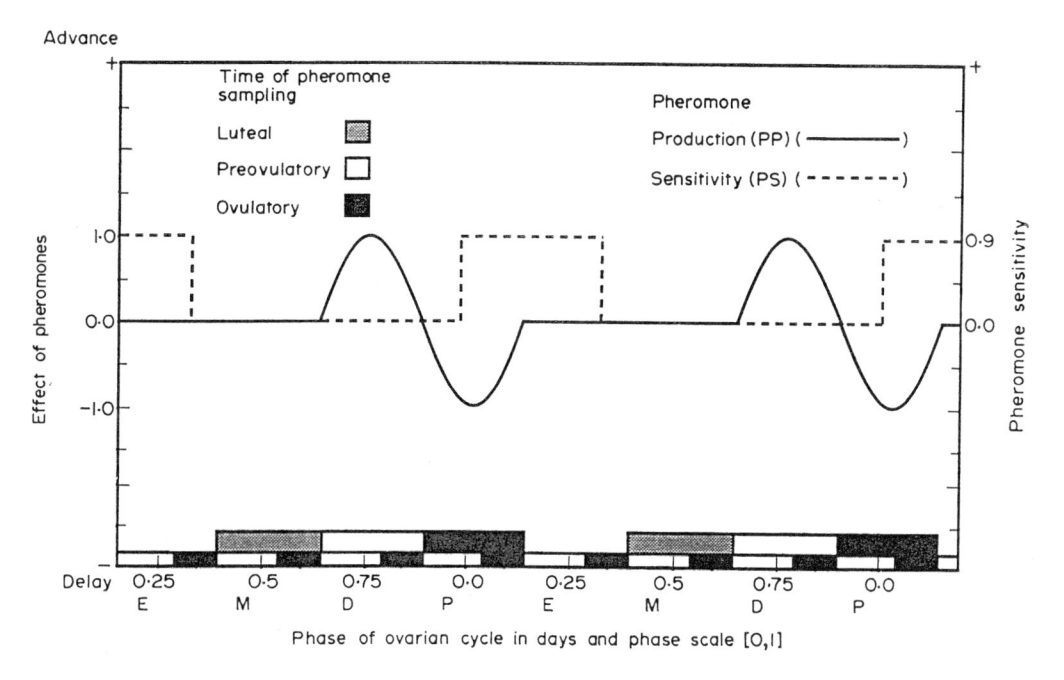

Figure 12.3 (continued)

ation in cycle length typical for this age group (Treloar et al. 1967), we created within-subjects controls by quantifying the effect of the compounds in terms of a change in length from a baseline menstrual cycle. Thus, a baseline cycle preceded the experimental conditions.

Then, in a crossover experimental design during the next four consecutive cycles, axillary compounds were applied daily. Half of the recipients received *follicular* compounds for two menstrual cycles and were then switched to exposure to *ovulatory* compounds for the next two cycles. The other half of the recipients received the same compounds, in the reverse order. In addition, the women who were donors served as a between-subjects control group. They received only the carrier above their upper lip each day: 70% isopropyl alcohol. Thus the experiment included both within- and between-subject controls. As expected, follicular and ovulatory compounds had opposite effects, together producing a 4-day change in cycle length in comparison to the recipient's own baseline cycle and those of the control group (figure 12.4). These data provide the first strong evidence that humans produce pheromones, and more specifically, priming pheromones that regulate ovarian function.

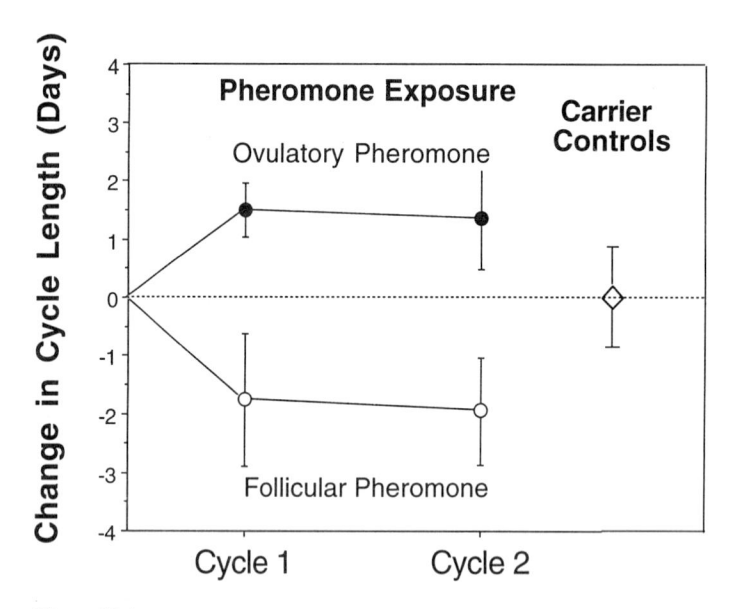

Figure 12.4
Primer pheromones collected from the axillae of women during the follicular and ovulatory phases have opposite effects on the length of the recipients' menstrual cycles. (From Stern and McClintock 1998.)

Regulating Follicular Phase Length

Once we demonstrated that human pheromones could shorten or lengthen the ovarian cycle, we needed to identify which phase of the cycle was affected in order to identify the specific neuroendocrine mechanism for these pheromones. In other mammalian species, a myriad of neuroendocrine mechanisms and phases have been hypothesized or demonstrated to regulate ovarian cycle length. It is easiest to organize a discussion of these potential mechanisms in terms of the phase of the cycle in which they operate.

The luteal phase of the ovarian cycle is the time after ovulation in which the corpus luteum produces progesterone, which prepares the uterus for implantation. In rodents, altering the life span of the corpus luteum directly changes cycle length. Keverne (1983b) has hypothesized that all of the pheromonal priming effects on ovarian function of the mouse are mediated by dopamine and prolactin which maintain the corpus luteum. Medical texts state that the human luteal phase is relatively fixed in length. Indeed, Cutler et al. (1980) hypothesized that axillary compounds from men prevent "defects" in the luteal phase and thus "stabilize" the cycle (Schweiger et al. 1990), perhaps mediated by changes in LH hormone pulse frequency (Suh and Betz 1993).

This luteal phase hypothesis, however, is based on two assumptions. The first is that regular ovarian cycles have a 14-day luteal phase in humans. The second assumption is that the pheromones "rescue" a defective corpus luteum, preventing it from involuting prematurely.

Neither concept is consistent with the data. The luteal phases of our subjects ranged from 5 to 18 days in length (based on onset time of LH surge and time of ovulation, pinpointed within 12 hours) (Stern and McClintock 1998). This range is well beyond the presumed normal life span of the corpus luteum in humans. This variance cannot be considered pathologic as our subjects had normal ovulatory cycles, no history of reproductive abnormality, and used only barrier contraception. They were 25 to 35 years of age, in their "reproductive prime." Thus, we must reevaluate our assumption that variation in the luteal phase beyond 14 ± 2 days is atypical or abnormal. Nonetheless, because luteal phase length did correlate with menstrual cycle length, it was a candidate for a target for pheromonal action.

Another phase of the cycle in which pheromones could operate is menses. Our original observation focused on the timing of menses because it is the ovarian cycle event most easily noted and recorded (McClintock 1971). The menses phase also varies in length and therefore correlates with cycle length (Stern and McClintock 1998). The exact neuroendocrine mechanisms for the onset or duration of menstrual flow are not known. There is a surge of follicle-stimulating hormone (FSH) right at the end of menstrual flow, which may affect the time of menses (Bancroft et al. 1983). And, in a large time frame, onset of menses follows a drop in progesterone levels, which have supported proliferation of the uterine endometrium. But the link between this drop and the specific onset time is not known, particularly when measured in terms of hours or a day or two. Since the changes in menstrual cycle length had been on the order of 2 days (McClintock 1971), it was possible that pheromones regulated cycle length by changing the onset or duration of menses.

Finally, pheromones may act on the length of follicular development, defined as the interval between the end of menses to the preovulatory surge of LH. The rate of follicular development determines the timing of the preovulatory surge of LH that ruptures the follicle, releases the egg, and enables it to transform from an organ producing estrogen to one producing predominately progesterone. The follicular phase has long been recognized as the most variable of the three phases. Moreover, pheromonal regulation of ovulation would certainly have a selective advantage if it enabled females to optimize the environments in which they could conceive and become pregnant (McClintock 1981b; Muller-Schwarze 1997).

Our data demonstrated unequivocally that the follicular phase mediated the effects of ovarian-dependent pheromones on length of the menstrual cycle and that no role was played by variation in the menses or luteal phases (figure 12.5). These other two phases were also variable and predicted variation in menstrual cycle length, but it was only in the follicular phase that that variance could be attributed to exposure to axillary compounds from other women.

Regulation of the Preovulatory Luteinizing Hormone Surge and Timing of Ovulation

Changes in cycle length are accomplished specifically by regulating the surge of LH that triggers ovulation. We now need to determine whether the putative pheromones accelerate or retard the rate at which the recruited ovarian follicle ripens (as in our model) or whether it changes the sensitivity of the brain to the estrogen priming required for the preovulatory LH surge and ovulation. As in other mammals, regulation could also be achieved by altering the neural signal directly regulating the pulsatile pattern of luteinizing hormone–releasing hormone (LHRH) and LH release, as well as the preovulatory LH surge (Wilson et al. 1984; Filicori et al. 1986; Schweiger et al. 1990; Veldhuis et al. 1989). Finally, our work in rats suggests another mechanism. In this species, pheromones mediate ovarian cycle length in two ways: making partial surges of LH more or less likely in addition to modulating prolactin and progesterone levels (Schank and McClintock 1992; Stern and McClintock 1998; Beltramino and Taleisnik 1983, 1985; Coquelin et al. 1984; Cohen-Tannoudji et al. 1994).

Implications for Future Research

Gene Expression, Infertility Treatment, and Contraception

Pheromones affect neuroendocrine function and thus regulate gene expression in the sensory systems that carry pheromonal information, as well as in the cells of the hypothalamic-pituitary-ovarian axis. Thus identifying the mechanism by which ovarian cycle pheromones act will be a case study in the social regulation of gene expression. Identifying the specific molecular mechanisms in this naturally occurring pathway could lead to the development of infertility treatment or contraceptives targeted at regulation of the preovulatory LH surge. The pheromonal system could be used to stimulate ovulation in women without spontaneous ovulation but with an intact hypothalamic-pituitary-ovarian axis. By taking advantage of an evolved pathway for regulating ovulation, the effects might be easier to control precisely, avoiding the multiple ovulations currently triggered by systemic administration of LH ago-

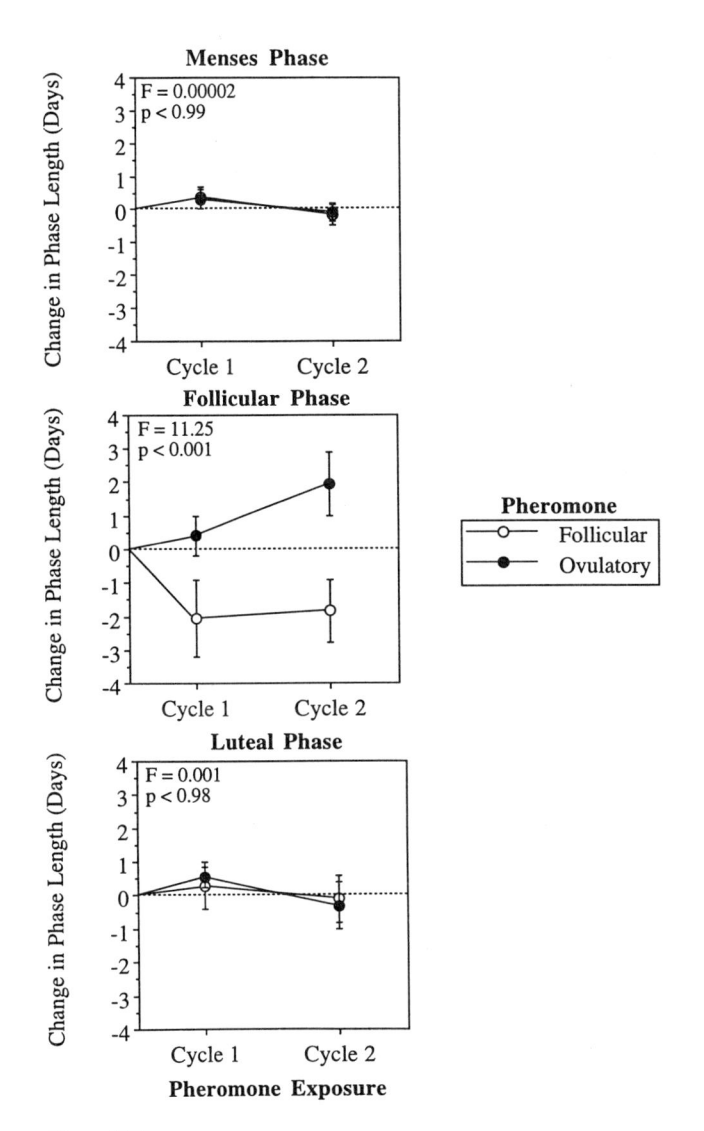

Figure 12.5
Axillary compounds regulated the length of the follicular phase (day after menses to the day before the preovulatory surge) and did not affect the menses phase (duration of menses) or the luteal phase (life span of the corpus luteum: day after ovulation to the day before menses).

nists. Likewise, a contraceptive effect might be achieved by manipulating the phase-delay pheromonal system to not only delay but inhibit ovulation.

Ovarian Dependence

The women in our study produced two functionally distinct pheromones at different phases of their ovarian cycles—the follicular phase and the ovulatory phase. The phase in which the pheromone is produced provides powerful clues to the neuroendocrine mechanism of pheromone production. There are many neuroendocrine differences between the follicular and ovulatory phases: estrogen predominates in the follicular phase, whereas progesterone, pregnenalone, androgens, FSH, and many other hormones rise precipitously at the time of the preovulatory LH surge.

The pheromones could be a direct metabolite of these particular hormones, particularly the steroids. Indeed there is an increase in androgens in the periovulatory period (Bancroft 1989; Bancroft et al. 1983), which may affect the amount of airborne steroids released in the armpit. On the other hand, these hormones may act on target tissue to change production of compounds that bear no direct relationship to the steroid hormones of the hypothalamic-pituitary-ovarian axis. This has proved to be the case in mouse pheromones (Novotny et al. 1998)

Bacterial populations of the armpit vary over the cycle. For example, lipophilic cutaneous diphtheroids play a role in cleaving apocrine secretions from their carrier proteins, and if the size of these populations changes between the follicular and ovulatory phases, then they could produce different types or profiles of pheromonal compounds in the axillae that are linked to the time of ovulation (Zeng et al., submitted).

Pheromone Transduction and Responses

The human priming pheromones are functionally similar to the pheromonal systems which regulate ovulation in hamsters, mice, and rats (Wysocki and Meredith 1987). These species have an accessory olfactory system that mediates pheromone responses (see chapter 13). This system is composed of a neural pathway linking sensory receptors to the neuronal structures of the hypothalamic-pituitary-ovarian axis. The functional parallel has suggested to some that humans may also have an accessory olfactory system. This controversial hypothesis is discussed in detail below under Pheromone Receptor Systems.

Although a tantalizing idea, an accessory olfactory system is not necessary to explain the pheromonal effects. In pigs, pheromonal modulation of estrus in sows is mediated by the main olfactory system (Dorries 1998). Another hypothesis is that the axillary compounds are absorbed directly through the skin of the upper lip or nasal mucosa.

Period of Sensitivity

The discovery of two types of pheromones, consistent with a coupled-oscillator model of social regulation of the ovarian cycle, implies that a female cannot be equally sensitive to these pheromones throughout her ovarian cycle. In other words, a female's sensitivity to two pheromone signals must vary at some point in the ovarian cycle. If females were equally sensitive throughout the cycle (i.e., if their sensitivity did not vary), then every member of a group would tend to respond in the same way at all times to the aggregate pheromone signal. Therefore, we hypothesized, there should be one phase of the ovarian cycle when females are most sensitive.

The literature on odor sensitivity in other mammals points to the periovulatory phase as a good candidate. Sprague-Dawley rats have cyclic changes in odor sensitivity, with peak olfactory sensitivity to odors occurring during a 32-hour period of the ovarian cycle centered at 0600 hours of vaginal estrus (Pietras and Moulton 1974). In addition, estradiol enhances the ability of female rats to recognize the olfactory signature of conspecifics. We hypothesized that peak sensitivity to odors coincides with peak sensitivity to ovarian pheromones. This hypothesis gains theoretical plausibility as a result of the successful match of simulation data with empirical data (Schank and McClintock 1992).

We verified our prediction by conducting simulation experiments with the computer model. For example, we systematically varied the time of pheromone sensitivity to determine when rats are sensitive to pheromones; we changed the hypothesized time of sensitivity by 1-hour increments across the entire 4-day ovarian cycle, running 1000 simulations at each setting and measuring the resultant distribution of synchrony scores and cycle lengths. Our model fit the empirical synchrony data when sensitivity was centered at a range of times between 1400 hours of proestrus to 2200 hours of estrus but fits the cycle length data *only* when it is centered at 0600 on estrus (figure 12.6). Thus, this simulation experiment precisely confirms our initial prediction based on Pietras and Moulton's study of olfactory sensitivity (Pietras and Moulton 1974). Moreover, finding that there is only one time when the model produces realistic results both for synchrony of the group and length of the ovarian cycle demonstrates the power of having built the model with multiple levels of analysis.

Women, too, are more sensitive to olfactory compounds during the periovulatory phase, particularly to steroids such as Exaltolide (Vierling and Rock 1967). Therefore, the parallels between the rat, computer, and human systems suggest that women may also be most responsive to pheromonal compounds at this time. In women, estrogen enhances responsiveness to environmental input through the sensory system.

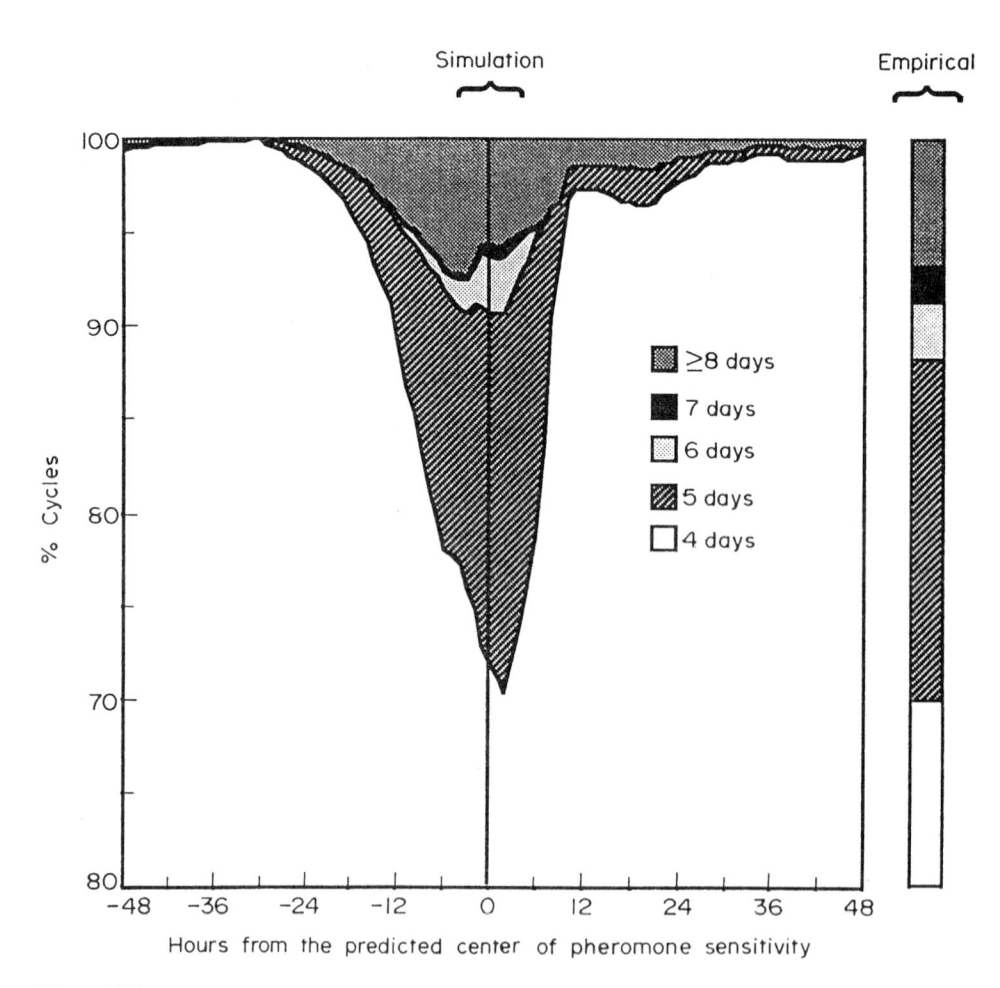

Figure 12.6
Results of a computer simulation experiment to measure the effect of the time of sensitivity to pheromones on the distribution of ovarian cycle lengths produced during the interaction. The simulation data match the observed data only when sensitivity to pheromones is centered during an 8-hour window, covering the time of the preovulatory surge and ovulation.

Sensory thresholds are lowest during ovulation (Henkin 1974; Messant 1976; Diamond et al. 1972). Moreover, estrogen enhances responsivity to sexual stimuli in the social environment (Englander-Golden 1976; Luschen and Pierce 1972). It has also been suggested that estrogen promotes other-directedness (Benedek and Rubenstein 1942; Reinisch 1977), social affiliation (vs. reclusiveness) (Luschen and Pierce 1972), and positive feelings toward others (Ivey and Bardwick 1968; Paige 1971).

Progesterone, the hormone that characterizes the postovulatory phase of the human menstrual cycle, has also been implicated in an individual's responsiveness to the environment. In contrast to estrogen, which enhances responsiveness to the environment, progesterone has been implicated in attenuating environmental stimuli. For example, Benedek and Rubenstein (1942) reported that their subjects became more self-oriented during the luteal phase of the cycle when progesterone is predominant. Benedek and Rubenstein argued that under the influence of progesterone, an individual becomes more oriented toward the body and its welfare, which would serve an important function during pregnancy when progesterone levels are particularly high. It has also been found that children exposed to abnormal levels of progesterone prenatally have personalities characterized as more inner-directed and individualistic than children exposed to abnormal amounts of estrogen prenatally (Reinisch 1977).

Odorless Compounds

We do not yet know the structure of the compounds that regulate ovulation without being detectable as odors. There are many compounds that change over the menstrual cycle (Preti and Wysocki 1998) and exhibit strong individual differences (e.g., producers and nonproducers of short-chain aliphatic acids; Labows and Preti 1991). Preti and his colleagues identified steroids from the pooled alcohol extracts of male and female sweat, yet did not detect an effect on menstrual cycle length (Spielman et al. 1995). Axillary compounds, such as 5α-androst-16-en-3β-ol or (E)-3-methly-2-hexenoic acid, are not likely to be the active compounds because they have a strong odor. On the other hand, the latter compound is odorless when bound tightly with its carrier protein, which would be the case in natural secretions (Spielman et al. 1995). Thus it could be that the larger molecule is odorless, even when the smaller component, which has a pheromonal action, does have an odor.

Individual Differences in Sensitivity and Production

Our experiments with a computer simulation of entrainment in rats revealed that large individual differences in sensitivity to pheromones are an essential aspect of the process. When the computer model did not include individual differences in pheromone sensitivity, the levels of synchrony achieved were unnaturally high, although stasis and asynchrony were still detectable (see figure 12.2B). When there was individual variation in this sensitivity parameter, the levels of synchrony matched the empirical data.

Therefore, we also expected individual differences in sensitivity to the pheromonal compounds in our human study. Indeed, only two thirds of the subjects responded in each condition. It was striking that many of the insensitive women had upper respiratory tract infections that might have prevented the compounds presented on their upper lip from reaching the receptors—a straightforward mechanical explanation for their lack of response.

Our research in this area is based on the working hypothesis that hormones affect sensitivity to environmental cues. This is a functional approach to understanding what role hormones play in emotional responses to social signals and behavior. We start with the assumption that the function of the endocrine system is to aid in an individual's interface with the environment. That is, at the level of the central nervous system, hormones are responsible for modulating the influence of external information on the organism, so that hormonal activity can enhance or depress an organism's receptivity to certain environmental stimuli (Kopell 1969; Selye 1946). For example, estrogen (which rises in the first half of the cycle, just prior to ovulation) has been implicated in female sexual responsivity to the male in rats and in primates (Kopell 1969; Leshner 1975). Estrogen has also been implicated in responsivity to food. Birke and Andrew (1979) found that female rats during estrus (when estrogen is predominant) are more focused and oriented toward a variety of experimentally manipulated stimuli than they are at other phases of the estrous cycle. Finally there are strong individual differences in sensitivity to androstenone that are not related to genetic differences in major histocompatibility complex (MHC) locus (Dorries et al. 1989; Wang et al. 1993; Wysocki and Beauchamp 1984; Pollack et al. 1982).

Among those women who did respond, the effect ranged from a 1-day change in cycle length to a 12-day change. Again, the presence of this variation was predicted by the computer simulation; future research is necessary to determine its cause. There are many possible explanations for this variability: receptor insensitivity to pheromones, resistance of the hypothalamic-pituitary-gonadal axis, limitations imposed by ovarian physiology; and other environmental signals or experiences that may have been more prepotent in controlling ovulation (i.e., environmental stressors, social interactions with men, or sexual activity).

Interactions with Males Women attending an all-women's college were more likely to have shorter and less variable cycles if they were in proximity to men at least three times a week (McClintock 1971); a parallel association has been reported between sleeping with a man and ovulatory cycles (Veith et al. 1983). Obviously, this correlation could be explained in many ways. It does parallel the pheromonal system in many mammals whereby male pheromones stimulate ovulation, both in spontaneously cycling species such as rats and mice and in induced ovulators such as voles (Carter et al. 1990; McClintock 1983a; see also chapter 14). In many mammals, males trigger ovulation, typically by modulating gonadotropin-releasing hormone (GnRH) and LH secretion (Vandenbergh 1994; Dluzen and Ramirez 1983; Dluzen et al. 1981).

Cutler et al. (1986) claimed evidence for male pheromones by reporting that four of seven women who thought they typically had aberrant cycle lengths then had cycles close to 29 days in length when they were exposed to a solution containing androgenic steroids (in contrast to three of nine such women exposed to a placebo; 57% vs. 33% of subjects). If a larger sample size reveals that this difference is statistically significant and if the baseline cycle lengths are measured, not simply recalled, such an effect would indeed warrant further testing for the presence of male pheromones that regulate the menstrual cycle in humans. In order to increase the power of future studies and their interpretation, however, it must recognized that 29.3 days is not the average length of a woman's menstrual cycle until age 32 (Treloar et al. 1967; Vollman 1977) which is a significantly older age than the average age of the women in the study of Cutler et al.

Until appropriate studies are done, there are alternative explanations for the association between shorter cycles and time spent in the proximity of men: less stress from isolation, more romantic interactions, and more sexual activity, to name but a few. Moreover, the direction of causality might be the reverse: women with regular ovulatory cycles might be more likely to seek out the company of men since the hormonal changes underlying ovulatory cycles influence female sexual motivation (see chapter 10). These hypotheses and alternative explanations must be tested rigorously before we conclude that men produce pheromones that regulate ovulation.

Testicular Function

In mammals, testosterone levels in males are increased by odors and pheromones from female conspecifics, particularly when the females are in heat (Vandenbergh 1994). Little research has focused on the pheromonal regulation of testicular function in humans. An anonymous case study reported that a man's beard grew more when he returned to the mainland to see his lover after living isolated on a remote island

(Anonymous 1970). However, it is unlikely that these effects were pheromonal, because the change occurred *in anticipation* of seeing his lover, not just being physically near her. On the other hand, there is an unpublished report that compounds from ovulating women increase testosterone in men (Jutte 1996), an effect that could possibly explain the higher levels of testosterone reported in married men (Persky et al. 1978).

Berliner et al. (1996), in an experiment with men that paralleled animal studies, exposed men to PDD (pregna-4,20-diene-3,6-dione), which is a progesterone derivative and hence an analogue of female compounds. Exposure of men to PDD produced a drop, not a rise, in plasma LH and FSH, followed by a decline in testosterone levels. Unfortunately, PDD is an artificial compound, which is asserted to be an agonist to an unidentified female pheromone. It certainly is a steroid derivative, but we await evidence of receptor binding that it functions as a pheromone agonist.

Berliner and colleagues propose a new term for this synthetic compound, "vomeropherin," because it is purported to change the surface potential of the human vomeronasal organ. However, no controls were run to see if similar effects were produced when compounds were applied to the olfactory epithelium or to demonstrate that the steroid was not absorbed directly into the bloodstream via the respiratory epithelium of the nasal cavity (Preti and Wysocki 1998). Moreover, it is not clear why a unique term is needed for human compounds, unless it applies only to synthetic analogues and not to naturally occurring compounds. The existing term for compounds that activate a VNO is "vomodor," a term applicable to compounds in any vertebrate system (Cooper and Burghardt 1990), although the term implies that such compounds are detected consciously as odors.

This preliminary finding necessitates further investigation before the broad interpretation of the data can be accepted. If PDD is indeed an analogue of a female pheromone, then humans would be the only species in which a female pheromone diminishes, rather than increases, testicular function (Meredith and Howard 1992; Meredith and Fernandez-Fewell 1994). An alternative explanation is that PDD mimics a male pheromone or odor that has a function in the context of sexual competition between males. Again, further research will determine whether it asserts its effects through conscious odor detection, in which case the effects should be interpreted in terms of sexual attraction, rather than competition between males (McClintock 1998). Clearly, tightly controlled studies are needed to establish unequivocally the function and meaning of these artificial compounds.

Behavior and Psychological State

Releasing Pheromones

Releasing pheromones produce some of the most dramatic behavioral effects. A male silkworm moth can detect a female half a mile away and appears compelled to follow the concentration gradient of the bombykol which she releases until he finds her and mates (Schneider 1974). The male pig exhales into the face of the sow, showering her with androstenol, a pheromone that triggers a lordosis reflex so strong that she is said to "stand," immovable, while the male circles around to mount her and mate. The behavioral effects are so prepotent that a breeder need only spray the female with an aerosol and stimulate her flanks to trigger the lordosis that facilitates artificial insemination (see figure 12.7). Pheromones have equally prepotent effects on male sexual behavior. Vaginal secretions of hamsters contain a sexual attractant that immediately triggers the male's mating behavior (Darby 1975; Singer et al. 1986, see also chapter 13) and female rats in heat emit a pheromone that is sufficient to induce erection in a male rat, without physical contact (Sachs 1997a).

Perfumers have long played on the hope and fantasy that there will be a human pheromone that is a potent aphrodisiac. But what are the behavioral data on which these claims are based and is there any aspect of human behavior which is triggered or "released" in such a stereotyped way? Michael and colleagues promoted this idea with their term "copulin" (Michael et al. 1971), vaginal aliphatic acids which putatively "released" sexual behavior in male rhesus monkeys. Goldfoot et al. (1976, 1980) critiqued this interpretation, pointing out that these compounds were neither sufficient nor necessary for copulation in this species (see chapter 10, for a more extensive critique of these studies). Likewise, when Morris and Udry (1978) tested these compounds in humans, they failed to find that aliphatic acids were sufficient to increase the frequency of intercourse in married couples, although there may have been an effect in a subset of couples. Paralleling the trend in the subset, Cutler et al. (1987) and Cutler and Stine (1988) also reported that women were more likely to have intercourse if they wore an extract of women's axillary secretions taken from a pool collected across all phases of the menstrual cycle. As in the Morris and Udry study, it is not clear who was being affected, and whether that effect was pheromonal or mediated by associations with high compound concentrations perceived consciously as odors.

Berliner, Monti-Bloch, and colleagues (Monti-Bloch et al. 1998) have studied androstadienone and estratetraene as putative human pheromones which release a relaxation response and increase self-confidence (Monti-Bloch and Grosser 1991;

Figure 12.7
The boar pheromones, Δ^{16}-androstenol and Δ^{16}-androstenone (left), which breeders spray on sows to induce the "standing" reflex (right), facilitating artificial insemination.

Taylor 1994). They have patented Realm, a series of fragrance products purported to contain "a number of known human pheromones" (Jennings-White 1995), although they are likely synthetic derivatives of naturally occurring compounds (see Preti and Wysocki 1998 for a detailed critique). Their support for this hypothesis did not include any quantified change in behavior or psychological state. Instead, they focused on the putative human VNO, claiming that these compounds change its surface potential (although the VNO is not a necessary part of pheromonal systems). In their marketing claims for these compounds, the authors assert that "[Realm] ... [is] the result of a scientific breakthrough which proves that the benefits of human pheromones go far beyond mere romance" (Jennings-White 1995). However, to date, the only behavioral data known to substantiate such assertions are unpublished daily diaries kept by the subjects while taking these compounds. Thus it is not possible to determine the experimental design, the detectability of the putative pheromones as odors, the exact nature of the response, or the statistical significance of the effect.

Cutler and colleagues also market sex-specific perfumes that contain unidentified compounds that they claim increase attractiveness and self-confidence (Pheromone 1013 and Pheromone 10X); reported to be dehydroepiandrosterone (DHEA), androstenone, and androstenol based on analysis by gas chromatography and mass spectrophotometry (Preti and Wysocki 1998). Recently Cutler et al. (1998) reported that they have made a synthetic compound (also unidentified for proprietary reasons) that increases the sociosexual behavior of men. However, they interpret their data not in terms of the compound's effect on men, but rather in terms of an inferred indirect effect on women. They state that the synthetic compound increased men's sexual activity, but only sexual interactions with women, for example, sexual intercourse or sleeping with a romantic partner; it had no effect on sexual activity that did not involve women, for example, masturbation. Therefore, the authors conclude that the compound was affecting the women partners of the wearer, and not the men themselves. This circuitous logic must be tested directly before these data can be cited as definitive evidence for the existence of human pheromones affecting social or sexual behavior.

There are several other critical aspects of the study that need clarification before it is appropriate to conclude that human pheromones exist which directly make women more likely to have sexual intercourse. (1) Do men naturally produce any compound for which this is the synthetic analogue, and under what circumstances? (2) Did the man's aftershave smell noticeably different when it contained the compound or was the compound not consciously detectable? If it was detectable, then the pheromone concept need not be invoked. (3) A reanalysis of the experimental design revealed that the men in the pheromone group, who were at least dating, were significantly more likely to be married or in a steady relationship ($P = .05$; reanalysis of Cutler et al. 1998). Would the same results be obtained if the men exposed to putative pheromones were not already in a social relationship that makes intercourse and sleeping together more likely? (4) Is the effect behaviorally or biologically meaningful? Many of the men who were counted as increasing their rate of sexual intercourse in fact went from having no intercourse in 2 weeks to once in 6 weeks, and even then, that one occurrence did not happen until they had been using the compounds for 5 to 6 weeks. (5) Would a statistically significant effect have been found had the experimental condition not been three times longer than the baseline?

A more subtle form of behavioral releaser has been hypothesized for volatile steroids such as androstenol, particularly when they occur at concentrations high enough to be easily detected as odors (Doty 1981; Schaal and Porter 1991). Although androstenol has been called a male sex attractant in the marketing claims of various fragrance companies, in the popular press, and in textbooks, most studies are interpreted most parsimoniously in terms of competition among men, not their sexual attractiveness to women. Androstenol made men avoid bathroom stalls and theater seats (Gustavson et al. 1987). Paradoxically, it made men rate other men as more sexually attractive than when androstenol was not present. Women did not show these effects (Black and Biron 1982; Filsinger et al. 1985). Androstenol also decreased the subjects' sense of their own attractiveness and depressed their moods, as does living near a pig farm—a natural source of the easily detectable strong odor of androstenol (Filsinger et al. 1984; Gustavson et al. 1987; Schiffman et al. 1995).

Given the high odor concentrations used in these studies, it is likely that these effects were mediated via odor associations and not pheromones. Moreover, there were significant individual differences in response: if the subject found the odor of androstenol pleasant, it had opposite and atypical effects on behavior (Kirk-Smith et al. 1978; Filsinger et al. 1985, 1990; Gilbert and Wysocki 1987). Thus, it would be important to determine how pleasant androstenol smelled to the college women in the study by Cowley et al. (1991). Women who wore a tube necklace overnight containing 0.25 mg of androstenol reported having had more conversations with men than did women wearing a control odor in the necklace.

The context of this effect is important for its interpretation. Typically, these young women had fewer conversations with men than with women. After wearing high concentrations of androstenol, however, they were more likely to report having conversations with men, attenuating the reported sex difference (Cowley et al. 1991). Note that the measure was reporting having had a conversation—there were no recorded data of actual social behavior. Moreover, from these data one cannot generate specific hypotheses about the mechanisms mediating this association. Did the compound affect the woman herself or the men around her? (Cowley et al. 1991).

Signaling Pheromones

Signaling pheromones have a much less dramatic effect on behavior because they only provide a recipient with information about the condition of the sender, and thus indirectly affect behavior. Signaling pheromones convey information about the sender's environmental, biological, or psychological state. Nonetheless, their effects are more specific than those of other semiochemicals (odors that carry information such as the odor of a person's breath, which is sufficient to identify his or her gender (Labows and Preti 1991; Muller-Schwarze 1997; Doty et al. 1982). Moreover, the way that signaling pheromones affect behavior depends entirely on how other brain systems utilize this information as well as information from other sensory information. In animals, such signaling pheromones convey information such as the dominance status of a male elephant in musth, the individual identity of hamsters or mice, or the fact that food is not poisonous (Galef and Heiber 1976 Yamazaki et al. 1994; Rasmussen et al. 1994).

Marriage Partners and the Major Histocompatability Complex

Ober and colleagues (Weitkamp et al., in press) have provided the most compelling evidence that humans have signaling pheromones containing information about the MHC of the sender (the genes that encode cell surface proteins necessary for recognition as self). Their data indicate that this genetic information may play a role in selecting a marriage partner with a low risk of miscarriage (or perhaps more precisely, to avoid a partner associated with a high risk of miscarriage; Ober 1992). This remarkable example of nonrandom marriage partners was found in the Hutterites Brethren, a small, ethnically homogeneous, socially closed religious community in South Dakota. The population originated in the Tyrolean Alps (1528 C.E.) and has a limited number of HLA alleles, one class of MHC molecules. Despite this low genetic variability, married couples are likely to have different HLA types, indicating that they avoid marrying partners with the same maternally inherited MHC haplotype. Not suprisingly, large populations and unconstrained mating patterns have precluded detecting this pattern in some populations (Rosenberg et al. 1983; Hedrick and Black 1997 Beauchamp and Yamazaki 1997).

That pheromones might mediate the nonrandom selection of marriage partners among the Hutterites is suggested strongly by the ability of mice to detect MHC differences in a single locus (Yamazaki et al. 1988; Egid and Brown 1989; Potts et al. 1991). Moreover, humans (as well as rats) can also discriminate the odors of urine of mice with different MHC types (Beauchamp et al. 1985; Gilbert et al. 1985). People can also discriminate human MHC type from compounds left on worn T-shirts (Wedekind et al. 1995; reviewed in Foidart et al. 1994), as can rats (Ferstl et al. 1991). Indeed, women rated the odors of MHC-dissimilar men as more pleasant than those with an MHC similar to their own, suggesting a psychological mechanism that could mediate the nonrandom marriage patterns observed in the Hutterites (Ober et al. 1997).

It remains to be determined which member of the Hutterite couple is responsible for choosing a marriage partner nonrandomly. It may be the woman. Urban women, who do not live in a closed religious community, report that they use smell as the single most important sensory modality for choosing a sexual partner and that body odor is the most potent sensory inhibitor of sexual arousal (Herz and Cahill 1997). Men, in contrast, use visual appearance as much as odor in choosing a partner and did not report that strong odors dampen their sexual arousal. On the other hand, the effectiveness of a signaling pheromone need not require conscious detection as an odor.

Parent-Infant Bonding

It has been suggested that human pheromones facilitate infant attachment because mothers readily detect their newborn's odors (Russell 1976; Porter et al. 1986). This ability, however, may be mediated by conscious odor detection. For example, the odors of babies are certainly consciously detectable as such, and the taste of compounds in mother's milk affects later feeding (Mennella and Beauchamp 1991), so the pheromone concept need not be invoked. However, signaling pheromones commonly provide redundant information and thus another sensory system could mediate the same response to the same information. On the other hand, they may be associated with compounds that do serve as pheromones that modulate a state of parental attachment.

Modulators—A New Concept and Type of Pheromone

We propose the term "modulator" to describe a heretofore unrecognized type of behavioral pheromone. This third type of behavioral pheromone has functional effects different from those of primer and signaling pheromones. State modulator pheromones change how an individual behaves or reacts to a particular situation. Its behavioral effects are less stereotyped than those triggered by a releaser pheromone, but more specific than the general effect of simply acquiring information about another's condition via a signaling pheromone. They are a part of the mechanism that guides ongoing behavior that is appropriate for a given context. State modulators are the type of pheromone most likely to guide human behavior (*if* indeed pheromones do have a nonvestigial role in human psychology).

Human behaviors and psychological states are complex and determined by the interplay of a wide variety of stimuli. Thus, it is unlikely that a pheromone alone would be sufficient to trigger stereotyped behavior when a person is reacting to a rich social and physical environment (McClintock 1998b). Pheromones would be unique among sensory systems if they could trigger complex behavior in a "fixed action pattern."

Contextual stimuli typically have a strong role in determining behavior. For example, being informed that an ambiguous sound is indeed language enables humans to recognize meaning instead of just random noise; the social setting determines whether insulting a person will make that person laugh or take umbrage. This contextual information is encoded by the state of the nervous and endocrine systems by altering such diverse processes as attention, arousal, and emotional state. It is via modulating these states that pheromones can modulate behavior. Alternatively, the other factors creating the behavior may be more prepotent and pheromones will not have an effect on observable behavior or conscious perception even though they are modulating the state of some part of the nervous or endocrine system (see chapter 10 for a parallel discussion of steroid hormones as modulators of sexual behavior).

We hypothesize that the effects of modulating pheromones are specific, for example, altering neural activity in a specific layer of the amygdala or nucleus in the hypothalamus. One major effect of these state changes is a change in sensitivity to sensory inputs. Thus, pheromonal effects can be measured in terms of sensitivity to and interpretation of stimuli. In addition, modulators may change the integration of sensory information or its integration with a particular response, changing either the probability of a behavior or a psychological state.

Perception of stimuli, sensorimotor integration, and psychological states are regulated by hormones, as well as by specific neural activity. So, as the neural and endocrine mechanisms of modulators are identified, the distinction between them may seem blurred. But modulators act within a short time frame, minutes and hours, and change behavior relatively quickly. In contrast, priming pheromones operate over days and weeks to produce long-term changes in endocrine function or development (Albone 1984).

The effects of modulating pheromones have a specific and unique neuroendocrine effect; other sensory signals cannot be substituted, that is, a word, sound, or image will not produce exactly the same response. Nonetheless, they cannot drive or create behavior de novo. It is the context and ongoing behavioral interactions that are the prepotent, primary determinants of behavior. State-modulating pheromones modulate this process. Moreover, a person's immediate situation and behavioral interactions will determine whether the state change is detectable, either as an observable behavior or in conscious awareness.

In sum, modulating pheromones affect (1) regulation of all sensory inputs during exposure and (2) the state or mood of the individual. Together, these contribute to the final common pathways that will determine the actual action or perception of other sensory information. In many situations, the changes induced by modulating pheromones may indeed be unconscious changes.

Pheromone Receptor Systems

Pheromones are not odorants because they do not exert their effects via the cortical structures that process conscious smell detection, that is, the primary cortex, temporal lobes, or the piriform and right orbitofrontal cortex (Zatorre et al. 1992). Therefore, pheromones have a variety of potential receptor systems other than those of the primary olfactory system. Some of these are likely to operate by the same principles of the olfactory neurons (Bartoshuk and Beauchamp 1994), namely, binding at a receptor protein and changing neural function of the receptor cell. Some pheromones, however, may operate via neuroendocrine principles and be carried in the blood to receptor cells, changing their genetic or molecular function. Thus, although some pheromones may have an odor at high concentrations, that is, be processed by primary olfactory cortex and downstream neocortical processes of the main olfactory system, this processing is not necessary for their function.

Vomeronasal System

In many mammals, pheromones bind at receptors in a specialized chemoreceptor, the vomeronasal organ (Wysocki and Meredith 1987; Johnson et al. 1985; Meredith and Fernandez-Fewell 1994; see also chapter 13). The mammalian VNO is typically located rostrally in the nasal cavity, near its anterior opening, within the first third of the cavity. Therefore it is much closer to the nares than to the olfactory epithelium, which is found at the most caudal and superior portion of the nasal cavity, tucked back up along its roof under the cribiform plate. Thus, airborne pheromones and odorants pass by the opening of the VNO and then a large surface area of nonsensory respiratory epithelium before they reach receptors for the main olfactory system's epithelium.

The VNOs rostral location on the septum, near the nares, means that a sniff can bring large chemical compounds to the VNO, which could not travel back up to the olfactory epithelium (e.g., nonvolatile compounds over 66,000 daltons). In some species, the opening of the VNO is adjacent to the nasopalatine duct, which connects the nasal and oral cavities through the roof of the mouth. In other species, such as snakes and elephants, the VNO receives chemosignals only via the oral cavity and the nasopalatine duct, not the nares and nasal cavity (Meredith and Burghardt 1978; Halpern 1987; Rasmussen et al. 1998). When viewed from the comparative perspectives of evolution and embryology, it is easy to hypothesize that these various locations evolved from the same location in a common ancestral species. Indeed, the organ is evolutionarily quite primitive, and is even found in the lamprey (Dyer and Berghard 1998).

This evolutionary analysis, however, cannot be used to argue that there must be a functional human VNO. The human VNO could well be vestigial, equivalent to the wings of the flightless cormorants in the Galapagos. Although clearly present in New World primates, we do not know whether the Old World primates most closely related to man, that is, chimpanzees and orangutans, have functional VNOs (Zingeser 1984; Preti and Wysocki 1998)

Humans do have a set of vomeronasal receptor genes analogous to two different families expressed in the VNO of rats and mice (Dulac and Axel 1995; Herrada and Dulac 1997; Matsunami and Buck 1997; Ryba and Tirindelli 1997). However, at least one contains a stop codon in the encoding region, suggesting that they may not be expressed in humans (Preti and Wysocki 1998). Moreover, the ligand for these receptors has yet to be identified in rodents, let alone humans—tantalizing problems yet to be solved. Equally tantalizing is the weak expression of immunoreactivity to vomeromodulin in the mucociliary complex of the vomeronasal epithelium of humans (Krishna et al. 1994). Vomeromodulin is hypothesized to be a transporter protein, capable of presenting a pheromone to a receptor in a mucous or aqueous solution, as has been so elegantly demonstrated in elephants (Rasmussen et al. 1998).

In human fetuses, there are two "septal pits" purported to open into the canal of the VNO. These structures form in the first month after conception, begin logarithmic growth in the second trimester, and attain their adult length of approximately 10 mm after birth (Boehm and Gasser 1993; Smith et al. 1997; but see Kjaer and Fischer Hansen, 1996a,b). Female fetuses appear to have a longer VNO, with a larger volume, than do males. Boehm and Gasser (1993), however, failed to find LHRH-staining evidence for a vomeronasal nerve in 36-week-old fetuses, suggesting that the developing nerve atrophies before birth. Thus, the functional activity of these clearly recognizable and sexually dimorphic structures has not been tested experimentally in the fetus or neonate, and any functional hypotheses are purely speculative at this time.

The human VNO was first described by Frederic Ruysch in case reports of soldiers with facial wounds (Ruysch 1703). Subsequently, Jacobson described the organ in animals, giving it his name (Jacobson 1811). Potiquet (1891) studied adult human cadavers (figure 12.8) and reported that the VNO was not easy to visualize in all adults (Potiquet was using only sunlight and did not have the benefit of a speculum with magnification!). In the 1930s several anatomists declared it completely nonexistent (Pearlman, 1934; Crosby and Humphrey 1939), as did others later (for reviews, see Pearson 1941; Wysocki 1979).

Nonetheless, recent histological and immunohistochemical studies of adult tissues have revealed that tissue from the putative organ is distinct from its adjacent respiratory epithelium (see review in Moran et al. 1994). Indeed, several different groups (Moran et al. 1991; Garcia-Velasco and Garcia-Casas 1995; Garcia-Velasco and Mondragon 1991; Stensaas et al. 1991) have recently concluded that, based on macroscopic examination, it is present in the majority of adults.

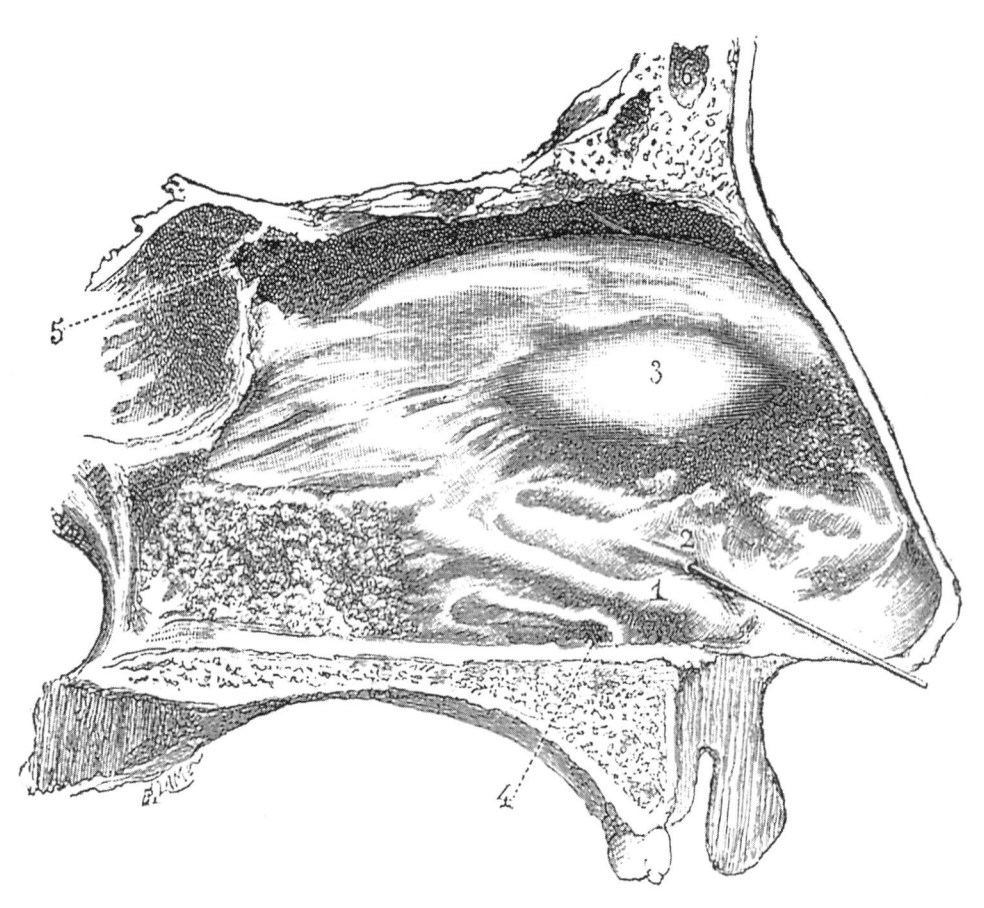

Figure 12.8
Potiquet's 1891 engraving of the septal structures not depicted in modern anatomy texts, which typically draw the septum as a smooth tissue. The caption reads: 1. Barrel made in part by Jacobson's cartilage. 2. Opening of Jacobson's canal, into which a stylus has been inserted. 3. Septal tubercle. 4. Nasopalatine infundibulum connecting the squelette to the nasopalatine canal. 5. Opening of the sphenoidal sinus. 6. Frontal sinus.

Nonetheless, in our own simple macroscopic examinations of normal young adults, we saw "septal dimples" in over 90% of subjects but not where Moran and colleagues (1994) described the opening of a putative VNO (figure 12.1 depicts it at the junction between the bottom of the septal cartilage and the top of the vomer bone). We found it, usually bilaterally, at the bottom of the vomer bone, near the floor of the nasal cavity and the nasal crest of the maxilla.

This confusion over its location in adults and the fetus, as well as the use of markedly different observation techniques and anatomical criteria, is a likely source of a controversy that has persisted for three centuries. We have concluded that there are clearly several different types of "septal dimples," anatomical structures in adults which must be studied further. Some may be associated with the incisive canal (Netter 1968), or a duct associated with a vomeronasal structure (Proctor and Anderson 1982). Others may be related to a nasopalatine canal (see figure 12.8; Potiquet 1891) or to large respiratory glands (Takami et al. 1993) We need to determine the morphological features that define these structures and make them reliably recognizable on macroscopic examination by direct visualization or endoscopy. Only then will it be possible to quantify individual differences in the presence and shape of the putative human VNO and the potentially related structures of the anterior nasal septum.

Then comes the question of function. Mammalian VNOs with well-established function have unciliated bipolar receptor cells on the medial wall of the lumen. The axons of the VNO receptor cells penetrate the basement membrane of the sensory epithelium, traverse the nonsensory respiratory epithelium, penetrate the cribiform plate to synapse in the accessory olfactory bulb (AOB) in the brain, having bypassed the main olfactory bulbs (Wysocki 1979).

The epithelium of the human VNO has bipolar cells containing two molecular markers characteristic of neurons: neuron-specific enolase and protein gene product 9.5 (note, however, that these markers are not unique to sensory neurons and are found in neuroendocrine cells as well; Takami et al. 1993; Preti and Wysocki 1998). Sensory neurons with axons that leave the VNO have yet to be identified in humans, as has a distinct AOB. These nerves would be unmyelinated and hence particularly difficult to find; moreover, the accessory and main olfactory bulbs can commingle in Old World primates, precluding easy identification of distinct pathways. All of this anatomical evidence is required before we can conclude that the human VNO is functional.

Monti-Bloch, Berliner, and their colleagues stated that specific compounds raise the surface potential of the VNO epithelium and that men and women have different response profiles (Berliner et al. 1996; Monti-Bloch and Grosser 1991). Indeed, a

0.3-pg pulse of specific steroids raised the surface potential of women threefold and produced a six-fold increase in men (relative to similar effects of clove oil or propylene glycol, which also produced statistically significant changes). However, the results were published without identifying the compounds, making it difficult to evaluate and interpret them until a patent application revealed they were androgens and estrogens (Berliner 1994).

Androstadienone was particularly effective in women, producing a six fold increase to 10 mV x s in contrast to estratetraenol, which has an eightfold effect in men. None of the steroids changed the surface potential of the olfactory epithelium, making the results hard to interpret. Given that these steroids have an odor at these concentrations, why was there no change in the main olfactory epithelium if these measurements are appropriately interpreted as a summated receptor potential? Preti and Wysocki (1998) point out that it is possible these potential changes could have occurred in the free nerve endings of the trigeminal nerve or the muscle layer of the nasal cavity, which is sexually dimorphic.

These data are potentially consistent with the hypothesis that the VNO mediates response to human pheromones, but they are far from sufficient proof. We need to determine if the VNO is a neuroendocrine structure or strictly neural. If it is a neural system, then we must show that the potential change on the epithelium is mediated by binding at specific receptors and changes in gene expression and molecular function, resulting in a change in the neural activity of receptor cells. Then we need to determine if there are neurons that project from these receptor cells and the synaptic pathways that project to the brain, the existence of which has been called into question (Preti and Wysocki 1998). Finally, some of these pathways must project to brain structures other than the main olfactory system—perhaps the amygdala or hypothalamus, as do the vomeronasal systems in other mammals.

The Nervus Terminalis

All vertebrates have another, very simple, chemosensory system: the nervus terminalis or terminal nerve (Wysocki and Meredith 1987) (figure 12.9). The terminal nerve originates from the olfactory placode in the nasal cavity where the neurons have chemosensory receptors in the submucosal plexus of the nasal septum. In humans and primates, the fibers then traverse the septum and should travel along with the vomeronasal nerve. Then, with the olfactory nerve on the inferior surface of the brain, it enters near the anterior perforated substance and projects to the midline forebrain (Brookover 1914; Pearson 1941). The fibers and cells of the nervus terminalis system contain LHRH and synapse in the hypothalamus (Schwanzel-Fukuda et al. 1996), which regulates gonadal function, along with the accessory and main

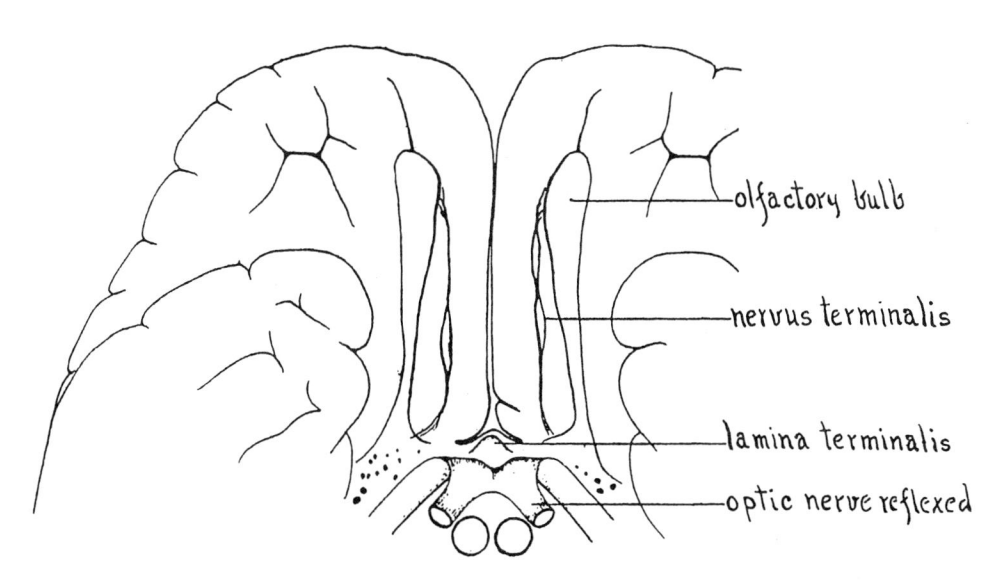

Figure 12.9
The nervus terminalis in a variety of species.

olfactory systems. This little-studied system deserves further attention to determine whether it mediates the effects of ovarian-dependent, or any other human pheromone.

Main Olfactory System

Pheromones can be detected consciously as odors if concentrations are high. These effects are presumably mediated by binding at specific receptors in the main olfactory system, which then project to the primary olfactory cortex (Gower and Ruparelia 1993; Sullivan 1997). Presumably they would act similarly to odors, activating immediate early genes such as *c-fos*, setting in motion a cascade that affects behavior such as the lordosis reflex in rats and mating behavior in male hamsters (Sagar et al. 1988; Pfaus et al. 1994; Pfaus and Heeb 1997; Fiber and Swann 1996).

5α-androst-16en-3α-ol (as well as its ketone, androstenone) is typically rated as unpleasant by women and affects their moods adversely (Filsinger et al. 1985; Grammer 1993; Benton 1982; Schiffman et al. 1995; Gower and Ruparelia 1993). Kohl and Francoeur (1995) demonstrated that women can detect the odor of androstadieneone at lower concentrations than can men; they are more sensitive and describe the odor variously as musky, sweet, tart, or "like sweat"—remarkably different responses by women to the same stimulus. Gonadal state affects sensory thresholds (e.g., see Doty 1981; Vierling and Rock 1967). That the putative pheromones described above can be detected consciously makes it imperative that studies of response be carefully designed to control for and rule out effects that are mediated by preconscious or unconscious associations with odors.

Perfumers have long known that humans perceive animal pheromones as odors. Civetone and muskone have been used in perfumes since 2000 B.C. Civetone is produced by a specialized gland in cats and has a pheromonal effect on their dominance behavior (Agosta 1992). Likewise, muskone, produced by the tarsal gland of the musk deer of Asia, acts as a pheromone to mediate dominance and as a sex attractant. Even now, perfumes are sold containing androstenol, the pheromone that the boar forcibly exhales in the face of the sow, immobilizing her so that he can walk around her, mount her from the rear, and copulate. Not exactly the same behavior claimed in the marketing of these compounds for humans!

Interestingly, these pheromonal effects in pigs can be mediated by the main olfactory bulb and do not require a VNO or AOB (Dorries et al. 1997). Likewise, the main olfactory system fine-tunes pheromonal responses in hamsters (Fiber and Swann 1996). Why, then, are they still considered pheromonal systems? It is because a pheromonal system is not defined anatomically, but functionally. Micromolar amounts of a compound are sufficient to instantly trigger a highly stereotyped behavior, standing immobile in a lordosis posture, suggesting a direct neural projection to a neural circuit underlying the behavior that is not dependent on complete conscious processing by the primary olfactory system. The behavior does not develop through a learned association with an odor; it is elicited the very first time that a pubertal sow comes into heat and is sprayed in the face with androstenol from an aerosol can (Boar-Mate).

Pheromonal signals may also be transduced by the main olfactory bulbs in Old World primates and in humans. In rhesus monkeys (*Macaca mulatta*), the fibers of the primary and accessory olfactory systems are so intermingled that they cannot be distinguished. Indeed, Turner et al. (1978) report the absence of an AOB in the rhesus monkey (Turner et al., 1978). Thus, immunocytochemistry will be needed in these species to determine if the putative VNO projects to an AOB, the main olfactory bulb, or to specialized nuclei that are embedded in the matrix of the main olfactory bulb. These systems are not readily distinguishable without single-cell recording or Dye I tracing techniques.

Other Chemosensory Systems

Salamanders deliver pheromones directly (Houck et al. 1997). In the mountain dusky salamander (*Desmognathus ochrophaeus*), the male scratches the female with his premaxillary teeth and then inoculates the female by rubbing his mental gland directly on the wound. Thus there is a precedent for the skin as a site of pheromonal reception (note that this is not exactly skin absorption because the male "injects" the pheromone into the female's circulatory system).

In humans, however, water-soluble proteins do not penetrate the lipid barrier of cutaneous epithelial cells. For example, topical anesthetics such as xylene are only effective if they are emulsified with a glycoprotein such as Emla (lidocaine; prilocaine) that enables skin penetration, and even so a depth of a few millimeters is achieved only with an hourlong exposure protected by an occlusive dressing.

Steroids pass through the skin more readily. Testosterone and estradiol can be administered via a skin patch, significantly raising blood levels of these steroids after several hours. Thus direct absorption through the skin is an unlikely mode of pheromone reception except possibly for steroidal pheromones, which are primers and act over a long time frame.

The nasal cavity, mouth, vagina, and penis are sites of receptors or direct absorption through the mucosal membrane. Steroids and GnRH agonists applied to the nasal mucosa raise blood levels of hormones within minutes. Finally, there are well-understood chemoreceptors on the tongue. But these mediate the sensation of taste and do not project to either the main or accessory olfactory systems. There is no species known to have collateral projections from the tongue to the brain that could have pheromonal effects by directly altering neural state, behavior, or endocrine function. Thus, although the tongue is specialized for chemoreception, it is unlikely to mediate pheromonal effects (see review in Diamond et al. 1996).

Pheromone Release

Pheromones are released into the air and other species members via a wide variety of body fluids. They are also released by specialized glands and may even be carried by the millions of skin cells sloughed daily into the environment. All studies of human priming pheromones (Russell et al. 1980; Cutler et al. 1986; Preti et al. 1986; Stern and McClintock 1998) collected compounds from the armpit (axilla), where exudates, glandular secretions, and sloughed skin cells are produced in a microenvironment containing colonies of commensal bacteria, which may further metabolize these compounds.

Exudates

The animal literature provides precedents for pheromone release from every human body exudate. Eccrine sweat contains pheromones which mediate sexual attraction in horses and synchrony of ovulation in cows (Izard and Vandenbergh 1979). Because eccrine sweat reflects blood concentrations of many molecules, it is likely that compounds carried in eccrine sweat are also released via tears, nasal secretions, saliva, and breath (Doty et al. 1982; Spielman et al. 1995). Detection of testosterone in saliva indicates that this, or breath, may be another source.

Urine is the fluid best known for pheromone release. Dogs mark their territory by urine carrying pheromones, as do elephants in musth (Rasmussen et al. 1993). Feces too contain pheromones. Female rats release a pheromone in their feces that helps their pups first orient to their natal nest, and, later in development, learn what foods are edible (Leon and Moltz 1972). While middle-class Americans living in the 1990s do not find urine or feces to be a likely source of human chemosignals, people in cultures that use communal toilets are exposed to them many times a day. In naked mole rats, reproduction is regulated by pheromones concentrated in the communal toilet area of their underground burrow (Jarvis 1981). Exposure to urine and feces is not restricted to so-called primitive cultures. The smell of urine very much captured the French social imagination in the late 1800s (Corbin 1986). Indeed, today a large percentage of the world's population does not use flush toilets. Thus, when considered in evolutionary time, exposure to compounds released in urine and even feces is an integral part of humanity and our social life.

Vaginal secretions (or the more poetic Mandarin term, *yinshui*, "hidden waters") are a ubiquitous source of female sexual attractants. The female hamster is normally solitary, living in the desert by vigorously defending the meager resources of her territory. Marking the borders of her territory with attractive vaginal secretions is one of the earliest signs that a female is coming into heat; they attract the male to her territory and she leads him into the burrow to mate (Darby 1975; Singer et al. 1997). Cervical mucus from cows changes the ovarian function of other cows (Izard and Vandenbergh 1979, 1982).

Male ejaculate can also contain pheromones. The male pheromone androstenol is found in the seminal fluid of boars (Gower 1972). In the guinea pig, it is compounds in the ejaculate that are absorbed through the female's vaginal epithelium and brings her out of heat, reducing the risk of sperm competition from subsequent mating with a rival male (Roy and Wilson 1981).

Specialized Glands

Many species have specialized glands from which pheromones are released, either into the air or by rubbing onto a surface. These glands are located near the vagina, testes, and anus of rodents, the ventrum of hamsters, the chin of gerbils, and the parotid glands in the mouth of boars.

Humans have no such obvious glands. However, we do have glands whose function is unknown and it has been argued that they are potentially a site of human pheromone release, particularly after an emotional stimulus (reviewed in Cohn 1994). Most commonly discussed are the apocrine glands, which are located in the armpit and anogenital area, along with hair and the eccrine sweat glands (Shellery et al. 1960), as well as the area around the nipples (areolae), and even the forehead, eyes, and ears (Shellery et al. 1960; Craigmyle 1984; Montagna and Parakkal 1974; Ebling 1989). Also mentioned are the sebum glands of the mouth and lips (Nicholson 1984) and the newly discovered apoeccrine glands, also localized in the axillae.

Chimpanzees (*Pan troglodytes*), which exhibit menstrual synchrony (Wallis 1985), have apocrine glands comparable to those in humans, as do gorillas (Cohn 1994). It is not known whether these glands occur in the other primates that exhibit ovarian synchrony while living in captivity—*Macaca fascicularis*; *Papio anubis*, and *Papio cynocephalus* (Wallis 1985, 1989; Wallis et al. 1986); they do not in a wild population of *P. cynocephalus* (J. Altmann, personal communication, 1997). They are also present in cows, which also exhibit ovarian synchrony (Cohn 1994; Izard and Vandenbergh 1979, 1982).

Apocrine glands are more common in humans than in any of these other primates. They are present diffusely over the body of fetuses, but disappear before birth in all but the aforementioned locations (phylogenetically, they may have evolved to serve as thermoregulation in species with hair Cohn 1994). Their secretions are viscous and milky, a complex mixture of fatty acids and steroids, as are the substances of other mammalian pheromone glands. Apocrine glands open directly into the hair follicle from which their compounds are dispersed along the hair shaft. In contrast, the adjacent apoeccrine glands open directly onto the skin and produce relatively large amounts of clear sweat (Sato et al. 1987). Both apocrine and apoeccrine glands become active at adolescence and inactive toward the end of the reproductive life span (Cohn 1994).

Most exciting is the identification of a lipocalin in apocrine secretions, identified as apolipoprotein D, an $\alpha_{2\mu}$-microglobulin carrier protein similar to those used in other mammalian odor and pheromone systems (Preti et al. 1997). A mixture of straight-chain, branched, and unsaturated acids (C_6–C_{11}) is also synthesized and secreted by apocrine glands. These compounds produce a characteristic underarm odor, but only after being separated from their carrier proteins by microorganisms living in the armpit. The potential pheromonal effects of these compounds have yet to be studied, along with identification of the many other compounds in apocrine secretions.

Women and men have areolar (surrounding the nipple) apocrine glands. Their products are little studied, although theoretically they are candidates for pheromone release. Equally neglected in studies are the products of Bartholin's glands in the vagina and Cowper's gland in the male reproductive tract. It will be difficult to determine whether either of these glands releases putative pheromones without having identified their chemical structures or developed a bioassay-based functional effect that satisfies the definition of a pheromone.

The Skin

With the growing practice of bathing daily, many assert that our culture has us remove putative pheromones produced in body products, exudates, and secretions. Despite modern hygiene, however, a person still sloughs 40 million skin cells each day and these cells may contain putative pheromones or carry the exudates and glandular products described above (Cohn 1994). Indeed, androstadienone and estratetraene, the putative releasing pheromones discussed earlier, have been isolated from skin cells, as well as been found in peripheral blood and semen (although the method of isolation and identification has not been described; Berliner et al. 1991; Preti and Wysocki 1998). There is little research on other compounds contained in cells sloughed from the skin, scalp, or the hair itself (reviewed in Labows and Preti 1991).

Certainly, skin cells are the primary source of the putative signaling pheromones containing information about the MHC complex. This information is encoded in the genome and expressed in skin cells as different classes of protein moieties on the cell surface. Humans, mice, and dogs can discriminate individual differences in MHC, although it is not clear whether this is done through unconscious or conscious odor detection or a pheromonal response (Wedekind et al. 1995; reviewed in Foidart et al. 1994; Ferstl et al. 1991; Yamazaki et al. 1988; Egid and Brown 1989; Potts et al. 1991; Beauchamp et al. 1985; Gilbert et al. 1985).

Potential Chemical Structures: Production, Activation, and Vagility

Structure and Production

Animal pheromones have strikingly different chemical structures. It is as if this type of communication is so ancestral that many different molecules have evolved or been co-opted to serve pheromonal functions. Some pheromones also serve hormonal or physiological functions within the individual, others are metabolites of hormonal and physiological pathways, and still others are unique compounds with no known function other than pheromonal communication.

Some pheromones are a single compound (Schneider 1974) while others are a cocktail of multiple compounds in exactly the correct proportions (e.g., ketones in Asian elephants; Rasmussen and Schulte 1997). The first identified was the silkworm moth's female sex attractant, bombykol, a complex alcohol compound (Agosta 1992).

It was long assumed that "simpler" species such as invertebrates would have simple compounds and that "complex" species such as mammals would have complex cocktails. This has proved false. Androstenone makes sows receptive (Dorries et al., in press; Galef and Heiber 1976). Salamander pheromones are composed of two or more glycoproteins (Houck et al. 1997; Houck and Reagan 1990). The single decapeptide sodefrin is a male sex attractant in red-bellied newts (*Cynops pyrroghaster*; Takahashi et al. 1997). Female goldfish produce $17\alpha,20\beta$-dihydroxy-4-pregnen-3-one, which increases gonadotropin II in the male along with milt volume and sexual behavior (Zheng and Stacey 1997). Male mice produce several different androgen-dependent volatile metabolites, each of which independently accelerates puberty in female mice: 6-hydroxy-6-methyl-3-heptanone; 2-*sec*-butyldihydrothiazole, 3,4-dehyro-exo-breicomin; α-and β-farnesenes. In contrast, the same chemosignal, 2,5-dimethlypyrazine, an adrenal-mediated urinary metabolite, mediates both delay of puberty onset in mice and the suppression of ovarian cycles in groups of adult females (Ma et al. 1997; Novotny et al. 1997). Therefore, in considering the kinds of human pheromones, one cannot know a priori what their chemical structures will be.

Moreover, human pheromones may not even be produced by humans, but rather by commensal bacteria. For example, the preputial gland of boars contains high concentrations of the bacteria necessary for production of their sex pheromone, although pheromones are not secreted by that gland (Gower 1972). Likewise, it may be bacterial action that interconverts the steroids that have been hypothesized to be human pheromones. Aerobic bacteria produce testosterone metabolites: 4,16-androstadienone, androstadienol, 5α-androstenone, and 3α-androstenol (Nixon et al. 1986; Rennie et al. 1989a,b; Labows and Preti 1991). Axillary hair contains the same steroids (Nixon et al. 1988) and ethanol extracts of axillary compounds contain dehydroepiandrosterone sulfate (DHEAS), as well as aliphatic acids (Preti et al. 1986). The natural product chemistry of the axillae is indeed complex and great headway in understanding it has been made by Preti and colleagues (Preti and Wysocki 1998).

Activation

Many pheromones cannot travel to receptor sites or exert their effects without being bound to a carrier protein. Carrier proteins play an essential role in the pheromonal signals of hamsters (Singer 1991), pigs (Booth and White 1988), mice (Robertson et al. 1993, 1997; Novotny et al. 1997; Mucignat-Caretta and Caretta 1997), and rats (Hinkens et al. 1997). Indeed, before the existence and role of carrier proteins was fully understood, they confounded identification of pheromones. For example, it was known that the active fraction of mouse urine contained large proteinaceous compounds. It is only recently that Novotny (pesonal communication 1998) determined that the pheromone itself is a small volatile compound, and appeared to be large because it is carried in urine bound to a large carrier protein.

In humans, Preti and colleagues (Spielman et al. 1995) have demonstrated that the component of underarm odor, (E)-3-methyl-2-hexenoic acid, is transported within the apocrine gland bound to 45- and 26-kDa proteins, and that it does not have an odor until it is liberated from the carrier proteins by axillary bacteria which is likely the case for human pheromones as well. Although these compounds have not been demonstrated to be pheromones, the functional principle likely holds.

Vagility

Finally, because so many pheromones are airborne, it is often assumed incorrectly that the chemical must be volatile. This is not the case. They need only be vagile, that is, capable of being dispersed.

We know well that antigens causing allergies are airborne, without being volatile chemicals (e.g., pollen and dust mite feces). All manner of compounds can become airborne by conjugation with another chemical, being aerosolized, carried on dust particles, or simply dispersed short distances through the air. Indeed, some pheromones are conveyed only by direct contact: the sex attractant in snakes and dominance pheromones in elephants (Houck et al. 1997; Rasmussen et al. 1998). In these cases the carrier proteins enable transport of lipophilic compounds in an aqueous medium. Thus, in considering the structure of putative human pheromones, there is no reason to assume they must be volatile, only sufficiently vagile to allow another human to receive them.

Context Dependence: Convergence of Environmental, Social, and Developmental Information

The criteria for vagility raise an essential question: In what contexts might human pheromones operate and what is the role of context in determining whether there will be a response and what will it be? If putative pheromones operate in the context of intimate sexual interactions, their vagility need not be great. On the other hand, if they signal the presence of an alarming and potentially dangerous situation, there would have been a strong selection pressure for highly vagile compounds that could signal danger well before the recipient was too close to escape.

Pheromone Type

In humans, context is likely to play quite different roles for the different types of pheromones. The ovarian-dependent priming pheromones that we have identified (Stern and McClintock 1998) are likely to produce the special case of menstrual synchrony under only specific circumstances. Weller and Weller (1997) found menstrual synchrony among Bedouin women living in Israel. These women live highly structured lives, eat similar food, and spend time in proximity, both while socializing and sleeping. This life style is similar to life in a women's dormitory in the late 1960s! Although social regulation of the human menstrual cycle has been detected in a variety of other contexts, for example, women working together in close quarters, it has not been detected when women live with men, have diverse living conditions, and interact relatively infrequently (Graham 1992; Weller and Weller 1993b).

The response to releasing pheromones operates, by definition, relatively independently of context, although context may mediate their release (Appelt and Sorensen 1997). Female pigs will respond to a spritz of Boar-Mate in a wide variety of breeding contexts. Male moths will respond to the fluorescent pheromonal signal of female moths, even when the signal is in fact coming from a candle flame and they fly up the gradient despite certain immolation. As discussed earlier, because human behavior is highly dependent on context, it is extremely unlikely that releaser pheromones will play a significant role in their behavior.

State-modulating pheromones, on the other hand, are more likely to affect human behavior or psychological state. With this type of pheromone, it is the context and meaning of a situation that determines if a person has a response and whether that response is conscious, preconscious, or unconscious. The initial or primary response to modulating pheromones may be similar across people. But then, it is their ongoing behavior, task, or psychological state that determines the specific downstream effects on the nervous system and, thereby, the specific form of their behavior. For example, pheromones with the potential for modulating sexual behavior may operate only in the presence of an appropriate sexual partner or a state of sexual arousal; putative pheromones that mediate attachment to a newborn may require knowledge of parental responsibility.

Signaling pheromones are the type most likely to be highly context-dependent and to have biologically significant effects only under specific circumstances. This is because signaling pheromones are but one source of information in a complex social interaction such as individual identity, kin recognition, or social status. Sheep recognize individual faces, individual male rats have different types of postejaculatory calls, and visual displays signal dominance in dogs along with pheromonal cues. Indeed, the response to a pheromone signaling high dominance status in a male wolf will lead to submission in one male and increased aggression and challenge to authority in another; the dramatically different forms of behavior depend on the environmental context and hormonal state of the recipient.

Learned Contextual Cues: A Pheromonal Function

Contextual information that shapes, elicits, or suppresses a pheromonal response is learned in many mammalian species, and thus is also likely to regulate human pheromonal systems. One of the most dramatic examples is pregnancy termination, caused when a pregnant female mouse is exposed to pheromones of a strange male, that is, a male with whom she has not mated (reviewed in Keverne 1983b). Only learning enables the female to distinguish strange and familiar males. When the strains of the stud and stranger are reversed, the pregnancy block is equally effective, that is, there is nothing inherent in the male's strain that prevents pregnancy disruption, just the fact that the female has mated with him. This pheromonal "imprinting" occurs even when the main olfactory system is removed, and Keverne has hypothesized that dopamine, associated with the experience of mating, enables retention of the pheromone memory by the accessory olfactory system.

Maternal attachment, which includes the response to a baby's odors, is determined by postpartum experience. While hormones of pregnancy do not affect growth of attachment to the fetus, they do predict the strength of attachment to the baby once it is born, when the baby is no longer just a concept or fantasy, but a real demanding organism that can be smelly, noisy, and not objectively attractive (Fleming et al. 1987, 1990, 1997).

These data suggest a way that modulating pheromones may act throughout the life span. Their function may be to create a state for learning sexual attraction to an appropriate partner in a supportive environment. They may not play a strong role once sexual behavior is experienced and fully developed. Instead, in humans, pheromones may serve as a scaffolding to guide developing attachment systems until cortical mechanisms are fully developed. Once the adult system is functional, loss of the pheromone-based system may have a negligible effect.

On the other hand, there may be aspects of sexual behavior and attachment that are modified early in the life span and not in adulthood, for example, olfactory recognition of androgens (Wysocki and Beauchamp 1984; Dorries et al. 1989; Wang et al. 1993), as well as their hedonic valence. If pheromones serve this developmental function, then ablation of this system in adulthood may indeed be deterministic (reviewed in Meredith 1991; Meredith and Fernandez-Fewell 1994; Wysocki and Lepri 1991). Such a system would contrast with odors, which are not innately pleasant or unpleasant (Engen 1982) and are amenable to conditioning throughout the life span (Hepper 1987; Van Toller and Kendal-Reed 1995).

Hormonal Context

Hormones cause differences between sexes in response to pheromones. Even within sexes, hormonal variation regulates sensitivity to pheromones. Therefore, the hormonal context of the recipient plays a critical role in shaping the response.

This principle has been documented beautifully at the level of single cells in the accessory olfactory system of hamsters (Wood 1996, 1997; Wood and Coolen 1997; see also chapter 13). Testosterone must be present in the very cells that are being driven by pheromonal stimuli in order to have an effect on behavior. In our study of rat pheromones, the output of our computer simulations best matched the empirical data when model rats are sensitive to pheromones only during the periovulatory phase (Schank and McClintock 1990, 1992). This makes functional sense given that, at least in humans, the pheromones are regulating the time of the preovulatory LH surge (Stern and McClintock 1998). Likewise, a functional approach leads to the hypothesis that women are sensitive to putative sexual pheromones only in the periovulatory phase when intercourse is most likely to result in conception. Indeed, the women rate the smell of androstenone as neutral at this time, whereas it is "unattractive" at other phases of the cycle (Grammer 1993).

Alteration of the timing of ovulation is likely to be but one of the many consequences of ovarian-dependent pheromones in women. Our work in rats and with computer simulations demonstrates that these same ovarian-dependent pheromones have qualitatively different effects that depend as well on the point in the reproductive life span when they occur (Schank and McClintock 1992). Further work in this area may well reveal that—as in rats—ovarian-dependent pheromones may have equal if not stronger effects on age at puberty, interbirth intervals, age at menopause, and level of chronic estrogen exposure during a woman's lifetime. Likewise, ovarian-dependent pheromones may be produced at stages of the reproductive life span other than during unfertilized cycles—for example, during pregnancy, lactation, and perimenopausally.

Hormones link social roles with different responses to pheromones, particularly during the intense social interactions that characterize social transitions. For example, during social transitions, males who are subordinate experience high corticosterone, lowered LH function, and lowered testosterone. This socially created hormonal profile reduces the subordinate male's response to female pheromones. Hormones may also regulate the pheromonal sensitivity of women. Women on birth control pills have a different sensitivity to the MHC than do women who are ovulating (Wedekind and Furi 1997).

The criteria for defining a "pheromone" were originally derived from insect studies. According to these definitions, a pheromone must be one or a few compounds with relative species-specificity, with clear and obvious behavioral or endocrine functions, and with effects that are minimally influenced by learning and more genetically programmed, as well as with effects that have been tested against control odors besides the diluent (Beauchamp et al. 1976). Additional criteria include not being a verbally conscious odor (Stern and McClintock 1998) and serving an evolutionary function (Meredith 1991). As in defining any chemical as having general functions (e.g., the criteria for being classified as a neurotransmitter), these criteria need to be revised as our knowledge increases.

Do Humans Use Pheromones?

Several key pieces are missing that must be discovered before we can fully answer the question, Do humans use pheromones? We began with the classic definition of a pheromone. They are chemicals, vagile signals, that are produced by one individual of a species and that trigger neuroendocrine responses which control the behavior, emotions, and fertility of another member of the same species (Karlson and Luscher 1959).

According to this definition, we have clear evidence for the existence of human priming pheromones—ovarian-dependent pheromones that regulate ovulation in women (Stern and McClintock 1998). Now we need to move away from our rigorously controlled experimental paradigm to determine whether these particular pheromones operate regularly in everyday life.

Conversely, Ober et al. (1997) have strong inferential evidence that humans use signaling pheromones based on MHC and HLA type to guide choice of marriage partner with whom they will have a lower risk of miscarriage. In this case, rigorously controlled experimental paradigms are now necessary to determine whether a signaling pheromonal system mediates this effect, which has been established in the everyday lives of the Hutterites Brethren.

Both of these cases meet the criterion that pheromones are not consciously detected; they exert their effects unconsciously. Localizing the neural pathways mediating these unconscious pathways presents a formidable challenge, yet sets the stage for identifying the neural pathways that would mediate pheromonal modulation of behavior and psychological state.

Species-specificity has traditionally been a key criterion for a system to be considered pheromonal. But now data indicate that although the dynamic system enables communication within a species, the actual compounds may well not be unique and may serendipitously exert effects on other species. For example, airborne chemosignals from female hamsters affect ovarian function in rats (Weizenbaum et al. 1977). Yet hamsters evolved in social isolation on the arid steppes of Asia, while the Norway rat likely evolved in the context of large social groups living commensally with humans. It is only in the artificial conditions of the laboratory that these chemosignals would in fact function as allomones, chemical communication between species.

Likewise, the pheromone recently identified in the urine of Asian elephants is chemically similar to that of butterflies (Lepidoptera). There is yet no evidence that this similarity coordinates the activity of the two species; their home ranges do not overlap and their dramatic size difference likely constrains the pheromonal system requirements for threshold, concentration, and vagility. The yeast pheromone, alpha-mating factor, is chemically similar to mammalian GnRH, so much so that it is chemically bound in the Norway rat's pituitary cells (Loumaye et al. 1982). Cat urine delays puberty in several rodent species (Vasilieva 1997). One of the most dramatic examples is part of European culture: humans use pigs to hunt truffle (*Tuber melanosporum*) because the fungus produces androstenol (figure 12.10). The pigs have to wear muzzles so they won't eat these fungi which are prized by humans as a delicacy. Thus, truffles (a fungus), pigs, and humans all produce the same compound.

If a compound is found in many species and functions as a pheromone in one of them, is the compound necessarily a pheromone in any of the other species? For example, androstenone and androstenol are pig pheromones, yet these steroids are produced not only by humans but also by most male mammals simply because they are part of the biosynthetic pathway for metabolizing androgen. Moreover, these steroids do affect human behavior at high concentrations and they have an odor processed by the main olfactory system. Thus, simply being present in humans and having behavioral effects cannot be used to infer that they are human pheromones. Perhaps the key to this species-specificity conundrum is the carrier protein. Proteins are more likely to be specific and locally present only in specific sensory tissues. Binding with a unique carrier protein may render a ubiquitous steroid both species-specific and odorless under natural conditions of everyday life.

Beauchamp and colleagues (1976) stipulated that pheromonal responses are not learned. Clearly, pheromonal effects should not depend on paired associations or

Figure 12.10
Pigs are still used today to hunt androstenol-producing truffles (*Tuber melanosporum*).

operant or classical conditioning. But this does not rule out the possibility that some environmental input is necessary for the development or fine-tuning of the pheromonal system. This is particularly likely for modulating pheromones that serve as scaffolding in the development of attachment systems.

Human behavioral pheromones, should they exist, would have evolved to affect behavior and psychological state only in specific contexts. Human pheromones are not likely to trigger a complex stereotyped behavior the first time they are encountered, nor even after many exposures. Instead, they are more likely to be modulators, regulating sensitivity and salience of stimuli, changing the way that sensory information is integrated, and altering sensorimotor integration and psychological states. Because they modulate a dynamic system, integrating many aspects of behavior and psychological state, and are highly dependent on environmental information and context, we anticipate that individual variability will be another hallmark of their effects on everyday human lives.

Acknowledgments

Suma Jacob, Charles Randles, and Bethanne Zelano provided invaluable assistance in preparing the manuscript. The research and this chapter were enabled by the support of NIMH R37 MH41788 and a grant from the Mind-Body Network of the Catherine A. and John T. MacArthur Foundation.

References

Agosta WC (1992). Chemical Communication: The Language of Pheromones. New York: Scientific American Library.

Albone ES (1984). Mammalian Semiochemistry. New York: John Wiley & Sons.

Altmann SA (1998). Foraging for Survival: Yearling Baboons in Africa. Chicago: University of Chicago Press.

Anonymous (1970). Effects of sexual activity on beard growth in man. Nature 226:869–870.

Appelt CW, Sorensen PW (1997). Female goldfish (*Carassius auratus*) signal their willingness to mate by changing their pattern of urinary pheromone release. *In* Chemical Signals in Vertebrates 8. Ithaca, NY: Cornell University Press.

Athena (1998). Advertisement, Cosmopolitan, January p 197.

Bancroft J (1989). Human Sexuality and Its Problems. New York: Churchill Livingstone.

Bancroft J, Sanders D, Davidson D, Warner P (1983). Mood, sexuality, hormones, and the menstrual cycle. III. Sexuality and the role of androgens. Psychosom Med 45:509–516.

Bartoshuk LM, Beauchamp GK (1994). Chemical senses. Annu Rev Psychol 45:419–449.

Bauchot R (1981). Etude comparative des volumes relatifs des bulbes olfactifs chez les vertébrés: L'homme est-il microsmatique? J Psychol Norm Pathol 1:71–80.

Beauchamp GK, Yamazaki K (1997). HLA and mate selection in humans: commentary. Am J Hum Genet 61:494–496.

Beauchamp GK, Doty RL, Moulton DG, Mugford RA (1976). The Pheromone Concept in Mammalian Chemical communication: A Critique. New York: Academic Press.

Beauchamp GK, Yamazaki K, Boyse EA (1985). The chemosensory recognition of genetic individuality. Sci Am 253:86.

Beltramino C, Taleisnik S (1983). Release of LH in the female rat by olfactory stimuli. Effect of the removal of the vomeronasal organs or lesioning of the accessory olfactory bulbs. Neuroendocrinology 36:53–58.

Beltramino C, Taleisnik S (1985). Ventral premammillary nuclei mediate pheromonal-induced LH release stimuli in the rat. Neuroendocrinology 41:119–124.

Benedek JF, Rubenstein G (1942). The Sexual Cycle in Women: The Relations Between Ovarian Functions and Psychodynamic Processes. Washington, DC: National Research Council.

Benton D (1982). The influence of androstenol—a putative pheromone—on mood throughout the menstrual cycle. Biol Psychol 15:249–256.

Berliner DL (1994). Fragrance compositions containing human pheromones. Erox Corporation Annual Report January 11.

Berliner DL, Jennings-White C, Lavker RM (1991). The Human skin: fragrances and pheromones. J Steroid Biochem Mol Biol 39:671–679.

Berliner DL, Monti-Bloch L, Jennings-White C, Diaz-Sanchez V (1996). The functionality of the human vomeronasal organ (VNO): evidence for steroid receptors. J Steroid Biochem Mol Biol 58:259–265.

Birke LI, Andrew RJ (1979). Distractibility changes during the oestrous cycle of the rat. Anim Behav 27:597–601.

Black SL, Biron C (1982). Androstenol as a human pheromone: no effect on perceived physical attractiveness. Behav Neural Biol 34:326–330.

Blumberg M, Mennella J, Moltz H, McClintock MK (1991). Facultative sex ratio adjustment in Norway rats: litters born asynchronously are female biased. Behav Ecol Sociobiol 31:401–408.

Boehm N, Gasser B (1993). Sensory receptor-like cells in the human foetal vomeronasal organ. Neuroreport 4:867–870.

Booth WD, White CA (1988). The isolation, purification and some properties of pheromaxein, the pheromonal steroid-binding protein, in porcine submaxillary glands and saliva. J Endocrinol 118:47–57.

Broca P (1888). Mémoirs d'anthropologie. Paris: Reinwald.

Brookover C (1914). The nervus terminalis in adult man. J Comp Neurol 24:131–135.

Carter CS, Williams JR, Witt DM (1990). The biology of social bonding in a monogamous mammal. In Advances in the Study of Behavior. New York: Academic Press.

Clark L, Mason J (1987). Olfactory discrimination of plant volatiles by the European starling. Anim Behav 35:227–235.

Cohen-Tannoudji J, Enhorn J, Signoret JP (1994). Ram sexual pheromone: first approach of chemical identification. Physiol Behav 56:955–961.

Cohn BA (1994). In search of human pheromones. Arch Dermatol 130:1048–1051.

Cooper WE, Burghardt GM (1990). Vomerolfaction and vomodor. J Chem Ecol 16:103–105.

Coquelin A, Clancey AN, Macrides F, Noble EP, Gorski RA (1984). Pheromonally induced release of luteinizing hormone in male mice: involvement of the vomeronasal system. J Neurosci 4:2230–2236.

Corbin A (1986). The Foul and the Fragrant. Cambridge, MA: Harvard University Press.

Cowley JJ, Johnson AL, Brooksbank BWL (1991). Human exposure to putative pheromones and changes in aspects of social behavior. J Steroid Biochem Mol Biol 39:647–659.

Craigmyle MBL (1984). The apocrine glands and the breast. New York: John Wiley & Sons.

Crosby, E, Humphrey, T (1939). Studies of the vertebrate telencephalon. I. The nuclear configuration of the olfactory and accessory olfactory formation and of the nucleus olfactorius anterior of certain reptiles, birds, and mammals. J Comp Neurol 71:121–213.

Cutler WB, Stine R (1988). Female essence increases heterosexual activity of women. Presented at the Annual Meeting of the American Fertility Society, Atlanta, Oct. 8–13 1988.

Cutler WB, Garcia CR, Krieger AM (1980). Sporadic sexual behavior and menstrual cycle length in women. Horm Behav 14:163–172.

Cutler WB, Preti G, Krieger A, Huggins GR, Garcia CR, Lawley HJ (1986). Human axillary secretions influence women's menstrual cycles: the role of donor extract from men. Horm Behav 20:463–473.

Cutler WB, Schleidt WM, Friedmann E, Preti G, Stine R (1987). Lunar influences on the reproductive cycle in women. Hum Biol 59:959–972.

Cutler WB, Friedmann E, McCoy NL (1998). Pheromonal influences on sociosexual behavior in men. Arch Sex Res 27:1–13.

Darby EM (1975). A presumptive sex pheromone in the hamster: some behavioral effects. J Comp Phsyiol Psychol 88:496–502.

Diamond M, Diamond AL, Mast M (1972). Visual sensitivity and sexual arousal levels during the menstrual cycle. J Nerv Ment Dis 155:170–176.

Diamond M, Binstock T, Kohl JV (1996). From fertilization to adult sexual behavior. Horm Behav 30:333–353.

Dluzen DE, Ramirez VD (1983). Localized and discrete changes in neuropeptide (LHRH and TRH) and neurotransmitter (NE and DA) concentrations within the olfactory bulbs of male mice as a function of social interaction. Horm Behav 17:139–145.

Dluzen DE, Ramirez VD, Carter CS, Getz LL (1981). Male vole urine changes luteinizing hormone-releasing hormone and norepinephrine in female olfactory bulb. Science 212:573–575.

Dorries KM (1998). Olfactory coding: time in a model. Neuron 20:7–10.

Dorries KM, Adkins Regan E, Halpern BP (1997). Sensitivity and behavioral responses to the pheromone androsterone are not mediated by the vomeronasal organ in domestic pigs. Brain Behav Evol 49:53–62.

Dorries KM, Schmidt HJ, Beauchamp GK, Wysocki CJ (1989). Changes in sensitivity to the odor of androstenone during adolescence. Dev Psychobiol 22:423–435.

Dorries KM, Adkins-Regan E, Halpern BP (in press). The vomeronasal organ of the pig: structure and role in female sexual behavior. Brain Behav Evol

Doty RL (1981). Olfactory communication in humans. Chem Senses 6:351–376.

Doty RL, Green PA, Ram C, Yankell SL (1982). Communication of gender from human breath odors: relationship to perceived intensity and pleasantness. Horm Behav 16:13–22.

Dulac C, Axel R (1995). A novel family of genes: encoding putative pheromone receptors in mammals. Cell 83:195–206.

Dyer L, Berghard A (1998). A novel family of ancient vertebrate odorant receptors in the lamprey. Presented at the Association for Chemoreception Sciences, April 23–26, Sarasota, FL.

Ebling FJG (1989). Apocrine glands in health and disorder. Int J Dermatol 28:508–511.

Egid K, Brown JL (1989). The major histocompatibility complex and female mating preferences in mice. Anim Behav 38:548–550.

Engen T (1982). The Perception of Odors. New York: Academic Press.

Engen T (1991). Odor Sensation and Memory. New York: Praeger.

Englander-Golden P (1976). Intellectual performance as a function of repression and the menstrual cycle. Presented at American Psychological Association Meeting, Sept 3–7 Wash DC.

Erox Corporation (1995). Fremont, CA, Realm advertisement.

Ferstl R, Eggert F, Westphal E, Zavazava N, Muller-Ruchholtz W (1991). MHC-Related odors in humans. In Chemical Signals in Vertebrates 6. (Doty RL, ed). New York: Plenum Press.

Fiber JM, Swann JM (1996). Testosterone differentially influences sex-specific pheromone-stimulated Fos expression in limbic regions of syrian hamsters. Horm Behav 30:455–473.

Filicori M, Santoro N, Merriam GR, Crowley WF (1986). Characterization of the physiological pattern of episodic gonadotropin secretion throughout the human menstrual cycle. J Clin Endrinol Metab 62:1136–1144.

Filsinger EE, Braun JJ, Monte WC, Linder DE (1984). Human (*Homo sapiens*) responses to the pig (*Sus scrofa*) sex pheromone 5-alpha-androst-16-en-3-one. J Comp Psychol 98:219–222.

Filsinger EE, Braun JJ, Monte WC (1985). An examination of the effects of putative pheromones on human judgments. Ethol Sociobiol 6:227–236.

Filsinger EE, Braun JJ, Monte WC (1990). Sex differences in response to the odor of alpha androstenone. Percept Mot Skills 70:216–218.

Fleming AS, Steiner M, Anderson V (1987). Hormonal and attitudinal correlates of maternal behaviour during the early postpartum period in first-time mothers. J Reprod Infant Psychol 5:193–205.

Fleming AS, Ruble DN, Flett GL, Van Wagner V (1990). Adjustment in first-time mothers: Changes in mood and mood content during the early postpartum months. Dev Psychol 26:137–143.

Fleming AS, Ruble D, Krieger H, Wong PY (1997). Hormonal and experiential correlates of maternal responsiveness during pregnancy and the puerperium in human mothers. Horm Behav 31:145–158.

Foidart A, J. J. Legros, J. Balthazart (1994). Les phéromones humaines: vestige animal ou realité non reconnue. Rev Med Liege 49:662–680.

Galef BG, Heiber L (1976). Role of residual olfactory cues in the determination of feeding site selection and exploration patterns of domestic rats. J Comp Physiol Psychol 90:727–738.

Garcia-Velasco J, Garcia-Casas S (1995). Nose surgery and the vomeronasal organ. Aesthetic Plast Surg 19:451–454.

Garcia-Velasco J, Mondragon M (1991). The incidence of the vomeronasal organ in 1000 human subjects and its possible clinical significance. J Steroid Biochem Mol Biol 39:561–563.

Gilbert AN, Wysocki CJ (1987). The National Geographic smell survey results. National Geographic 172:514–524.

Gilbert AN, Rosenwasser AM, Adler NT (1985). Timing of parturition and postpartum mating in Norway rats: interaction of an interval timer and a circadian gate. Physiol Behav 34:61–63.

Glass L, Mackey MC (1988). From Clocks to Chaos. Princeton NJ: Princeton University Press.

Goldfoot DA, Kravetz MA, Goy RW, Freeman SK (1976). Lack of effect of vaginal lavages and aliphatic acids on ejaculatory responses in rhesus monkeys: behavioral and chemical analyses. Horm Behav 7:1–27.

Goldfoot DA, Westerborg-Van Loon H, Groeneveld W, Slob AK (1980). Behavioral and physiological evidence of sexual climax in the female stump-tailed macaque (*Macaca arctoides*). Science 208:1477–1479.

Gower DB (1972). 16-Unsaturated C19 steroids. A review of their chemistry, biochemistry and possible physiological role. J Steroid Biochem 3:45–103.

Gower DB, Ruparelia BA (1993). Olfaction in humans with special reference to odorous 16–androstenes: their occurrence, perception, and possible social, psychological, and sexual impact. J Endocrinol 137:167–187.

Graham CA (1992). Menstrual synchrony: an update and review. Hum Nat 2:293–311.

Grammer K (1993). 5α-Androst-16en-3βα-on: a male pheromone? A brief report. Ethol Sociobiol 14:201–207.

Gustavson AR, Dawson ME, Bonett DG (1987). Androstenol, a putative human pheromone, affects human (*Homo sapiens*) male choice performance. J Comp Psychol 101:210–212.

Halpern, M (1987). The organization and function of the vomeronasal system. Annu Rev Neurosci 10:325–362.

Hedrick PW, Black FL (1997). HLA and mate selection: no evidence in South Amerindians. Am J Hum Genet 61:505–511.

Henkin RI (1974). Sensory changes during the menstrual cycle. *In* Biorhythms and Human Reproduction. (Ferin M, Halberg F, Richart RM, Vande Wiele, RL, eds). New York: John Wiley & Sons.

Hepper P (1987). The amniotic fluid: an important priming role in kin recognition. Anim Behav 35:1343–1346.

Herrada G, Dulac C (1997). A novel family of putative pheromone receptors in mammals with a topographically organized and sexually dimorphic distribution. Cell 90:763–773.

Herrick CJ (1924). Neurological Foundations of Behavior. New York: Holt.

Herz R, Cahill E (1997). Differential use of sensory information in sexual behavior as a function of gender. Hum Nat 8:275–289.

Hinkens D, Zidek L, Weidong M, Shank J, Alberts J, Novotny MV (1997). Rat urinary volatile constituents, their endocrine dependencies and binding to the urinary protein. *In* Chemical Signals in Vertebrates 8. Ithaca, NY: Cornell University Press.

Houck LD, Reagan NL (1990). Male courtship pheromones increase female receptivity in a plethodontid salamander. Anim Behav 39:729–734.

Houck L, Feldhoff RC, Rollmann SM (1997). Courtship pheromones in terrestrial salamanders. *In* Chemical Signals in Vertebrates 8. Ithaca, NY: Cornell University Press.

Ivey ME, Bardwick JM (1968). Patterns of affective fluctuation in the menstrual cycle. Psychosom Med XXX: 336–345.

Izard K, Vandenbergh J (1979). Pheromonal synchronization of bovine estrus. Presented at Conference on Reproductive Behavior.

Izard MK, Vandenbergh JG (1982). Priming pheromones from oestrous cows increase synchronization of oestrus in dairy heifers after $PGF_{2\alpha}$ injection. J Reprod Fertil 66:189–196.

Jacob S, Zelano B, Gungor A, Abbot D, Naclerio R, McClintock MK (1996). Variation in a human septal structure: the vomeronasal controversy. Society for Neuroscience, Nov 7–12, 1998, Los Angeles.

Jacobson L (1811). Description anatomique d'un organe observe dans la mammifères. Ann Museum Hist Nat. Paris 18:412–424.

Jarvis JUM (1981). Eusociality in a mammal: cooperative breeding in naked mole-rat colonies. Science 212:571–573.

Jennings-White C (1995). Perfumery and the sixth sense. Perfumer Flavorist 20:1–7.

Johnson A, Josephson R, Hawke M (1985). Clinical and histological evidence for the presence of the vomeronasal (Jacobson's) organ in adult humans. J Otolaryngol 14:71–79.

Johnston RE (1977). The causation of two scent-marking behaviour patterns in female hamsters (*Mesocricetus auratus*). Anim Behav 25:317–327.

Jolly A (1966). Lemur Behavior, Chicago: University of Chicago Press.

Jutte A (1996). Cognitive and physiological responses of men to female pheromones. Presented at 13th International Conference of the International Society for Human Ethology, Aug 5–10, Vienna.

Karlson P, Butenandt A (1959). Pheromones (ectohormones) in insects. Annu Rev Entomol 4:39.

Karlson P, Luscher M (1959). "Pheromones": a new term for a class of biologically active substances. Nature 183:55–56.

Katz RA, Shorey HH (1979). In defense of the term "pheromone." J Chem Ecol 5:299–301.

Keverne EB (1983a). Chemical communication in primate reproduction. *In* Pheromones and Reproduction in Mammals. Vandenbergh JG, ed), pp 79–92. New York: Academic Press.

Keverne EB (1983b). Phermonal influences on the endocrine regulation of reproduction. Trends Neurosci 6:381–384.

Kirk-Smith M, Booth DA, Carroll D, Davies P (1978). Human social attitudes affected by androstenol. Res Commun Psychol Psychiatr Behav 3:379–384.

Kjaer I, Fischer Hansen B (1996a). The human vomeronasal organ: prenatal developmental stages and distribution of luteinizing hormone–releasing hormone. Eur J Oral Sci 104:34–40.

Kjaer I, Fischer Hansen B (1996b). Luteinizing hormone-releasing hormone and innervation pathways in human prenatal nasal submucosa: factors of importance in evaluating Kallmann's syndrome. APMIS 104:680–688.

Kohl JV, Francoeur RT (1995). The Scent of Eros: Mysteries of Odor in Human Sexuality. New York: Continuum.

Kopell B (1969). Variations in some measures of arousal during the menstrual cycle. J Nerv Ment Dis 148:180–187.

Krishna NSR, Getchell ML, Getchell TV (1994). Expression of the putative pheromone and odorant transporter vomeromodulin mRNA and protein in nasal chemosensory mucosae. J Neurosci Res 39:243–259.

Labows JN, Preti G (1992). Human semiochemicals. *In* Fragrance: The Psychology and Biology of Perfume (S van Toller and GS Dodd, eds), pp 69–90. London: Elsevier Applied Science Publications.

Leon M, Moltz H (1972). The development of the pheromonal bond in the albino rat. Physiol Behav 8:683–686.

Leshner AI (1975). A model of hormones and agonistic behavior. Physiol Behav 15:225–235.

Loumaye E, Thorner J, Catt KJ (1982). Yeast mating pheromone activates mammalian gonadotrophs: evolutionary conservation of a reproductive hormone? Science 218:1323–1325.

Luschen ME, Pierce DM (1972). Effect of the menstrual cycle on mood and sexual arousability. J Sex Res 8:41–47.

Ma W, Miao Z, Novotny MV (1997). The role of adrenal-mediated chemosignals in estrus suppression in the house mouse: the Lee-Boot effect revisited. *In* Chemical Signals in Vertebrates 8. Ithaca, NY: Cornell University Press.

Macdonald DW (1985). The carnivores: order Carnivora. *In* Social Odours in Mammals. (Brown RE, Macdonald DW, eds), p 882. Oxford: Clarendon Press.

Martin IG (1980). "Homeochemic," intraspecific chemical signal. J Chem Ecol 6:517–519.

Matsunami H, Buck LB (1997). A multigene family encoding a diverse array of putative pheromone receptors in mammals. Cell 90:775–784.

McClintock MK (1971). Menstrual synchrony and suppression. Nature 229:244–245.

McClintock MK (1981a). Social control of the ovarian cycle. Biosci Rep 31:138–139.

McClintock MK (1981b). Social control of the ovarian cycle and the function of estrous synchrony. Am Zool 21:243–256.

McClintock MK (1983a). Pheromonal regulation of the ovarian cycle: enhancement, suppression and synchrony. *In* Pheromones and Reproduction in Mammals. (Vandenbergh JG, ed), pp 113–149. New York: Academic Press.

McClintock MK (1983b). Synchronizing ovarian and birth cycles by female pheromones. *In* Chemical Signals in Vertebrates III. (Muller-Schwarze D, Silverstein RM, eds), pp 159–178. New York: Plenum Publishing.

McClintock MK (1984a). Estrous synchrony: modulation of ovarian cycle length by female pheromones. Physiol Behav 32:701–705.

McClintock MK (1984b). Group mating in the domestic rat as a context for sexual selection: consequences for analysis of sexual behavior and neuroendocrine responses. Adv Stud Behav 14:1–50.

McClintock MK (1998a). Whither menstrual synchrony? Annu Rev Sex Res 9: In

McClintock Mk (1998b). On the nature of mammalian and human pheromones. Ann NY Acad Sci 855:390–392.

McClintock MK, Adler NT (1978). Induction of persistent estrus by airborne chemical communication among female rats. Horm Behav 11:414–418.

Mennella J, Blumberg M, McClintock MK, Moltz H (1990). inter-litter competition and communal nursing among Norway rats: Advantages of birth synchrony. Behav Ecol Sociobiol 27:183–190.

Mennella JA, Beauchamp GK (1991). Maternal diet alters the sensory qualities of human milk and the nursling's behavior. Pediatrics 88:737–743.

Meredith M (1991). Vomeronasal damage not nasopalatine duct damage behavior deficits in male hamsters. Chem Senses 16:155–168.

Meredith M, Burghardt G (1978). Electrophysiological studies of the tongue and accessory olfactory bulb in garter snakes, Physiol Behav 21: 1001–1008.

Meredith M, Fernandez-Fewell G (1994). Vomeronasal system, LHRH, and sex behaviour. Psychoneuroendocrinology 19:657–672.

Meredith M, Howard G (1992). Intracerebroventricular LHRH relieves behavioral deficits due to vomeronasal organ removal. Brain Res Bull. 29:75–79.

Messant PR (1976). Female hormones and behavior. *In* Exploring Sex Differences. (Lloly B, Archer J, eds). London: Academic Press.

Michael RP, Keverne EB, Bonsall RWI-BRH, Institute of Psychiatry, Beckenham, England (1971). Pheromones: Isolation of male sex attractants from a female primate. Science 172:964–966.

Moehlman PD (1985). The odd-toed ungulates: order Perrisodactyla. *In* Social Odours in Mammals. (Brown RE, Macdonald DW, eds) p 882. Oxford: Clarendon Press.

Montagna W, Parakkal P (1974). Structure and Function of Skin. New York: Academic Press.

Monti-Bloch L, Grosser BI (1991). Effect of putative pheromones on the electrical activity of the human vomeronasal organ and olfactory epithelium. J Steroid Biochem Mol Biol 39:573–582.

Monti-Bloch L, Grosser BI, Jennings-White C, Berliner DL (1998). Behavioral effect of androsta-4,16-dien-3-one (androstadienone). Presented at Association for Chemoreception Sciences, Sarasota, FL April 23–26, 1998.

Moran DT, Jafek BW, Rowley III JC (1991). The vomeronasal (Jacobson's) organ in man: ultrastructure and frequency of occurrence. J Steroid Biochem Mol Biol 39:545–552.

Moran DT, Monti-Bloch L, Stensaas LJ, Berliner DL (1994). Structure and function of the human vomeronasal organ. *In* Handbook of Olfaction and Gustation. (Doty RL, ed), pp 793–820. New York: Marcel Dekker.

Morris NM, Udry JR (1978). Pheromonal influences on human sexual behaviour: an experimental search. J Biosoc Sci 10:147–157.

Mucignat-Caretta C, Caretta A (1997). Protein bound odorants as flags of male mouse presence. *In* Chemical Signals in Vertebrates 8. Ithaca, NY: Cornell University Press.

Muller-Schwarze D (1997). Signal, sending, sensitivity: which evolves most in mammals? *In* Chemical Signals in Vertebrates 8. Ithaca, NY: Cornell University Press.

Netter FH, ed. (1989). Atlas of Human Anatomy. consulting ed. Colacino, S Summit, NJ: CIBA-GEIGY.

Nevitt GA, Reid K (1998). Evidence for different olfactory foraging strategies in Antarctic seabirds. Presented at Association for Chemoreception Sciences, Sarasota, FL

Nicholson B (1984). Does kissing aid human bonding by semiochemical addiction? Br J Dermatol 3:623–627.

Nixon A, Mallet A, Jackman P, Gower D (1986). Testosterone metabolism by isolated human axillary *Corynebacterium* spp.: a gas-chromatographic mass-spectrometric study. J Steroid Biochem Mol Biol 24:887–892.

Nixon A, Mallet A, Gower D (1988). Simultaneous quantification of 5 odorous steroids (16-androstenes) in the axillary hair of men. J Steroid Biochem Mol Biol 29:505–510.

Novotny MV, Ma W, Zidek L (1997). New biochemical insights into puberty acceleration, estrus induction and puberty delay in the house mouse. *In* Chemical Signals in Vertebrates 8. Ithaca, NY: Cornell Unviersity Press.

Novotny M, Ma W, Zidek L (1998). The chemosignals causing puberty acceleration in the house mouse: natural stimuli and their structural analogs. Presented at the 27th Annual Meeting, 1998, Association for Chemical Senses, Sarasota, FL

Ober C (1992). The maternal-fetal relationship in human pregnancy: an immunogenetic perspective. Exp Clin Immunogenet 9:1–14.

Ober C, Weitkamp LR, Cox N, Dytch H, Kostyu D, Elias S (1997). HLA and mate choice in humans [see comments]. Am J Hum Genet 61:497–504.

Paige KE (1971). Effects of oral contraceptives on affective fluctuations associated with the menstrual cycle. Psychosom Med 33:515–537.

Pearlman S (1934). Jacobson's organ (organon vomeronasale Jacobsoni): its anatomy, gross, microscopic and comparative, with some observations as well on its function. Ann Otol Rhinol Laryngol 43:740–768.

Pearson AA (1941). The development of the nervus terminalis in man. J Comp Neurol 75:39–66.

Persky H, Lief HI, Strauss D, Miller WR, O'Brien CP (1978). Plasma testosterone level and sexual behavior of couples. Arch Sex Behav 7:157–173.

Pfaus JG, Heeb MM (1997). Implications of immediate-early gene induction in the brain following sexual stimulation of female and male rodents. Brain Res Bull 44:397–407.

Pfaus JG, Jakob A, Kleopoulos SP, Gibbs RB, Pfaff DW (1994). Sexual stimulation induces Fos immunoreactivity within GnRH neurons of the female rat preoptic area: interaction with steroid hormones. Neuroendocrinology 60:283–290.

Pietras RJ, Moulton DG (1974). Hormonal influences on odor detection in rats: changes associated with the estrous cycle, pseudopregnancy, ovariectomy, and administration of testosterone propionate. Physiol Behav 12:475–491.

Pollack MS, Wysocki CJ, Beauchamp GK, Braun D Jr, Callaway C, Dupont B (1982). Absence of HLA association or linkage for variation in sensitivity to the odor of androstenone. Immunogenetics 15:579–589.

Porter RH, Balogh RD, Chernoch JM, Franchi C (1986). Recognition of kin through characteristic body odors. Chem Senses 11:389–395.

Potiquet M (1891). Le Canal Jacobson. Rev Laryngol Otol Rhinol 2:737–753.

Potts WK, Manning CJ, Wakeland EK (1991). Mating patterns in semi-natural populations of mice influenced by MHC genotype. Nature 352:619.

Preti G, Huggins GR (1975). Cyclic changes in volatile acidic metabolites of human vaginal secretions and their relation to ovulation. J Chem Ecol 1:361–376.

Preti G, Wysocki CJ (1998). Human pheromones: releasers or primers—fact or myth? *In* Advances in Chemical Signals in Vertebrates. New York: Plenum Press.

Proctor DF, Anderson IB (1982). The Nose: Upper Airway Physiology and the Atmospheric Envirnment, Amsterdam: Elsevier Biomedical.

Preti G, Cutler WB, Garcia CR, Huggins GR, Lawley HJ (1986). Human axillary secretions influence women's menstrual cycles: the role of donor extract of females. Horm Behav 20:474–482.

Preti G, Spielman AI, Leyden JJ (1997). Human axillary secretions and odors: a human chemical signal source? *In* Chemical Signals in Vertebrates 8. Ithaca, NY: Cornell University Press.

Randall JA (1993). Behavioral adaptations of desert rodents (Heteromyidae). Anim Behav 45:263–287.

Rasmussen LEL, Schulte BA (1997). Ecological and biochemical constraints on pheromonal signaling systems in Asian elephants: evolutionary implications especially in comparison to insects. *In* Chemical Signals in Vertebrates 8. Ithaca, NY: Cornell University Press.

Rasmussen LEL, Lee TD, Daves GD Jr, Schmidt MJ (1993). Female-to-male sex pheromones of low volatility in the Asian elephant, *Elephas maximus*. J Chem Ecol 19:2115–2128.

Rasmussen LEL, Perris TE, Gunawardena N (1994). Isolation of potential musth-alerting signals from the temporal gland secretions of male Asian elephants (*Elephas maximus*): a new method. Chem Senses 19:540.

Rasmussen JL, Greenwood D, Feng L, Prestwich G (1998). Initial characterizations of secreted proteins from Asian elephants that bind the sex pheromone (Z)-7-dodecenyl acetate. Presented at Association for Chemical Senses, Sarasota, FL

Reinisch JM (1977). Prenatal exposure of human foetuses to synthetic progestin and oestrogen affects personality. Nature 266:561–562.

Rennie PJ, Holland KT, Mallet AI, Watkins WJ, Gower DB (1989a). Interconversion of androst-16-ene steroids by human axillary aerobic coryneform bacteria. Biochem Soc Trans 17:1027–1028.

Rennie PJ, Holland KT, Mallet AI, Watkins WJ, Gower DB (1989b). Testosterone metabolism by human axillary bacteria. Biochem Soc Trans. 17:1017–1018.

Robertson DHL, Benyon RL, Evershed RP (1993). Extractions, characterization and binding analysis of two pheromonally active ligands associated with major urinary proteins of house mouse (*Mus musculus*). J Chem Ecol 19:1405–1416.

Robertson DHL, Pes D, Gaskell SJ, Hurst JL, Beynon RJ (1997). Characterisation of major urinary proteins in wild mice. *In* Chemical Signals in Vertebrates 8. Ithaca, NY: Cornell University Press.

Rosenberg LT, Cooperman D, Payne R (1983). HLA and mate selection. Immunogenetics 17:89–93.

Roy EJ, Wilson MA (1981). Diurnal rhythm of cytoplasmic estrogen receptors in the rat brain in the absence of circulating estrogens. Science 213:1525–1527.

Russell MJ (1976). Human olfactory communication. Nature 260:520–522.

Russell MJ, Switz GM, Thompson K (1980). Olfactory influences on the human menstrual cycle. Pharmacol Biochem Behav 13:737–738.

Ruysch F (1703). Thesaurus anatomicus. Amsterdam.

Ryba NJP, Tirindelli R (1997). A new multigene family of putative pheromone receptors. Neuron 19:371–379.

Sachs B (1997a). Erection evoked in male rats by airborne scent from estrous females. Physiol Behav 62:921–924.

Sachs BD (1997b). Airborne aphrodisiac odor from estrous rats: implication for pheromonal classification. *In* Chemical Signals in Vertebrates 8. Ithaca, NY: Cornell University Press.

Sagar SM, Sharp FR, Curran T (1988). Expression of c-fos protein in brain: metabolic mapping at the cellular level. Science 240:1328–1330.

Sato K, Leidal R, Sato F (1987). Morphology and development of an apoeccrine sweat gland in human axillae. Am J Physiol 252:R166–R180.

Schaal B, Porter RH (1991). "Microsmatic humans" revisited: the generation and perception of chemical signals. *In* Advances in the Study of Behavior (Peter JB, Slater JSR, Beer C, Milinski M, eds), pp 135–199. San Diego: Academic Press.

Schank JC (1991). Model-building, computer simulation and experimental design in biology. Dissertation. The Committee on the Conceptual Foundations of Science, University of Chicago.

Schank J, McClintock MK (1990). A computer simulation of ovarian cycle synchrony in female rats: an object oriented integration of computer simulation and laboratory experiment. *In* Modelling and Simulation (Schmidt B, ed), pp 590–595. San Diego: SCS Press.

Schank JC, McClintock MK (1992). A coupled-oscillator model of ovarian-cycle synchrony among female rats. J Theor Biol 157:317–362.

Schank JC, McClintock MK (1993). A model of ovarian-cycle synchrony in groups of Norway rats. Anim Behav Society, UC Davis July 24–30.

Schank JS, Tomasino CI, McClintock MK (1995). The development of a pheromone isolation and delivery (PID) system for small animals. Anim Tech 46:103–113.

Schiffman SS, Sattely Miller EA, Suggs MS, Graham BG (1995). The effect of environmental odors emanating from commercial swine operations on the mood of nearby residents. Brain Res Bull 37:369–375.

Schneider D (1974). The sex-attractant receptor of moths. Sci Am 231:28–35.

Schwanzel-Fukuda M, Crossin KL, Pfaff DW, Bouloux PMG, Hardelin JP, Petit C (1996). Migration of luteinizing hormone–releasing hormone (LHRH) neurons in early human embryos. J Comp Neurol 366:547–557.

Schweiger U, Tuschl R, Broocks A, Pirke K-M (1990). Gonadotrophin secretion in the second half of the menstrual cycle: a comparison of women with normal cycles, luteal phase defects and disturbed follicular development. Clin Endocrinol 32:25–32.

Selye H (1946). The general adaptation syndrome and the diseases of adaptation. J Clin Endocrinol Metab 6:117.

Shapiro JA (1998). Thinking about bacterial populations as multicellular organisms. Annu Rev Microbiol 52:81–104.

Shelley WB, Hurley HJ, Nichols AC (1960). Axillary odor: experimental study of the role of bacteria, apocrine sweat, and deodorants. Arch Dermatol 430–446.

Singer AG (1991). A chemistry of mammalian pheromones. J Steroid Biochem Mol Biol 39:627–632.

Singer AG, Macrides F, Clancy AN, Agosta WC (1986). Purification and analysis of a proteinaceous aphrodisiac pheromone from hamster vaginal discharge. J Biol Chem 261:13323–13326.

Singer AG, Beauchamp GK, Yamazaki K (1997). Volatile signals of the major histocompatibility complex in male mouse urine. Proc Natl Acad Sci U S A 94:2210–2214.

Smith TD, Siegel MI, Mooney MP, Burdi AR, Burrows AM, Todhunter JS (1997). Prenatal growth of the human vomeronasal organ. Anat Rec 248:447–455.

Spielman A, Zeng X-N, Leyden J, Preti G (1995). Proteinaceous precursors of human axillary odor: isolation of two novel odor-binding protein. Experientia 51:40–46.

Stensaas L, Lavker RM, Monti-Bloch L, Grosser B, Berliner D (1991). Ultrastructure of the human vomeronasal organ. J Steroid Biochem Mol Biol 39:553–560.

Stephan H, Bauchot R, Andy OJ (1970). Data on size of the brain and of various brain parts in insectivores and primates. *In* The Primate Brain. (Noback CR, Montagna W, eds), pp 289–297. New York: Appleton-Century-Crofts.

Stern KN (1992). Ovulation in Women: Methods of Detection, Behavioral Correlates, and Pheromonal Regulation. Chicago: Department of Psychology, Committee on Biopsychology, University of Chicago.

Stern K, McClintock MK (1995). Biological rhythms and individual variation: accurate measurement of menstrual cycle phase and the preovulatory LH surge. *In* Psychopharmacology of Women. (Jensvold MF, Halbreich U, Hamilton J, eds), pp 393–413. Washington, DC: American Psychiatry Press.

Stern K, McClintock M (1998). Regulation of ovulation by human pheromones. Nature 392:177–179.

Suh BY, Betz G (1993). Altered luteinizing hormone pulse frequency in early follicular phase of the menstrual cycle with luteal phase defect patients in women. Fertil Steril 60:800–805.

Sullivan SL (1997). Information coding in the mammalian olfactory system. *In* Chemical Signals in Vertebrates 8. Ithaca, NY: Cornell University Press.

Takahashi N, Iwata T, Umezawa K, Miura S, Toyoda F, Kikuyama S (1997). Molecular cloning of a sodefrin, a female-attracting pheromone in red-bellied newt. *In* Chemical Signals in Vertebrates 8. Ithaca, NY: Cornell University Press.

Takami S, Getchell M, Chen Y, Monti-Bloch L, Berliner D, Stensaas L, Getchell TV (1993). Vomeronasal epithelial cells of the adult human express neuron-specific molecules. Neuroreport 4:375–378.

Taylor R (1994). Brave new nose: sniffing out human sexual chemistry. J Natl Inst Health Res 6:47–51.

Treloar AE, Boynton RE, Behn DG, Brown BW (1967). Variation of the human menstrual cycle through reproductive life. Int J Fertil 12:77–126.

Turner BH, Gupta KC, Mishkin M (1978). The locus and cytoarchitecture of the projection areas of the olfactory bulb in *Macaca mulatta*. J Comp Neurol 177:381–396.

Vandenbergh JG (1994). Pheromones and mammalian reproduction. *In* The Physiology of Reproduction. (Knobil E, Neill JD, eds), pp 343–359. New York: Raven Press.

Van der Lee S, Boot LM (1956). Spontaneous pseudopregnancy in mice II. Acta Physiol Pharmacol (Neerl) 5:213–214.

Van Toller S, Kendal-Reed M (1995). A possible protocognitive role for odor in human infant development. Brain Cogn 29:275–293.

Vasilieva NY (1997). Influence of cats' urinary chemosignals on reproduction and sexual maturation in some rodent species. *In* Chemical Signals in Vertebrates 8. Ithaca, NY: Cornell University Press.

Veith JL, Buck M, Getzlaf S, Van Dalfsen P, Slade S (1983). Exposure to men influences the occurrence of ovulation in women. Physiol Behav 31:313–315.

Veldhuis JD, O'Dea LSL, Johnson ML (1989). The nature of the gonadotropin-releasing hormone stimulus–luteinizing hormone secretory response of human gonadotrophs in vivo. J Clin Endocrinol Metab 68:661–670.

Vierling JS, Rock J (1967). Variations in olfactory sensitivity to exaltolide during the menstrual cycle. J Appl Physiol 22:311–315.

Vollman RF (1977). The menstrual cycle. *In* Major Problems in Obstetrics and Gynecology 7. Philadelphia: WB Saunders.

Wallis J (1985). Synchrony of estrous swelling in captive group living chimpanzees (*Pan troglodytes*). Int J Primatol 6:335–350.

Wallis J (1989). Synchrony of menstrual cycles in captive group-living baboons (*Papio cynocephalus*). Am J Primatol 18:167–168.

Wallis J, King BJ, Roth-Meyer C (1986). The effect of female proximity and social interaction on the menstrual cycle of crab-eating monkeys (*Macaca fascicularis*). Primates 27:83–94.

Wallraff H (1986). Directional components derived from initial-orientation data of inexperienced homing pigeons. J Comp Physiol [A] 159:143–159.

Wang HW, Wysocki CJ, Gold GH (1993). Induction of olfactory receptor sensitivity in mice. Science 260:998–1000.

Wedekind C, Furi S (1997). Body odour preferences in men and women: do they aim for specific MHC combinations or simply heterozygosity? Proc R Soc Lond B Biol Sci 264:1471–1479.

Wedekind C, Seebeck T, Bettens F, Paepke AJ (1995). MHC-dependent mate preferences in humans. Proc R Soc Lond B Biol Sci 260:245–249.

Weitkamp LR, Cox N, Ober C (In Press). Five, eleven, and sixteen locus HLA haplotypes: influence of HLA on mate choice. Immunogetics.

Weizenbaum F, McClintock MK, Adler NT (1977). The effect of the presence of hamsters on the estrous cycle of the rat. Horm Behav 8:342–347.

Weller A, Weller L (1993a). Menstrual synchrony between mothers and daughters and between roommates. Physiol Behav 53:943–949.

Weller A, Weller L (1997). Menstrual synchrony under optimal conditions: Bedouin families. J Comp Psychol 111:143–151.

Weller L, Weller A (1993b). Human menstrual syncrhony: å critical assessment. Neurosci Behav Rev 17:427–439.

Wilson HC (1987). Female axillary secretions influence women's menstrual cycles: a critique. Horm Behav 21:536–546.

Wilson RC, Kesner JS, Kaufman JM, Uemura T, Akema J, Knobil E (1984). Central electrophysiologic correlates of pulsatile luteinizing hormone secretion in the rhesus monkey. Neuroendocrinology 39:256.

Wood RI (1996). Functions of the steroid-responsive neural network in the control of male hamster sexual behavior. Trends Neurosci 7:338–344.

Wood RI (1997). Thinking about networks in the control of male hamster sexual behavior. Horm Behav 32:40–45.

Wood RI, Coolen LM (1997). Integration of chemosensory and hormonal cues is essential for sexual behaviour in the male Syrian hamster: role of the medial amygdaloid nucleus. Neuroscience 78:1027–1035.

Wysocki CJ (1979). Neurobehavioral evidence for the involvement of the vomeronasal system in mammalian reproduction. Neurosci Biobehav Rev 3:301–341.

Wysocki CJ, Beauchamp GK (1984). Ability to smell androstenone is genetically determined. Proc Natl Acad Sci U S A 81:4899–4902.

Wysocki CJ, Lepri JJ (1991). Consequences and Removing the vomeronasal organ. J Steroid Biochem Mol Biol 39:661–670.

Wysocki CJ, Meredith M (1987). The Vomeronasal System. New York: John Wiley & Sons.

Yamazaki K, Beauchamp GK, Imai Y, Curran M, Bard J, Boyse EA (1994). Odor types determined by the major histocompatibility complex in mice. *In* Olfaction and Taste XI. (Kurihara K, Suzuki N, Ogawa H, eds). Berlin: Springer-Verlag.

Yamazaki K, Beauchamp GK, Kupniewski D, Bard J, Thomas L, Boyse EA (1988). Familial imprinting determines H-2 selective mating preferences. Science 240:1331–1332.

Zeng X-N, Leyden JJ, Spielman AI, Preti G (submitted) Analysis of the characteristic female axillary odors: qualitative comparison to males. J Chem Ecol

Zheng W, Stacey N (1997). A steroidal pheromone and spawning stimuli act via different neuroendocrine mechanisms to increase gonadotropin and milt volume in male goldfish *Carassius auratus*. Gen Comp Endocrinol 105:228–238.

Zingeser M (1984). The nasopalatine ducts and associated structures in the rhesus monkey (*Macaca mulatta*): topography, prenatal development, function, and phylogeny. Am J Anat 170:581–595.

IIB SOCIAL CONTEXT AND MECHANISMS UNDERLYING BEHAVIOR

The previous three chapters focused on how social context affects the expression of reproductive behavior. Martha McClintock's chapter addressed, in addition, one possible mechanism by which social context might manipulate behavior under some conditions. The first two chapters in this section delve more deeply into mechanisms that might modulate social contextual influences on behavior.

Chapter 13, by Ruth Wood and Jennifer Swann, develops the pathway by which olfactory information can modulate sexual behavior in Syrian hamsters (*Mesocricetus auratus*), a species in which olfactory cues play an extensive role in sociosexual behavior. The authors illustrate a complex interaction between the gonadal state of the male and his capacity to both display sexual behavior and respond to odor cues from the female. These studies develop the notion that gonadal hormones, in addition to affecting sexual motivation, also modulate perceptual systems altering the stimuli that an animal detects and responds to in its environment. Ultimately, such an interplay can result in hormonally modulated cognition as a regulator of sexual activity.

Patricia Schiml, Scott Wersinger, and Emilie Rissman contribute, in chapter 14, a perspective on how social experience can modify gene expression within the brain, in this case the expression of genes for the production of gonadotropin-releasing hormone (GnRH). Their review of the comparative literature demonstrates a tight linkage between sexual behavior and the activation of the neuroendocrine axis in species where either ovarian function or ovulation is dependent on behavioral input. Studies of the musk shrew, an insectivore with induced ovarian function, demonstrate that activation of the GnRH system in females is reliably induced by sexual interactions with a male. Even visual contact with a male through a transparent barrier affects regional distribution of GnRH, though in a direction opposite to that produced by sexual interactions. This chapter illustrates the power of using social manipulations to alter internal physiology to elucidate the processes underlying behavior.

Chapter 15, by David Crews, brings us back to the importance of both social and environmental context in understanding behavior. His chapter emphasizes how sexuality, the individualistic pattern of physical and behavioral characteristics associated with reproductive behavior, arises from contextual inputs. In this chapter the principle social context consideration is the human social context which affects our conceptualization of sex and sexuality. Dr. Crews argues that the social history of investigating sexuality limits the factors considered to be important in understanding sexuality. Using results from species with nonchromosomal systems of sex determination, he brings this book full circle by demonstrating the crucial role the environment plays in producing individual differences in sexual differentiation.

The central message of all of these chapters is that organisms face complex and difficult environments, posing critical challenges to survival, much less reproduction. Elucidating the principles and mechanisms that have evolved to deal with this unpredictable changing environment entails consideration of the complexity of the challenges facing individuals and creating approaches that introduce essential control into meaningful social contexts. Behavior allows organisms to adapt to a range of environments and to seek environments to which they are better suited. Similarly, studying a range of environments and using environmental change to challenge the animals under study provides the mechanism to eventually understand the grand sweep of behavioral adaptation.

13 Neuronal Integration of Chemosensory and Hormonal Signals in the Control of Male Sexual Behavior

Ruth I. Wood and Jennifer M. Swann

Reproduction is profoundly influenced by social context, as discussed in the preceding chapters. The effects of social context on fertility and sexual behavior are often mediated by odor cues from conspecifics in many species, including our own (see chapter 12). However, odor cues seldom act alone. Instead, the response to odors is influenced by the endocrine milieu. McClintock has summarized evidence that reproductive hormones interact with the olfactory and reproductive system to affect sex and social behavior in species throughout the animal kingdom (see chapter 12). Her laboratory has provided the first unequivocal evidence in support of the notion that reproductive processes are influenced by pheromones in human females. However, very little is known about the specific neuroanatomical and physiological mechanisms that underlie these phenomena in human beings and most other mammals.

We are beginning to understand the pathways that mediate chemosensory influences on sex behavior in male Syrian hamsters (*Mesocricetus auratus*). Our recent studies investigating the neural circuitry through which chemosensory and hormonal cues are combined in the brain to facilitate male sexual behavior are extensions of the pioneering work of Sarah W. Newman. Male hamsters are similar to males of most other species in that gonadal steroids are essential to mating. If circulating gonadal steroids are eliminated by castration or suppressed by short day lengths, mating gradually declines (Morin and Zucker 1978). Male hamsters are also dependent upon chemosensory stimuli for the initiation of copulation (Murphy and Schneider 1970). When chemosensory cues are interrupted, mating is immediately and permanently abolished. Through studies of neural processing of sexually relevant odors in a hormone-sensitive circuit for male sexual behavior, we are beginning to unravel mechanisms underlying the influence of social context on reproduction.

We begin by introducing the behavioral endocrinology of male Syrian hamsters by way of comparisons with males of other laboratory species. Using this comparative approach we discuss the effects of chemosensory stimuli on gonadal hormone secretion as well as the influence of the hormonal milieu on the behavioral responsiveness to odor cues. Next, we discuss in detail the neural pathways that mediate odor-hormone interactions that lead to mating behavior in Syrian hamsters, including the experimental evidence that led to the elucidation of these neural pathways. These data include both descriptive studies of brain steroid receptors and neural activation in response to odor cues, as well as experimental studies to determine the functional significance of specific brain areas. We establish the precise groups of steroid-sensitive neurons that are critical for sex behavior. We provide strong evidence that a site-specific interaction between neurons stimulated by odor cues and those stimulated by gonadal steroids is required for mating in male Syrian hamsters. We go on to discuss the effects of gonadal steroid-odor interactions on neurotransmitters, neuronal morphology, and neural connectivity. Finally, we briefly discuss the significance of hormone-odor interactions within the context of other environmental and social cues that affect reproduction in other mammals.

Internal and External Cues Stimulate Male Mating Behavior

Males of many species express sexual behavior only when conditions in the internal and external environment indicate that reproductive activity is likely to be successful. Although the energetic investment in production of sperm and seminal fluid is small, a male may expend considerable energy in seeking out receptive females or in intermale competition (Bronson 1989). Wasted reproductive effort could compromise his chances of survival. Thus, sexual behavior is usually not expressed when successful copulation is unlikely.

Gonadal steroid hormones are one important humoral cue to the brain which signify the state of the internal milieu. Concentrations of gonadal steroids in circulation are influenced by many different factors, including nutrition, stress, and state of maturity (Bronson 1989). Thus, steroids provide an integrated signal to the brain about the internal state of reproductive readiness. The synthesis and release of testicular steroids is governed by luteinizing hormone (LH) from the anterior pituitary, which in turn is under control of pulsatile gonadotropin-releasing hormone (GnRH) from the preoptic area and hypothalamus (reviewed for hamsters in Bartke 1985). Steroid hormones are essential to spermatogenesis. However, they also have effects throughout the body which facilitate sexual activity. In particular, the brain is an important target for steroid hormones. Testosterone and estrogen organize sex-specific patterns of behavior during development and are required for the activation of these patterns in adulthood (Bartke 1985).

Likewise, males respond to external sensory cues to assess availability and sexual receptivity of female conspecifics. Depending on the species, males may rely on a combination of visual, auditory, tactile, or chemosensory cues from females (Silver 1992). Unlike visual or tactile communication, odors can be broadcast widely. In contrast to auditory signals, chemosensory cues from body secretions (urine, vaginal fluid, skin glands) persist in the environment. For these reasons, chemosensory stimuli influence male sexual behavior in many mammalian species. This is particularly true of solitary animals, such as Syrian hamsters.

Odor-Hormone Interactions

In a natural environment, a male receives information from different sources simultaneously. Both internal blood-borne signals and external sensory cues must be integrated to cause an appropriate behavioral response. Accordingly, it is reasonable to expect that steroid and chemosensory signals for sexual activity are not independent, but interact reciprocally. Receipt of chemosensory cues can stimulate activity of the hypothalamic-pituitary gonadal axis, and hormonal status influences responsiveness to chemosensory cues (figure 13.1).

Chemosensory Cues Stimulate Hormone Release

Odor cues can rapidly trigger the release of gonadal steroids. This has been demonstrated in several rodent species, including rats (Kamel et al. 1977), mice (Coquelin and Bronson 1980), and hamsters (Macrides et al. 1974; Pfeiffer and Johnston 1992). In hamsters, testosterone increases in circulation within 30 minutes after presentation of a receptive female (Macrides et al. 1974). The odor-induced release of testosterone is presumably mediated by increases in pulsatile GnRH from the hypothalamus which drives LH secretion. GnRH cannot be measured in systemic circulation, but changes in the secretion of this peptide can be inferred from LH release. Male rats and mice show a dramatic rise in LH within 5 minutes of exposure to a receptive female or her odors (Coquelin and Bronson 1980; Graham and Desjardins 1980; Coquelin and Desjardins 1982). The rapidity of this response suggests that the effect of odors on GnRH is neuronal, rather than a response to steroid feedback. In addition, the odor-driven LH response can be classically conditioned to nonsexual chemical cues (Graham and Desjardins 1980) and habituates following repeated exposure to the same female (Coquelin and Bronson 1979).

Interestingly, neuroendocrine responses to odor cues are only evident in sexually active animals. Gonadally intact male guinea pigs (Harding 1981) and rams (Perkins et al. 1992) that refuse to mate with a receptive female also fail to show the female-induced testosterone increases displayed by mating conspecifics. Although the nature of the mating deficit is unknown, this finding suggests fundamental differences in the neuroendocrine organization of mating and nonmating males. It also implies a behavioral role for the acute activation of the hypothalamic-pituitary-gonadal axis.

Hormonal Modulation of Responsiveness to Chemosensory Cues

Just as odor cues induce neuroendocrine release, the activity of the neuroendocrine system also stimulates the response to chemosensory cues. While castrated males

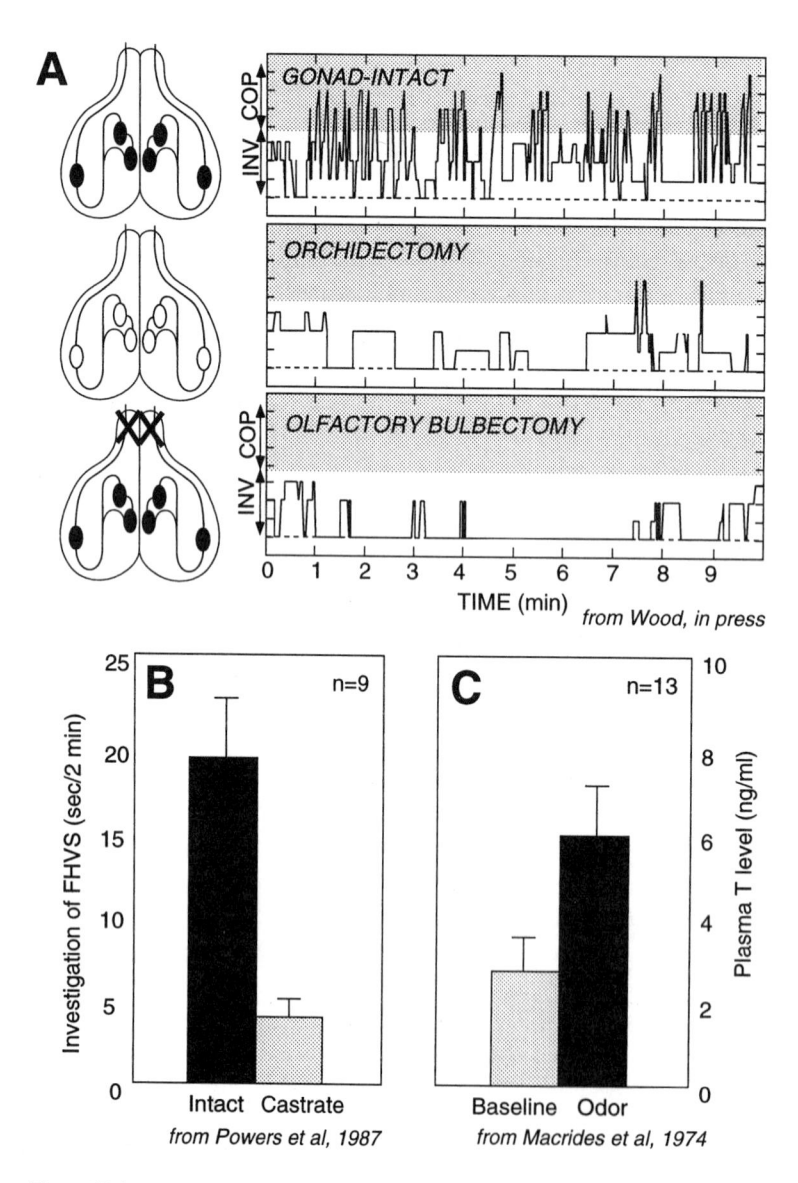

Figure 13.1
Interaction of odors and hormones in the control of male hamster sexual behavior. (A) Sexual behavior of a normal male, a long-term castrate, and a male with bilateral olfactory bulbectomy during a 10-minute exposure to a receptive female. Black circles indicate steroid binding; dotted line indicates no interaction with the female; shading indicates copulatory behavior. COP, mounts, intromissions, ejaculation; INV, investigatory behavior (self-grooming, investigation of the female, anogenital investigation). (Redrawn from Wood, 1998). (B) Gonadal steroids facilitate investigation of female hamster vaginal secretion (FHVS). (Redrawn from Powers et al. 1985). (C) Chemosensory stimuli enhance testosterone (T) concentrations in circulation. (Redrawn from Macrides et al. 1974.)

have no apparent difficulty in detecting odors, their interest in investigating sexually relevant odors is greatly reduced (Gandelman 1982; Powers et al. 1985). Responsiveness to female hamster vaginal secretion (FHVS) is restored by exogenous testosterone treatment. This suggests that steroid hormones act on central neural pathways that process chemosensory stimuli.

Such odor-hormone interactions have significance for a male hamster in a natural environment. Although it has been difficult to study hamsters in their natural habitat, we know that hamsters are solitary, seasonally breeding animals. During the spring and summer breeding season, it is thought that a male encounters a female by following her trail of vaginal marks back to her burrow (Johnston 1985). Each encounter with a mark would stimulate testosterone release, enhancing both sexual readiness and interest in chemosensory cues. Since the interactions of odors and hormones are both stimulatory, the net effect is to increase the likelihood that the male will engage in copulatory behavior. However, during fall and winter, the testes regress and circulating concentrations of testosterone fall to basal levels (Bartke 1985). In terms of mating behavior, this seasonal testicular regression is equivalent to castration (Campbell et al. 1978). At the same time, interest in chemosensory cues from females is reduced. A male conserves reproductive effort by not pursuing anestrous females during the nonbreeding season.

Additional evidence indicates that hypothalamic GnRH also has a direct effect on chemosensory responsiveness, independent of its role in steroid release. Intracerebral injections of GnRH increase the reproductive activity of intact and castrated rats (Moss and McCann 1973). More recently, Meredith and Howard (1992) have shown that exogenous GnRH restores sexual behavior in nonmating hamsters (reviewed in Meredith and Fernandez-Fewell 1994). As discussed below, removal of the vomeronasal organ (VNO) eliminates mating in sexually inexperienced male hamsters. Intracerebral injections of GnRH restore mating in these animals within 30 minutes. Moreover, behavior is enhanced in both VNx and intact males receiving AcLHRH-5-10, a GnRH analogue that does not stimulate LH release. In view of the peculiar ontogeny of GnRH neurons from the olfactory placode (Schwanzel-Fukuda and Pfaff 1989; Wray et al. 1989), perhaps it is not surprising that GnRH is closely tied to chemosensory stimuli throughout life.

Neural Pathways for Odor-Hormone Interactions

The interaction of odors and hormones in regulating behavior is achieved through interactions of neural pathways which transduce and transmit chemosensory and steroid stimuli. This is illustrated in figure 13.2. Chemosensory cues detected in the

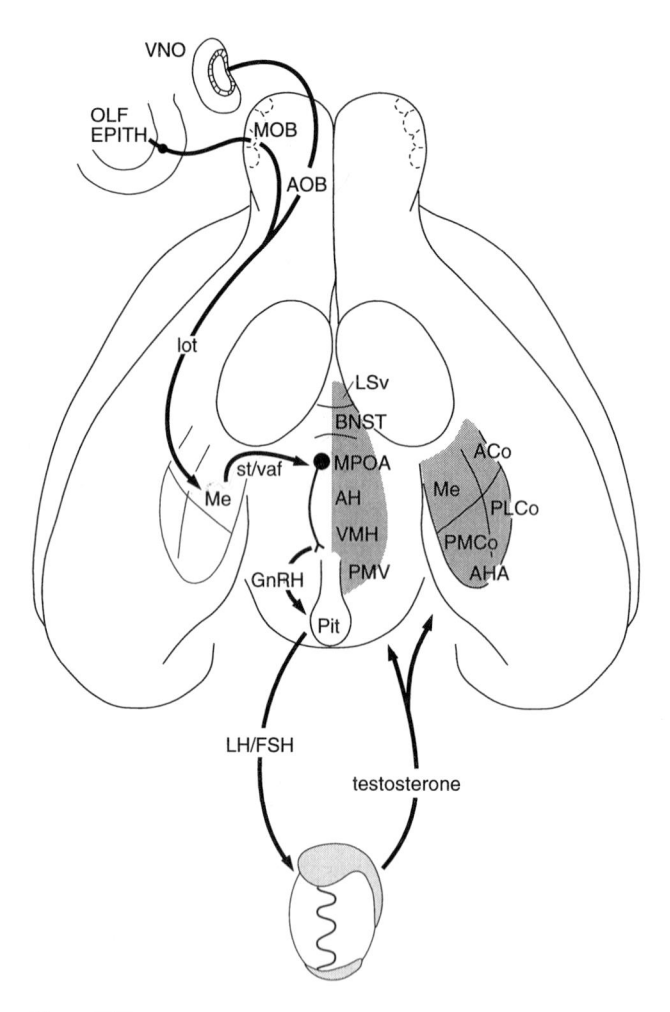

Figure 13.2
Schematic diagram of the ventral surface of the hamster brain. (Left) Chemosensory cues detected in the olfactory mucosa and vomeronasal organ can stimulate sexual behavior and neuroendocrine release in the medial preoptic area (MPOA) through a multisynaptic pathway that includes the olfactory bulbs and the medial amygdaloid nucleus (Me). (Right) Conversely, steroids modulate responsiveness to odor cues by binding to their receptors (shaded areas) along the chemosensory pathway. ACo, anterior cortical amygdaloid nucleus; AH, anterior hypothalamus; AHA, amygdalohippocampal area; AOB, accessory olfactory bulb; BNST, bed nucleus of the stria terminalis; FSH, follicle-stimulating hormone; GnRH, gonadotropin-releasing hormone; LH, luteinizing hormone; lot, lateral olfactory tract; LSv, ventral lateral septum; Me, medial amygdaloid nucleus; MOB, main olfactory bulb; MPOA, medial preoptic area; OLF EPITH, olfactory epithelium; Pit, anterior pituitary; PLCo, posterolateral cortical nucleus; PMCo, posteromedial cortical amygdaloid nucleus; PMV, ventral premammillary nucleus; vaf, ventral amygdalofugal pathway; VMH, ventromedial hypothalamic nucleus; VNO, vomeronasal organ; st, stria terminalis.

VNO and olfactory mucosa (OM) are transmitted through a multisynaptic, steroid-sensitive pathway to reach the medial preoptic area (MPOA). Chemosensory cues are essential to male hamster sexual behavior. Ablation (Murphy and Schneider 1970) or deafferentation (Winans and Powers 1977) of the olfactory bulbs prevents mating. Likewise, the integrity of MPOA is essential to expression of male sexual behavior (Powers et al. 1987), and the majority of GnRH neurons are located in MPOA in hamsters (Jennes and Stumpf 1980) and most other species.

The pathways that connect these critical areas have been highly conserved during evolutionary history (reviewed in Wood and Newman 1995a). Beginning in the nasal cavity, sensory neurons in the OM and VNO project through the cribiform plate to the main (MOB) and accessory olfactory bulbs (AOB), respectively. These anatomically distinct sensory systems play distinct roles in the regulation of male sexual behavior and steroid hormone release (Winans and Powers 1977; Pfeiffer and Johnston 1994). Destruction of the VNO causes severe deficits in copulatory behavior (Winans and Powers 1977; Meredith 1986) and eliminates FHVS-induced androgen surges (Pfeiffer and Johnston 1994). In contrast, destruction of the main olfactory system does not affect these measures (Winans and Powers 1977; Pfeiffer and Johnston 1994). Nonetheless, even in the absence of the VNO, exposure to an estrous female increases androgen levels (Pfeiffer and Johnston 1994). Thus, the accessory olfactory system plays the greater role in mediating responses to FHVS, but other sensory systems suffice to stimulate androgen release in the presence of a female.

The MOB and AOB send efferents through the lateral olfactory tract (lot) to several nuclei in the amygdala. Of these, the medial amygdaloid nucleus (Me) is the critical link between chemosensory targets and the hypothalamus. Me integrates information from the main and accessory olfactory systems. In particular, the anterior portion of this nucleus (MeA) receives a substantial projection from AOB, and modest input from MOB (Davis et al. 1978). MeA is essential to copulation (Lehman and Winans 1982), whereas destruction of posterior Me (MeP) selectively inhibits chemoinvestigatory behavior (Lehman et al. 1983). Me relays critical chemosensory information to the bed nucleus of the stria terminalis (BNST) and MPOA via the stria terminalis (st) and ventral amygdalofugal pathway (vaf, Lehman 1980). BNST and MPOA are also functionally distinct. Destruction of BNST decreases interest in sexually relevant odors, while lesions centered in MPOA immediately and permanently abolish copulation (Powers et al. 1987).

Steroid hormones can increase responsiveness to odor cues via binding to androgen and estrogen receptors in central chemosensory nuclei. In this regard, Me, BNST, and MPOA each contain abundant steroid receptors (see Wood and Newman 1995a). Steroid receptors have also been described in both the olfactory bulbs

(Keliher, personal communication, 1997) and the VNO (Getchell et al. 1996), but these are sparse compared with the central nuclei of the mating behavior circuit. This distribution is consistent with the behavioral effects of steroids on odor responsiveness. Castration does not appear to reduce the threshold for detection of odors in general (Carr et al. 1962), as would be expected if steroids acted preferentially on OM, VNO, or the olfactory bulbs. Instead, removal of gonadal steroids seems to diminish interest in sexually relevant chemosensory cues (Carr et al. 1966), consistent with the distribution of steroid receptors in central components of the mating behavior circuit.

Neural Activation Reveals the Mating Behavior Circuit

The functions and connections of nuclei in the mating behavior circuit were originally defined by lesions and tract tracing (reviewed in Wood and Newman 1995a). Recently, immunoreactivity for Fos, the protein product of the *c-fos* gene, has been used to refine this circuit by identifying specific groups of neurons activated in response to sexually relevant odors and copulatory behavior. Fos immunocytochemistry is a valuable tool for tracing functional activation along a multisynaptic pathway (Dragunow and Faull 1989) such as that controlling sexual behavior. The pattern of Fos activation in the olfactory bulbs, Me, BNST, and MPOA in response to FHVS is depicted in figure 13.3. When a male is exposed to FHVS, Fos is expressed in each component of the mating behavior circuit (Fiber et al. 1993; Fiber and Swann 1996). However, Fos-immunoreactive neurons are not distributed uniformly throughout these regions. In MOB and AOB, odors induce Fos in mitral and granule cells, the output and feedback neurons of the olfactory bulbs, respectively (see Shipley and Ennis 1996). Granule cell feedback inhibition may be particularly important in the regulation of responses to diverse chemosensory stimuli. Both granule and mitral cells are stimulated by sexually relevant, as well as nonsexual chemosensory cues (Fiber et al. 1993). However, selective processing of chemosensory cues determines whether these stimuli are relayed to central target areas. Nonspecific cues stimulate areas involved in general olfaction, such as the piriform cortex. Only FHVS from Syrian hamsters stimulates Fos expression in Me, BNST, and MPOA.

In Me and BNST, FHVS-stimulated Fos is concentrated in MeP and the posteromedial subdivision of BNST (BNSTpm, Fiber et al., 1993). These regions are densely interconnected through st and vaf, receive direct projections from AOB, and play similar roles in the regulation of behavior (Gomez and Newman 1992). In MPOA, FHVS induces Fos expression in both the medial preoptic nucleus (MPN) and its more lateral magnocellular subdivision (MPNmag). These areas are differentially connected to Me and BNST and may play different roles in the regulation of

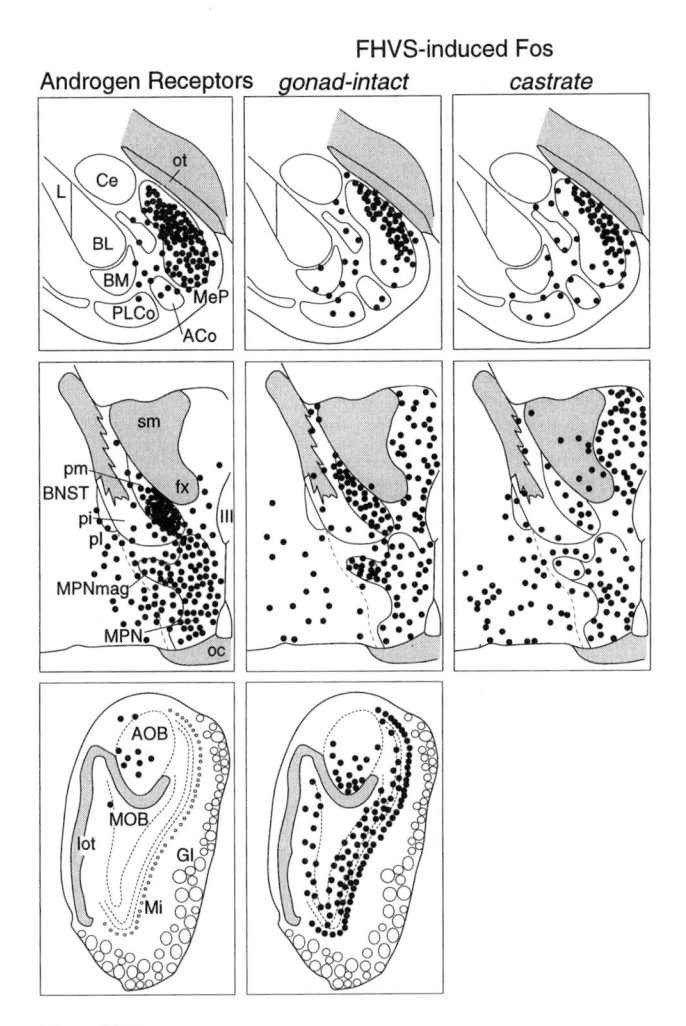

Figure 13.3
Androgen receptor immunoreactivity and odor-induced Fos activation in the olfactory bulbs, medial amygdaloid nucleus (Me), bed nucleus of the stria terminalis (BNST) and medial preoptic area of a representative gonad-intact male hamster (left, middle), and in a long-term castrate (right). III, third ventricle; ACo, anterior cortical amygdaloid nucleus; AOB, accessory olfactory bulb; BL, basolateral amygdaloid nucleus; BM, basomedial amygdaloid nucleus; BNSTpi, posterointermediate BNST; BNSTpl, posterolateral BNST; BNSTpm, posteromedial BNST; Ce, central amygdaloid nucleus; fx, fornix; Gl, glomeruli; L, lateral amygdaloid nucleus; lot, lateral olfactory tract; MeP, posterior Me; Mi, mitral cell layer; MOB, main olfactory bulb; MPN, medial preoptic nucleus, mag, magnocellular subdivision; oc, optic chiasm; ot, optic tract; PLCo, posterolateral cortical nucleus; sm, stria medullaris.

behavior. Odor-induced Fos expression in MPN is dependent on experience and can be induced by the testing arena alone in sexually experienced hamsters, gerbils, and rats (Kollack-Walker and Newman 1997; Heeb and Yahr 1996; Coolen et al. 1997b). Odor-induced Fos in MPNmag is independent of experience and specific to FHVS alone (Fiber et al. 1993; Fiber and Swann 1996; Kollack-Walker and Newman 1997). Destruction of MPNmag eliminates copulatory behavior (Powers et al. 1987). Thus, MPN may regulate behavior in response to conditioned cues, while MPNmag is hard-wired to specific sensory stimuli.

Fos is also expressed in MPOA, BNST, and Me in response to copulatory behavior. However, the distribution of mating-induced Fos-immunoreactive neurons is substantially different from that induced by chemosensory cues alone. BNSTpm, MPN, and MeP each show gradual increases in Fos that are correlated with increases in sexual activity (Kollack-Walker and Newman 1997). These changes presumably reflect sensory feedback from genital areas (Baum and Everitt 1992). In addition, copulation induces small clusters of Fos-positive neurons in lateral MeP and in dorsal and medial BNSTpm of hamsters, rats, and gerbils (Heeb and Yahr 1996; Coolen et al. 1996). The clusters of Fos-positive neurons appear in rats that ejaculate without intromission, suggesting that they are ejaculation-specific (Coolen et al. 1997a). Moreover, ejaculation-induced clusters may signal an end to mating behavior, because they are expressed only when males approach sexual satiety (Parfitt and Newman 1995). In support of this hypothesis, destruction of ejaculation-specific clusters in MeP prolongs copulation (Parfitt et al. 1996).

Fos and Odor-Hormone Interactions

Steroids Affect Odor-Induced Fos Immunoreactivity Fos is a useful tool for investigating the neural mechanisms of odor-hormone interactions. Because FHVS-induced Fos is expressed in the same brain regions that contain abundant steroid receptors, it is logical to expect that there is an interaction of Fos and steroid receptor–containing neurons which mirrors the interaction of odors and hormones in controlling sexual behavior and neuroendocrine function. Figure 13.3 compares the distribution of FHVS-stimulated Fos in gonad-intact vs. gonadectomized males. Importantly, removal of gonadal steroids does not cause a widespread reduction in Fos expression. There were no differences in the number of Fos-positive neurons in the olfactory bulbs, Me, or BNST after castration (Swann 1997). Only in MPNmag was Fos significantly reduced in orchidectomized males.

At face value, this result could imply that steroids are important for Fos expression only in MPNmag. However, this explanation is overly simplistic. Current evidence suggests that Fos does not reflect the electrical output of a neuron (Dragunow and

Faull 1989). Rather, it is thought to indicate a change in the afferent input. Loss of Fos immunoreactivity in MPNmag after castration implies that input to this subnucleus is reduced in the gonadectomized male. Me and BNST are important sources of afferent projections to MPNmag in rats (Simerly and Swanson 1986). Thus, we hypothesize that the loss of steroid stimulation in Me, BNST, and MPOA prevents transmission of odor cues along the chemosensory pathway. In this manner, steroid binding in Me, BNST, and MPOA may act as a gating signal to permit or facilitate transmission of chemosensory cues. According to this model, Fos immunoreactivity is unchanged in the distal portions of the chemosensory pathway (Me and BNST) of castrated males because afferent input to these regions from the olfactory bulbs is not regulated by gonadal steroids.

Fos Activation of Steroid Receptor–Containing Neurons In view of the overlapping distribution of Fos and steroid receptors, it should not be surprising that Fos is expressed in steroid-responsive neurons of Me, BNST, and MPOA after mating (Wood and Newman 1993). This suggests that steroid receptor–containing neurons are activated during sexual behavior. To our knowledge, no one has determined if Fos co-localization with steroid receptors is due to chemosensory stimuli or copulatory behavior. However, the distribution of these double-labeled neurons is consistent with both odor- and mating-induced activation. It is unlikely that odor-induced steroid secretion is responsible for Fos expression in steroid-responsive neurons because the pattern of Fos is unchanged in gonadectomized male hamsters and rats replaced with exogenous gonadal steroids (Baum and Wersinger 1993; Swann 1996).

Fos Activation of GnRH Neurons Although chemosensory cues are known to stimulate testosterone secretion (Macrides et al. 1974), presumably via stimulation of GnRH release, it was surprising to find that GnRH neurons do not express Fos either during mating or in response to odor cues (Fernandez-Fewell and Meredith 1994). Odor cues can reach GnRH neurons in the rostral preoptic area by projections directly from MeP or via BNST (Gomez and Newman 1992). In this regard, the projection from BNSTpm to the anteroventral periventricular nucleus (AVPV) in the MPOA is sexually dimorphic in rats, with a substantially larger projection in males (Hutton et al. 1998). The AVPV regulates the preovulatory LH surge in female rats (Weigand and Terasawa 1982), and may contribute to odor-induced hormonal release in males.

The absence of Fos in GnRH neurons highlights a troublesome aspect of using Fos as a marker of neuronal activation. Although we may presume that neurons which express Fos are indeed active in response to a given stimulus, the converse is not necessarily true. That is to say, lack of Fos expression does not imply that a neuron was inactive or unaffected. In the case of GnRH, we know that GnRH neurons of female rats and sheep are capable of expressing Fos in response to a surge-inducing dose of estradiol (Hoffman et al. 1990; Moenter et al. 1993). However, the preovulatory GnRH surge represents a substantial activation above baseline conditions. Other stimuli that are known to increase the activity of the GnRH neurosecretory system, such as removal of steroid inhibition to increase pulsatile GnRH release, fail to cause Fos expression in GnRH neurons (Moenter et al, 1993). Based on the increment of testosterone secretion in hamsters in response to sexually relevant odors (Macrides et al. 1974), the odor-induced release of GnRH represents a rather modest stimulus compared to the massive activation required for Fos induction.

Sites of Steroid Action along the Chemosensory Circuit

Lesions along the mating behavior pathway that eliminate mating have established the importance of Me, BNST, and MPOA in male sexual behavior. Our understanding of this circuit is enhanced through the use of Fos to identify activated neurons, including steroid-responsive neurons. However, these techniques alone do not determine the significance of steroid action in the chemosensory circuit. Castration eliminates mating, and replacement with exogenous testosterone restores sexual activity. Me, BNST, and MPOA each contain large numbers of steroid receptor–containing neurons (see Wood and Newman 1995a), and it is reasonable to expect that these areas transduce hormonal cues for sexual behavior. However, other limbic nuclei also have abundant steroid receptors. How do we establish which groups of steroid-sensitive neurons are important for sexual behavior? One way to tackle this question is by using intracerebral steroid implants in castrated males to locally stimulate groups of steroid-responsive neurons in the brains of castrated males. The results of several such studies are summarized in figure 13.4.

Intracerebral implants of testosterone in either Me or BNST/MPOA enhance sexual activity above levels in castrates (Lisk and Bezier 1980; Rasia-Filho et al. 1991; Wood and Newman 1995c). These effects are mediated by aromatization to estrogen, for implants of the nonaromatizable androgen, dihydrotestosterone, in either area are without effect (Lisk and Greenwald 1983; Wood 1996). Steroid implants not only facilitate male copulatory behavior, they also enhance the response to odor cues. Specifically, male mice with intracerebral implants in MPOA will make ultrasonic vocalizations to female urine (Sipos and Nyby 1996). Likewise, intracerebral testosterone implants in Me or BNST MPOA of male hamsters stimulate anogenital investigation (Wood and Newman 1995c). Together, these studies suggest

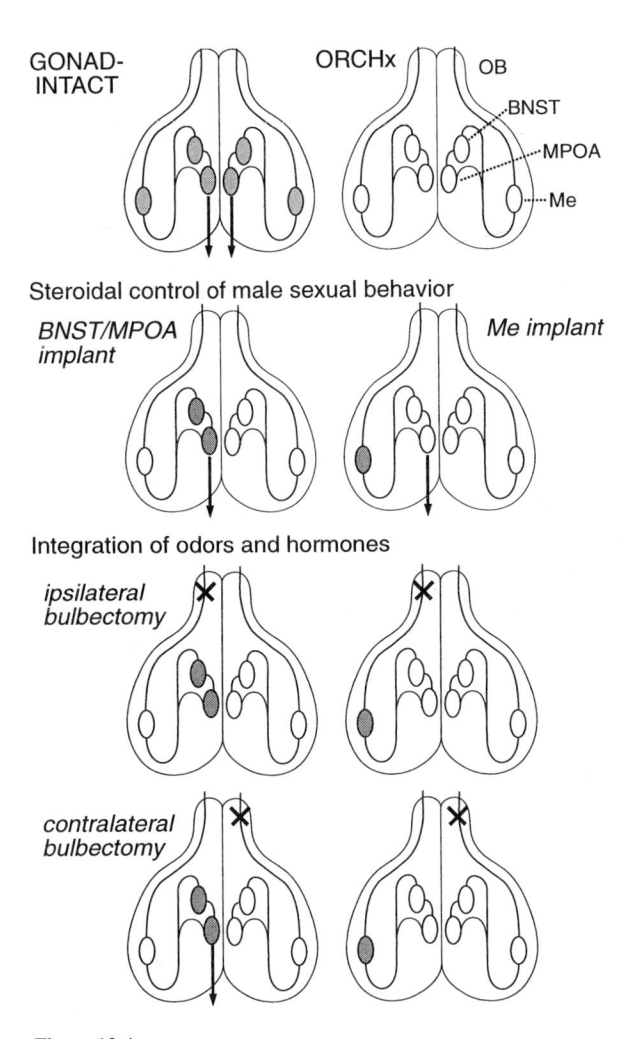

Figure 13.4
Schematic diagram of the chemosensory circuit for male mating behavior in the Syrian hamster summarizing the steroidal control of sexual activity. (Top) Castration abolishes mating. Middle, Focal stimulation of steroid receptor–containing neurons using intracerebral implants of testosterone in medial amygdaloid nucleus (Me) or bed nucleus of the stria terminalis (BNST) medial preoptic area (MPOA) increases sexual activity in castrated (ORCHx) males. (Bottom) Combining an intracerebral testosterone implant with unilateral removal of an olfactory bulb (OB) determines if steroid interaction with odors is essential to mating. Ipsilateral bulbectomy prevents interaction of odors and hormones, and therefore blocks sexual activity. Contralateral bulbectomy permits mating in males with testosterone in BNST and MPOA, but not in males with testosterone in Me.

that steroid action along the chemosensory pathway facilitates sexual behavior and responsiveness to sexually relevant chemosensory stimuli.

This leads us to investigate the neuroanatomical basis of odor-hormone interactions. In particular, we know that the same brain regions transmit chemosensory cues and transduce hormonal signals. Moreover, we know that odors and hormones interact to facilitate behavior and neuroendocrine function. Where does this interaction take place in the brain? What is the significance of odor-hormone interaction in the control of male sexual behavior?

Recently, we tested the hypothesis that an interaction of steroids and odor cues is required for mating by combining an intracerebral testosterone implant with removal of a single olfactory bulb, either ipsilateral or contralateral to the steroid implant (see figure 13.4). This approach takes advantage of the predominantly ipsilateral projections of the olfactory bulbs (Davis et al. 1978). If an interaction of steroids and odors is required, males with ipsilateral bulbectomy will fail to mate because the odor and hormonal cues for sexual behavior are segregated in opposite hemispheres of the brain. Contralateral bulbectomy is intended as a control. The sexual behavior of castrated males with a testosterone implant in BNST MPOA supported our hypothesis (Wood and Newman 1995b). That is, males with ipsilateral bulbectomy failed to mate, whereas those with contralateral bulbectomy responded to the steroid implant. The conclusion is that male hamsters require an interaction of odors and hormones for mating.

The behavior of males with a testosterone implant in Me was somewhat more difficult to interpret (Wood and Coolen 1997). In those males, both ipsilateral and contralateral bulbectomy abolished sexual activity. Based on this result, it appears that steroid stimulation of mating through Me is more sensitive than BNST MPOA to disruption of chemosensory input.

Cellular Mechanisms of Odor-Hormone Interaction

The studies described above support the concept that an interaction of chemosensory and hormonal stimuli is essential to sexual behavior. The challenge now is to understand the cellular mechanisms for this interaction. Steroid hormones have a fundamentally different mode of action from chemosensory stimuli. Odors facilitate mating by activating neurons along a defined neural pathway. This type of stimulus is particularly amenable to study by techniques such as Fos that reflect acute changes in neural function. However, steroids have a more long-term, permissive action in the brain to facilitate behavior. Mating behavior persists for weeks after castration, and reinstatement of copulation with exogenous testosterone requires chronic steroid exposure (Noble and Alsum 1975). These effects of steroids on mating behavior cannot be explained by simple activation of neurons through the binding of steroids to their receptors in the brain. Steroids have diverse effects on neural structure and function, which are summarized in figure 13.5. Although steroids are capable of rapid neuronal activation, they also have more long-term effects on neurotransmitters, connectivity, and neuronal morphology. The time course of hormonal control of sexual behavior is consistent with these latter mechanisms.

Steroid Effects on Neurotransmitters in the Chemosensory Pathway

There is considerable evidence for hormonal modulation of neurotransmitters and their receptors, including in the mating behavior pathway. Male sexual behavior is controlled by many different neurotransmitter systems, nearly all of which are sensitive to steroids. Castration reduces immunoreactivity for several neurotransmitters in the hamster mating behavior circuit, including substance P, nitric oxide synthase, and the catecholamine synthetic enzyme, tyrosine hydroxylase (reviewed in Wood and Newman 1995a). Moreover, castration also decreases neurotransmitter receptors in the same circuit. Thus, steroids may reduce synaptic efficiency in the transmission of odor cues by inhibiting release of neurotransmitters or suppressing postsynaptic receptors.

Steroid Effects on Neuronal Morphology in the Chemosensory Pathway

Gonadal steroids also promote growth and branching of neurons. In general, these effects are most dramatic during development, and are responsible for sex differences in the morphology of brain nuclei and neurons along the chemosensory pathway. However, steroids also stimulate neuronal growth in the adult, as detailed by Garcia-Segura et al. (1994). In the hamster mating behavior pathway, castration reduced the mean somal area, highest dendritic branch, and percentage of neurons in Me with tertiary branches (Gomez and Newman 1991).

Steroid effects on neuronal structure are not limited to the cell soma and dendrites. Recently, VanderHorst and Holstege (1997) have shown that estrogen can induce axonal growth in adult female cats. Steroid-induced growth occurs in the pathway to the lumbosacral motor group, which may be the final common pathway for lordosis. Since female cats do not exhibit lordosis except when in heat, these results suggest that axonal growth mediates steroidally modulated behavior.

Steroid Effects on Connectivity in the Chemosensory Pathway

Although steroid exposure in the adult has only modest effects on neuronal morphology, the effects on synaptic connections are much more dramatic. Changes in

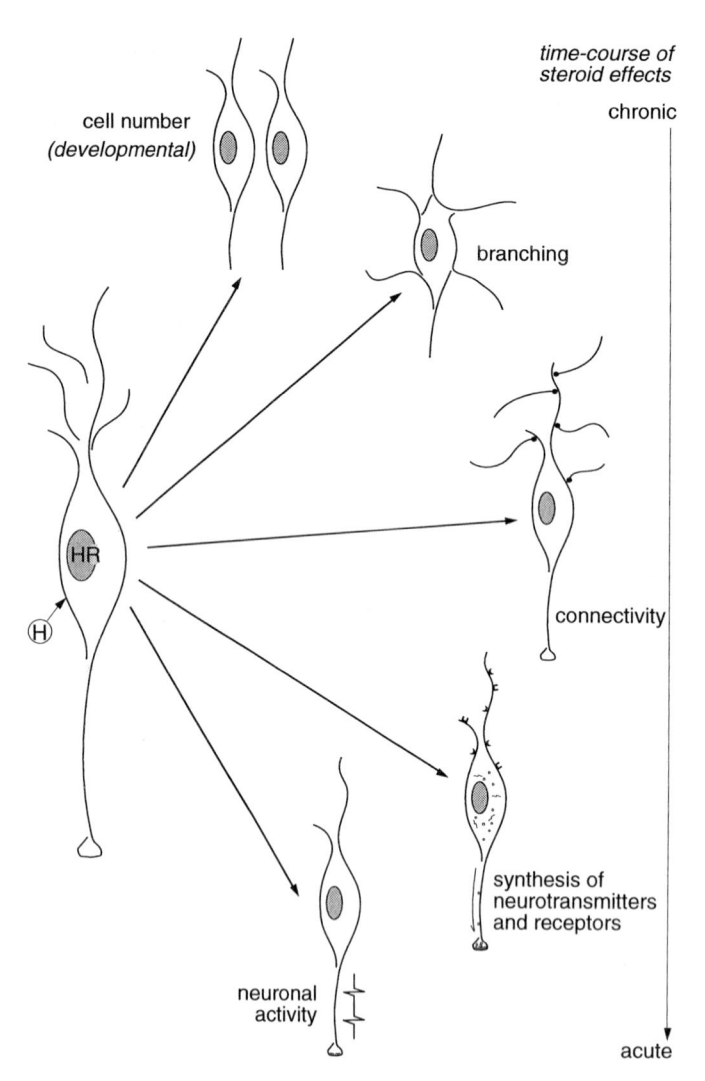

Figure 13.5
Examples of steroid action in the brain. Steroids influence neuronal structure and function through binding to their receptors. These actions can include long-term effects on dendritic branching and connectivity, as well as rapid stimulation of transmitter function and neuronal activity. (From Wood 1998.)

connectivity can take place rapidly in the absence of substantial alterations in gross neuronal structure. This has been best studied in females, where synapses in the arcuate nucleus of the hypothalamus are remodeled over the course of the 4-day estrous cycle (Olmos et al. 1989). In the male, loss of synaptic connections after castration could have a major impact on transmission of chemosensory cues through Me, BNST, and MPOA.

In sum, the net effect of castration at the cellular level is to attenuate neural connections and reduce synaptic efficiency through loss of neurotransmitters and receptors. It is likely that both structural and neurochemical mechanisms contribute to the decline in sexual behavior after castration. These effects are consistent with steroids acting as a gating mechanism to permit or enhance transmission of chemosensory cues through steroid-sensitive brain nuclei, as shown in figure 13.6. According to our current model, steroids act in Me, BNST, and MPOA to enhance both the number and activity of synapses that relay chemosensory cues from the olfactory bulbs to the preoptic area. When steroids are present, odor cues from a receptive female activate neurons in MPOA essential to sexual behavior. In addition, chemosensory stimuli activate GnRH neurons in MPOA to further increase circulating testosterone. When steroids are no longer present, synaptic connections attenuate and odor cues are no longer transmitted to MPOA. While the available evidence supports our hypotheses, this model remains to be tested.

Significance of Odor-Hormone Interactions

A great deal of research is directed at understanding the signals and pathways through which individual sensory and endocrine stimuli influence mating behavior. However, sensory and humoral cues are seldom presented in isolation. Instead, an organism in a natural setting is confronted with multiple signals from its internal and external environment that together are interwoven to yield an experience richer than the sum of the individual modalities (Stein and Meredith 1993).

To understand complex motivated behaviors, we must determine how different cues are weighed, prioritized, and integrated. The specific signals that guide sexual activity are tailored to the biology of the species. Whereas hamsters rely principally on chemosensory stimuli to initiate behavior, other species use visual, auditory, or somatosensory stimuli, or a combination of sensory cues (Silver 1992). Nonetheless, multimodal integration of sensory and humoral signals is ubiquitous, and further studies may uncover common mechanisms that guide the behavior of diverse species.

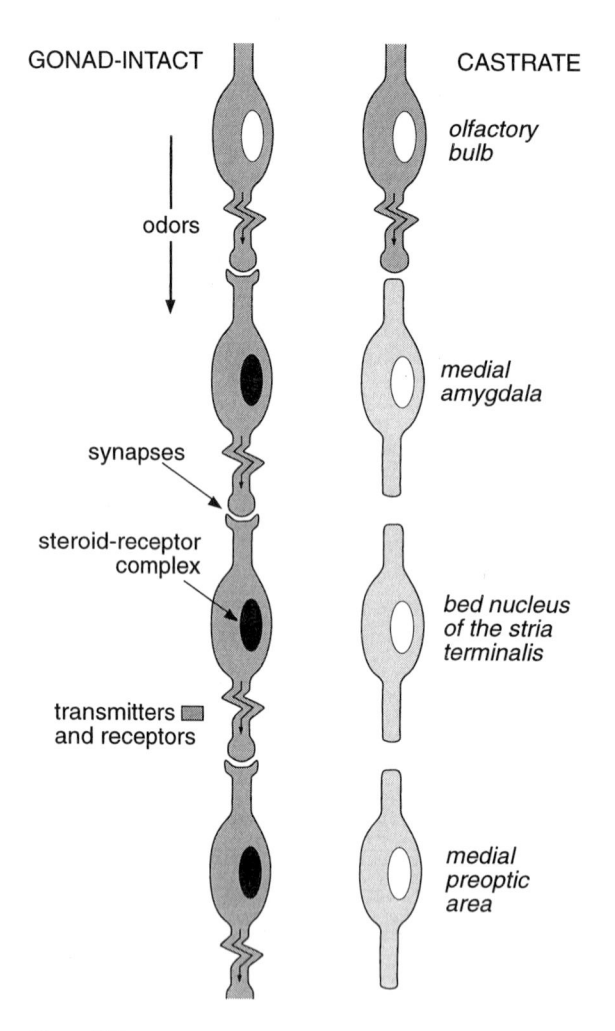

Figure 13.6
Potential cellular mechanisms of odor-hormone interactions. In the gonad-intact male (left), steroids act on hormone-responsive neurons (black nuclei) to facilitate transmission of chemosensory cues by promoting synaptic contacts and enhancing neurotransmitter production (shading). In the absence of steroids (right), connections between neurons attenuate, and transmission of odor cues is impaired.

Acknowledgments

We thank Sarah W. Newman for training and guidance. This work was supported by grants from the NIH (HD-03269 and MH-05534 to R.I.W., HD-28467 to J.M.S.), and NSF (IBN-9753021 to J.M.S.).

References

Bartke A (1985). Male hamster reproductive endocrinology. *In* The Hamster: Reproduction and Behavior (Siegel HL, ed) pp 74–98. New York: Plenum Press.

Baum MJ, Everitt BJ (1992). Increased expression of c-fos in the medial preoptic area after mating in male rats: role of afferent inputs from the medial amygdala and midbrain central tegmental field. Neuroscience 50:627–646.

Baum MJ, Wersinger SR (1993). Equivalent levels of mating-induced neural c-fos immunoreactivity in castrated male rats given androgen, estrogen, or no steroid replacement. Biol Reprod 48(6):1341–1347.

Bronson FH (1989). Mammalian Reproductive Biology. Chicago: University of Chicago Press.

Campbell CS, Finkelstein JS, Turek FW (1978). The interaction of photoperiod and testosterone on the development of copulatory behavior in castrated male hamsters. Physiol Behav 21:409–415.

Carr WJ, Solberg B, Pfaffman C. (1962). The olfactory threshold for estrus female urine in normal and castrated male rats. J Comp Physiol Psychol 55:415–417.

Carr WJ, Loeb LS, Wylie NR (1966). Responses to feminine odors in normal and castrated male rats. J Comp Physiol Psychol 62:336–338.

Coolen LM, Peters HJ, Veening JG (1996). Fos immunoreactivity in the rat brain following consummatory elements of sexual behavior: a sex comparison. Brain Res 738(1):67–82.

Coolen LM, Olivier B, Peters HJ, Veening JG (1997a). Demonstration of ejaculation-induced neural activity in the male rat brain using 5-HT1A agonist 8-OH-DPAT. Physiol Behav 62(4):881–891.

Coolen LM, Peters HJ, Veening JG (1997b). Distribution of Fos immunoreactivity following mating versus anogenital investigation in the male rat brain. Neuroscience 77:1151–1161.

Coquelin A, Bronson FH (1979). Release of luteinizing hormone in male mice during exposure to females: habituation of the response. Science 206(4422):1099–1101.

Coquelin A, Bronson FH (1980). Secretion of luteinizing hormone in male mice: factors that influence release during sexual encounters. Endocrinology 106(4):1224–1229.

Coquelin A and Desjardins C (1982). Luteinizing hormone and testosterone secretion in young and old male mice. Am J Physiol 243(3):E257–263.

Davis BJ, Macrides F, Young WM, Schneider SP, Rosene DL (1978). Efferents and centrifugal afferents of the main and accessory olfactory bulbs in the hamster. Brain Res Bull 3:59–72.

Dragunow M, Faull R (1989). The use of c-fos as a metabolic marker in the neuronal pathway tracing. J Neurosci Methods 29:261–265.

Fernandez-Fewell GD, Meredith M (1994). c-Fos expression in vomeronasal pathways of mated or pheromone-stimulated male golden hamsters: contributions from vomeronasal sensory input and expression related to mating performance. J Neurosci 14(6):3643–3654.

Fiber JM, Swann JM (1996). Testosterone differentially influences sex-specific pheromone-stimulated Fos expression in limbic regions of Syrian hamsters. Horm Behav 30:455–473.

Fiber JM, Adames P, Swann JM (1993). Pheromones induce c-fos in limbic areas regulating male hamster mating behavior. Neuroreport 4:871–874.

Gandelman R (1983). Gonadal hormones and sensory function. Neurosci Biobehav Rev 7:1–17.

Garcia-Segura LM, Chowen JA, Parducz A, Naftolin F (1994). Gonadal hormones as promoters of structural synaptic plasticity: cellular mechanisms. Prog Neurobiol 44:279–307.

Getchell ML, Kulkarni-Narta A, Marcinek R, Getchell TV (1996). Estrogeu receptors are expressed in olfactory and vomeronasal receptor neurons. Soc Neurosci Abstr 22(2):1594.

Gomez DM, Newman SW (1991). Medial nucleus of the amygdala in the adult Syrian hamster: a quantitative Golgi analysis of gonadal hormonal regulation of neuronal morphology. Anat Rec 231:498–509.

Gomez DM, Newman SW (1992). Differential projections of the anterior and posterior regions of the medial amygdaloid nucleus in the Syrian hamster. J Comp Neurol 317:195–218.

Graham JM, Desjardins C (1980). Classical conditioning: induction of luteinizing hormone and testosterone secretion in anticipation of sexual activity. Science 210(4473):1039–1041.

Harding CF (1981). Social modulation of circulating hormone levels in the male. Am Zool 21:223–231.

Heeb MM, Yahr P (1996). C-Fos immunoreactivity in the sexually dimorphic area of the hypothalamus and related brain regions of male gerbils after exposure to sex-related stimuli or performance of specific sexual behaviors. Neuroscience 72(4):1049–1071.

Hoffman GE, Lee W-S, Attardi B, Yann V, Fitzsimmons MD (1990). Luteinizing hormone–releasing hormone neurons express c-fos after steroid activation. Endocrinology 126:1736–1741.

Hutton LA, Gu G, Simerly RB (1998). Development of a sexually-dimorphic projection from the bed nuclei of the stria terminalis to the anteroventral periventricular nucleus in the rat. J Neurosci 18(8):3003–3013.

Jennes L, Stumpf WE (1980). LHRH-systems in the brain of the golden hamster. Cell Tissue Res. 209:239–256.

Johnston RE (1985). Communication. *In The Hamster: Reproduction and Behavior* (Siegel HI, ed), pp 121–154. New York: Plenum Press.

Kamel F, Wright WW, Mock EJ, Frankel AI (1977). The influence of mating and related stimuli on plasma levels of luteinizing hormone, follicle stimulating hormone, prolactin, and testosterone in the male rat. Endocrinology 101:421–429.

Kollack-Walker SK, Newman SW (1997). Mating-induced expression of c-fos in the male Syrian hamster brain: role of experience, pheromones, and ejaculations. J Neurobiol 32(5):481–501.

Lehman MN (1982). Neural pathways of the vomeronasal and olfactory systems controlling sexual behavior in the male golden hamster. Dissertation, University of Michigan.

Lehman MN, Winans SS (1982). Vomeronasal and olfactory pathways to the amygdala controlling male hamster sexual behavior. Brain Res 240:27–41.

Lehman MN, Powers JB, Winans SS (1983). Stria terminalis lesions alter the temporal pattern of copulatory behavior in the male golden hamster. Behav Brain Res 8:109–128.

Lisk RD, Bezier JL (1980). Intrahypothalamic hormone implantation and activation of sexual behavior in the male hamster. Neuroendocrinology 30:220–227.

Lisk RD, Greenwald DP (1983). Central plus peripheral stimulation by androgen is necessary for complete restoration of copulatory behavior in the male hamster. Neuroendocrinology 36:211–217.

Macrides F, Bartke A, Fernandez F, D'Angelo W (1974). Effects of exposure to vaginal odor and receptive females on plasma testosterone in the male hamster. Neuroendocrinology 15:355–364.

Meredith M (1986). Vomeronasal organ removal before sexual experience impairs male hamster mating behavior. Physiol Behav 36:737–743.

Meredith M, Fernandez-Fewell G (1994). Vomeronasal organ, LHRH, and sex behavior. Psychoneuroendocrinology 19(5–7):657–672.

Meredith M, Howard G (1992). Intracerebroventricular LHRH relieves behavioral deficits due to vomeronasal organ removal. Brain Res Bull 29(1):75–79.

Moenter SM, Karsch FJ, Lehman MN (1993). Expression of c-fos in GnRH cells of the ewe during increased secretory activity. Endocrinology 133:896–903.

Morin L and Zucker I (1978). Photoperiodic regulation of copulatory behavior in the male hamster. J Endocrinol 77:249–258.

Moss RL and McCann SM (1973). Induction of mating behavior in rats by luteinizing hormone releasing factor. Science 181:177–179.

Murphy MR, Schneider GE (1970). Olfactory bulb removal eliminates mating behavior in the male golden hamster. Science 167:302–304.

Noble RG, Alsum PB (1975). Hormone dependent sex dimorphisms in the golden hamster (*Mesocricetus auratus*). Physiol Behav 14:567–574.

Olmos G, Naftolin F, Perez J, Tranque PA, Garcia-Segura LM (1989). Synaptic remodelling in the rat arcuate nucleus during the estrous cycle. Neuroscience 32:663–667.

Parfitt DB and Newman SW (1995). Specific neuronal activation in the medial amygdala after mating is correlated with the onset of sexual satiety. Soc Neurosci Abstracts 21(1):702.

Parfitt DB, Coolen LM, Newman SW, Wood RI (1996). Lesions of the posterior medial nucleus of the amygdala delay sexual satiety. Soc Neurosci Abstracts 22(1):155.

Perkins A, Fitzgerald JA, Price EO (1992). Luteinizing hormone and testosterone response in sexually active and inactive rams. J Anim Sci 70:2086–2093.

Pfeiffer CA, Johnston RE (1992). Socially stimulated androgen surges in male hamsters: the roles of vaginal secretions, behavioral interactions, and housing conditions. Horm Behav 26(2):283–293.

Pfeiffer CA, Johnston RE (1994). Hormonal and behavioral responses of male hamster to females and female odors: roles of olfaction, the vomeronasal system, and sexual experience. Physiol Behav 55(1):129–138.

Powers JB, Bergondy ML, Matochik JA (1985). Male hamster sociosexual behaviors: effects of testosterone and its metabolites. Physiol Behav 35:607–616.

Powers JB, Newman SW, Bergondy ML (1987). MPOA and BNST lesions in male Syrian hamsters: differential effects on copulatory and chemoinvestigatory behaviors. Behav Brain Res 23:181–195.

Rasia-Filho AA, Peres TMS, Cubilla-Gutierrez FH, Lucion AB (1991). Effect of estradiol implanted in the corticomedial amygdala on the sexual behavior of castrated male rats. Braz J Med Biol Res 24:1041–1049.

Schwanzel-Fukuda M, Pfaff DW (1989). Origin of luteinizing hormone–releasing hormone neurons. Nature 338(6211):161–164.

Shipley MT, Ennis M (1996). Functional organization of olfactory system. J Neurobiol 30(1):123–176.

Silver RA (1992). Environmental factors influencing hormone secretion. *In* Behavioral Endocrinology (Becker J, Breedlove M, Crews D, eds), pp 401–422. Cambridge, MA: MIT Press.

Simerly RB, Swanson LW (1986). The organization of neural inputs to the medial preoptic nucleus of the rat. J Comp Neurol 246:312–342.

Sipos ML, Nyby JG (1996). Concurrent androgenic stimulation of the ventral tegmental area and medial preoptic area: synergistic effects on male-typical reproductive behaviors in house mice. Brain Res 729(1):29–44.

Stein BE, Meredith MA (1993). The Merging of the Senses. Cambridge: MIT Press.

Swann JM (1997). Gonadal steroids regulate behavioral responses to pheromones by actions on a subdivision of the medial preoptic nucleus. Brain Res 750:189–194.

VanderHorst VG, Holstege G (1997). Estrogen induces axonal outgrowth in the nucleus retroambiguous–lumbrosacral motoneuronal pathway in the adult female cat. J Neurosci 173:1122–1136.

Weigand SJ, Terasawa E (1982). Discrete lesions reveal functional heterogeneity of suprachiasmatic structures in regulation of gonadotropin secretion in the female rat. Neuroendocrinology 34:395–404.

Winans SS, Powers JB (1977). Olfactory and vomeronasal deafferentation of male hamsters: histological and behavioral analyses. Brain Res 126:325–344.

Wood RI (1996). Estradiol, but not dihydrotestosterone, in the medial amygdala facilitates male hamster sexual behavior. Physiol Behav 59(4/5):833–841.

Wood RI (in press). Integration of chemosensory and hormonal input in the male Syrian hamster brain. Ann N Y Acad Sci

Wood RI, Coolen LM (1997). Integration of chemosensory and hormonal cues is essential for mating behavior in the male Syrian hamster: role of the medial amygdala. Neuroscience 78(4):1027–1035.

Wood RI, Newman SW (1993). Mating activates androgen receptor–containing neurons in chemosensory pathways of the male Syrian hamster brain. Brain Res 614:65–77.

Wood RI, Newman SW (1995a). Hormonal influence on neurons of the mating behavior pathway in male hamsters. *In* Neurobiological Effects of Sex Steroid Hormones (Micevych P, Hammer RP, eds), pp 3–39. Cambridge: Cambridge University Press.

Wood RI, Newman SW (1995b). Integration of chemosensory and hormonal cues is essential for mating behavior in the male Syrian hamster. J Neurosci 15(11):7261–7269.

Wood RI, Newman SW (1995c). The medial amygdaloid nucleus and medial preoptic area mediate steroidal control of sexual behavior in the male Syrian hamster. Horm Behav 29:338–353.

Wray SA, Nieburges A, Elkabes S (1989). Spatiotemporal cell expression of luteinizing hormone–releasing hormone in the prenatal mouse: evidence for an embryonic origin in the olfactory placode. Brain Res Dev Brain Res 46(2):309–318.

14 Behavioral Activation of the Female Neuroendocrine Axis

Patricia A. Schiml, Scott R. Wersinger, and Emilie F. Rissman

In this chapter, we focus on the social regulation of one critical aspect of mammalian female reproduction: ovulation. In order to study social control of ovulation, it is important to keep in mind the underlying realities of gamete production in both males and females. In females relatively few, and in some species as few as one, gametes mature during each ovulatory cycle. By contrast, in males, gamete maturation is a continuous and ongoing process; millions of spermatozoa are produced daily. In addition, the timing of gamete production is more critical in females than in males, because the energetic costs of reproduction are greater to mammalian females. Timing of reproduction is regulated by many factors, including genetic, experiential, and environmental variables. Ultimately, even if all other factors are favorable, if an appropriate mate is not available, reproduction cannot occur. Thus, in females of many species, environmental and social influences, as well as neuroendocrine mechanisms, tend to time mating behavior so that it coincides with ovulation.

The preceding chapters documented the importance of social context in the control of reproduction, and also provided clues to the specific types of signals that may mediate social influences. We are interested specifically in the sensory and neuroendocrine mechanisms that control ovulation in mammals.

We review work on social control of ovulation that has been conducted in several mammalian species. These examples have been selected because they serve as model systems for studying sociobehavioral regulation of the activation and maintenance of the functional neuroendocrine axis. In these species, social manipulations can be used in a carefully controlled manner to activate the neuroendocrine axis. Thus, the events that precede and follow ovulation can be observed from start to finish.

The pivotal player in the neuroendocrinology of ovulation is the ten–amino-acid neuropeptide, gonadotropin-releasing hormone (GnRH). We present data collected in several species of mammals in which the activation of the GnRH system can be stimulated, to a greater or lesser extent, by social interactions with male conspecifics.

In the second half of the chapter, we review data from our laboratory on behavioral activation of the neuroendocrine axis and ovulation in the musk shrew (*Suncus murinus*). In conclusion, we present an evolutionary argument that behavioral activation of the GnRH system represents the primitive mammalian condition.

Neuroendocrinology of Ovulation

The Hypothalamic-Pituitary-Gonadal Axis

Ovulation requires the maturation and synchronization of the ovary, pituitary, and brain (see figure 1.1). In the most commonly studied mammals, ovulation occurs spontaneously at a specific time during the menstrual or estrous cycle, regardless of the presence of males, in response to steroid hormones. The timing of ovulation is regulated by a series of changing feedback relationships between the ovaries, brain, and pituitary, as described in the General Introduction to this book (see figure 1.1). In the beginning of the cycle, only immature follicles are present in the ovaries and these produce little estradiol. Follicle maturation is promoted by the pulsatile secretion of hypothalamic GnRH into the portal blood supply, which upon reaching the anterior pituitary promotes the pulsatile secretion of luteinizing hormone (LH), as well as the secretion of follicle-stimulating hormone (FSH) (Karsch et al. 1997). As follicles mature, their synthesis and secretion of ovarian steroids increases, and plasma levels of estradiol also increase. Just prior to ovulation, a dramatic increase in estradiol acts on the brain to trigger GnRH release and increases pituitary sensitivity to the GnRH signal (Levine 1997). The resulting LH surge is the trigger for ovulation.

In other species, ovulation does not occur spontaneously in a cyclic manner. In some of these species, ovarian maturation leads to increased secretion of estradiol, which in turn promotes the display of sexual behavior. During mating, the vaginocervical stimulation received during coitus triggers a neuroendocrine reflex that results in the preovulatory LH surge (Sawyer and Markee 1959; Spies et al. 1997). These two patterns are referred to as spontaneous and induced ovulation, respectively. However, some investigators have argued that these are not dichotomous variables, but instead two extremes on a continuum (Conaway 1971).

Distribution of GnRH Neurons

Regardless of whether the LH surge is induced by positive feedback from estradiol or by mating, it is controlled in all mammalian species by the same neural releasing factor, GnRH. A small subset of neurons (e.g., about 1200 in the rat,) express GnRH gene and produce GnRH peptide. Although the adult distribution of GnRH neurons is species-specific, there are a few generalizations that can be made (figure 14.1). The GnRH neurons are distributed in a relatively diffuse continuum along the mediobasal part of the forebrain. They project heavily to the median eminence (ME) which lies just above the pituitary-portal system. In rats and mice, the GnRH neurons are concentrated in a continuum from the lateral septum through the diagonal band of

Behavioral Effects on the Female Neuroendocrine System 447

RAT

MUSK SHREW

RHESUS MONKEY

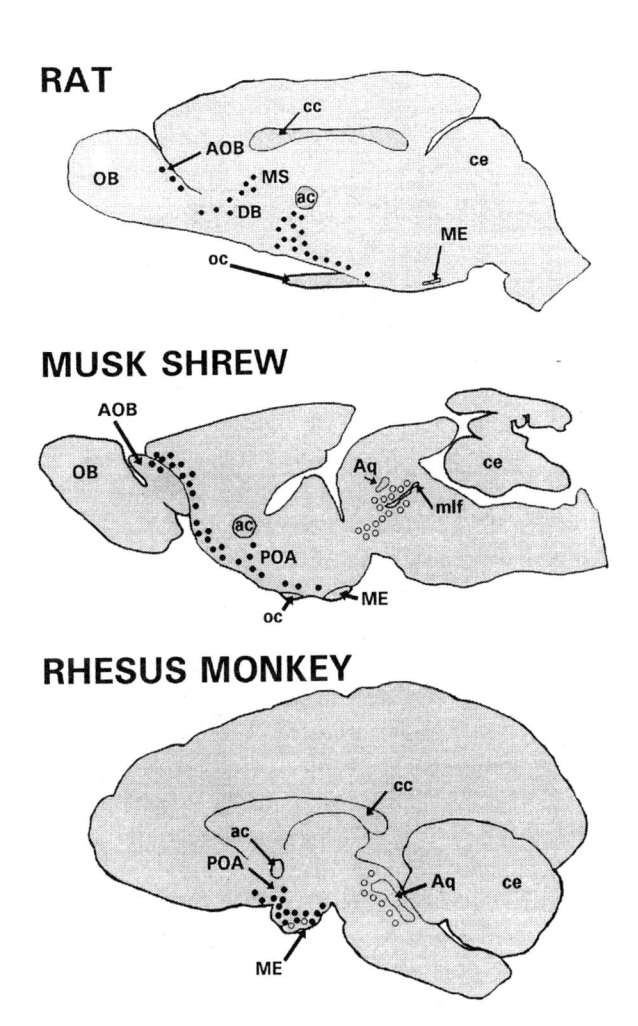

Figure 14.1
The distribution of gonadotropin-releasing hormone (GnRH) neurons depicted in parasagittal views in the rat, the musk shrew, and the rhesus monkey. Solid circles represent neurons containing the mammalian form of GnRH. Open circles represent neurons containing chicken II GnRH. The brains are not drawn to scale. The number of circles represent the relative distribution and density of GnRH neurons, not specific numbers of GnRH neurons (Silverman et al. 1982; Witkin et al. 1982; Dellovade et al. 1993; Lescheid et al. 1997). ac, anterior commissure; AOB, accessory olfactory bulb; Aq, cerebral aqueduct; cc, corpus callosum; ce, cerebellum; DB, diagonal band of Broca; ME, median eminence; mlf, medial longitudinal fascicularis; POA, preoptic area; MS, medial septum; OB, main olfactory bulb; oc, optic chiasm.

Broca (DB) and into the rostral medial preoptic area (MPOA; Silverman et al. 1994). In musk shrews and opossums, the majority of the GnRH cells in the forebrain are present in the most anterior regions, including the accessory olfactory bulb, areas associated with the terminal nerve, and in the anterior medial septum and DB (Schwanzel-Fukuda et al. 1988; Dellovade et al. 1993). In other species, such as the ferret and rhesus monkey, the GnRH neurons remain in a diffuse continuum, but this continuum extends further caudal and may extend back as far as the mammillary bodies (Sisk et al. 1988; Bibeau et al. 1991: Silverman et al. 1994). No functional significance for these different distributions has been found.

A few noteworthy features distinguish the GnRH neurons from other neuronal populations that produce other pituitary releasing factors, such as corticotropin-releasing factor (CRF), somatostatin, and growth hormone–releasing hormone (GHRH). First, although only a small number of cells make these other peptides, they are typically organized into a few discrete cell nuclei. For example, the cells that make CRF and project to the ME reside in the paraventricular nucleus. In addition, like GnRH, these other releasing factors may function secondarily as neurotransmitters in the brain, but unlike GnRH do so in a site-specific manner. For example, somatostatin-containing neurons in the periventricular nuclei project to the ME, but in the ventro- and dorsomedial hypothalamus somatostatin cells project widely to other regions (Zaborsky 1982; Krieger 1983). Another unusual aspect of the neuroanatomy of GnRH neurons is their unique origin. In contrast to all other known brain neurons, the GnRH neurons originate outside of the brain, in the olfactory placode. During embryonic development the GnRH cells migrate through the cribiform plate and along the terminal nerve, at which time they are already producing GnRH peptide (Schwanzel-Fukuda et al. 1988; Silverman et al. 1994). They then migrate along the mediobasal forebrain to their adult positions.

A second molecular form of GnRH has been reported to be present in brains of a subset of placental mammals, including humans, rhesus and stumptail macaques, tree and musk shrews, and golden moles (Dellovade et al. 1993; Kasten et al. 1996; King et al. 1994; Urbanski et al. 1999; White et al. 1998). Using several methods, including immunocytochemistry (ICC), high-performance liquid chromatography (HPLC), and radioimmunoassay (RIA), the molecular form has been identified as chicken GnRH II (cGnRH-II). This peptide is 70% homologous to the original GnRH isolated in mammalian brain and has been identified in many other nonmammalian vertebrates (Muske 1993; King et al. 1994). In adult musk shrews, tree shrews, and rhesus monkeys, the cGnRH-II–containing cell bodies reside in the midbrain, along the midline, ventral to the midbrain central gray (Dellovade et al. 1993: Kasten et al. 1996; Lescheid et al. 1997; see figure 14.1). It is tempting to speculate that these two cell populations have different functions. One appealing hypothesis is that the midbrain GnRH acts as a neurotransmitter (Muske 1993).

Subpopulations of GnRH Neurons?

Given the many different types of hormonal and sensory inputs the GnRH system integrates, it seems logical that the neurons would be organized into functional, if not distinct neuroanatomical, subgroups. The clearest anatomical distinction present in the GnRH neuronal system is that introduced above. The GnRH-producing cells in the forebrain that produce mammalian GnRH contrast with the GnRH cells that reside in the midbrain and express cGnRH-II. In the musk shrew enough information is available that we can make a few basic comparisons between the two systems (Dellovade et al. 1993; Rissman et al. 1995) (figure 14.2). The size and shape of the GnRH cells in the two populations are different. Whereas the GnRH cells in the forebrain have small cell somata and extensive fibers, the cGnRH-II cell bodies appear larger, are more oblong in shape, and only the proximal fiber is typically visible. The cGnRH-II fibers have a fine punctate appearance and are diffusely located throughout the forebrain and brain stem. The major terminal field for the cGnRH-II cells is the medial habenula and in contrast with forebrain GnRH fibers, few cGnRH-II fibers are visible in the ME. As expected from the localization of fibers, the concentration of cGnRH-II outside of the midbrain is highest in the medial habenula. Interestingly, cGnRH-II synapses have been identified in this region by electron microscopy (Rissman et al. 1995).

Preliminary data show that the two forms of GnRH are products of two different genes in musk shrew brains, as they are in the tree shrew and human (Kasten et al. 1996; White et al. 1998; E. F. Rissman and R. D. Fernald, unpublished data). In the musk shrew embryo, forebrain GnRH cells can be detected crossing out of the olfactory placode and into the brain at day 15 (Gill et al. 1995). However, we have never detected cGnRH-II–containing cells during these early stages of migration. Likewise, in the fetal rhesus monkey brain, GnRH-ir (immunoreactive) cells migrating from the olfactory placode do not contain cGnRH-II. Thus, it appears that in mammalian brain, as in amphibian brain (Muske and Moore 1990), the cGnRH-II and forebrain GnRH cells arise from different embryonic origins (Gill et al. 1995; Lescheid, et al. 1997).

In the forebrain, primarily in rodents, retrograde tract tracers have been used to attempt to identify the GnRH cells that specifically regulate the preovulatory LH surge. Several studies have used tract tracers given either systemically or applied directly to the ME. In the rat, between 40% and 80% of the forebrain GnRH cells

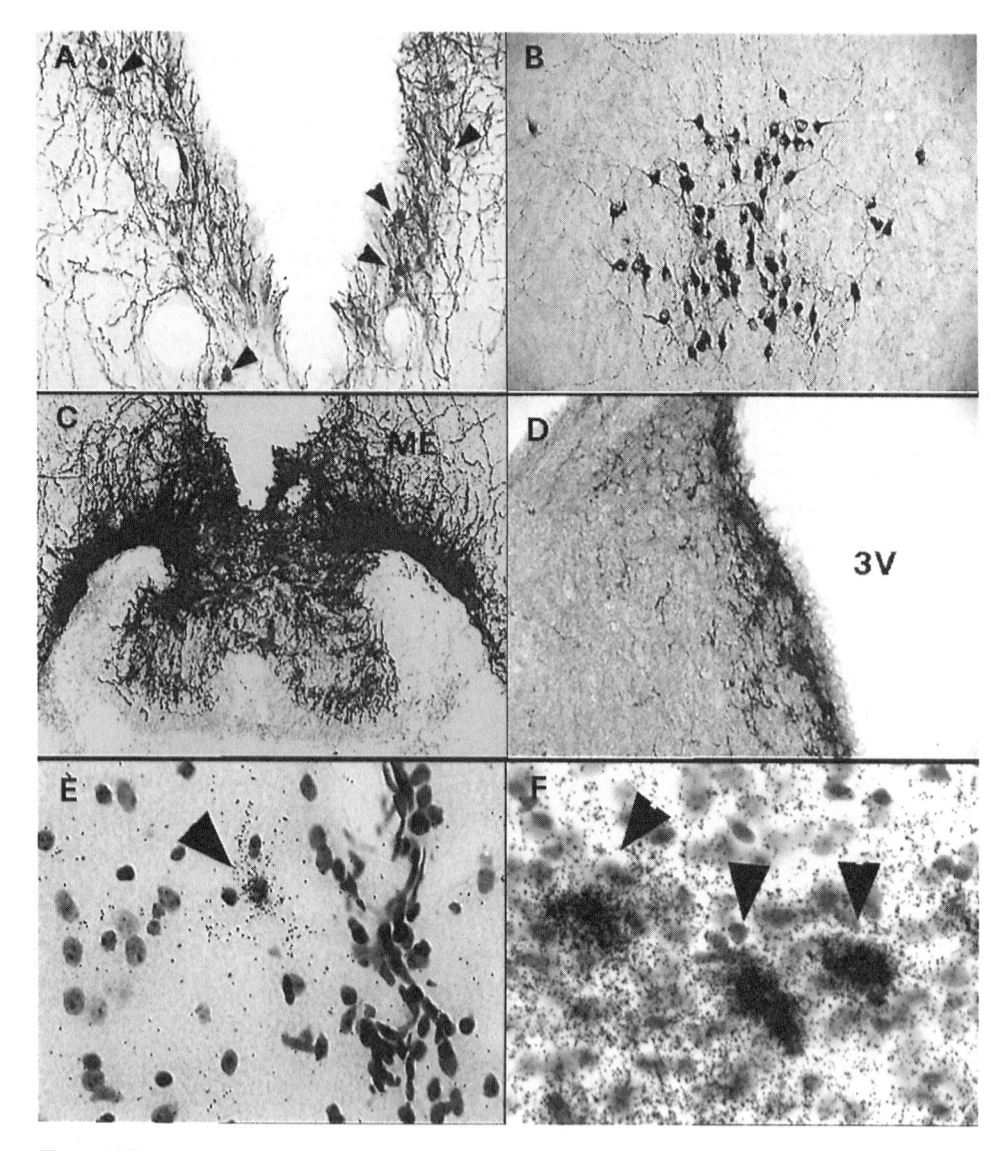

Figure 14.2
Photomicrographs comparing the two forebrain gonadotropin-releasing hormone (GnRH) systems. Pictures in the left column are cells and fibers containing mammalian GnRH. Those in the right column depict chicken GnRH II (cGnRH-II) cells and fibers. (A and B) Examples of the mammalian GnRH and cGnRH-II cell bodies, respectively. The mammalian cells (A, indicated by arrowheads) are small and embedded in extensive fiber networks. The cGnRH-II somata are larger and appear together in distinctive clusters with few fibers. (C and D) The terminal fields of the mammalian GnRH cells (C, median eminence) and cGnRH-II cells (D, medial habenula). There is little overlap in the terminal field of these two populations of GnRH neurons. (E) Silver grains in the medial septum after in situ hybridization with a ribopropbe specific for mammalian GnRH. (F) Silver grains in the midbrain after in situ hybridization using a riboprobe to a portion of the musk shrew cGnRH-II sequence.

project to the ME (Silverman et al. 1987; Merchenthaler et al. 1989). Using systemic injections of fluorogold, about the same percentage of GnRH cells were dual-labeled (between 33% and 88%). Interestingly, the number of fluorogold-filled GnRH cells was elevated in hormone-primed females as compared with ovariectomized animals (Silverman et al. 1990b), suggesting that there were more active terminals in the hormone-primed animals since fluorogold uptake is enhanced in active nerve terminals. In all of these studies, the location of the GnRH cell bodies that project to the ME appears to be evenly distributed throughout in the organum vasculosum of the lamina terminalis (OVLT), DB, and POA. In other words, GnRH neurons that project to the ME are not conveniently concentrated in one small brain nucleus.

Researchers have used dual-labeled ICC and in situ hybridization to try to detect subpopulations of GnRH neurons. The elucidation of obvious subpopulations, however, has been elusive. Initial attempts to determine whether estrogen receptors were present in GnRH cells all proved negative (Shivers et al. 1983; Watson et al. 1992). However the recent discovery of a second form of estrogen receptor (ER_β) makes it likely that these studies will have to be conducted again using specific probes and antiseum for ER_β (Kuiper et al. 1996; Mosselman et al. 1996; Tremblay et al. 1997). In the guinea pig, a subpopulation of GnRH cells that coexpress progesterone receptor (PR) has been reported (King et al. 1995). Finally, the presence of the type II glucocorticoid receptor in over 30% of the GnRH neurons (Ahima and Harlan 1992) in rat brain suggests that the adrenocorticoids are involved in GnRH regulation.

Only one neurotransmitter is extensively coexpressed in subpopulations of GnRH cells (Coen et al. 1990, Merchanthaler et al. 1990). In the rat medial septum, OVLT, POA, and hypothalamus (Merchenthaler et al. 1991; Marks et al. 1992), some of the GnRH cells also contain galanin. Sex, hormonal condition, and age influence the degree of coexpression of galanin in GnRH-containing neurons (Marks et al. 1993; Ceresini et al. 1994; Rossmanith et al. 1994). In young male and female rats, the amount of galanin messenger RNA (mRNA) per GnRH cell increases during puberty (Rossmanith et al. 1994). This increase is blocked by gonadectomy prior to puberty (Rossmanith et al. 1994). In adult females, gonadectomy causes a reduction in galanin mRNA levels in GnRH cells, and estradiol treatment (after ovariectomy) blocks this effect. Progesterone further augments the increase in galanin (Rossmanith et al. 1996). In all brain regions, but particularly in the OVLT and POA, females have many more galanin-GnRH double-labeled (with ICC) cells than males (Merchenthaler et al. 1991). Males, however, have more galanin mRNA per GnRH cell than do females (Ceresini et al. 1994).

Galanin content in the ME and POA varies with cycle stage in females. These regions contain more galanin peptide during proestrus than estrus (Coen et al. 1990). In situ hybridization reveals that the amount of galanin grains per GnRH cell increases between 1200 and 1800 hours on the day of proestrus and stays elevated for the next 24 hours. By contrast, galanin expression in non-GnRH cells in the same area (POA and hypothalamus) remains constant during the cycle (Marks et al. 1993). Finally, infusion of pharmacological amounts of galanin into the lateral ventricles causes LH release (Coen et al. 1990), and a galanin antagonist blocks LH secretion (Sahu et al. 1994). Galanin may play a role in the regulation of the preovulatory LH surge, since hormonal status and galanin expression are related, galanin is coexpressed in GnRH cells, and galanin infusions can stimulate LH release (Marks et al. 1994). Since little work on the relationship between galanin and GnRH has been conducted in animals other than rats, we must wait to see if this is a general mechanism.

To date, immunocytochemical localization of Fos, the protein product of the immediate early gene *c-fos*, has yielded some progress on the question of subpopulations of GnRH cells. The induction of Fos protein and mRNA have been used as markers of neuronal activation in neuroendocrine systems (reviewed in Hoffman et al. 1993). When used in conjunction with a marker for GnRH, individual GnRH cells that do and do not express Fos under different types of experimental conditions can be identified. These studies are reviewed in the next section.

An alternative approach to this question of subpopulations has employed the spontaneous mutant *hpg* (hypogonadal) mouse (reviewed in Silverman et al. 1994). These animals have a mutation in the GnRH gene that truncates transcription (Mason et al. 1986). They are infertile as adults, but are responsive to exogenous administration of GnRH, LH, or sex steroids. Fetal grafts of medial basal hypothalamic tissues into the third ventricle have been used to partially restore fertility in male and female *hpg* mice (Gibson et al. 1984a,b). Grafts containing as few as one immunoreactive GnRH neuron can support ovulation in *hpg* mice (Gibson et al. 1984a). Thus, we can conclude that only a very small subset of GnRH neurons is absolutely required to support a preovulatory LH surge. In these studies, females were maintained under optimal laboratory conditions. It would be interesting to determine if relevant environmental factors that are known to affect LH release in mice, such as male pheromones or nutrition, can be transduced by these few GnRH neurons.

Production of Gonadotropin-Releasing Hormone

As we have outlined above, GnRH plays the pivotal role in the interaction between internal and external cues that influence reproduction. There are several steps in

GnRH production and secretion that may be regulated by these cues. First, transcription or translation of the GnRH gene can be modified. Many studies have examined the ability of steroid hormones to alter GnRH gene expression (Sagrillo et al. 1996). The results of these studies are not uniform. The hormonal state of the animals and the methods employed to quantify gene expression have varied, but these studies tend to show that ovariectomy results in decreased GnRH mRNA (Kim et al. 1989; Roberts et al. 1989). Conversely, treatment with estradiol enhances GnRH mRNA (Kim et al. 1989; Petersen et al. 1989; Roberts et al. 1989; Rothfield et al. 1989; Rosie et al. 1990). The effects of social cues on GnRH mRNA have not yet been examined in any mammal, but we would predict that given the right selection of animals, social cues could enhance GnRH mRNA, perhaps in an even more dramatic manner than estrogen treatment.

Like many neuropeptides the active GnRH peptide is processed from a much larger (ninety-two–amino-acid) prohormone. The prohormone molecule contains three regions: a signaling portion, the GnRH sequence, and a gonadotropin-associated peptide (GAP). Several protyolytic processing steps are required to cleave the GnRH molecule from its prohormone. A few studies have asked if steroids regulate processing of mature GnRH peptide from the prohormone precursor. In the rat, long-term ovariectomy results in decreased pro-GnRH and GnRH content (Kelley et al. 1989; Roberts et al. 1989). Conversely, levels of pro-GnRH and GnRH increase after estradiol administration (Roberts et al. 1989). In the ferret, an electron microscopy study showed more Golgi complexes in GnRH cells in brains of ovariectomized, estradiol-treated females that received vaginocervical stimulation 75 minutes prior to sacrifice, as compared with females stimulated 2 days earlier (Bibeau et al. 1991). This same hormone and stimulation treatment leads to an LH surge and thus mating-induced vaginoacervical stimulation may increase intracellular stores of pro-GnRH.

If rapid release and subsequent processing of mature GnRH peptide is activated by ovulation, social cues that trigger ovulation may also act through this pathway. Little is know about the availability of the processing enzymes and if any of these are ever rate-limiting.

Behavioral Activation of GnRH

Use of Fos to Assess Status of GnRH Neurons

About 10 years ago, a number of laboratories began employing dual-label ICC localization of Fos and GnRH to investigate which GnRH neurons are activated on the evening of proestrus and therefore might underlie the preovulatory LH surge.

Some neurons express Fos, a transcription factor, in response to specific stimuli. Thus, if cells express Fos after an animal has been presented with a stimulus, but not in the absence of the stimulus, then some cellular modification (activation, inhibition, and/or adaptation) can be assumed to have occurred in this subpopulation. This technique has proved particularly important for understanding the neuroanatomy of GnRH cells, because the usual lesion and pharmacological studies have been difficult to employ given that the GnRH neurons reside in a diffuse continuum rather than a discrete nucleus. Using ovariectomized, steroid-primed rats, Hoffman et al. (1990) demonstrated that Fos-like immunoreactivity (Fos-ir) is increased in GnRH neurons after exogenous administration of steroid hormones at the time of an expected LH surge. Interestingly, this increase occurred in an anatomical subset of GnRH neurons, those in the OVLT and POA. In intact, cycling female rats, Lee et al. (1990) found virtually the same thing; at the time of the expected LH surge, there was Fos-ir within the same populations of OVLT and POA GnRH neurons. In addition, GnRH mRNA content is greater in GnRH cells that coexpress Fos during the LH surge (Wang et al. 1995). These results suggest that, at least in the rat, GnRH neurons in the OVLT, POA, and anterior hypothalamus (AH) drive the preovulatory LH surge.

This correlation between Fos-ir and GnRH neurons also holds in a variety of mammalian species. Moenter et al. (1993) showed that in the ewe, Fos-ir in GnRH neurons was correlated with the generation of the LH surge, but not the ovariectomy-induced increase in LH. In female Syrian hamsters, Fos-ir is augmented in GnRH neurons around the time of the preovulatory LH surge (Doan and Urbanski 1994) in more rostral forebrain regions (DB, medial and rostral POA). Interestingly, in hamsters, GnRH neurons in the caudal POA expressed high levels of Fos-ir throughout the cycle. Food deprivation, which among many other things delays puberty and suppresses ovulatory cycles (see chapters 2 and 3), prevented the periovulatory increase of Fos-ir in GnRH neurons in the rostral GnRH neurons and significantly reduced the percentage of Fos-ir and GnRH double-labeled neurons in the caudal POA (Berriman et al. 1992). These data are consistent with the idea that the GnRH neurons that control tonic pulsatile secretion of LH are located more caudally than GnRH neurons that control the surge. Other investigators have also suggested that the surge results from the activity of a subpopulation of the neurons that are responsible for pulsatile secretion. In response to increasing levels of estradiol, this subpopulation of neurons may increase amplitude and frequency of pulses to produce the surge pattern that is observed.

In many species, pheromonal cues from a conspecific can affect reproductive physiology. For example, puberty can be advanced in the female mouse or vole by contact with a male (Bronson 1989; Vandenbergh 1994). Again, using Fos as a marker of neural activation, a number of laboratories have shown that pheromonal cues activate neurons in a number of regions in the forebrain that are involved in reproduction. But no one has yet reported that pheromones induce Fos in GnRH neurons (Fernandez-Fewell and Meredith 1994; Wersinger and Baum 1997b), although clearly pheromones affect the GnRH system. Thus, the mechanisms that activate GnRH neurons via estrogen, vs. nontactile environmental cues such as pheromones, may not be equivalent.

Induced Ovulation and GnRH Activity

The Ferret Ovulation is induced by copulation in species as diverse as rabbits, cats, ferrets, musk shrews, and camels, to name a few. In these species, mating per se triggers ovulation. The musk shrew and ferret are excellent examples of induced ovulating species. The musk shrew will be described in detail later. The ferret comes into behavioral estrus once per year in response to a long-day photoperiod (reviewed in Baum et al. 1990). During this time, estrogen levels are elevated and the female becomes behaviorally receptive. Plasma LH remains undetectable, and the female will not ovulate unless mating occurs (Baum et al. 1990). Once mating ensues, plasma LH levels begin to rise within 1.5 hours of receiving an intromission (Carroll et al. 1985, 1987). This preovulatory surge of LH results in ovulation approximately 24 to 36 hours later. Treatment with a number of exogenous steroid regimens has never successfully resulted in ovulation in this species (Baum et al. 1990). Thus, it represents an example of a species in which only social cues can trigger ovulation.

How does mating induce the LH surge? Theoretically there are several possibilities. One possibility is that the mechanical stimulation of the cervix could trigger a classic neuroendocrine reflex arc that results in the release of the preovulatory surge of LH from the pituitary. Alternatively, pheromonal cues associated with the male may induce the release of LH. In addition, visual or tactile cues, other than cervical stimulation, may regulate the LH surge. Finally, combinations of some or all of these may have synergistic effects on LH. Pheromonal cues alone are insufficient to induce the preovulatory LH surge in the ferret (Wersinger and Baum 1997b). Yet attempts to induce ovulation in ferrets by simply using a glass rod have been unsuccessful. However, if the female is in physical contact with a male, and the vaginocervical region is stimulated with a glass rod, GnRH is released and presumably an LH surge occurs (Bibeau et al. 1991). Clearly, the cues that affect the GnRH system and the interaction between them need to be investigated further.

The correlation of Fos in GnRH neurons and the preovulatory LH surge described above also holds in the ferret (Lambert et al. 1992a; Wersinger and Baum 1996).

Mating (but not exposure to chemosensory cues alone) results in a large increase in plasma LH in female ferrets. Intromission results in a large increase in Fos-containing GnRH neurons in the caudal mediobasal hypothalamus (MBH) in the female. This suggests that this subpopulation of GnRH neurons is involved in the LH surge. However, the neurotransmitters and neuropeptides through which sensory cues trigger GnRH activity are not known. Data generated in the rabbit have addressed this question.

The Rabbit Female rabbits come into behavioral estrus, mate, and ovulate during the breeding season. In females, LH levels rise within 15 minutes after the first intromission (reviewed in Ramirez and Beyer 1988). Increased GnRH in microdialysates sampled from the MBH precedes this rise in LH (Kaynard et al. 1990) and a rise in norepinephrine (NE) in the MBH precedes the increase in GnRH by about 7 minutes (Yang et al. 1996). Intraventricular infusion of NE can facilitate LH release in hormone-primed females (Sawyer 1952). Prazosin, an α_1-adrenergic receptor blocker, prevents the mating-induced rise of GnRH in the MBH (Spies et al. 1997). Finally, tyrosine hydroxylase (TH)–immunoreactive neurons in the locus coeruleus which project to the MBH show increased optical density 15 minutes after mating (Spies et al. 1997). This suggests that mating upregulates the production of TH mRNA in midbrain neurons that may regulate GnRH release. Thus, in the rabbit, it appears that mating results in the release of NE from midbrain neurons, which triggers the release of stored GnRH in the MBH and results in a surge of LH.

There are major similarities between rabbits and ferrets in the mechanism of their induced ovulation. First, mating appears to activate the NE system in both species (Spies et al. 1997; Wersinger et al. 1997a). Second, high levels of estrogen are associated with the LH surge (Pau and Spies 1986; Baum et al. 1990). Also, the release of GnRH occurs within minutes after mating and ovulation occurs many hours later. Although the mechanisms may be similar, the cues that are required to trigger ovulation are different, thus emphasizing the importance of studying these cues in more than one species. Female rabbits will ovulate in response to glass rod stimulation of the cervix. Thus, they do not require the pheromonal-visual-tactile cues that female ferrets require. However, descriptions of courtship in wild rabbits involve the male scent marking his mate with chin gland secretions, and spraying urine on the female (Southern 1947). Perhaps as the result of generations of domestication, environmental cues that modified GnRH release in wild rabbits are now vestigial traits in laboratory populations. It is also possible that under less optimal laboratory conditions rabbits would be sensitive to these external factors.

The Musk Shrew

Natural History and Evolution

There is very little published information about musk shrew social behavior in the field. To our knowledge, no detailed analyses exist. Baker (1946) estimated that the density of musk shrew populations on Guam was 6.2 animals per acre. This number is low, when compared with ranges of between 20 and 300 voles per acre in some studies (Lidicker 1973). Our best guess is that in nature musk shrews have infrequent but occasional social contact with one another. Scent-marking behavior and male-male interactions in the laboratory (P.A.S., personal observations) suggest that males are territorial, although no field data exist to verify this observation. The only complex social behavior that has been studied in the laboratory (besides sexual behavior studied in our laboratory) is caravaning behavior (Tsuji and Ishikawa 1984). This occurs when preweanling-age pups and their mothers are placed in an unfamiliar or disturbed area. During caravaning, young shrews grasp the mother's rump with their open mouths and follow her, forming a chain of shrews. This behavior is associated with a distinct vocalization emitted by the pups, which is similar to the vocalizations produced by receptive females toward males (Gould 1969). Additionally, musk shrews produce a wide variety of vocalizations which could be used during intraspecific encounters (Gould 1969). Although the precise function of caravaning and shrew vocalizations is unknown at this time, these findings indicate that social communication and social interactions are not absent in this species and would argue against this species being completely solitary.

In our laboratory, we have been investigating behavioral influences on the female neuroendocrine axis in the musk shrew. All shrews (order Insectivora) are considered to be primitive eutherian mammals (figure 14.3). Musk shrews originated in central Asia, but have adapted to and spread rapidly through other locations after being accidentally introduced (Barbehenn 1962, 1974). Shrews live in areas populated by humans as well as in areas relatively void of human activity (Barbehenn 1974). Moreover, musk shrews can be found in a wide range of environments, including subtropical regions (Harrison 1955) and deserts (Advani and Rana 1981).

Because shrews are considered an example of an early-evolved placental mammal, studying musk shrews may shed light on modes of reproduction utilized by ancestral mammalian species. Musk shrews are not seasonal breeders. In the field, pregnant musk shrews are found at all times of the year. Seasonal peaks in pregnancy rates are modest, at best (Barbehenn 1962; Harrison 1955; Louch et al. 1966). Changes in female and male fertility measures do not always correspond with one another

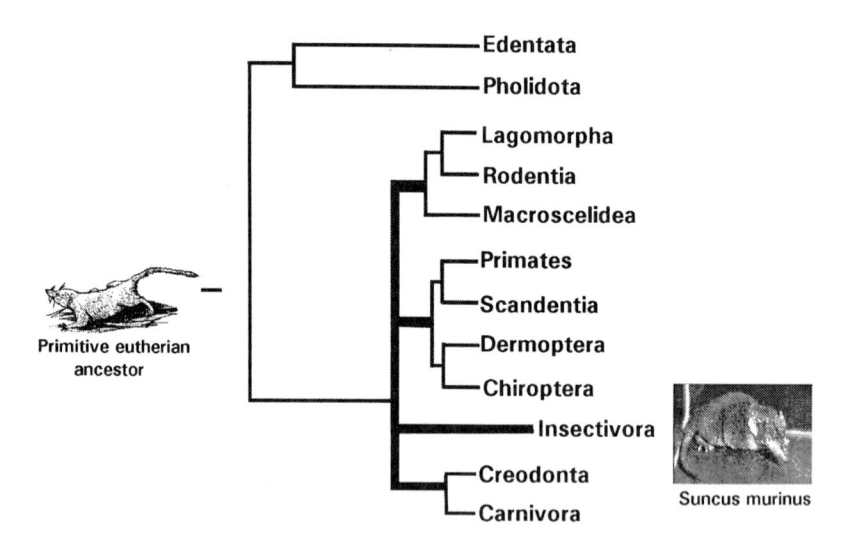

Figure 14.3
Phylogenetic tree showing relationships between orders of eutherians evolved from a common ancestor. The species discussed in this chapters are included in the orders Lagomorpha (rabbits), Rodentia (rats, mice, hamsters), Primates (monkey, human), Scandentia (tree shrew), Insectivora (musk shrew, mole), and Carnivora (ferret).

(Harrison 1955; Louch et al. 1966). Moreover, changes in gonadal status in both males and females cannot be correlated with any specific environmental factor, such as photoperiod, rainfall, or temperature (Beg et al. 1986). In the laboratory, manipulations of photoperiod have modest effects on various reproductive variables. Short day lengths are associated with reduced weights of androgen-sensitive tissues in males (Wayne and Rissman 1990, 1991) and females are less likely to mate after they have resided on short days (Rissman et al. 1987). However, musk shrews inhabit areas with relatively little seasonal variation in day length. When ecologically relevant day lengths are used in laboratory studies, the effects of photoperiod, at least in males, are further reduced (Wayne and Rissman 1991). On the other hand, a 50% reduction in calories for a 48-hour period significantly reduced the expression of sexual behavior in females (Gill and Rissman 1997). Therefore, given their natural habitat and the variability in the availability of their insect diet, it seems likely that reproduction in musk shrews is opportunistic and regulated by local food availability.

General Characteristics of Musk Shrew Reproductive Behavior

Female musk shrews do not exhibit an ovarian cycle or behavioral estrus, and sexual behavior is not tied to any ovarian event (Dryden 1969). We have tried to determine which cues from the male, or those derived from behavioral interactions with a male, activate behavioral receptivity in the female. Furthermore, we have studied the hormonal and neural systems, altered by behavioral interactions that mediate female sexual behavior.

Musk shrews exhibit a stereotyped pattern of sexual behavior that is more similar to that displayed by other species with induced ovulation than to spontaneous ovulating species such as the rat. Female musk shrews are initially very aggressive during interactions with males. This aggressive phase can last anywhere from 5 to 60 minutes in virgin females, and is much shorter in sexually experienced females (Rissman 1987). Gradually, the female becomes sexually receptive, a state characterized by rump presentation and tail-wagging. Male musk shrews follow females once tail-wagging begins and start mounting them shortly thereafter. Females do not exhibit lordosis during mounting, but rather continue to walk or run around the arena while wagging their tail. Males can mount up to a hundred times prior to ejaculation. Only 30% to 50% of the male's mounting attempts do not include penile intromission, probably because female shrews do not stand still during mounting attempts. Females continue to be receptive to males for several days, even after they have received multiple ejaculations (E.F.R., unpublished observations).

Steroid Hormones Regulate Female Sexual Behavior

Because female musk shrews do not have a behavioral estrous cycle, and will mate prior to, during, and after ovulation, it would seem likely that the neural control of copulation is disengaged from ovarian hormones. Further, when virgin shrews begin to copulate, estradiol levels in plasma are relatively low (5 pg/mL). Yet, the ovaries are required for display of receptivity (Rissman and Bronson 1987). Hormone replacement after ovariectomy with supraphysiological doses of estradiol can reinstate sexual behavior, but in a physiological dose range, only testosterone is effective (Rissman and Crews 1988; Rissman et al. 1990). To further elucidate the mechanisms through which testosterone has its effect on female sexual behavior, we examined the roles of the androgen vs. the estrogen receptor. Ovariectomized, testosterone-treated females given aromatase inhibitors show reduced sexual behavior relative to controls (Rissman et al. 1990, 1996). Furthermore, treatment with the antiestrogen tamoxifen blocks female sexual behavior, but treatment with the antiandrogen flutamide has no significant effect (Rissman 1991).

Adrenal steroids also are involved in sexual behavior. When females become sexually receptive there is a concurrent increase in plasma levels of progesterone (Rissman and Crews 1988) and cortisol (Schiml and Rissman 1999). Because the levels of these steroids increase in less than 15 minutes in the intact female, we assume that they are secreted by the adrenal glands. Either surgical or pharmacological adrenalectomy (Fortman et al. 1992; Rissman and Bronson 1987; Schiml and Rissman 1999) reduces female receptivity. In long-term dexamethasone-treated, ovary- and adrenal-intact animals, testosterone treatment restores sexual behavior (Fortman et al. 1992) and cortisol treatment can reinstate some aspects of sexual behavior in metyrapone-treated females (Schiml and Rissman 1999). Therefore, testosterone, of both adrenal and ovarian origin, and cortisol from the adrenal are important for the display of female sexual behavior. The effects of these steroid hormones on GnRH and ovulation need to be explored.

Sexual Behavior Affects Steroid Levels and Ovulation

Adult virgin musk shrews have immature ovaries, which do not contain preovulatory follicles. The production of mature follicles begins after the first mating (Fortune et al. 1992). Fewer than one third of females, however, ovulate after the initial ejaculatory bout (Rissman 1992). Yet, when ovaries removed from virgins are cultured for several hours with gonadotropins, they are fully capable of steroidogenesis (Fortune et al. 1992). In addition, administration of human chronic gonadotropin or GnRH promotes ovulation in vivo (Rissman et al. 1995, Freeman et al. 1998).

Plasma estradiol concentrations are uniformly low at various time points (between 1 and 8 hours) after an initial mating. However, estradiol levels can increase by up to ten-fold 15 hours after mating in females that have ovaries containing preovulatory follicles (Fortune et al. 1992). Estradiol is also elevated in females 24 or 40 hours after mating and ovulation (Dellovade et al. 1995b). While multiple matings given in a single day do not dramatically increase the likelihood of pregnancy, multiple matings over a period of several days increase the likelihood that a female will ovulate and become pregnant (Clendenon and Rissman 1990). Thus, we hypothesized several years ago that the first mating serves to prime the hypothalamic-pituitary-gonadal axis, making it more responsive to subsequent matings (Clendenon and Rissman 1990; Rissman 1992).

Methods to Study the GnRH System

Most of the information we have collected to date on the neuroendocrine system in the musk shrew has relied on ICC employing antibodies to GnRH. Despite differences in GnRH gene sequence, the ten amino acids that make up the peptide are highly conserved across species (reviewed in King and Millar 1995). Thus, there are many antibodies available that recognize GnRH peptide in musk shrew brain. In most of our work, we employed a generalist GnRH antiserum (LR1, generously provided by Dr. R. Benoit). This antibody recognizes amino acids 3,4, and 7 through 10 of the mammalian GnRH peptide and has been used with success in many species of birds and mammals (Silverman et al. 1994). LR1 also reacts with both the mature GnRH peptide and the pro-GnRH hormone (Silverman et al. 1990a). In order to distinguish between GnRH peptide and pro-GnRH protein in cells, we employed two other antisera. One of these reacts with amidated mature GnRH peptide. This antiserum is a commercially available monoclonal antibody (SMI-41, produced by Sternberger Monoclonals Inc., Baltimore, MD) made against the five amino acids adjacent to the C-terminal of the mammalian GnRH peptide, and the amidation site. Because the amine complex is formed after the final cleavage step, only mature peptide is recognized by this antiserum.

Additionally, an antiserum was used to visualize cells containing immunoreactive pro-GnRH protein. This antiserum was made against a peptide sequence that spans from the end of the GnRH peptide into the GAP portion of the rat GnRH prohormone (amino acids 6 through 16, antiserum no. 1947, made by Dr. R. P. Millar, Cape Town, S. Africa). Using these three antisera, we have shown that the distribution of neurons that contain mature GnRH peptide, pro-GnRH hormone, or both is identical (Rissman et al. 1997). In dual-labeling studies with the GnRH prohormone and peptide antisera, we found that most GnRH-ir cells cross-react with both antisera. The appearance of these GnRH-ir cells is such that pro-GnRH immunoreactivity is located in the soma, and the cell bodies are surrounded by fibers containing GnRH peptide (Rissman et al. 1997).

Ovulation Affects Gonadotropin-Releasing Hormone

Given the data in other species, we did not predict that we would be able to detect a measurable change in GnRH cells by simply using immunocytochemistry. However, we discovered that both numbers of GnRH-ir cells and content of GnRH, as measured by enzyme immunoassay, are correlated with ovulatory status. Females that experience one mating but do not ovulate and are sacrificed 24 to 48 hours later have significantly more GnRH-ir cells in the forebrain as compared with unmated controls (Dellovade et al. 1995b). However, in the females that ovulated as a result of a single mating, GnRH-ir cell numbers were low (Dellovade et al. 1995b). This suggests that mating increases either production or storage, or both, of GnRH and that ovulation reduces GnRH content in cell bodies. In addition, in all regions of the forebrain examined, GnRH content increased after mating, reaching a peak just before ovulation (figure 14.4). After ovulation, GnRH content declined in brains of females that ovulated but continued to increase in animals that did not ovulate. These results

Figure 14.4
The gonadotropin-releasing hormone (GnRH) content in the ventral forebrain of female musk shrews at various times after mating. The mated females were divided into two groups: those that ovulated (as determined by the presence of corpora lutea on the ovaries), and those that did not. The GnRH content was compared in ovulating and nonovulating females. Females that did not ovulate had significantly more GnRH in the ventral forebrain than did ovulating females.

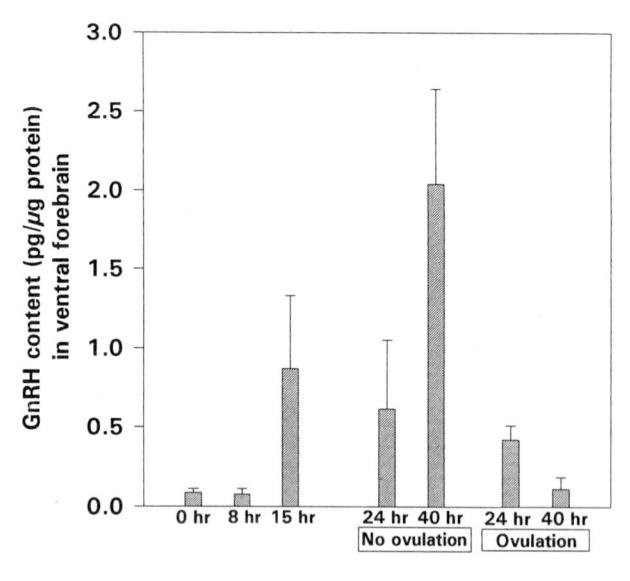

indirectly support our hypothesis that the first mating, which typically does not trigger ovulation, primes the GnRH system. Current studies are underway to assess the relationship between multiple matings, ovulation, and GnRH-ir.

Social Interactions with Males Affect GnRH

To isolate the type(s) of social stimuli that activate GnRH production, we have conducted a series of studies to examine the immunoreactivity of GnRH cells in brains of females sacrificed after engaging in various types of encounters with males, or male-related social cues. We have shown that mating has rapid, dramatic effects on the number of GnRH-ir neurons in the forebrain. The number of GnRH-ir neurons in the most anterior portion of the GnRH system increase in the time it takes females to exhibit tail-wagging, relative to controls, and numbers decrease a few minutes later, following the receipt of an ejaculation (Dellovade et al. 1995a). Less dramatic increases in GnRH-ir cells are seen when females are exposed to male-soiled bedding for as little as 25 minutes. This finding suggests that social cues and pheromones have synergistic effects on the GnRH system.

In a recent set of studies, the GnRH prohormone and peptide antisera were employed, side by side, on tissues from the same brains. In the first study, females were sacrificed after they expressed certain behaviors during their first mating bout. In this study, the numbers of cells that contained GnRH peptide did not vary. However, the pro-GnRH-ir cells increased when females began tail-wagging. A second, significant rise in pro-GnRH cell numbers was noted 1 hour after an ejaculation (Tai et al. 1997).

In a companion experiment, virgin females were housed in one of four conditions. To control for the amount of time spent with a male, all of these conditions lasted 1 hour. Females resided either in a cage containing male-soiled bedding with an active adult male on the other side of a screen barrier, under identical conditions with an anesthetized male, or in a clean cage with a male on the other side of a transparent solid barrier. Control animals resided in a clean cage without a male present. When mating was prevented, behavioral interactions with an active male, or his olfactory cues, did not affect the numbers of pro-GnRH-ir cells, but did result in a decrease in the numbers of GnRH peptide–containing cells in the caudal forebrain. In addition, we noted an increase in the area of GnRH peptide immunoreactivity in the ME in brains of females that interacted with males across a solid transparent barrier. We hypothesize that social contact with a male depletes stores of mature GnRH peptide in cell bodies, while increasing GnRH peptide stores in fibers, perhaps also stimulating LH release if male pheromonal cues are present. On the other hand, copulation may stimulate new transcription and translation of pro-GnRH. Current studies are underway testing these hypotheses.

Musk Shrew Future Directions

The musk shrew provides us with a valuable model to study neuroendocrine control of ovulation. There are several aspects of the interaction between behavior and the GnRH system that we are addressing currently in our laboratory. First, we would like to know, as specifically as possible, which aspects of the mating interaction activate GnRH production. We have found that a brief exposure (less than 30 minutes) to male-soiled bedding increase the numbers of GnRH-ir cells in the anterior forebrain. Yet, preexposure to male-soiled bedding does not facilitate ovulation when it precedes a single mating bout (Rissman 1992), but it can reduce the latency with which females display receptivity (Rissman 1987,1989). Perhaps the GnRH produced in cell bodies in the anterior forebrain is not involved in ovulation directly, but instead acts as a neurotransmitter and affects behavior.

GnRH infusion studies are being conducted to examine whether GnRH facilitates sexual behavior in the musk shrew, as has been reported in rats (Moss and McCann 1973; Pfaff 1973). We have employed three doses of GnRH, or vehicle, and examined sexual behavior 15 minutes post infusion. Virgin females fitted with indwelling cannulas directed at the lateral ventricle receiving a 100-ng dose of GnRH show tail-wagging and rump presentations sooner than controls (Schiml et al. 1998).

One-hour interactions with males across a screen barrier stimulate a drop in number of GnRH-ir cells containing mature GnRH peptide, and the same effect is seen if the interaction occurs across a solid, but clear barrier (Tai et al. 1997). These behavioral interactions are noteworthy for the amount and intensity of the vocal communication that occurs. These data suggest that auditory communication promotes release of GnRH peptide from a subset of GnRH cell bodies. Perhaps under normal conditions, when pairs have both auditory and olfactory contact, these two sensory cues have synergistic actions on GnRH. Interestingly, complete bilateral olfactory bulbectomy does not affect the onset or display of female mating behavior (E.F.R., unpublished data). Studies in which females are exposed to auditory signals emitted by breeding pairs are presently underway.

Another behavioral input to GnRH that warrants further investigation is the finding that the number of cells immunoreactive for pro-GnRH increases 1 hour after ejaculation. This finding could indicate that GnRH is released in association with mating and that new transcription and translation of GnRH occur subsequent to mating. This hypothesis is based in part on work in rats, in which careful analysis reveals that Fos induction in GnRH neurons may occur after the LH surge (Finn et al. 1998). We are presently employing in situ hybridization to quantify GnRH gene expression. GnRH genes have been cloned and sequenced from many species of fish, a few birds, and many mammals (Seeberg and Adelman 1984; Lovejoy et al. 1992; Kasten et al. 1996; White et al. 1998). We have located one riboprobe made by Adelman from the human GnRH gene which cross-reacts with GnRH in musk shrew forebrain (Rissman and Finn, unpublished data). This probe is being employed to examine gene expression before, during, and after mating.

Of course, one of the most important and obvious questions that we are addressing in the musk shrew is, What is the function of the second form of GnRH? We have noted that the numbers of immunoreactive cGnRH-II cells in the midbrain decline several hours after ovulation (Dellovade et al. 1995b; P.A.S., unpublished data). We also have found a robust sex difference in both immunoreactive cGnRH-II cells and the fiber terminals in the habenula. The direction of this difference is that females have less immunoreactivity than males. However, this difference is abolished by ovariectomy (Rissman and Li 1998). Thus, the second form of GnRH is regulated by ovarian hormones. We have just completed cloning and have a partial sequence for the cGnRH-II gene present in the musk shrew midbrain (Rissman and Fernald, GenBankIt 237405 AF107315). We will use this tool to determine if ovarian hormones regulate mRNA for this GnRH form and whether behavioral interactions affect transcription of cGnRH-II mRNA.

Evolution of Behavioral Control of GnRH Activity

Mammals have evolved many diverse reproductive strategies. One example is the diversity that underlies the mechanisms that stimulate the preovulatory LH surge. Classically, mammals have been divided into species that show mating-induced ovulation and those that are spontaneous ovulators. According to the traditional definitions, induced-ovulating species are those in which the preovulatory LH surge is triggered by external (e.g., mating-associated) stimuli. If this stimulus is not present, the animal does not ovulate. Spontaneous ovulators, in contrast, ovulate in response to an endogenous signal (e.g., rising estrogen and progesterone levels) and do not require external stimuli. Although this dichotomy is useful for the study of mechanisms triggering ovulation, we believe that it is also important to consider the effect of the environment on reproductive physiology.

Many mammalian females, including both induced and spontaneous ovulators, display some degree of behavioral regulation of their neuroendocrine axis. For example, the onset of the breeding season in ewes can be accelerated by exposure to a ram (Martin 1984; Wayne et al. 1989). Puberty in female mice can be advanced by the presence of an adult male and delayed if the female is group-housed with other females (Bronson 1989; Vandenbergh 1994). The rat, often used as the prototype of a spontaneous ovulating species, can exhibit characteristics of an induced ovulator. Typically, under laboratory conditions, rats have a 4-day estrous cycle. Interestingly, persistent estrus can be induced in rats by placing them on a regimen of constant light illumination, and also occurs in aged females (Brown-Grant et al. 1973; LeFevre and McClintock 1992). Very high levels of estrogen characterize the state of persistent estrus. These animals will never ovulate spontaneously. However, if rats in persistent estrus mate and receive an intromission, ovulation will occur. Thus, although rats normally ovulate spontaneously, they are capable of induced ovulation (Brown-Grant et al. 1973). Another case in point is the *hpg* mouse referred to earlier. When fetal hypothalamic tissue is grafted into the brains of females that lack GnRH, estrus ensues. However, estrous cycles do not commence; instead these animals undergo induced ovulation after mating (Gibson et al. 1984a). Therefore, the neural substrate that underlies induced ovulation, although not normally used, is present and can be functional in laboratory rodents that normally display spontaneous ovulation.

In addition, the converse situation has never been reported. No one has succeeded in getting an induced-ovulating species to ovulate in response to steroid hormones alone. This strongly suggests that whatever the neural substrate that allows an animal to ovulate in response to steroids, it is absent in induced-ovulating species. Taken together, the data suggest to us that induced ovulation is the ancestral mammalian trait and that spontaneous ovulation is the derived condition. The taxonomic distribution of induced-ovulating mammalian species is broad (reviewed in Jochle 1975). There are induced ovulating species in virtually every order: the Marsupialia, Insectivora, Chiroptera, Rodentia, Lagomorpha, Carnivora, Pinnipedia, and Arteriodactyla (Jochle 1975). In addition, there are examples in all mammalian orders of species in which social cues can affect some aspect of female reproductive physiology (Bronson 1989). We hypothesize that these adaptive responses to social cues reflect the remnants of the ancestral condition that included reliance on copulation to trigger ovulation. Thus, understanding the mechanism by which social cues can alter reproductive physiology will not only give us an appreciation of the incredible diversity of reproductive strategies adopted by mammals but also help us to understand the universal mechanisms that govern reproduction.

Acknowledgments

We thank NIH for the following support: NRSA MH11534 (P.A.S.), NRSA NS10444 (S.R.W.), KO2 MH401349 and RO1 NS35429 (E.F.R.).

References

Advani R, Rana BD (1981). Food of the house shrew, *Suncus murinus sindensis*, in the Indian desert. Acta Theriologica 26:133–134.

Ahima RS, Harlan RE (1992). Glucocorticoid receptors in LHRH neurons. Neuroendocrinology 56:845–850.

Baker RH (1946). A study of rodent populations on Guam, Mariana Islands. Ecol Monogr 16:393–408.

Barbehenn, KR (1962). The house shrew on Guam. Pacific Island Ecology (Storer T, ed), pp 247–256. Honolulu, Bernice P. Bishop Museum.

Barbehenn, KR (1974). Recent invasions of Micronesia by small mammals. Micronesia 10:41–50.

Baum MJ, Carroll RS, Cherry JA, Tobet SA (1990). Steroidal control of behavioural, neuroendocrine and brain sexual differentiation: studies in a carnivore, the ferret. J Neuroendocrinol 2:401–418.

Beg, MA, Kauser, S, Hassan, MM, Khan, AA (1986). Some demographic and reproductive parameters of the house shrew in Punjab (Pakistan). Pakistan J Zool 18:201–208.

Berriman SJ, Wade GN, Blaustein JD (1992). Expression of Fos-like proteins in gonadotropin-releasing hormone neurons of Syrian hamsters: effects of estrous cycles and metabolic fuels. Endocrinology 131:2222–2228.

Bibeau CE, Tobet SA, Anthony ELP, Carroll RS, Baum MJ, King JC (1991). Vaginocervical stimulation of ferrets induces release of luteinizing hormone–releasing hormone. J Neuroendocrinol 3:29–36.

Bronson, F (1989). Mammalian Reproductive Biology. Chicago: University of Chicago Press.

Brown-Grant K, Davidson JM, Greig F (1973). Induced ovulation in albino rats exposed to constant light. J Endocrinol 57:7–22.

Carroll RS, Erskine MS, Doherty PC, Lundell LA, Baum MJ (1985). Coital stimuli controlling luteinizing hormone secretion and ovulation in the female ferret. Biol Reprod 32:925–933.

Carroll RS, Erskine MS, Baum MJ (1987). Sex difference in the effect of mating on the pulsatile release of luteinizing hormone in a reflex ovulator, the ferret. Endocrinology 121:1349–1359.

Ceresini G, Merchenthaler A, Negro-Villar A, Merchenthaler I (1994). Aging impairs galanin expression in luteinizing hormone–releasing hormone neurons: effects of ovariectomy and/or estradiol treatment. Endocrinology 134:324–330.

Clendenon, AL and Rissman, EF (1990). Prolonged copulatory behavior facilitates pregnancy success in the musk shrew. Physiol Behav 47:831–835.

Coen CW, Montagnese C, Opacka-Juffry J (1990). Coexistence of gonadotropin-releasing hormone and galanin: immunohistochemical and functional studies. J Neuroendocrinol 2:107–111.

Conaway CH (1971). Ecological adaptation and mammalian reproduction. Biol Reprod 4:239–247.

Dellovade TL, King JA, Millar RP, Rissman EF (1993). Presence and differential distribution of distinct forms of immunoreactive gonadotropin-releasing hormone in the musk shrew brain. Neuroendocrinology 58:166–177.

Dellovade TL, Hunter E, Rissman EF (1995a). Interactions with males promote rapid changes in gonadotropin-releasing hormone immunoreactive cells. Neuroendocrinology 62:385–395.

Dellovade TL, Ottinger MA, Rissman EF (1995b). Mating alters gonadotropin-releasing hormone cell number and content. Endocrinology 136:1648–1657.

Doan A, Urbanski HF (1994). Diurnal expression of Fos in luteinizing hormone releasing hormone neurons of Syrian hamsters. Biol Reprod 50:301–308.

Dryden GL (1969). Reproduction in *Suncus murinus*. J Reprod Fertil Suppl 6:377–396.

Fernandez-Fewell GD, Meredith M (1994). C-fos expression in vomeronasal pathways of mated or pheromone-stimulated male golden hamsters: contributions from vomeronasal sensory input and expression related to mating performance. J Neurosci 14:3643–3654.

Finn PD, Steiner RA, Clifton DR (1998). Temporal patterns of gonadotropin-releasing hormone (GnRH), c-fos, and galanin gene expression in GnRH neurons relative to luteinizing hormone surge in the rat. J Neurosci 18:713–719.

Fortman M, Dellovade TL, Rissman EF (1992). Adrenal contribution to the induction of sexual behavior in the female musk shrew. Horm Behav 26:76–86.

Fortune JE, Eppig JJ, Rissman EF (1992). Mating stimulates estradiol production by ovaries of the musk shrew (*Suncus murinus*). Biol Reprod 46:885–891.

Freeman LM, Arora T, Rissman EF (1998). Neonatal androgen affects copulatory behavior in the female musk shrew. Horm Behav 34:231–238.

Gibson MJ, Krieger DT, Chalrton HM, Zimmerman EA, Silverman AJ, Perlow MJ (1984a). Mating and pregnancy can occur in genetically hypogonadal mice with preoptic area brain grafts. Science 225:949–951.

Gibson MJ, Perlow MJ, Chalrton HM, Zimmerman EA, Davies TF, Krieger DT (1984b). Preoptic area brain grafts in hypogonadal (*hpg*) female mice abolish effects of congenital hypothalamic gonadotropin releasing hormone (GnRH) deficiency. Endocrinology 114:1938–1940.

Gill CJ, Rissman EF (1997). Female sexual behavior is inhibited by short- and long-term food restriction. Physiol Behav 61:387–394.

Gill CJ, King JA, Millar RP, Rissman EF (1995). Development of two GnRH systems in neuronal and immune cells of a mammal. Soc Neurosci Abstracts 21(3):1897.

Gould E (1969). Communication in three genera of shrews (Soricidae): *Suncus*, *Blarina*, and *Cryptotis*. Commun Behav Biol 3A:11–31.

Harrison, JL (1955). Data on the reproduction of some Malayan mammals. Proc Zool Soc London 125:445–460.

Hoffman GE, Lee WS, Attardi B, Yan V, Fietzsimmons MD (1990). Luteinizing hormone–releasing hormone neurons express c-fos antigen after steroid activation. Endocrinology 126: 1736–1741.

Hoffman GE, Smith MS, Verbalis JG (1993). C-fos and related immediate early gene products as markers of activity in neuroendocrine systems. Front Neuroendocrinology 14:173–213.

Jochle W (1975). Current research in coitus-induced ovulation: a review. J Reprod Fertil Suppl 22:165–207.

Karsch FJ, Bowen JM, Caraty A, Evans NP, Moenter SM (1997). Gonadotropin-releasing hormone requirements for ovulation. Biol Reprod 56:303–309.

Kasten TL, White SA, Norton TT, Bond CT, Adelman JP, Fernald RD (1996). Characterization of two new preproGnRH mRNAs in the tree shrew: first direct evidence for mesencephalic GnRH gene expression in a placental mammal. Gen Comp Endocrinol 104:7–19.

Kaynard AH, Pau KYF, Hess DL, Spies HG (1990). Gonadotropin-releasing hormone and norepinephrine release from the rabbit mediobasal and anterior hypothalamus during the mating-induced luteinizing hormone surge. Endocrinology 127:1176–1185.

Kelly MJ, Garrett J, Bosch MA, Roselli CE, Douglass J, Adelman JP, Ronnekleiv OK (1989). Effects of ovariectomy on GnRH mRNA, proGnRH and GnRH levels in the preoptic hypothalamus of the female rat. Neuroendocrinology 49:88–97.

Kim K, Lee BJ, Park Y, Cho WK (1989). Progesterone increases messenger ribonucleic acid (mRNA) encoding luteinizing releasing hormone (LHRH) level in the hypothalamus of ovariectomized estradiol-primed prepubertal rats. Mol Brain Res 6:151–158.

King JA, Millar RP (1995). Evolutionary aspects of gonadotropin-releasing hormone and its receptor. Cell Mol Neurobiol 15:5–23.

King JA, Steneveld AA, Curlewis JD, Rissman EF, Millar RP (1994). Identification of chicken GnRH II in brains of metatherian and early-evolved eutherian species of mammals. Regul Peptides 54:467–477.

King JC, Tai DW, Hanna IK, Pfeiffer A, Haas P, Ronsheim PM, Mitchell SC, Turcotte JC, Blaustein JD (1995). A subgroup of LHRH neurons in guinea pigs with progestin receptors is centrally positioned within the total population of LHRH neurons. Neuroendocrinology 61:265–275.

Krieger DT (1983). Brain peptides: what, where, and why? Science 222:975–985.

Kuiper GGJM, Enmark E, Pelto-Huikko M, Nilsson S, Gustafsson JA (1996). Cloning of a novel estrogen receptor expressed in rat prostate and ovary. Proc Natl Acad Sci U S A 93:5925–5930.

Lambert GM, Rubin BS, Baum MJ (1992a). Sex difference in the effect of mating on c-fos expression in luteinizing hormone–releasing hormone neurons of the ferret forebrain. Endocrinology 131:1473–1480.

Lee WS, Smith S, Hoffman GE (1990). Luteinizing hormone–releasing hormone neurons express fos protein during the proestrous surge of luteinizing hormone. Proc Natl Acad Sci U S A 87:5163–5167.

LeFevre JA, McClintock MK (1992). Social modulation of behavioral reproductive senescence in female rats. Physiol Behav 52:603–608.

Lescheid DW, Terasawa E, Abler LA, Urbanski HF, Warby CM, Millar RP, Sherwood NM (1997). A second form of gonadotropin-releasing hormone (GnRH) with characteristics of chicken GnRH-II is present in the primate brain. Endocrinology 138:5618–5629.

Levine JE (1997). New concepts in the neuroendocrine regulation of gonadotropin surges in rats. Biol Reprod 56:293–302.

Lidicker WZ (1973). Regulation of numbers in an island population of the California vole, a problem in community dynamics. Ecol Monogr 43:271–302.

Louch CD, Ghosh AK, Pal BC (1966). Seasonal changes in weight and reproductive activity of *Suncus murinus* in West Bengal, India. J Mammal 47:73–78.

Lovejoy D, Fischer W, Ngamvongchon S, Craig A, Nahorniak C, Peter R (1992). Distinct sequence of gonadotropin-releasing hormone (GnRH) in dogfish brain provides insight into GnRH evolution. Proc Natl Acad Sci U S A 89:6373–6377.

Marks DL, Lent KL, Rossmanith WG, Clifton DK, Steiner RA (1994). Activation-dependent regulation of galanin gene expression in gonadotropin-releasing hormone neurons in the female rat. Endocrinology 134:1991–1988.

Marks DL, Smith MS, Vrontakis M, Clifton DK, Steiner RA (1993). Regulation of galanin gene expression in gonadotropin-releasing hormone neurons during the estrous cycle of the rat. Endocrinol 132:1836–1844.

Marks DL, Wiemann JN, Burton KA, Lent KL, Clifton DK, Steiner RA (1992). Simultaneous visualization of two cellular mRNA species in individual neurons by use of a new double in situ hybridization method. Mol Cell Neurosci 3:395–405.

Martin GB (1984). Factors affecting the secretion of luteinizing hormone in the ewe. Biol Rev 59:1–87.

Mason AJ, Hayflick JS, Zoeller RT, Young WS, Phillips HS, Nikoliks K, Seeburg P (1986). A deletion truncating the GnRH gene is responsible for hypogonadism in the hpg mouse. Science 234:1366–1371.

Merchenthaler I, Setalo G, Csontos C, Petrusz P, Flerko B, Negro-Vilar A (1989). Combined retrograde and immunocytochemical identification of luteinizing hormone–releasing hormone and somatostatin-containing neurons projecting to the median eminence of the rat. Endocrinology 125:2812–2821.

Merchenthaler I, Lopez FJ, Lennard DE, Negro-Vilar A (1990). Colocalization of galanin and luteinizing hormone-releasing hormone in a subset of preoptic hypothalamic neurons: anatomical and function correlates. Proc Natl Acad Sci U S A 87:6326–6330.

Merchenthaler I, Lopez FJ, Lennard DE, Negro-Villar A (1991). Sexual differences in the distribution of neurons co-expressing galanin and luteinizing hormone–releasing hormone in the rat brain. Endocrinology 127:2431–2436.

Moenter SM, Karsch FJ, Lehman MN (1993). Fos expression during the estradiol-induced gonadotropin-releasing hormone (GnRH) surge of the ewe: induction in GnRH and other neurons. Endocrinology 133:896–903.

Moss R, McCann SM (1973). Induction of mating behavior in rats by luteinizing hormone–releasing factor. Science 181:177–179.

Mosselman S, Polman J, Dijkema R (1996). ERβ: identification and characterization of a novel human estrogen receptor. FEBS Lett 92:49–53.

Muske LE (1993). Evolution of gonadotropin-releasing hormone (GnRH) neuronal systems. Brain Behav Evol 42:215–230.

Muske LE, Moore FL (1990). Ontogeny of immunoreactive gonadotropin-releasing hormone neuronal systems in amphibians. Brain Res 534:177–187.

Pau KYF, Spies HG (1986). Estrogen-dependent effect of norepinephrine on hypothalamic gonadotropin-releasing hormone release in the rabbit. Brain Res 399:15–23.

Petersen SL, Cheuk C, Hartman RD, Barraclough CA (1989). Medial preoptic microimplants of the antiestrogen, keoxifene, affect luteinizing hormone–releasing hormone mRNA levels, median eminence luteinizing hormone–releasing hormone concentrations and luteinizing hormone–release in ovariectomized, estrogen-treated rats. J Neuroendocrinol 1:279–283.

Pfaff DW (1973). Luteinizing hormone-releasing factor potentiates lordosis behavior in hypophysectomized ovariectomized female rats. Science 182:1148–1149.

Ramirez VD, Beyer C (1988). The ovarian cycle of the rabbit: its neuroendocrine control. In The Physiology of Reproduction (Knobil E, Neill JD, eds), pp 1873–1892. New York: Raven Press.

Rissman EF (1987). Social variables influence female sexual behavior in the musk shrew (Suncus murinus). J Comp Psychol 101:3–6.

Rissman EF (1989). Male-related chemical cues promote sexual receptivity in the female musk shrew. Behav Neural Biol 51:114–120.

Rissman EF (1991). Evidence that neural aromatization of androgen regulates the expression of sexual behavior in female musk shrews. J Neuroendocrinol 3:441–448.

Rissman EF (1992). Mating induces puberty in the female musk shrew. Biol Reprod 47:473–477.

Rissman EF, Bronson FH (1987). Role of the ovary and adrenal glad in the sexual behavior of the musk shrew, Suncus murinus. Biol Reprod 36:664–668.

Rissman EF, Crews D (1988). Hormonal correlates of sexual behavior in the female musk shrew: The role of estradiol. Physiol Behav 44:1–7.

Rissman EF, Li X (1998). Sex differences in mammalian and chicken-II gonadotropin-releasing hormone immunoreactivity in musk shrew brain. Gen Comp Endocrinol 112:346–355.

Rissman EF, Nelson RJ, Blank JL, Bronson FH (1987). Reproductive response of a tropical mammal the musk shrew (Suncus murinus) to photoperiod. J Reprod Fertil 81:563–566.

Rissman EF, Clendenon AL, Krohmer RW (1990). Role of androgens in the regulation of sexual behavior in the female musk shrew. Neuroendocrinology 51:468–473.

Rissman EF, Alones V, Craig-Veit CB, Millam JR (1995). Distribution of chicken II gonadotropin-releasing hormone in mammalian brain. J Comp Neurol 357:524–531.

Rissman EF, Harada N, Roselli CE (1996). Effect of vorozole, an aromatase enzyme inhibitor, on sexual behavior, aromatase activity and neural immunoreactivity. J Neuroendocrinol 8:199–210.

Rissman EF, Li X, King JA, Millar RP (1997). Behavioral regulation of GnRH production. Brain Res Bull 44:459–464.

Roberts JL, Dutlow CM, Jakubowski M, Blum M, Millar RP (1989). Estradiol stimulated preoptic area-anterior hypothalamic pro-GnRH-GAP gene expression in ovariectomized rats. Mol Endocrinol 6:127–134.

Rosie R, Thomson E, Fink G (1990). Oestrogen positive feedback stimulates the synthesis of LHRH mRNA in neurones of the rostral diencephalon of the rat. J Endocrinol 124:285–289.

Rossmanith WG, Marks DL, Clifton DK, Steiner RA (1994). Induction of galanin gene expression in gonadotropin-releasing hormone neurons with puberty in the rat. Endocrinology 135:1401–1408.

Rossmanith WG, Marks DL, Clifton DK, Steiner RA (1996). Induction of galanin mRNA in GnRH neurons by estradiol and its facilitation by progesterone. J Neuroendocrinol 8:185–191.

Rothfield JM, Hejtmancik JF, Conn PM, Pfaff DW (1989). In situ hybridization for LHRH mRNA following estrogen treatment. Mol Brain Res 6:121–125.

Sagrillo CA, Grattan DR, McCarthy MM, Selmanoff M (1996). Hormonal and neurotransmitter regulation of GnRH gene expression and related reproductive behaviors. Behav Genet 26:241–277.

Sahu A, Xu B, Kalra SP (1994). Role of galanin in stimulation of pituitary luteinizing hormone secretion as revealed by a specific receptor antagonist, galantide. Endocrinology 134:529–536.

Sawyer CH (1952). Stimulation of ovulation in the rabbit by the intraventricular injection of epinephrine or norepinephrine. Anat Rec 112:385.

Sawyer CH, Markee JE (1959). Estrogen facilitation of release of pituitary ovulating hormone in the rabbit in response to vaginal stimulation. Endocrinology 65:614–621.

Schiml PA, Rissman EF (1999). Cortisol facilitates induction of female sexual behavior in the musk shrew. Behav Neurosci 113:166–175.

Schiml PA, Li X, Rissman EF (1998). The effects of intraventricular infusion of GnRH on female sexual behavior. Soc Neurosci Abstracts 24:948.

Schwanzel-Fukuda M, Fadem BH, Garcia MS, Pfaff DW (1988). Immunocytochemical localization of luteinizing hormone–releasing hormone (LHRH) in the brain and nervus terminalis of adult and early neonatal gray short-tailed opossum (*Monodelphis domestica*). J Comp Neurol 276:44–60.

Seeburg P, Adelman J (1984). Characterization of cDNA for precursor of human luteinizing hormone releasing hormone. Nature 311:666–668.

Shivers BD, Harlan RE, Morrell JI, Pfaff DW (1983). Absence of oestradiol concentration in cell nuclei of LHRH immunoreactive neurones. Nature 304:345–347.

Silverman AJ, Antunes JL, Abrams GM, Nilaver G, Thau R, Robinson JA, Ferin M, Krey LC (1982). The luteinizing hormone–releasing hormone pathways in rhesus (*Macaca mulatta*) and pigtailed (*Macaca nemestrina*) monkeys: new observations on thick, unembedded sections. J Comp Neurol 211:309–317.

Silverman AJ, Jhamandas J, Renaud LP (1987). Localization of luteinizing hormone–releasing hormone (LHRH) neurons that project to the median eminence. J Neurosci 7:2312–2319.

Silverman AJ, Witkin JW, Millar RP (1990a). Light and electron microscopic immunocytochemical analysis of antibodies directed against GnRH and its precursor in hypothalamic neurons. J Histochem Cytochem 38:803–813.

Silverman AJ, Witkin R, Silverman C, Gibson MJ (1990b). Modulation of gonadotropin-releasing hormone neuronal activity as evidenced by uptake of fluorogold from the vasculature. Synapse 6:154–160.

Silverman AJ, Levine I, Witkin JW (1994). The gonadotropin-releasing hormone (GnRH) neuronal systems: Immunocytochemistry and in situ hybridization. *In* The Physiology of Reproduction (Knobil E, Neill JD, eds), pp 1683–1709. New York: Raven Press.

Sisk CL, Moss RL, Dudley CA (1988). Immunocytochemical localization of hypothalamic luteinizing hormone–releasing hormone in male ferrets. Brain Res Bul 20:157–161.

Southern HN (1947). Sexual and aggressive behaviour in the wild rabbit. Behaviour 1:173–194.

Spies HG, Pau KYF, Yang SP (1997). Coital and estrogen signals: a contrast in the preovulatory neuroendocrine networks of rabbits and rhesus monkeys. Biol Reprod 56:310–319.

Tai VC, Schiml PA, Li X, Rissman EF (1997). Behavioral interactions have rapid effects on immunoreactivity of prohormone and gonadotropin-releasing hormone peptide. Brain Res 772:87–94.

Tremblay GB, Tremblay A, Copeland NG, Gilbert DJ, Jenkins NA, Labrie F, Giguere V (1997). Cloning chromosomal localization and functional analysis of the murine estrogen receptor β. Mol Endocrinol 11:353–365.

Tsuji K and Ishikawa T (1984). Some of observations of the caravaning behavior in the musk shrew (*Suncus murinus*). Behaviour 90:167–183.

Urbanski HF, White RB, Fernald RD, Kohama SG, Garyfallou VT, Densmore US (1999). Regional expression of mRNA encoding a second form of gonadotropin-releasing hormone in the macaque brain. Endocrinology 140:1945–1948.

Vandenbergh JG (1994). Pheromones and mammalian reproduction. *In* The Physiology of Reproduction 2nd ed (Knobil E, Neill J, eds), pp 343–362 New York: Raven Press.

Wang HJ, Hoffman GE, Smith MS (1995). Increased GnRH mRNA in the GnRH neurons expressing cFos during the proestrous LH surge. Endocrinology 136:3673–3676.

Watson RE, Langub MC, Landis JW (1992). Further evidence that most luteinizing hormone–releasing hormone neurons are not directly estrogen-responsive: simultaneous localization of luteinizing hormone–releasing hormone and estrogen receptor immunoreactivity in the guinea-pig brain. J Neuoendocrinol 4:311–317.

Wayne NL, Rissman EF (1990). Effects of photoperiods and social variables on reproduction and growth in the male musk shrew (*Suncus murinus*). J Reprod Fertil 89:707–715.

Wayne NL, Rissman EF (1991). Tropical photoperiods affect reproductive development in the musk shrew, *Suncus murinus*. Physiol Behav 50:549–553.

Wayne NL, Malpaux B, Karsch FJ (1989). Social cues can play a role in timing onset of the breeding season of the ewe. J Reprod Fertil 87:707–713.

Wersinger SR, Baum MJ (1996). The temporal pattern of mating-induced immediate-early gene product immunoreactivity in LHRH and non-LHRH neurons of the estrous ferret forebrain. J Neuroendocrinol 8:345–359.

Wersinger SR, Baum MJ (1997a). Sexually dimorphic activation of midbrain tyrosine hydroxlyase neurons after mating or exposure to chemosensory cues in the ferret. Biol Reprod 56:1407–1414.

Wersinger SR, Baum MJ (1997b). Sexually dimorphic processing of somatosensory and chemosensory inputs to forebrain luteinizing hormone–releasing hormone neurons in mated ferrets. Endocrinology 138:1121–1129.

White, RB, Eisen, JA, Kasten, JL, and Fernald, RD (1998). Second gene for gonadotropin releasing hormone in humans. Proc Nat Acad Sci U S A 95:305–309.

Witkin JW, Paden CM, Silverman AJ (1982). The luteinizing hormone–releasing hormone (LHRH) systems in the rat brain. Neuroendocrinology 35:429–438.

Yang SP, Pau KYF, Hess D, Spies HG (1996). Sexual dimorphism in secretion of hypothalamic gonadotropin-releasing hormone and norepinephrine after coitus in rabbits. Endocrinology 137:2683–2693.

Zaborsky L (1982). Afferent connections of the medial basal hypothalamus. Adv Anat Embryol Cell Biol 69:1–107.

15 Sexuality: The Environmental Organization of Phenotypic Plasticity

David Crews

If we want to understand how a phenotype evolved, we have to understand how its regulation is organized, and how it might have been pieced together during evolution from preexisting traits.
—Mary Jane West-Eberhard (1992)

Sex is a concept that is divisible into genetic, gonadal, morphological, physiological, neural, and behavioral components. For example, the genetic sex of an individual mammal refers to its sex chromosome complement, that is, whether it has two X chromosomes, or one X and one Y chromosome. Gonadal sex refers to whether or not an individual has ovaries or testes. Behavioral sex refers to the display of male- or female-typical behaviors. In contrast to sex, sexuality is indivisible because, by definition, it refers to the unique composition of the various aspects of sex that identify the individual. An individual's sexuality is a mosaic of the components described above, as well as factors such as experience, socialization, partner preference, gender attributes, motivational state, performance capabilities, maturation, nutrition, and health. In addition, sexuality is the product of the evolutionary history of the species. Given that evolution depends upon inherited variation (the fabric of change) and environmentally regulated reproduction (the vehicle of change), as well as stochastic forces, how does an individual's sexuality develop? Answering this question requires that we study both ultimate and proximate mechanisms that determine sex and sexuality. In this chapter I outline experimental methodologies and theoretical approaches that may bring us closer to this goal.

Perspectives in biology and psychology often change with new discoveries. Usually this stimulates new research, but it can also limit our vision. This occurs when discoveries narrow our view of biological processes (Russert-Kramer 1989). It might be argued that our present understanding of the ultimate and proximate mechanisms underlying sexuality is an example of a perspective that has generated new data, but has also limited our vision.

Fifty years ago the prevailing view was that there were "male" (= androgens) and "female" (= estrogens) hormones that were responsible for "male" (= mounting and intromission) and "female" (= receptivity) sexual behaviors. The current view of sex differences in hormones and behavior can be traced to a paradigm shift that occurred 40 years ago. Phoenix and associates (1959) observed that exogenous testosterone propionate administered to pregnant guinea pigs will masculinize the behavior of female offspring. Their contribution was to show that the organizational effects of testosterone, which had been known to occur in the body, also occurred with regard to behavior, and by implication, the brain. This and subsequent work suggested that sex steroid hormones early in development organize the body and the brain such that these same hormones later in adulthood activate sex-typical mating behaviors. Since that time a considerable literature on the sexual differentiation of the brain and behavior has accumulated apace with advances in technology.

Behavioral endocrinology has provided us with an ample collection of tools to address sexuality. The vast majority of these experimental designs and outcomes have been interpreted within the context of the organizational hypothesis. It can be argued that this perspective has limited our understanding of sexuality. Because the emphasis has focused on population characteristics, such as the mean lordosis quotient or mean number of mounts made by an animal, we have not progressed very far in understanding the behavior of the individual. I will try to turn attention from group means to sources of individual variation. Along the way I introduce some concepts to guide the reader as we explore new directions in the study of sexuality. This is followed by a collection of recent findings that I hope will shift our focus from current typologies and stimulate new ways of thinking. Implementation of finer-grain analyses, which take into account the context in which the behavior occurs, as well as the nature and pattern of hormone secretion, may bring us closer to our ultimate goal.

The chapter begins with a brief discussion of the meaning of sex, sexuality, phenotype, and phenotypic plasticity. I argue that sexual differentiation can be viewed as one type of phenotypic plasticity. This is followed by a discussion of some of the more widely held paradigms in behavioral endocrinology concerning sex and sexual differentiation. I point out how these paradigms fail to incorporate individual variation, phenotypic plasticity, and environmental effects on sexuality. The remainder of the chapter examines alternative model systems and paradigms that are useful in the study of individual variation in sexuality, and are divided into two main themes.

First, I discuss environmental organization of sex differences, and second, I discuss the evolution of the neuroendocrine mechanisms that control sex behavior. The first theme, environmental organization of sex differences in behavior, is discussed in three subsections. In the first subsection, I introduce species that can be particularly useful for elucidating the sources of individual variation leading to sexuality. One characteristic common to these species is that they exhibit sex differences that do not arise from sex chromosomes. Rather, sex differences in these species arise from environmental cues, such as ambient temperature and social cues from conspecifics. After these model systems are introduced, I illustrate how these species can be used to examine the mechanisms that lead to sex differences. For example, I evaluate the notion that differences in the size of specific brain areas can lead to sex differences in behavior. In the last subsection under Environmental Organization of Phenotypic

Plasticity, I present some examples of nongenetic effects on sexuality that are passed from generation to generation. This is followed by the second main theme of this chapter, which is concerned with the evolution of neuroendocrine mechanisms controlling sex behavior. I conclude with a theme that has been echoed throughout this book, that a thorough understanding of sex and sexuality requires a synthesis of knowledge, in this case from classic developmental biology and embryology, neuroscience, endocrinology, and particularly those fields that incorporate the environment and evolution such as ethology, comparative animal behavior, and evolutionary ecology.

Some Guiding Concepts

The individual, or more accurately, the unique constellation of genes that constitutes the individual, is the unit of selection in the broad scope of evolution. Thus, selection does not act at the level of definable traits, which may be adaptive; it acts at the level of the whole organism. The genotype refers to the allelic complement of the individual and does not change during the life of the organism. However, each genotype can lead to more than one phenotype. The phenotype reflects both the individual's genotype, as well as the environment in which it develops, and thus the phenotype can change during the life span. This fact has become clear in the past decade as imaginative experiments in behavioral ecology and evolutionary biology have revealed the influence of life history and maturation on mate choice. Similarly, experiments in cellular and molecular biology have demonstrated that ontogenetic mechanisms constrain evolutionary change, particularly as they relate to the brain. Emerging from both areas has been the appreciation that the phenotype can change throughout different stages of an individual's life history and that every genotype can yield a variety of phenotypes. Since evolution selects for outcomes, and not mechanisms, it is important that behavioral endocrinologists consider both *individuals* and *context* in both *time* and *space* as important elements in the development of sexuality.

Like sex, phenotype is a concept that describes the constellation of morphological, physiological, and behavioral traits that, when considered together, characterize the individual. Thus, the phenotype concept refers to aspects of an individual's behavior, brain, body parts, physiology, or organs. Like the components of sex, these different phenotypic traits usually are concordant within time, space, and state of an individual, although this often has been a presumption without substantiation.

Phenotypic plasticity refers to the process by which the internal and external environments induce different phenotypes from a given genotype (McNamara and Houston 1996; Piglucci 1996). The range of possible responses from an individual is termed "the reaction norm." In natural populations, genetic variation in phenotypic plasticity is both trait- and environment-specific. The mechanisms underlying plasticity can be either committed and fixed, or labile and reversible. Sexual differentiation can be seen as a kind of phenotypic plasticity. For example, many vertebrate species lack sex chromosomes. In such species, gonadal sex can be fixed during embryonic development, as in the case of certain reptiles in which gonadal sex is plastic during a short period but is fixed by environmental temperature thereafter. In other cases, gonadal sex can change in adulthood, as in the case of certain coral reef fish in which gonadal sex is plastic and sex change is socially regulated in adulthood. Even in species with sex chromosomes, each individual possesses all of the genes necessary to develop the sexual phenotype of both sexes. These considerations together imply that the process of sexual differentiation represents a form of phenotypic plasticity.

Traditional Views in Behavioral Endocrinology Compared to the Study of Individual Variation in Sexuality

The individual emerges from the dynamic interactions within and across all levels of biological organization, from genes to molecules to cells to systems to organisms operating in environments both past and present. Thus, an individual's sexuality results from heritable traits, transmitted genomically or nongenomically, as well as environmental and social stimuli the individual experiences at different stages in its life history. Although behavioral endocrinology has revealed much about the proximate causes of sex differences in behavior, the prevailing hormonal organizational hypothesis of sexual differentiation is of limited value in the quest to understand sexuality.

The foremost reason is subtle but profound. Concepts become preconceptions when they narrow our perceptions, and the resulting orthodoxy thereby excludes alternative views, hypotheses, or preparations. Consider, for example, the "biological species concept" in evolutionary biology. Predicated as they are on sexual organisms, most species concepts are not readily applied to asexual organisms. Thus, the species concept excludes what can be learned about speciation from the study of asexual organisms. The same might be said of the hormonal organization hypothesis of sexual differentiation. While much has been learned about the physiology and, more recently, the molecular biology of sex determination and differentiation of the structures associated with sex in mammals, this information tells us little about the epigenetic process from which sexuality emerges. Genetic and gonadal sex are used to categorize individuals in a majority of vertebrates. These are generally discontinuous and discrete traits; the individual has either the male chromosome complement or the female complement. In contrast, sexuality is a continuously variable suite of traits uniquely expressed in each individual. Because sexual reproduction requires the production of viable complementary gametes and the associated behavior to efficiently combine them, the system of sex determination is biased toward the production of the polar outcomes we refer to as male and female. In contrast, sexuality is more variable, producing both reproductive and nonreproductive phenotypes

A second reason that the predominant behavioral endocrinology methods are of limited value is that we know only what we study, and we tend to study only what we know. In behavioral endocrinology, the majority of research is conducted on relatively few species, virtually all of them highly domesticated. This is particularly evident in studies of sex determination and sexual differentiation. In all of the preparations that have been studied in detail, sex chromosomes determine the type of gonad that will be formed. While this genetic difference facilitates the study of sex differences, it actually hinders our understanding of sexuality. That is, because genetic sex and gonadal sex are linked, it is difficult to distinguish epigenetic from genetic contributions to sexuality. Consider, for example, sociosexual behaviors displayed by both sexes, but at different frequencies or intensities. To what extent are the differences observed between adult males and females due to chromosomal constitution, differences in nongenomic yet heritable influences, maternal influences, the nature and pattern of gonadal steroid hormone secretion, or even sex-typical experiences?

Third, it must always be kept in mind that each individual is bipotential, arising as it does from a single cell. Even as adults both males and females are capable of displaying those behaviors characteristic of the opposite sex. The result is that the range of variation among individuals within a sex usually is greater than the difference between the averages for each of the sexes. In other words, although the sexes differ behaviorally in many ways, this difference is a statistical phenomenon. When trying to detect unifying principles it is often necessary to minimize individual differences. It is time to develop a set of approaches to studying individual differences that are as equally powerful as the mean-based approaches now used to reveal underlying principles.

Environmental Organization of Phenotypic Plasticity

Ideally, studies of the plasticity of the sexual phenotype would utilize animal models that exhibit sex-typical differences in the traits of interest, yet do not have the complications arising from sex chromosomes. In other words, a species that can illustrate how different environments can elicit different phenotypes from a particular genotype without the confounding of sex-limited genes. Do such organisms exist in nature? In species with genotypic sex determination, such as mammals, the inheritance of specific chromosomes, whether in type or in number, fixes the sex of the individual at the moment of fertilization. Scientists have known for many years that in certain plants and invertebrates, the sex ratio is skewed and, further, that these deviations are due to environmental conditions. However, it has only been in the last two decades that we have come to realize that many vertebrates also exhibit environmental sex determination. There are two basic types, behavior-dependent and temperature dependent, and in both instances gonadal sex is determined after conception.

Behavior-dependent Sex Determination

In behavior-dependent sex determination, it is the individual's perception of its social environment that establishes gonad type. Such species are sequentially hermaphroditic and exhibit one of three distinct patterns (Grober 1998). In protogyny, individuals mature and reproduce first as females and then, as they age, turn into males. In protandry, individuals develop and reproduce first as males, and then later turn into females. The third pattern is serial sex change in which the individual functions first as a female, changes into a functional male, and then reverts to female, repeating this process over and over again. This latter pattern is functionally equivalent to simultaneously hermaphroditic species in that individuals alternate their behavior and the type of gamete that is shed in successive matings. However, in alternating behavior-dependent sex determination, the gonads undergo a complete morphological change, producing exclusively the gonad-typical gamete during each successive phase, whereas in simultaneous hermaphroditism the gonads are ovotestes.

When sex change in coral reef fish was first discovered, the prevailing view was that the behavioral changes observed were a consequence of the changes in the gonads. Recently this idea has been turned on its head. In the bluehead wrasse, a single (type I) male dominates a coral head with the rest of the individuals being females or type II males. Within minutes of the dominant male's removal, the largest female will begin to behave as a male, aggressively defending the coral head and soliciting other females. Morphological changes in the gonad are observed many days later and physiological changes are seen several days after that. In other words, it takes approximately 2 weeks after a new fish takes over before its gonads are transformed. Even if the largest bluehead female in a harem is ovariectomized, she will still exhibit all of these behavioral changes and, although muted, color changes following removal of the dominant male (Godwin et al. 1996). Indeed, even though these neutered females do not shed gametes, they entice other females to spawn with them. In most instances this occurs immediately, as would occur when an intact female assumes the dominant role in the harem. Recent studies indicate that the transcript of the gene coding for arginine vasotocin changes in parallel with these behavioral changes (Godwin et al. 1998).

Temperature-Dependent Sex Determination

In many reptiles the temperature of the incubating egg determines an individual's gonadal sex, a process known as temperature-dependent sex determination (TSD). Species with TSD lack sex chromosomes and have little or no genetic predisposition to respond to temperature in particular ways. For example, research with the red-eared slider turtle indicates that the physical stimulus of temperature is transduced in the midtrimester of development to modulate expression of the genes coding for steroidogenic enzymes and sex steroid hormone receptors (Crews 1996). This, in turn, alters the hormonal milieu, and the temperature-specific sex-determining cascades appropriate for that temperature stimulate and inhibit such that individuals develop as gonadal males or females. At intermediate incubation temperatures, hermaphrodites are not formed; rather the ratio of males to females changes relative to that of the higher and lower temperatures. Thus, in TSD species each individual has an equal ability to become a male or a female, and temperature serves as the trigger activating and suppressing the cascades that lead to the development of testes or ovaries. Environmental sex-determining mechanisms, of which TSD is one, are believed to be the evolutionary precursor to genotypic sex-determining mechanisms (Bull 1983; Janzen and Paukstis 1991).

Embryonic Temperature Shapes the Adult Phenotype

In the leopard gecko, incubation of eggs at 26°C produces only female hatchlings, whereas incubation at 30°C produces a female-biased sex ratio, and 32.5°C produces a male-biased sex ratio; incubation of 34° to 35°C again produces virtually all females. Hence, females from eggs incubated at 26°C are referred to as low-temperature females, whereas those females from eggs incubated at 34°C are referred to as high-temperature females; the two intermediate incubation temperatures are referred to as female-biased (30°C) and male-biased (32.5°C) temperatures.

By incubating eggs at these various temperatures and then following individuals as they age, we have found that incubation temperature accounts for much of the phenotypic variation seen among adults both between and within the sexes. For example, adult leopard geckos are sexually dimorphic, with males having open secretory pores anterior to the cloaca. In low-temperature females these pores are closed, whereas in females from a male-biased temperature they are open (Crews 1988). Head size is also sexually dimorphic, with males having wider heads than females, yet within females, those from a male-biased temperature have wider heads than do those from a low temperature (Crews 1988). Similarly, although males are the larger sex, incubation temperature has a marked effect on growth within a sex. Females from a male-biased temperature grow faster and larger than do females from a female-biased temperature, and become as large as males from a female-biased temperature (Tousignant and Crews 1995).

Circulating concentrations of testosterone in both newborn and adult males are approximately 100 times higher than in adult females (Gutzke and Crews 1988; Tousignant and Crews 1995). However, the endocrine physiology of the adult varies in part due to the temperature experienced during incubation (Coomber et al. 1997; Tousignant et al. 1995). For example, plasma estrogen levels are significantly higher in males from a female-biased temperature than in males from a male-biased temperature. Among females, circulating estrogen levels are significantly higher, and androgen levels significantly lower, in low-temperature females than in females from a male-biased temperature. Whether this also is the case in hatchlings from different incubation temperatures is not yet known.

Incubation temperature also has a major influence on the nature and frequency of the behavior displayed by the adult leopard gecko. For example, females usually respond aggressively only if attacked, whereas males will posture and then attack other males but rarely females (Gutzke and Crews 1988; Flores et al. 1994). However, males from a female-biased temperature are less aggressive than males from the higher, male-biased temperature and, although not as aggressive as males from that same incubation temperature, females from a male-biased temperature are significantly more aggressive toward males than are females from a low or female-biased temperature. These same females show the male-typical pattern of offensive aggression. Incubation temperature also influences the ability of exogenous testosterone to restore aggression. Following ovariectomy and testosterone treatment, low-temperature females do not exhibit increased levels of aggression toward *male* stimulus animals, whereas females from male-biased temperatures return to the high levels exhibited while gonadally intact (Flores and Crews 1995). This suggests that incubation temperature influences how the individual responds to steroid hormones in adulthood.

Courtship is a male-typical behavior. In a sexual encounter, the male will slowly approach the female, touching the substrate or licking the air with his tongue. Males also have a characteristic tail vibration, creating a buzzing sound, when they detect a female. Intact females have never been observed to exhibit this tail-vibration behavior, regardless of their incubation temperature. However, if ovariectomized females from low and male-biased temperatures are treated with testosterone, they will begin to tail-vibrate toward female, but not male, stimulus animals; males appear to regard such females as male because they are attacked (Flores and Crews 1995).

Attractiveness in the leopard gecko is a female-typical trait and is measured by the intensity of a sexually active male's courtship behavior toward the female. Females from a male-biased temperature are less attractive than are females from lower incubation temperatures (Flores et al. 1994). Interestingly, attractiveness in high-temperature females is greater than that of females from male-biased temperatures and not different from that of low-temperature females. Long-term castrated males are attractive and initially courted by intact males, but on olfactory inspection they are attacked. This suggests that both sexes can produce both a female-typical attractiveness pheromone and a male-typical recognition pheromone as do red-sided garter snakes (Mason et al. 1989). As is the case with females, incubation temperature influences sensitivity to exogenous hormones in males. Estrogen treatment will induce receptive behavior in castrated males if they are incubated at a female-biased temperature, but not if they are incubated at a male-biased temperature.

The nature and pattern of an individual's growth, hormone secretion, and behavior are expressions of brain activity. Thus, it is reasonable to expect that sex differences in growth hormone secretion and behavior arise from sex differences in the brain, that is, neural phenotypes (Balaban 1990, 1997; Bass 1996, 1998). It stands to reason, therefore, that the morphological, physiological, and behavioral phenotypes we have discovered in the leopard gecko might be the reflection of neural phenotypes. Much to our surprise, we discovered that there is no statistically significant sexual dimorphism in the preoptic area (POA) and ventromedial hypothalamus (VMH) between males and females at those incubation temperatures that produce both sexes (Coomber et al. 1997). However, there are consistent differences across incubation temperatures. For example, females from the male-biased incubation temperature have a larger POA than females from low and female-biased incubation temperatures. Further, the volume of the POA is larger in both males and females from the male-biased temperature than in animals from the female-biased temperature. Similarly, the volume of the VMH is larger in low-temperature females than in females from the male-biased temperature. These data suggest that the incubation temperature of the embryo may directly organize the brain independent of gonadal sex.

Metabolic activity in particular brain areas was measured using the cytochrome-c oxidase (COX) method. We found that males on average have greater COX activity in the POA, whereas females on average have greater COX activity in the VMH (Coomber et al. 1997). Again, incubation temperature is an important determinant, but so too is gonadal sex (i.e., hormones). Males and females from the male-biased temperature have greater COX activity in the POA than animals from the other incubation temperatures, whereas females from the female-biased temperature have greater COX activity in the VMH than females from the male-biased temperature.

As mentioned, there is a significant increase in aggression in females with increasing incubation temperature (Flores et al. 1994). In reptiles, the nucleus sphericus and external amygdala are homologous to the medial and basolateral amygdala of mammals, respectively; as in mammals, both areas are involved in the control of aggression. Analysis of females from different incubation temperatures reveals that COX activity increases in this and other brain areas as a function of incubation temperature in a manner that parallels these differences in aggression within females.

Are these differences in the volume and COX activity of brain nuclei a consequence of a direct action of temperature, or an indirect result of temperature's sex-determining function? This has been tested using the classic gonadectomy and hormone replacement therapy approach (Crews et al. 1996a). As might be expected, androgen treatment following gonadectomy stimulates courtship behavior in both males and females, as well as in females from the all-female–producing temperature (Crews et al. 1996a; Flores and Crews 1995). Following gonadectomy, the volume of the POA decreases and that of the VMH increases in males as well as in females from the male-biased temperature, but not in females from the all-female–producing temperatures. Similarly, androgen treatment results in an increase in metabolic capacity in the POA and in the VMH in males and both types of females. In summary, androgen treatment increases courtship behavior and metabolic activity in the brain area controlling male-typical courtship behavior without a concordant increase in volume of this brain area. Thus, it is more likely that it is the activity, not the size, of a structure that is important for sex differences in behavior. This latter result is also similar to previous findings that females from different incubation temperatures have different behavioral sensitivities to exogenous testosterone (Flores and Crews 1995), which in other lizards reflects steroid hormone receptor abundance (Crews et al. 1996b; Young and Crews 1995). Thus, it appears that male and female geckos from the same incubation temperature respond to sex steroids in the same way, but that within a sex, geckos from different incubation temperatures respond to sex steroid hormone manipulation differently. These behavioral and neurobiological differences indicate that incubation temperature is the primary stimulus in the differentiation of the brain areas involved in sociosexual behaviors.

Although we do not know at this time if hormones differ in the yolk, during embryogenesis, or among same-sex neonates from different incubation temperatures, steroid hormones cannot explain the fact that the level and intensity of aggressive behavior is a function of incubation temperature, not gonadal hormones. Similarly, the rate of growth appears to be organized directly by temperature, perhaps via a temperature effect on growth hormone secretion. Males grow faster and larger than females from the same incubation temperature. If estrogen is administered to eggs incubating at a male-producing temperature, the individual will be a female (rather than a male), yet its rate of growth will be accelerated. Finally, if ovariectomized at birth, low-temperature females grow at a rate characteristic of males from the male-biased temperature, whereas females from the male-biased temperature do not show an acceleration of growth or an increase in aggression (Tousignant and Crews 1995). Although prehatching organizational effects with a postnatal hormonal contribution could explain these results, such evidence suggests that incubation temperature plays an important role in determining an individual's response to gonadectomy and hormone replacement therapy.

Cause vs. Consequence, Size vs. Activity

An important question in the study of brain-behavior relationships is whether brain differences predispose individuals to behave in a certain way, or if individual differences in behavior result in corresponding differences in the brain. This issue of cause vs. consequence is of critical importance in the interpretation of data on the size of sexually dimorphic areas in the brain. A dramatic example of sexual dimorphism in brain structure is found in some species of songbirds. The size of the brain areas involved in the production of song changes as a function of gonadal hormones and, further, individuals with more complex songs have larger and more complex song control brain nuclei (e.g., the higher vocal center or HVC) (Arnold 1992). Similarly, a number of investigations have established that the brain of male and female mammals can differ in a variety of ways, including the size of nuclei implicated in the control of sexual behavior. Do individuals having larger song nuclei learn more complex songs, or do individuals that learn large song repertoires develop large song nuclei?

To answer this question, Brenowitz et al. (1996) hand-reared marsh wrens and exposed them to one of two song repertoires having a fivefold difference in size. Although the two groups showed corresponding differences in the size of their learned song repertoires, they did not differ in the volume of the HVC or in the size, number, or density of neurons within this nucleus. These results suggest that the relationship between song complexity and brain space develops independently of early or later experience, either as a result of inherent differences among individuals in the size of song nuclei, or of other epigenetic influences such as individual differences in hormones in the yolk, steroid metabolism, or allocation of the type or quality of nutrition. In the last instance the nutritional status of the embryo and neonate (Nowicki et al. 1998) and perhaps even the presence of nonsteroidal estrogens in the food (e.g., phytoestrogens or grain fungi) or xenobiotics resulting from industrial contamination must also be considered.

There are examples in which the size of a brain area and its metabolic activity are concordant. In sexual whiptail lizards, the anterior hypothalamus (AH)–POA is larger in males than in females, whereas the VMH, which controls female-typical receptivity, is larger in females (Crews et al. 1990). During hibernation or following castration, the AH-POA shrinks and the VMH enlarges (i.e., both brain areas become female-like); a similar relationship applies to the neurons within these areas (Wade and Crews 1992). These results indicate that in the ancestral sexual species structural dimorphisms develop in the adult and, further, that testicular androgens control the seasonal growth of these areas. Studies using metabolic markers of brain activity reveal that the AH-POA is more active in male, and the VMH more active in female, whiptails during mating (Rand and Crews 1994). The metabolic activity of brain nuclei can also vary according to uterine position. In gerbils, the sexually dimorphic area of the preoptic area (SDA-POA) is responsible for copulatory behavior in males (Yahr 1995) and, as females differ in their sexual behavior according to intrauterine position, the SDA-POA is likely to be involved in their behavior as well. Cytochrome oxidase histochemistry reveals long-term changes in the metabolic capacity in the SDA-POA, with 2M (located between two males) female fetuses having greater activity than 2F (located between two females) female fetuses (Jones et al. 1997). There also is a difference in COX activity in the posterior anterior hypothalamus, an area replete with neurons containing gonadotropin-releasing hormone (GnRH) (Silverman et al. 1994), which may explain the physiological differences between 2M and 2F females.

This relationship between the size of a structure and its activity is not universal, however. For example, because the unisexual whiptail displays both malelike and female-like pseudosexual behavior, it would seem reasonable to predict that its brain would also be bisexual, resembling both the male and the female of the ancestral sexual species. To my great surprise, the AH-POA and a VMH of the unisexual whiptail is not significantly different in size from those seen in females of the sexual ancestral species (Crews et al. 1990; Wade and Crews 1991). There also is no difference in neuron somata size in those individuals exhibiting malelike pseudosexual behavior compared with those exhibiting female-like pseudosexual behavior (Wade and Crews 1992). Even if the parthenogen is treated with androgen so that it exhibits only malelike behavior, the brain remains unchanged. The size of the AH-POA is not correlated with male- or female-like pseudosexual typical behavior. Another example is the red-sided garter snake. In this species only the males court. Despite the sex-specificity of behavior, the AH-POA and the VMH are not dimorphic in area (Crews et al. 1993).

These findings raise questions as to the meaning of sexual dimorphisms in the vertebrate brain. For example, we hear today that homosexuals behave as they do because their brain is different from that of heterosexuals. However, the parthenogenic lizard described above clearly retains the ability to express malelike behaviors, but not because it has a masculinized AH-POA. It retains the ability to express malelike behaviors because it has co-opted the progesterone surge to trigger the masculine behavioral potential that *remains* in a brain that is "feminized" with regard to the size of particular brain areas. In this case, "feminized" refers to a definition related to the size of these brain areas in the heterosexual species. In summary, behavioral differences need not be paralleled by structural differences in the brain.

The Importance of Experience and Age

Throughout the 1950s and 1960s ethologists, as well as comparative and physiological psychologists, were preoccupied with questions of the development of behavior and, in particular, the role of experience. Many experiments demonstrated that certain experiences during specific periods of development were critical to the formation of those behavioral suites characteristic of the species (e.g., displays typically performed by males during courtship) or of a sex (e.g., maternal behavior). For example, the young male white-crowned sparrow must hear his species song during a particular period of development if he is to sing in adulthood. In rats, rabbits, and monkeys, multiparous females are more skilled mothers than are nulliparous females (Lehrman 1962).

Experience is cumulative, depending upon preceding events and, at the same time, setting the stage for future experiences. Social deprivation, environmental complexity, and nourishment can alter the physiology and structure of the brain (Bhide and Bedi 1984; Kraemer et al. 1984; Turner and Greenough 1985). Although we have known for years that sociosexual experience and age influence an individual's sexual behavior, there has been only limited work relating these effects to brain morphology and activity (e.g., see Keverne et al. 1993; Kollack-Walker and Newman 1997; Witkin 1994; Witkin and Romero 1995). Engaging in sexual behavior brings about a decrease in the size of the sexually dimorphic nucleus of the bulbocavernosus in the spinal cord of male rats (Breedlove 1997). In rhesus monkeys, rearing individuals without mothers or allowing only brief periods of interaction with peers results in elevated levels of aggression in males, and submissive behavior in females, in adulthood; if reared in same-sex vs. mixed-sex groups, males showed less of the male-typical foot-clasping mounting behavior as adults, whereas females exhibited more of this behavior as adults (Wallen 1996). Hormones can combine with the social environment early in life to shape adult sexuality. If a female zebra finch is treated with estrogen shortly after birth so as to masculinize its song control nuclei, and then reared with other females from adolescence to adulthood, it will prefer females as sex partners when given androgen in adulthood (Mansukhani et al. 1996).

Since organisms age as they gain sociosexual experience, but do not necessarily gain sociosexual experience as they age, it is important to separate the effects of age from those attributable to experience. For example, neuron density in the hippocampus decreases with age in humans, monkeys, and rats, but it is not clear to what extent epigenetic factors such as experiential or environmental factors could contribute to these age-related changes in the volume of the hippocampus. In rats, aging is associated with a decline in the density of synaptic input to GnRH neurons in the POA, but reproductive experience will counter this trend and maintain synaptic input to GnRH neurons at young adult levels (Witkin 1992; Witkin and Romero 1995).

To assess the relative effects of age and sociosexual experience, we incubated leopard gecko eggs at temperatures that produce either all females or a male-biased sex ratio (26°C or 32.5°C, respectively) and then raised the animals in isolation for at least 1 year before housing some lizards together in breeding groups. In this way it was possible to obtain animals that were either young (1 year of age) or old (2 to 3 years of age), as well as socially experienced or inexperienced.

Sociosexual experience and age can have different effects on the volume and COX activity of brain areas (Crews et al. 1997). For example, the volume of the POA increases with sexual experience in low-temperature females, but not in females from the male-biased temperature, whereas COX activity in the VMH increases in females from the male-biased temperature, but not in low-temperature females. In males as they age, the POA becomes smaller and more active. Such results indicate that the volume and metabolic capacity of specific brain regions are (1) dynamic in adulthood, changing as leopard geckos age and gain sociosexual experience, (2) that the size and activity of brain areas can be independent, and (3) the embryonic environment influences the nature and degree of these changes.

Hormonal Inheritance, Transgenerational Effects, and Individual Variation

Transmission of traits across generations occurs by the inheritance of genes, but characteristics can also be transmitted across generations by nongenomic means, such as culture in humans. Phenomena similar to cultural transmission have been described in animals, such as tits stealing cream from milk bottles in England, black rats feeding on pine nuts in Israel, and washing of rice by macaque monkeys in Japan.

Consider what is involved in genomic vs. nongenomic patterns of inheritance. The first involves individual-to-individual transmission, whereas the latter involves context-to-context transmission that, in turn, influences individuals. These contexts may derive from the culture or local population or from properties of physiological processes. Take, for example, stimuli experienced during fetal development. Forty years ago Christian and LeMunyan (1958) demonstrated that the stress of crowding on pregnant mice adversely affected the physiology and behavior of two generations of progeny. Similar stress effects can be induced by handling pregnant females or housing them in socially unstable conditions; the mechanism of action appears to be due to activation of adrenocortical responses of the mother and the consequent effects on fetal endocrine physiology (Herrenkohl 1979; Sachser and Kaiser 1996; Ward 1972; Ward and Weisz 1980). Moore (1995) and colleagues (Moore et al. 1997) discovered that mother rats behave differently toward male and female pups and, further, these differences reinforce and accentuate subsequent sex differences when the pup reaches adulthood.

Nongenomic inheritance can also occur as a consequence of the uterine environment during pregnancy. Among mammals that bear litters, the fetuses are arranged like peas in a pod, with flanking neighbors except for those at the ends. Clemens was the first to propose that the position of the female rat fetus in the uterus relative to its siblings influences its morphology, physiology, and behavior in adulthood (Clemens 1974; Clemens and Coniglio 1971). Evidence for the intrauterine-position effect has now been found in rats, gerbils, mice, and even humans. One of the features of this phenomenon is that a 2M female is exposed to higher levels of androgen produced by the neighboring males than a 2F female (Clark et al. 1991). As adults these 2M females have lower estrogen and higher testosterone levels, have a masculinized phenotype, are less attractive to males and more aggressive to females, and produce litters with significantly greater male-biased sex ratios relative to 2F females (cf. Clark and Galef 1995; Vom Saal 1991).

Exposure to excessive hormone levels during pregnancy as a result of genetic or metabolic disorders, or hormone therapy of the mother, can lead to abnormal development of secondary and accessory sex characters, including psychosocial alterations in gender identity and gender role in adulthood (Reinisch et al. 1991). Hormonally induced organization also occurs normally in human beings. For example, the cochlea produces spontaneous otoacoustic emissions, with females producing more than males. This sexual dimorphism in emissions is not evident in the female of opposite-sex fraternal twins, suggesting that the prenatal environment created by the male fetus decreases these emissions in females with male co-twins (McFadden 1993; McFadden and Loehlin 1995; see also McFadden and Pasanen 1998). As will become evident, it is not insignificant that otoacoustic emissions occur in lizards and turtles and are temperature-dependent in adults (Köppl 1995). The circulating levels of androgen and sex hormone–binding globulin during the mother's pregnancy may also have an organizing effect on both the hormone profiles and psychosocial test scores of young adult women (Udry et al. 1995). In this remarkable study, a substantial portion of the variance in the women's "gendered" behavior was accounted for by measurements of androgen exposure only during the midtrimester of development and by their circulating levels of androgen as adults.

Environmental conditions can also influence the psychosexual differentiation of embryos or neonates. Cooling newborn rats to 18°C (the normal temperature of the mother is 34°C) for 2 hours delays the testosterone surge that occurs normally about 2 hours after birth which, in turn, affects sexual behavior in adulthood (Roffi et al. 1987). Finally, in Siberian hamsters, as discussed extensively in chapter 7, the photoperiod experienced by the mother during pregnancy influences the circadian rhythms of the offspring (Stetson et al. 1989).

All of these examples indicate that environmental factors influence brain organization. Does a similar phenomenon occur in egg-laying vertebrates? The shelled egg is not immune from maternal or uterine influences (Crews and Bull 1987), and classic studies in embryology used chicken eggs to establish that steroid hormones influence the development of sexually dimorphic structures and behavior (reviewed in Adkins-Regan 1981, 1987; Shumacher and Balthazart 1985; Sayag et al. 1991). Like milk (Koldovsky 1980; Kacsoh et al. 1989), yolk is a significant repository of circulating hormones and reflects the hormonal profile of the mother at the time of its deposition. In fish, the circulating concentrations of a variety of steroid and other hormones in the female are greatest at the time the eggs are yolking and these hormones have an important impact on the development of the fry (Bern 1990; Schreck et al. 1991). In the Japanese quail, circulating steroid levels in the female are correlated with steroid levels in the yolk of eggs (Adkins-Regan et al. 1995).

In the zebra finch and the canary, the testosterone content in the yolk varies predictably across eggs within a clutch and these differences correlate with subsequent behavioral differences in the adults (Schwabl 1993; see also Schwabl et al. 1997). Eggs laid on later days have higher testosterone levels than eggs laid earlier which, in turn, correlates positively with the subsequent growth and social rank of the individual; that is, males hatched from eggs laid later tend to grow faster and achieve higher social status (Schwabl 1993, 1996b). As would be expected, circulating levels of testosterone, which increase during yolking, vary among females and under different environmental regimens (Schwabl 1996a). The spotted hyena displays similar, but more pronounced, maternal effects on offspring phenotype, as discussed in chapter 9.

Environmental Control of Sexuality

In this species the female's clitoris is hypertrophied, resembling the male's penis. This unique pattern of female urogenital development results from maternal ovarian hormone production that is transformed and transported to the fetus by the placenta. Other aspects of the female's phenotype, such as body size and aggressive behavior, are also masculinized such that females dominate males in the social group (Glickman et al. 1997). I have classified this process, in which the endocrine state of the mother influences the offspring via placental transfer or through hormones deposited in the egg yolk, as "hormonal inheritance" (Crews et al. 1989). Study of monotremes would be particularly important in this regard.

The Evolution of Neuroendocrine Mechanisms Controlling Sexual Behavior

Can evolutionary history be used as a lens to approach issues of molecular action in relation to reproductive behavior? Beginning with a chance observation 20 years ago (Crews and Fitzgerald 1980), we have made considerable progress toward this goal. Along the way, we have made some important discoveries that have been extended to mammals.

Because the ancestors of most animals are extinct, most approaches to the evolution of neuroendocrine mechanisms regulating behavior have yielded approximations. The whiptail lizards present a rare case in which representatives of both the ancestral sexual and the descendant unisexual species still exist, representing a "snapshot" of evolution and enabling study of the evolutionary process directly (Crews 1989). Further, because the parthenogenetic species exists only as females, yet still exhibits both malelike and female-like sexual behaviors, we are able to probe the fundamental nature of sexuality without the complication of gender (Crews 1988). Finally, they enable us to examine two fundamental issues in behavioral neuroscience from a new perspective: First, how might the cellular mechanisms that control sexual behaviors have evolved? Second, how do the neural circuits that subserve male-typical and female-typical sexual behaviors differ?

When considered together with the sex-changing fish, the experiments with the unisexual whiptail lizard make it clear that sexuality resides in the brain and not in the gonad. Individual unisexual whiptails show primarily female-like pseudosexual behavior during the preovulatory stage when 17β-estradiol (E_2) concentrations are relatively high and progesterone levels are relatively low; just the opposite is seen during the display of malelike pseudocopulatory behavior in the postovulatory phase when concentrations of E_2 are low and progesterone levels have increased. Androgens (either testosterone or dihydrotestosterone) are not detectable in the circulation of the parthenogen. Changes in behavior commonly occur at transitions in circulating levels of hormones, and the close parallel between the severalfold rise in progesterone levels at ovulation suggests that it may be the hormone responsible for the expression of pseudocopulatory behavior. Supporting evidence is that (1) subcutaneous implants of progesterone elicit pseudocopulatory behavior in ovariectomized parthenogens housed with ovariectomized, estrogen-treated parthenogens; (2) intrahypothalamic implantation of progesterone elicits mating behavior in castrated males of the sexual species; and (3) progesterone upregulates androgen receptor (AR) gene expression in the AH-POA.

Although the unisexual whiptail descended directly from a sexual whiptail, the two species differ in an important aspect of their reproductive biology; namely, circulating concentrations of E_2 in reproductively active parthenogenetic whiptails are approximately fivefold lower than in reproductively active female sexual whiptails (Young and Crews 1995). Since changes in the circulating concentrations of sex steroid hormones can have dramatic effects on endocrine physiology and behavior, one might expect the severalfold difference in E_2 between the parthenogenetic whiptail and the sexual whiptail to be accompanied by species differences in estrogen-dependent phenomena. This is evident in comparisons of ER messenger RNA (mRNA) content in the POA of unisexual whiptails and female whiptails. Thus, an inverse relationship exists between sex steroid receptor gene expression in the POA and circulating sex steroid hormone concentration. The increased level of ER gene expression in the POA may result in a greater sensitivity to the circulating concentrations of E_2, which could in turn result in lower levels of E_2 through feedback effects.

Why is ER mRNA expression in the POA higher in the parthenogenetic whiptail than in its maternal ancestral species? One possibility is that it is linked to the increased gene dosage resulting from the triploid nature of the genome. It has been suggested that one reason that polyploid species differ physiologically and ecologically from their diploid relatives is that the increased gene dosage may result in higher enzyme and hormone levels. This phenomenon has been well documented for a number of enzymes in several plant species. Allozyme analysis of a number of diploid and polyploid whiptail lizards, including the parthenogenetic whiptail, demonstrate that each of the three sets of chromosomes are actively transcribing genes at rates proportional to the gene dosage, rather than one chromosome set becoming inactivated, as might be expected. Triploidy, therefore, could result in increased sensitivity to E_2, not only by increasing the basal rate of ER production but also by increasing estrogen-dependent gene transcription rates as the target gene number is also increased.

Consistent with this hypothesis, the species differences in circulating concentrations in E_2 and ER mRNA are accompanied by differences in sensitivity to E_2. Dose-response studies reveal that lower dosages of estradiol benzoate are required to induce receptive behavior, as well as changes in gene expression, in the VMH of parthenogenetic whiptails compared with sexual whiptails. As in other vertebrates, the VMH is involved in the hormonal induction of receptive behavior in whiptail lizards (Wade and Crews 1991; Kendrick et al. 1995). Thus, species differences in reproductive physiology (e.g., brief follicular phases, as in the rat and mouse, compared to extended follicular phases, as in whiptail lizards and rabbits) may explain species differences in neuroendocrine controlling mechanisms. Thus, at least three factors may contribute to species differences in endocrine physiology and behavior: (1) sensitivity to sex steroid hormones, (2) hormone-dependent regulation of sex steroid hormone receptor gene expression, and (3) neuroanatomical distribution of steroid receptor gene expression, especially in nonlimbic structures.

How could an androgen-dependent male-typical mating behavior of the sexual ancestral species evolve to become a progesterone-dependent malelike pseudosexual behavior in the unisexual descendant species? The courtship behavior in males of the sexual ancestral species depends upon testicular androgens. Gould and Vrba (1982) pointed out that existing features can be produced by two distinct historical processes. One of these is adaptation, or the gradual selection of traits resulting in improved functions. Some traits, however, evolved from features that served other roles, or had no function at all, and were co-opted for their current role because they enhanced current fitness. This latter process may be termed *exaptation*. In adaptation, traits are constructed by selection for their present functions, while exaptations are co-opted for a new use.

Evolution depends upon *individual variation*; without it there would be no basis on which to evolve. This question is of biomedical importance because individual variation in response to standardized stimuli is fundamental and affects every aspect of healthcare. How does this relate to the whiptail lizards?

Some males of the sexual ancestral species are sensitive to progesterone; that is, in about one third of castrated males, administration of exogenous progesterone *restores* the complete repertoire of male-typical sexual behavior. Progesterone is probably exerting its stimulatory action as a progestagen and *not* via conversion to other sex steroid hormones. Other data suggest that progesterone acts via the PR and not via the AR. Thus, in the sexual ancestral species, individual variation in sensitivity to progesterone appears to have served as the substrate for the evolution of the novel hormone-brain-behavior relationship observed in the parthenogen. That is, the elevation of progesterone following ovulation presented a reliable stimulus that, given the low circulating concentrations of androgens, was co-opted to trigger mounting behavior in the parthenogen.

Is this behavioral responsiveness to progestin in males specific only to reptiles? What little is known of the physiology of progesterone in males points to a functional role. For example, $17\alpha, 20\beta$-dihydroxyprogesterone stimulates spawning behavior in castrated rainbow trout (Mayer et al. 1994). In male rats there is a pronounced diurnal rhythm in progesterone secretion, with the peak in progesterone levels coinciding with the period of greatest copulatory activity (Kalra and Kalra 1977). When administered in physiological dosages (rather than the pharmacological dosages usually used) progesterone causes some castrated male rats to mate with receptive females; when combined with subthreshold dosages of testosterone, all of the males respond (Witt et al. 1994, 1995). Further, this progesterone response is blocked by the progesterone antagonist RU486. Finally, recent experiments with transgenic mice indicate that males with a null mutation for the PR not only exhibit deficits in sexual behavior while intact, but also have a severely impaired capacity for maintaining sexual behavior following castration (Phelps et al. 1998). Thus, although progesterone has long been thought of as a "female" hormone involved in the control of female-typical sexual behavior, this evidence points to a previously unsuspected role of progesterone in the control of male sexual behavior.

Concluding Remarks

Variation results from the inherited genotype, environmental context, and the interaction between genotype and environment. Traditionally separate, molecular and developmental biologists and ecological and evolutionary biologists have recently focused on the epigenetic properties of the genotype-environment interaction. This is the well-traveled domain of psychobiology. Psychobiology offers a bridge between perspectives by recognizing that two levels exist in the organization of the adult sexual phenotype. Primary organization is the process of sexual differentiation that follows the determination of the gonad and is manifest as the morphological, physiological, behavioral, and neural traits that characterize the sexual phenotype. Secondary organization follows and is the basis of an individual's sexuality. Sexuality results from heritable genetic variation as well as nongenomic factors that include, but are not limited to, sex steroid hormones. Some of the other factors include (1) the environments encountered throughout life, (2) age and sociosexual experiences, and (3) the psychological and physiological condition of the mother.

This conceptual separation of the mechanisms that determine gonadal and morphological sex from those that shape sexuality is central to any study of individual differences. However, in species in which gonadal sex is determined at fertilization by sex chromosomes, primary and secondary organization are linked, thereby becoming a confounding variable in studies of reproductive traits and sexuality. Thus, assessing the relative contributions of genetic and environmental influences requires alternative animal model systems in which the determination of gonadal sex is independent of genetic sex. Such a comparative approach generates new ideas about cause vs. consequence and size vs. activity in brain-behavior relationships, the importance of experience and age, and how hormonal inheritance, transgenerational effects, and individual variation are interrelated. With such information it becomes possible to study directly the evolution of the neuroendocrine mechanisms that control sexual behavior.

Because individual variation is the substance of evolutionary change, understanding its organization will require both new paradigms as well as alternative animal model systems that allow separation of the effects of genes and hormones from environmental and experiential stimuli. In general, much of the research on the proximate mechanisms underlying the establishment of the sexual phenotype has emphasized the role of gonadal sex hormones in the development of sexual dimorphisms. But it is time to refocus, move away from differences in mean values of males and females, and instead direct our attention to the individual variations within each sex. The diverse natural systems described herein suggest that factors other than sex chromosomes and the steroid hormones secreted following gonadal differentiation can play important roles in the development of within-sex differences in sexual behavior. Some possible factors discussed include (1) embryonic environment (broadly defined to include not only the hormones produced by neighboring fetuses or found in the yolk but also physical factors such as temperature or number and type of other conspecifics), (2) the psychological and physiological condition of the mother during pregnancy or egg-laying, (3) the sociosexual experiences during growth and adulthood, and (4) the aging process.

Along this line, it might be more profitable to view hormones and other factors during embryogenesis as allowing for the growth of neural connections, whereas participation in particular behaviors throughout life specifies and consolidates the unique functional integrity of the attendant neuroendocrine systems of each individual (Crews 1987). For example, an individual's sociosexual experience not only can modify its mating behavior, but experience also can change how it responds to sex steroid hormones. In guinea pigs, the amount of sexual activity experienced as a juvenile affects the level of sexual behavior displayed as an adult, as well as modifies the individual's response to castration and androgen replacement therapy (Valenstein and Young 1955). In cats, males that have had sexual experience before castration persist in displaying sexual behavior, whereas males that are sexually naive prior to castration show an abrupt cessation in sexual behavior; conversely, castrated, sexually naive male cats given exogenous testosterone take longer to exhibit sexual behavior than castrates that have had sexual experience (Rosenblatt 1965). In gulls, testosterone treatment of males reared in social groups induces sexual behaviors, but not in males reared in isolation; males reared in social groups until sexually mature and then subjected to social isolation for 2 weeks exhibit lower levels of sexual behavior in response to exogenous testosterone than similarly reared males exposed to unfamiliar conspecifics (Groothuis 1995).

One interpretation of these data is that sociosexual experience mediates the effects of androgen by affecting the individual's sensitivity to androgen, perhaps by influencing receptor density, the metabolism of the hormone, or alterations of neural circuitry, or a combination of these. Consistent with this hypothesis is the observation that the behavior of the partner can regulate the abundance of steroid receptor in the brain independently of the gonads (Hartman and Crews 1996). Thus, experience can have a permanent organizing effect on behavior and, hence, the brain. Lehrman (1962) noted that "the two sexes within the same species might differ with respect to the relative degree of dependence of their sexual behavior upon the presence of various hormones and upon various situation and experiential factors" (p. 142). This same logic can be extended to individual differences. The leopard gecko illustrates how such nongenetic factors can affect the differentiation of the adult phenotype without the confounding of genetically based sex differences. In this species, incubation temperature accounts for much of the variation in morphology, endocrine physiology, sociosexual behavior, and the size and metabolic activity of associated brain areas.

A final question is whether the relationship between individual differences in levels and intensity of sociosexual behavior is the cause or consequence of individual differences in relevant brain areas. Experience is continuous, punctuated by discrete events that shape subsequent behaviors. Recent studies suggest that the behavior, or the predisposition to behave, appears to be a consequence of brain activity, rather than the behavior causing the differences in brain activity.

Related to this issue is the finding that it is not the size, but the activity, of a brain area that matters. It may be then that differences are more likely to be reflected in the regulation of genes encoding receptor expression, metabolic activity, hormonal modulation of sensory processing, and so forth, rather than in neuronal circuitry, distribution of receptors, or the volume of brain areas. This has important implications for how we approach the problem of brain-behavior causality. How and why individuals differ is perhaps the original question in both psychology and biology. The ideas presented herein are not novel, and can be found in the writings of D. O. Hebb, Z. Y. Kuo, D. S. Lehrman, and C. H. Waddington. The plasticity of the sexual phenotype and how each individual emerges out of its own unique circumstances will require integrating classic developmental, ethological, and neuroscience concepts with more current perspectives. By exploiting preparations provided in nature and developing new approaches, the foundation for a paradigm shift leading to a better understanding of individual differences will be laid in much the same way that it was for sex differences four decades ago.

Acknowledgments

I thank E. A. Brenowitz, J. Godwin, S. B. Hrdy, D. McFadden, M. C. Moore, M. Rand, M. J. Ryan, J. Sakata, E. Vance, and J. Wingfield for reading earlier versions of the manuscript. The support of NIMH Research Scientist Award 00135, MH 41770, and MH 57874 is gratefully acknowledged.

References

Adkins-Regan E (1981). Early organization effects of hormones: An evolutionary perspective. *In* Neuroendocrinology of Reproduction (Adler AT, ed), pp 159–228. New York: Plenum Press.

Adkins-Regan E (1987). Hormones and sexual differentiation. *In* Hormones and Reproduction in Fishes, Amphibians, and Reptiles (Norris DO, Jones RE, eds), pp 1–30. New York: Plenum Press.

Adkins-Regan E, Ottinger MA, Park J (1995). Maternal transfer of estradiol to egg yolks alters sexual differentiation of avian offspring. J Exp Zool 271:466–470.

Arnold AP (1992). Developmental plasticity in neural circuits controlling birdsong: sexual differentiation and the neural basis of learning. J Neurobiol 23:1506–1528.

Balaban E (1990). Avian brain chimeras as a tool for studying species behavioural differences. *In* The Avian Model in Developmental Biology: From Organism to Genes (Dourarin N, Dieterlen-Lievrre F, Smith F, eds), pp 105–118. Paris: Editions du CNRS.

Balaban E (1997). Changes in multiple bran regions underlie species differences in a complex, congenital behavior. Proc Natl Acad Sci U S A 94:2001–2006.

Bass AH (1996). Shaping brain sexuality. Am Scientist 84:352–363.

Bass AH (1998). Behavioral and evolutionary neurobiology: a pluralistic approach. Am Zoologist 38:97–107.

Bern HA (1990). The "new" endocrinology: its scope and its impact. Am Zoologist 30:877–885.

Bhide PG, Bedi KS (1984). The effects of a lengthy period of environmental diversity on well-fed and previously undernourished rats. II. Synapse to neuron ratios. J Comp Neurol 227:305–310.

Breedlove SM (1997). Sex on the brain. Nature 389:801.

Brenowitz EA, Lent K, Kroodsma D (1996). Brain space for learning song in birds develops independently of song learning. J Neurosci 15:6281–6286.

Bull JJ (1983). Evolution of Sex Determining Mechanisms. Menlo Park, CA: Benjamin/Cummings.

Christian JJ, LeMunyan CD (1958). Adverse effects of crowding on reproduction and lactation of mice and two generations of progeny. Endocrinology 63:517–529.

Clark MM, Galef BG (1995). Prenatal influences on reproductive life history strategies. Trends Ecol Evol 10:151–153.

Clark MM, Crews D, Galef BG (1991). Circulating concentrations of sex steroid hormones in pregnant and fetal mongolian gerbils. Physiol Behav 49:239–243.

Clemens LG (1974). Neurohormonal control of male sexual behavior. In Reproductive Behavior (Montagna W, Sadler WA, eds), pp 23–54. New York: Plenum Press.

Clemens LG, Cognilio L (1971). Influence of prenatal litter composition on mounting behavior of female rats. Am Zoologist 11:617–618.

Coomber P, Gonzalez-Lima F, Crews D (1997). Effects of incubation temperature and gonadal sex on the morphology and metabolic capacity of brain nuclei in the leopard gecko (*Eublepharis macularius*), a lizard with temperature-dependent sex determination. J Comp Neurol 380:409–421.

Crews D (1987). Functional associations in behavioral endocrinology. In Masculinity/Femininity: Basic Perspectives (Reinisch JM, Rosenblum LA, Sanders SA, eds), pp 83–106. Oxford: Oxford University Press.

Crews D (1988). The problem with gender. Psychobiology 16:321–334.

Crews D (1989). Unisexual organisms as model systems for research in the behavioral neurosciences. In Evolution and Ecology of Unisexual Vertebrates (Dawley RM, Bogart JP, eds), pp 132–143. Albany, NY: State Museum.

Crews D (1996). Temperature-dependent sex determination: the interplay of steroid hormones and temperature. Zool Sci 13:1–13.

Crews D, Bull JJ (1987). Evolutionary insights from reptile sexual differentiation. In Genetic Markers of Sexual Differentiation (Haseltine FP, McClure ME, Goldberg EH, eds), pp 11–26. New York: Plenum Press.

Crews D, Fitzgerald KT (1980). "Sexual" behavior in parthenogenetic lizards (*Cnemidophorus*). Proc Natl Acad Sci U S A 77:499–502.

Crews D, Wibbels T, Gutzke WHN (1989). Action of sex steroid hormones on temperature-induced sex determination in the snapping turtle (*Chelydra serpentina*). Gen Comp Endocrinol 75:159–166.

Crews D, Wade J, Wilczynski W (1990). Sexually dimorphic areas in the brain of whiptail lizards. Brain Behav Evol 36:262–270.

Crews D, Robker R, Mendonça M (1993). Seasonal fluctuations in brain nuclei in the red-sided garter snake and their hormonal control. J Neurosci 13:5356–5364.

Crews D, Coomber P, Baldwin R, Azad N, Gonzalez-Lima F (1996a). Effects of gonadectomy and hormone treatment on the morphology and metabolic capacity of brain nuclei in the leopard gecko (*Eublepharis macularius*), a lizard with temperature-dependent sex determination. Horm Behav 30:474–486.

Crews D, Godwin J, Hartman V, Grammar M, Prediger EA, Shephard R (1996b). Intrahypothalamic implantation of progesterone in castrated male whiptail lizards (*Cnemidophorus inornatus*) elicits courtship and copulatory behavior and affects androgen- and progesterone receptor–mRNA expression in the brain. J Neurosci 16:7347–7352.

Crews D, Coomber P, Gonzalez-Lima F (1997). Effects of age and sexual experience on the morphology and metabolic capacity of brain nuclei in the leopard gecko (*Eublepharis macularius*), a lizard with temperature-dependent sex determination. Brain Res 758:169–179.

Flores DL, Crews D (1995). Effect of hormonal manipulation on sociosexual behavior in adult female leopard geckos (*Eublepharis macularius*), a species with temperature-dependent sex determination. Horm Behav 29:458–473.

Flores DL, Tousignant A, Crews D (1994). Incubation temperature affects the behavior of adult leopard geckos (*Eublepharis macularius*). Physiol. Behav. 55:1067–1072.

Glickman SE, C. J. Zabel CJ, Yoerg SI, Weldele ML, Drea CM, Frank LG (1997). Social facilitation, affiliation, and dominance in the social life of spotted hyenas. In The Integrative Neurobiology of Affiliation (Carter CS, Lederhendler II, Kirkpatrick B, eds), pp 175–185. New York: New York Academy of Sciences.

Godwin J, Crews D, Warner RR (1996). Behavioral sex change in the absence of gonads in a coral reef fish. Proc Soc Lond B Biol Sci 263:1683–1688.

Godwin J, Sawby R, Crews, D, Warner RR, Grober MS (1999). Hypothalamic arginine vasotocin mRNA abundance variation across sexes and with sex change in a coral reef fish. Brain, Behav., and Evolution (in press).

Gould SJ, Vrba ES (1982). Exaptation—a missing term in the science of form. Paleobiology 8:4–15.

Grober MS (1998). An integrative analysis of sex change. Acta Ethol 1:3–17.

Groothuis TGG (1995). Social experience and the development of sexual behavior. In Proceedings of the 24th International Ethological Conference Abstracts, p 6.

Gutzke WHN, Crews D (1988). Embryonic temperature determines adult sexuality in a reptile. Nature 332:832–834.

Hartman V, Crews D (1996). Sociosexual stimuli regulate ER- and PR-mRNA abundance in the hypothalamus of all-female whiptail lizards. Brain Res 741:344–347.

Herrenkohl LR (1979). Prenatal stress reduces fertility and fecundity in female offspring. Science 206:1097–1099.

Janzen FJ; Paukstis GL (1991). Environmental sex determination in reptiles: ecology, evolution, and experimental design. Q Rev Biol 66:149–179.

Jones D, Gonzalez-Lima F, Crews D, Galef BG, Clark MM (1997). Effects of intrauterine position on hypothalamic activity of female gerbils: a cytochrome oxidase study. Physiol Behav 61:513–519.

Kacsoh B, Terry LC, Meyers JS, Crowley WR, Grosvenor CE (1989). Maternal modulation of growth hormone secretion in the neonatal rat. I. Involvement of milk factors. Endocrinology 125:1326–1336.

Kalra PS, Kalra SP (1977). Circadian periodicities of serum androgens, progesterone, gonadotropins and luteinizing hormone–releasing hormone in male rats: the effects of hypothalamic deafferentiation, castration and adrenalectomy. Endocrinology 10:1821–1827.

Kendrick AM, Rand MS, Crews D (1995). Electrolytic lesions of the ventromedial hypothalamus abolish sexual receptivity in female whiptail lizards, *Cnemidophorus uniparens*. Brain Res. 680:226–228.

Keverne EB, Levy F, Guerara-Guzman R, Kendrick KM (1993). Influence of birth and maternal experience on olfactory bulb neurotransmitter release. Neuroscience 56:557–565.

Koldovsky O (1980). Hormones in milk. Life Sci 26:1833–1836.

Kollack-Walker S, Newman SW (1997). Mating-induced expression of *c-fos* in the male Syrian hamster brain: role of experience, pheromones and ejaculations. J Neurobiol 32:481–501.

Köppl C (1995). Otoacoustic emissions as an indicator for active cochlear mechanics: a primitive property of vertebrate auditory organs. In Advances in Hearing Research (Manley GA, Klump GM, Köppl C, Fastl H, Oeckinghaus H, eds), pp 207–218. Singapore: World Scientific.

Kraemer GW, Ebert MH, Lake, CR, McKinney WT (1984). Hypersensitive to *d*-amphetamine several years after early social deprivation in rhesus monkeys. Psychopharmacology 82:266–271.

Lehrman DS (1962). Interaction of hormonal and experiential influences on development of behavior. In Roots of Behavior (Bliss EL, ed), pp 142–156. New York: Harper & Row.

Mansukhani V, Adkins-Regan E, Yang S (1996). Sexual partner preference in female zebra finches: the role of early hormones and social environment. Horm Behav 30:506–511.

Mason RT, Fales HM, Jones TH, Pannell LK, Chin JS, Crews D (1989). Sex pheromones in snakes. Science 245:290–293.

Mayer I, Liley NR, Borg B (1994). Stimulation of spawning behavior in castrated rainbow trout (*Oncorhyncus mykiss*). by 17α, 20β-dihydroxy-4-pregnen-3-one, and not by 11-ketoandrostenedione. Horm Behav 28:181–190.

McFadden D (1993). A masculinizing effect on the auditory systems of human females having male co-twins. Proc Natl Acad Sci U S A 90:11900–11904.

McFadden D, Loehlin J (1995). On the heritability of spontaneous otoacoustic emissions: a twins study. Hear Res 85:181–198.

McFadden D, Pasanen EG (1998). Comparison of the auditory systems of heterosexuals and homosexuals: click-evoked otoacoustic emissions. Proc Natl Acad Sci U S A. 1998 95:2709–2713.

McNamara JM, Houston AI (1996). State-dependent life histories. Nature 380:215–220.

Moore CL (1995). Maternal contributions to mammalian reproductive development and divergence of males and females. *In* Advances in the Study of Behavior (Slater PJP, Rosenblatt JS, Snowdon CT, Milinski M, eds), pp 47–118. New York: Academic Press.

Moore CL, Wong L, Daum MC, Leclair OU (1997). Mother-infant interactions in two strains of rats: implications for dissociating mechanism and function of a maternal pattern. Dev Psychobiol 30:301–312.

Nowicki S, Peters S, Podos J (1998). Song learning, early nutrition and sexual selection in songbirds. Am Zoologist 38:179–190.

Phelps SM, Lydon J, O'Malley BW, Crews D (1998). Regulation of male sexual behavior by progesterone receptor, sexual experience and androgens. Horm Behav 34:294–302.

Phoenix CH, Goy RW, Gerall AA, Young WC (1959). Organizational action of prenatally administered testosterone propionate on the tissues mediating behavior in the female guinea pig. Endocrinology 65:369–382.

Pigliucci M (1996). How organisms respond to environmental changes: From phenotypes to molecules (and vice versa). Trends Ecol Evol 11:168–173.

Rand MS, Crews D (1994). The bisexual brain: sex behavior differences and sex differences in parthenogenetic and sexual lizards. Brain Res 663:163–167.

Reinboth R (1988). Physiological problems of teleost ambisexuality. Environ Biol Fishes 2:455–488.

Reinisch JM, Ziemba-Davis M, Sanders SA (1991). Hormonal contributions to sexually dimorphic behavioral development in humans. Psychoneuroendocrinology 16:213–236.

Rosenblatt JS (1965). Effects of experience on sexual behavior in male cats. *In* Sex and Behavior (Beach FA, ed), pp 416–439. New York: John Wiley & Sons.

Russert-Kramer L (1989). "I'll see it when I believe it!"; Investigating nervous system/reproductive system interactions in animal organisms. Am Zoologist 29:1141–1155.

Roffi J, Chami F, Corbier P, Edwards DA (1987). Influence of the environmental temperature on the postpartum testosterone surge in the rat. Acta Endocrinol 115:478–482.

Sachser N, Kaiser S (1996). Prenatal social stress masculinizes the females' behavior in guinea pigs. Physiol Behav 60:589–594.

Sayag N, Robinson B, Snapir N, Arnon E, Grimm VE (1991). The effects of embryonic treatments with gonadal hormones on sexually dimorphic behavior of chicks. Horm Behav 25:137–153.

Schreck CB, Fitzpatrick MS, Feist GW, Yeoh C-G (1991). Steroids: developmental continuum between mother and offspring. *In* Proceedings of the Fourth International Symposium on Reproductive Physiology of Fish (Scott AP, Sumpter JP, Kime DE, Rolfe MS, eds), pp 256–258. Sheffield, UK: University of East Anglia, Fish Symposia 91.

Schumacher M, Balthazart J (1985). Sexual differentiation is a biphasic process in mammals and birds. *In* Neurobiology (Gilles R, Balthazart J, eds), pp 203–219. Berlin: Springer-Verlag.

Schwabl H (1993). Yolk is a source of maternal testosterone for developing birds. Proc Natl Acad Sci U S A 90:11446–11450.

Schwabl H (1996a). Environment modifies the testosterone levels of a female birds and its eggs. J Exp Zool 276:157–163.

Schwabl H (1996b). Maternal testosterone in the avian egg enhances postnatal growth. Comp Biochem Physiol 114A:271–276.

Schwabl H, Mock DW, Cieg JA (1997). A hormonal mechanism for parental favouritism. Nature 386:231.

Silverman AJ, Livne I, Witkin JW (1994). The gonadotropin-releasing hormone (GnRH) neuronal systems: immunocytochemistry and in situ hybridization *In* The Physiology of Reproduction, 2nd ed, vol 1 (Knobil E, Neill JD, eds), pp 1683–1709. New York: Raven Press.

Stetson MH, Ray SL, Creyaufmiller N, Horton TH (1989). Maternal transfer of photoperiodic information in Siberian hamsters. II. The nature of the maternal signal, time of signal transfer, and the effect of the maternal signal on peripubertal reproductive development in the absence of photoperiodic input. Biol Reprod 40:458–465.

Tousignant A, Crews D (1995). Incubation temperature and gonadal sex affect growth and physiology in the leopard gecko (*Eublepharis macularius*), a lizard with temperature-dependent sex determination. J Morphol 224:159–170.

Tousignant A, Viets B, Flores D, Crews D (1995). Ontogenetic and social factors affecting the endocrinology and timing of reproduction in the female leopard gecko (*Eublepharis macularius*). Horm Behav 29:141–153.

Turner M, Greenough WT (1985). Differential rearing effects on rat visual cortex synapses. I: Synaptic and neuronal density and synapses per neuron. Brain Res 329:195–203.

Udry JR, Morris NM, Kovenock J (1995). Androgen effects on women's gendered behavior. J Biosoc Sci 27:359–368.

Valenstein ES, Young WC (1955). An experiential factor influencing the effectiveness of testosterone propionate in eliciting sexual behavior in male guinea pigs. Endocrinology 56:173–177.

Vom Saal FS (1991). Prenatal gonadal influences on mouse sociosexual behaviors. *In* Heterotypical Behavior in Man and Animals (Haug M, Brain PF, Aron C, eds), pp 42–70. London: Chapman & Hall.

Wade J, Crews D (1991). Relationship between reproductive state and "sexually" dimorphic brain areas in sexually reproducing and parthenogenetic whiptail lizards. J Comp Neurol 309:507–514.

Wade J, Crews D (1992). Sexual dimorphisms in the soma size of neurons in the brain of whiptail lizards (*Cnemidophorus* species). Brain Res 594:311–314.

Wallen K (1996). Nature needs nurture: the interaction of hormonal and social influences on the development of behavioral sex differences in rhesus monkeys. Horm Behav 30:364–378.

Ward IL (1972). Prenatal stress feminizes and demasculinizes the behavior of males. Science 175:82–84.

Ward IL, Weisz J (1980). Maternal stress alters plasma testosterone in fetal males. Science 207:328–329.

West-Eberhard MJ (1992). Behavior and evolution. *In* Molds, Molecules, and Metazoa: Growing Points in Evolutionary Biology. Grant, PR and Horn, H (eds.), pp 57–75. Princeton, NJ: Princeton University Press.

Witkin JW (1992). Reproductive history affects the synaptology of the aging gonadotropin-releasing hormone system in the male rat. J Neuroendocrinol 4:427–432.

Witkin JW, Romero M-T (1995). Comparison of ultrastructural chacteristics of gonadotropin-releasing hormone neurons in prepubertal and adult male rats. Neuroscience 64:1145–1151.

Witt DM, Young LJ, Crews D (1994). Progesterone and bisexual behavior in males. Psychoneuroendocrinology 19:553–562.

Witt DM, Young LJ, Crews D (1995). Progesterone modulation of androgen-dependent sexual behavior in male rats. Physiol Behav 57:307–313.

Yahr P (1995). Neural circuitry for the hormonal control of male sexual behavior. *In* Neurobiological Effects of Sex Steroid Hormones (Micevych PE, Hammer RP, eds), pp 40–56. Cambridge: Cambridge University Press.

Young LJ, Crews D (1995). Comparative neuroendocrinology of steroid receptor gene expression and regulation: relationship to physiology and behavior. Trends Endocrinol Metab 6:317–323.

Contributors

Gregory F. Ball
Department of Psychology
Johns Hopkins University
Baltimore, Maryland

George E. Bentley
Department of Psychology
Johns Hopkins University
Baltimore, Maryland

Franklin H. Bronson
Department of Zoology
Institute of Reproductive Biology
University of Texas at Austin
Austin, Texas

David Crews
Departments of Zoology and Psychology
University of Texas at Austin
Austin, Texas

Jeffrey A. French
Department of Psychology and
Nebraska
Behavioral Biology Group
University of Nebraska at Omaha
Omaha, Nebraska

Michael R. Gorman
Department of Psychology
University of California at San Diego
La Jolla, California

Kay E. Holecamp
Departments of Zoology and Psychology
Michigan State University
East Lansing, Michigan

Jerry D. Jacobs
Department of Zoology
University of Washington
Seattle, Washington

Sabra L. Klein
Department of Psychology,
Neuroscience, and
Population Dynamics
Johns Hopkins University
Baltimore, Maryland

Theresa M. Lee
Department of Psychology
University of Michigan
Ann Arbor, Michigan

Donna L. Maney
Department of Zoology
University of Washington
Seattle, Washington

Martha McClintock
Department of Psychology
The University of Chicago
Chicago, Illinois

Simone Meddle
Department of Zoology
University of Washington
Seattle, Washington

Randy J. Nelson
Department of Psychology,
Neuroscience, and
Population Dynamics
Johns Hopkins University
Baltimore, Maryland

Nicole Perfito
Department of Zoology
University of Washington
Seattle, Washington

Emilie Rissman
Department of Biology
University of Virginia
Charlottesville, Virginia

Colleen M. Schaffner
Department of Psychology
St. John's University
Collegeville, Minnesota

Patricia A. Schiml
Department of Biology
University of Virginia
Charlottesville, Virginia

Jill E. Schneider
Department of Biological Sciences
Lehigh University
Bethlehem, Pennsylvania

Rae Silver
Department of Psychology
Barnard College and Columbia University
New York, New York

Ann-Judith Silvermam
Department of Psychology
Barnard College and Columbia University
New York, New York

Laura Smale
Departments of Zoology and Psychology
Michigan State University
East Lansing, Michigan

Kiran Soma
Department of Zoology
University of Washington
Seattle, Washington

Jennifer M. Swann
Department of Biological Sciences
Lehigh University
Bethlehem, Pennsylvania

Anthony D. Tramontin
Department of Zoology
University of Washington
Seattle, Washington

George Wade
Center for Neuroendocrine Studies
University of Massachusetts
Amherst, Massachusetts

Kim Wallen
Department of Psychology
Emory University
Atlanta, Georgia

Scott R. Wersinger
Department of Biology
University of Virginia
Charlottesville, Virginia

John C. Wingfield
Department of Zoology
University of Washington
Seattle, Washington

Ruth I. Wood
Department of Obstetrics and Gynecology
Yale University School of Medicine
New Haven, Connecticut

Author Index

Baker, R. H., 457
Baker, R. R., 307
Ball, G. F., 83, 129, 149
Ball, J., 293
Baptista, L. F., 144
Beach, F., 296
Beauchamp, G. K., 407
Bellis, M. A., 307
Benedek, J. F., 379
Benoit, J., 133
Bentley, G. E., 129
Berliner, D. L., 382, 393
Bernstein, I., 296
Berthold, P., 83
Boehm, N., 391
Bonsall, R., 295
Brenowitz, E. A., 483
Broca, P., 357
Bronson, F. H., 13, 15, 21, 187, 208
Bunning, E., 134

Carpenter, C. R., 293
Cheng, M. F., 147
Christian, J. J., 487
Clemens, L. G., 487
Crews, D., 473
Cutler, W. B., 372, 381, 383, 385

Darwin, C., 5, 35
Dewsbury, D. A., 327
Dixson, A. F., 344
Dobzhansky, T., 3
Dresner, N., 304

Ellis, L., 274
Epple, G., 335
Evans, S., 334
Everitt, B., 295

Farner, D. S., 143
Fliess, W., 358
Follet, B. K., 133
Freidman, M. I., 37, 55
French, J. A., 288, 325
Freud, S., 358

Gasser, B., 391
Gautier, J. P., 332
Gautier-Hion, A., 332
Goldfoot, D. A., 295, 383
Gordon, T. P., 296
Gorman, M. R., 142, 187, 191
Gould, S. J., 5, 491
Goy, R. W., 294, 296
Groult, B., 289

Halasz, B., 161
Hamner, W. M., 134
Hartman, C., 293
Heape, W., 292
Hebb, D. O., 495
Herbert, J., 295
Hinde, R. A., 146
Holekamp, K. E., 187, 257
Howard, G., 427

Inglett, B. J., 345

Jacobs, J. D., 83, 85
Johnson, V. E., 296

Kappeler, P. M., 329
Kendrick, K. M., 344
Keverne, E. B., 295, 357, 372
Kinsey, A. C., 310
Kleiman, D. G., 342
Klein, S. L., 188, 219
Kuo, Z. Y., 492

Labows, J. N., 379
Lee, G. H., 44, 142
Lee, T. M., 187, 191
Lehrman, D. S., 83, 145, 296, 494

Malinowski, B., 291
Maney, D. L., 83, 85
Mason, W. A., 331
Masters, W. H., 296
McClintock, M. K., 288, 355, 423
Meddle, S., 83, 85, 133
Menendes-Pelaez, A., 176
Meredith, A., 427
Michael, R. P., 294, 296, 383
Miller, G. S., 292
Moenter, S. M., 454
Monti-Bloch, L., 383, 393
Moore, C. L., 487
Moore, M. C., 149
Moran, D. T., 393
Morin, L. P., 47
Morris, N. M., 383
Morton, M. L., 146, 149

Nelson, R. T., 187, 219
Newman, S. W., 423
Nicholls, T. J., 142

Ober, C., 387

Perfito, N., 83, 85
Phoenix, C. H., 473

Phoenix, C. P., 295
Pietras, R. J., 377
Pope, N. S., 302
Porter, R. H., 357
Potiquet, M., 392
Preti, G., 362, 379, 401

Resko, J., 295
Rice, F. J., 304
Ricklefs, R. E., 83
Rissman, E. F., 187, 445
Rose, R., 296
Rowan, W., 83, 160
Russell, M. J., 362
Ruysch, F., 391

Schaal, B., 357
Schaffner, C. M., 288, 325, 342
Scharrer, B., 160
Schiml, O. A., 445
Schneider, J. E., 14, 35
Seibt, U., 327
Selye, H., 237
Silver, R., 83, 159
Silverman, A-J., 159
Smale, L., 187, 257
Soma, K., 83, 85
Stanislaw, H., 304
Swann, J. M., 423

Tenaza, R. R., 331
Tilson, R. L., 331
Tinklepaugh, O. L., 303
Tramontin, A. D., 83, 85
Turner, B. H., 396

Waddington, C. H., 495
Wade, G. N., 14, 35, 187
Wallen, K., 289, 344
Weller, L., 403
Wersinger, S. R., 445
West-Eberhard, M. J., 473
Wickler, W., 327
Wingfield, J. C., 83, 85, 109, 188, 208
Wood, R. I., 423

Yokoyama, K., 143
Young, W. C., 294, 296

Zehr, J. L., 302
Zucker, I., 187
Zuckerman, S., 292, 317
Zumpre, D., 295

Subject Index

Accessory olfactory bulb, 393
  Fos activation, 430
  male sexual behavior, 429
Acylhomoserine lactone, signal molecule, 356
Adenohypophysis. *See* Anterior pituitary gland
Adenosine triphosphate (ATP)
  food intake, 38
  mast cells, 170, 174
  metabolism, 41
Adiposity
  fertility, 51
  signals, 18
Adipostatic hypothesis
  energy balance, 47
  functional diagram, 48
  reproduction, 47, 50
Adolescent sexuality, 310
Adrenal steroids, sexual behavior, 459
Adrenergic receptor blockers, 456
Adrenocorticotropic hormone (ACTH), 41, 243
Age and aging
  behavioral development, 485
  brain activity, 486
  vs. experience, 486
  GNRH neurons, 486
  importance, 485
Aggression. *See also* Territorial aggression
  immune responses, 246
  parental behavior, 107
  sexual dimorphism, 109
  temperature-dependent, 482
  testosterone, 240
Agriculturist society, puberty and energy, 26
Allozymes, gene transcription, 490
Amino acids
  metabolism, 40
  nutritional infertility, 37
Amygdala, 434
Androgens. *See also* Testosterone
  brain activity, 482
  dependent traits, 240
  immunoreactivity, 431
  sexual behavior, 489, 492
Androstadienone, gender differences, 394
Androstenol, 385, 407
Androstenone, sex-specific perfumes, 385
Anestrus, fasting induced, 45, 55, 60
Animal pheromones
  chemical structures, 400
  human perception, 396
  types, 359–361
Anorexia nervosa, 35
Anterior pituitary gland, 160

Antibody production, immune function, 226
Aphrodisiacs, pheromones
Apocrine glands
  animal and human, 399
  pheromones, 366, 399
Apolipoprotein D, 399
Area postrema, 68
Aromatization, estrogen, 434
Artificial insemination, 46, 384
Athena, perfume additive, 358
Athletes, 18, 35
Attractiveness, female-typical trait, 481
Auditory cues, GnRH effects, 143
Autosexual behavior, contraceptives, 305
Autumnal sexuality, 132

Baboons, interbirth interval, 275
Ballet dancers, puberty, 18
Bartholin's gland, pheromone release, 400
Basophils, courtship behavior, 162, 166
B-cells
  adaptive immunity, 224
  antibody production, 226
  brain physiology, 178
  function, 162
  immune system, 168
  social factors, 243
Bed nucleus of stria terminalis, 394
Behavior. *See also* Sexual behavior
  development, 485
  ethnology and psychology, 485
  mechanisms, 421
  sex differences, 257
  social context, 421
Biological determinism, 7
Bird reproduction
  breeding stage, 88, 94, 101
  ecological basis, 85–87
  environmental signals, 91–93
  food storage, 129
  GnRH control, 139–142
  hormone-behavior interactions, 85–87
  integrating information, 93
  life-history stages, 86
  modifying information, 92
  multiple brooding, 91
  nonphotoperiodic termination, 113–116
  photoperiodism, 112, 131, 139
  predictive cues, 92, 94
  seasonal, 129–131
  social interactions, 113
  supplementary factors, 92, 94
  synchronizing information, 93
  unpredictable events, 92, 94

Birth control pills, sexual desire, 305
Birth rate, seasonal variation, 27
Blood-brain barrier
　hematopoietic cells, 178
　mast cells, 172–174
　medial habenula, 173
　nervous system, 172–174
Bluehead wrasse, sex determination, 478
Boar pheromones, 384, 396
Body fat
　leptin, 57–58
　metabolic signal, 15, 28, 48, 50
　reproduction, 50
　thermoregulation, 24
Body size
　dimorphism index, 109
　puberty, 23, 28
　wild animals, 23, 28
Brain
　behavior relationship, 483
　breeding changes, 138
　environmental factors, 488
　experience and age, 486
　GnRH release, 138
　hematopoietic cells, 177–180
　immune-privileged, 177
　metabolism, 481
　organization, 488
　ovulation, 446
　physiology, 177–180
　ringdove studies, 177–180
　sex differences, 481
　sexual dimorphism, 483
　steroid effects, 437
Brain secretions
　brain drain, 177–179
　early studies, 160
　endocrine, 160
　immune system, 161–163
　sexual behavior, 159
Brain stem
　area postrema, 68
　metabolism, 68
　nucleus of solitary tract, 68
　vocal behavior, 148
Breeding stage. *See also* Bird reproduction
　chemosensory cues, 237
　development phase, 94–100
　expression and control, 93
　flexibility, 142
　food availability, 95
　gonadal development, 97, 100
　hormone control, 101
　maturation phase, 101

mortality rates, 239
phase diagram, 89, 91
predictability concept, 95–100
seasonal, 94
social cues, 100, 236
species differences, 142
strategies, 105–108
stress effects, 237–239
termination, 110–112
testosterone patterns, 105–108
Budgeriars, vocal behavior, 145
Bulbectomy, contralateral, 436

Callitrichidae, monogamy and taxonomy, 331, 333
Canaries
　courtship song, 146
　testosterone levels, 488
Caravaning behavior, musk shrew, 457
Carnitine palmitoyltransferase I, 62, 67
Carnivores, feeding strategy, 25
Carrier proteins, pheromones, 401
Castration, 311
Catecholamines, metabolism, 41
Cebidae, monogamy and taxonomy, 330
Central nervous system (CNS)
　blood-brain barrier, 173
　glucose oxidation, 62
　immune network, 179
　mast cells, 170, 176
　metabolic signals, 39, 41
Cercopithecidae, monogamy and taxonomy, 331
Cerebrospinal fluid (CSF), mast cells, 171, 174, 178
Cheirogaleidae, monogamy and taxonomy, 329
Chemical communication, 7, 355, 423
Chemosensory systems. *See also* Pheromones
　breeding cues, 237
　circuit diagram, 435
　environmental signals, 355
　female signals, 424
　hormonal cues, 425–427
　male signals, 423
　neurotransmitters, 437
　pathways, 437
　steroid action, 434–436
Chimpanzees, reproductive success, 274
Cholecystokinin, nutritional infertility, 37
Chromosomes, gene transcription, 490
Circadian rhythms, melatonin, 134
Circannual rhythms, bird reproduction, 112
Civetone, perfumes, 396
Cloud forest mouse, food availability, 22
Cocaine amphetamine related transcript (CART),
　leptin effects, 37, 64
Communication, chemical, 7

Subject Index

Competition, testosterone, 240
Concanavalin A, splenocyte proliferation, 235
Conception rates, spotted hyena, 265
Conditioned immune responses, 161
Connectivity, steroid effects, 437
Contextual influences, reproductive behavior, 340–346
Continuous breeders, 96, 100
Contraceptives
 human pheromones, 374
 intrusive and nonintrusive, 305
 nonhormonal, 305
 sexual desire, 305
 types, 305
Corticosteroid-binding globulin, breeding stress, 239
Corticotropin-releasing factor (CRF)
 energy balance, 37
 locus of effects, 44, 119
 ovulation, 448
Corticotropin-releasing hormone, stress response, 161
Courtship behavior
 androgens, 482
 animal, 9
 leopard gecko, 480
 male-typical, 480
 mast cells, 166
 nonneuronal cells, 164–167
 photoperiod effects on birdsong, 146
 ringdoves, 164–167
Cowper's gland, pheromone release, 400
Crowding stress, physiology and behavior, 487
Cyclophosphamide, immune suppression, 161
Cytochromic oxidase, brain activity, 481
Cytokines, male choice, 241

Dairy cattle, breeding, 46
Dancers, training programs, 35
Day length
 environmental cue, 195
 maternal influence, 201–208
 photoperiod history, 201–208
 seasonal breeding clock, 134
Degranulation, mast cells, 176
Dehydroepiandrosterone (DHEA), 385
Delayed hypersensitivity, skin tests, 226
2-deoxy-D-glucose, 51–52, 62
Determinism, biological, 7
Diabetes, hormonal hypothesis, 54
Dimorphism index, testosterone, 109, 111
Disease
 breeding stress, 239
 chemosensory cues, 237
 immune function, 242
 social interactions, 242
Double-matings, 308

Eating behavior. *See* Food intake
Eating disorders. *See* Anorexia nervosa
Eccrine glands, pheromones, 366, 399
Ecology
 bird reproduction, 85–87
 ecologists, 13
 female reproduction, 263
Egg-laying
 brain organization, 488
 environmental factors, 98
 log-linear analysis, 96
 predictability, 94, 99
 social cues, 102
 theoretical models, 98
Ejaculation, testosterone level, 313–315
Embryonic temperature, adult phenotype, 479–483
Emergency life-history stages (ELHS)
 bird reproduction, 86, 88
 corticosterone effects, 119
 hormonal changes, 118
 perturbation of reproduction, 116
Endoplasmic reticulum, mRNA content, 490
Energy
 functional diagram, 17
 GnRH control, 18–20
 homeostasis, 219
 immune function, 232–235
 metabolism, 40–41
 reproductive system, 15, 35
Energy balance
 adipostatic hypothesis, 47, 50
 environmental context, 13
 hormonal hypothesis, 47
 immune function, 219
 laboratory view, 16
 metabolic hypothesis, 49, 62
 negative, 42
 process characteristics, 18
 puberty, 15–17, 20, 23, 28
 reproduction, 13, 35–39
Energy reserves
 body size, 23, 28
 environmental aspects, 26
 feeding strategy, 25, 28
 hibernation, 24
 natural habitats, 20
 puberty, 15–17
Environment
 bird reproduction, 91–93
 energy balance, 13, 15–17, 26, 47, 50

Environment (cont.)
  immune function, 219–223, 227–235
  meadow voles, 192, 195
  photoperiod variables, 210
  regulating factors, 129–131
  reproduction timing, 191–195
  seasonal breeding, 129–131, 219–223
  types of signals, 91–94
Eosinophils, function, 162
Epithelial cells, pheromones, 366
Erox Corporation, 358
Estradiol
  birds, 135
  mast cell, 176
  musk shrew, 459, 460
  postovulatory, 9
  sexual behavior, 45, 297, 301–305, 459, 489
  social rank, 301
  territorial behavior, 113
Estrogen. *See also* Estradiol
  aromatization, 434
  environmental, 484
  follicle, 446
  immunoreactivity, 67
  mast cells, 175
  menstrual synchrony, 370
  nonsteroidal, 484
  receptor, 67
  sexual behavior, 292, 380
  sexual preference, 237
Ethology, behavioral development, 485
European starling, seasonal breeding, 131, 141
Evolution
  biology, 4, 476
  hyena, 257
  individual variation, 491
  reproduction theory, 5–7
  sex and sexuality, 5–7
Exaltolide, 377
Exaptation, 491
Experience
  vs. age, 486
  behavior, 485
  brain activity, 486
  sociosexual, 485
Extrapair copulations, gibbons, 333
Extraretinal photoreceptor, seasonal breeding, 132
Exudates, pheromone release, 397

Fashion models, infertility, 35
Fasting
  estrous cycle, 44–47, 53
  sex behavior, 40, 45

Fasting-induced anestrus, 45–46
Fat pads, insulin effects, 56
Feces, pheromone release, 398
Feeding behavior. *See* Food intake
Feisty females, spotted hyenas, 257–259
Female hamster vaginal secretions (FHVS), 424–434
Female reproduction
  adiposity, 50
  aggression, 109, 257
  birds, 144
  computer simulation, 362–366
  ecological basis, 263
  energy, 15, 35
  food availability, 15, 35, 264
  hyenas, 257
  life span, 266–268
  male vocalizations, 144–148
  musk shrew, 457
  neuroendocrine behavior, 445
  pheromones, 362–366
  risk, 289
  seasonal patterns, 263
  sexual behavior, 45, 289, 459
  sexuality, 303–310
  social influences, 265
  system development, 144–148
  testosterone patterns, 109
Ferret, pheromonal cues, 455
Fertility. *See also* Infertility
  adiposity, 51
  energy balance, 51
  goddesses, 35
  laboratory research, 35–39
  sculptures, 35
  sexual behavior, 45, 289
Fight-or-flight response, 117
Follicle-stimulating hormone (FSH)
  breeding stage, 94, 100
  locus of effects, 43
  menstrual cycle, 373
  ovulation, 446
  photoperiod response, 135
Follicular compounds, menstrual cycle, 371
Food intake
  brain stem, 68
  breeding stage, 95
  fertility, 35
  insulin effects, 56
  intake and storage, 56, 68, 129
  leptin effects, 68
  metabolic signals, 37
  photoperiod, 210, 211
  puberty, 16

Subject Index

reproduction, 13, 38, 220, 272
seasonal, 21
social rank, 272
Foraging activity, 17, 25
Forebrain, GnRH levels, 449, 462
Fos protein
  GnRH neurons, 453–455
  immunoreactivity, 70, 432, 454
  mating behavior, 430
  odor-hormone interactions, 432–434
  steroid effects, 432
Free fatty acids
  CPT-1 effects, 62
  energy balance, 37
  leptin effects, 64
  metabolism, 41
Free-living species, neuroendocrine mechanisms, 4

Galanin, gender differences, 451
Gambia, agriculturist society, 26
Gene
  expression, 374
  human pheromones, 374
  inheritance, 486
  transcription, 490
Genotype, 475
Gerbils, neural activation, 484
Gibbons, pair-bonding, 333
Glial cells, medial habenula, 172
Glucagon, 37, 41
Glucocorticoids, infertility, 37
Glucogenesis, liver, 41
Glucose
  brain uptake, 41
  energetic challenges, 51, 62, 63, 233–235
  infertility, 37
  metabolism, 40, 62
Goddess of Lespurge (sculpture), 36
Gonad. *See also* Ovary; Testis
  breeding stage, 97, 100
  development, 97, 100
  morphology, 478
  priming pheromones, 361
  sex determination, 478
Gonadal steroids. *See also* Estradiol; Progesterone; Testosterone
  internal cues, 424
  mast cells, 174
  neuronal morphology, 437
  odor cues, 425
  sexual desire, 299
Gonadectomy, 171
Gonadotropin, 8, 453

Gonadotropin-releasing hormone (GnRH)
  absorption, 168
  behavior activation, 453–456
  bird reproduction, 139–142, 166
  brain areas, 43, 138
  breeding stage, 94, 100, 138
  cell dynamics, 138, 451
  chemosensory response, 427
  chicken epitope, 167, 448
  courtship behavior, 170, 174
  cues and effects, 143
  energetic control, 18–20, 42–44
  forebrain levels, 449, 462
  Fos protein, 453–455
  future research, 463–465
  hyena, GnRH challenge, 267
  induced ovulation, 455
  locus of effects, 43
  male socialization, 462
  mast cells, 170, 174
  medial habenula, 166
  metabolic signal, 66–70
  molecular studies, 139
  musk shrew, 460
  neuroanatomy, 449
  neuronal system, 134–136, 446–455
  ovulation, 446, 448, 461
  peptide sequences, 164
  photoperiod reproduction, 139–142
  physiology, 136
  pituitary secretions, 161
  preoptic area, 135
  production, 452
  pulse generator, 66–70
  seasonal breeding, 134–136
  species differences, 142
  study methods, 460
  testosterone regulation, 103
Grain fungi, brain size, 484
Granulocytes
  feedback inhibition, 430
  function, 162
Grooming behavior, marmosets, 341
Ground squirrels, hibernation and puberty, 24
Growth. *See also* Puberty
  energy balance, 15
  laboratory view, 16
Guinea pigs, social modulation, 298

Habenula, ringdove courtship, 165
Hamsters, reproduction, 46, 132, 209, 423
Hematopoiesis, 168
Hematopoietic cells
  brain physiology, 177–180

Hematopoietic cells (cont.)
 courtship behavior, 166
 immune system, 168
 ringdove studies, 177–180
 type and function, 162
Hibernation, energy reserves, 24, 28
Higher vocal center, songbirds, 146, 483
Hippocampus, experience and age, 486
Histamine, mast cells, 166, 174
Hormone hypothesis
 energy balance, 47, 52
 functional diagram, 48, 52
 insulin, 52–57
 leptin, 57–61
Hormone replacement therapy, 482
Hormones
 aggression, 243
 bird reproduction, 85–87
 behavior, 2, 85–87
 brain secretions, 160
 breeding stage, 101
 chemosensory cues, 425–427
 context influence, 405
 control mechanisms, 101
 defining, 160
 early views, 291–293
 energy, 42–44
 female primates, 291–293
 indirect changes, 117
 infertility, 37
 male behavior, 423
 mast cells, 175–177
 monogamy, 344
 pheromones, 405
 renesting, 117
 sex differences, 473
 sexual behavior, 291, 297, 312
 sexual desire, 301–305
 sexuality, 289, 486–489
 signals, 423
 social modulation, 297–303
Housing conditions, immune function, 243
Huddling behavior, marmosets, 335
Human leukocyte antigens (HLA), 387
Human pheromones
 activation, 400–402
 bacteria source, 401
 chemical structures, 400
 communication, 358
 context dependence, 402–406
 developmental data, 402
 environmental factors, 402
 follicular regulation, 372
 future research, 374–382

 marriage partners, 387
 odor concentrations, 386
 olfactory system, 395
 overview, 355–361
 parent-infant bonding, 388
 production, 400–402
 reproductive behavior, 355–361
 samples, 366
 social information, 402–406
 synthetic compounds, 385
 testicular function, 381
 transduction and responses, 376
 vagility, 402
Humoral immunity, 224
Hutterites, marriage partners, 387
Hyenas. See Spotted hyenas
Hylobatidae, monogamy and taxonomy, 332
Hypersensitivity, skin tests, 226
Hypocretin. See Orexin
Hypothalamic-pituitary-adrenal axis (HPA)
 endocrinology, 446
 ovulation, 446
 population density, 247
Hypothalamus
 neural pathways, 161
 reproduction, 8
 vocal control, 148

Immune function
 antibody production, 226
 assessment, 223–227
 breeding stress, 237–239
 day length, 230
 energy challenges, 232–235
 environmental factors, 219–223, 227–235
 mate choice, 240
 measuring, 225
 photoperiod, 227–231
 population density, 247
 pregnancy, 239
 seasonal breeding, 220, 227, 233
 serology, 227
 social interactions, 237–248
 stress factor, 221
 temperature, 231
Immune responses
 adaptive, 224
 cell-mediated, 246
 conditioned, 161
 tail length, 240
Immune system
 brain secretions, 161–163
 components, 223
 hematopoietic cells, 168

Subject Index

humoral, 224
mast cells, 166
Immunoglobulin E (IgE), 168
Immunoglobulin G (IgG), 231, 244
Immunology. *See also* Immune function
  endocrinology, 163
  psychoneuroendocrine, 159, 163
Immunoreactive cells, 166
Immunosuppresant drugs, brain secretions, 161
Imprinting, pheromonal, 404
Individual variation, evolution, 491, 493
Indridae, monogamy and taxonomy, 329
Infant attachment, human pheromones, 388
Infertility. *See also* Fertility
  glucocorticoids, 37
  human pheromones, 374
  nutritional, 35–39
  treatment, 374
Ingestive behavior, 38. *See also* Food intake
Inheritance
  genomic, 487
  hormonal, 486–489
  stress effects, 487
Insect pheromones, 356
Insectivores, feeding strategy, 25
Insulin
  brain mechanisms, 54
  effector pathways, 18
  food intake, 54
  hormone hypothesis, 52–57
  infertility, 37
  metabolism, 41, 55, 63
Insulin-like growth factor (IGF), 20
Interbirth interval, social rank, 268, 270
Interferons, social interactions, 246
Interleukins, mate choice, 241
Interstitial fluid, brain physiology, 178
Intruder encounters, species differences, 339
Isolation, stress, 242
Isolectin, 166

Joint parental care, monogamy, 328

Kenaba, agriculturist society, 26
Kenya. *See* Spotted hyenas
Keyhole limpet hemocyanin (KLH), 233, 244

Labile perturbation factors (LPF)
  hormonal changes, 117
  indirect and direct, 117–119
  renesting behavior, 117
  reproductive function, 116
Lactation, 10
  immune function, 239
  metabolism, 41
  prolonged diestrus, 45
Learned cues, pheromonal function, 4
Leopard gecko
  embryonic temperature, 479
  experience and age, 486
  testosterone levels, 480
Leptin
  body fat, 58
  effector pathways, 18
  hormone hypothesis, 57–61
  infertility, 37
  metabolism, 41, 61, 63
  plasma levels, 60
  puberty, 15
Lesbian couples, sexual activity, 304
Leydig cells, leutinizing hormone, 135
Life-history stages
  bird reproduction, 86
  machine theory, 87
  termination phase, 110
Life span
  female, 266–268
  puberty, 21, 28
  reproduction, 45
  wild animals, 21, 28
Lipocalins, 399
Lipogenesis, 41
Lipolysis, 41
Lipostatic hypothesis. *See* Adipostatic hypothesis
Litter size
  birth rate, 268
  social rank, 274
  uterine component, 487
Liver, glucogenesis, 41
Lordosis, fasting effect, 45
Luteinizing hormone, 8
  breeding stage, 94, 100
  food availability, 51
  GnRH effect, 269
  insulin effects, 57
  locus of effects, 43
  male songbirds, 148
  mating behavior, 455
  ovulation, 374, 446
  pheromones, 367, 372
  photoperiod, 135, 148
  pulsing diagram, 19
  surge mechanism, 9, 374
Lymphocytes, types and proliferation, 168, 225

Macaques, reproductive success, 274
Macronutrients, types, 40
Macrophages, function, 162

Magnocellular neurons, 160
Main olfactory bulb, 429
Major histocompatibility complex (MHC)
 androsterone, 380
 human pheromones, 387
 marriage partners, 387
Male reproduction
 contextual influences, 276–279
 courtship behavior, 277
 genetic analysis, 279
 hamster, 423
 natal and immigrant, 277
 social rank, 277
 success rate, 277
Male sexual behavior
 aggression, 105
 brain schematic, 428
 chemosensory signals, 423
 desire, 310
 endocrinology, 147, 423
 hamster, 423
 homosexuality, 7
 hormonal signals, 423
 hyena, 276
 marmoset, 340–346
 neural pathways, 147, 423
 pheromones, 380
 risk, 310
 sexuality, 310–315
 songbirds, 144–148
 stimulating cues, 424
 vocalizations, 144–148
Mammalian Reproductive Biology (Bronson), 13
Mammals
 environmental factors, 187–189
 seasonal signals, 191–195
Marmosets
 contextual influences, 333
 monogamy, 331
 pair-bonding, 337
 reproductive behavior, 340–346
 same-sex conspecifics, 339
 sexual displays, 335
 sociosexual behavior, 333
 stranger responses, 334
 taxonomy, 331
Marriage partners, Hutterites, 387
Marsh wrens, brain size, 483
Masai Mara National Reserve (Kenya), 261
Mast cells
 antibody markers, 166
 blood-brain barrier, 172–174
 brain changes, 170–172
 CNS degranulation, 176

 courtship behavior, 166
 function, 162
 hormones, 175–177
 immigration and detection, 170–172
 immune system, 168
 medial habenula, 171
 mediators, 169
 nervous system, 172–175
 neuronal, 174
 number and distribution, 176
 overview, 169
 phenotypes, 169
 progenitors, 169
 reproductive organs, 176
 secretion, 170, 174
 signaling properties, 174
 tissue and organs, 169
Mate choice
 aggression, 118
 female traits, 241
 social interactions, 240
Maternal behavior
 day length, 201–208
 separation, 242
Mating behavior. *See also* Sexual behavior
 luteinizing hormone, 455
 male hamster, 435
 monogamy, 327
 neural circuit, 430–432
 pathway lesions, 434
 relationships, 328
 steroid effects, 436
 systems, 105–108
Maturation, breeding stage, 101
Meadow voles
 environment, 192, 195
 mate choice, 241
 odor preference, 237
Medial peoptic area (MPOA), 43, 67, 430, 448. *See also* Preoptic area
Medial preoptic nucleus, 432
Mediobasal hypothalamus, 456
Meek males, spotted hyenas, 257–259
Melatonin
 Circadian rhythms, 134
 photoperiod, 200, 231
 reproduction, 199
Menstrual cycle
 axillary compounds, 375
 Bedouin women, 403
 group patterns, 366
 individual lengths, 366
 neuroendocrinology, 373
 pheromones, 366–374

Subject Index 513

sexual desire, 298–305
synchrony, 403
Mercaptoacetate, metabolic effect, 63
Messenger RNA (mRNA)
  breeding stage, 94
  gene transcription, 58
  sexual behavior, 490
Metabolic hypothesis, energy balance, 49, 62
Metabolic signals
  blood-borne, 14, 28
  brain stem, 68
  fatty acids, 51–52, 62
  glucose, 51–52, 62
  GnRH pulse generator, 66–70
  vs. mediators, 39
  pathway diagram, 70
  source and nature, 15
Metabolism
  feeding and fasting, 40
  leptin effects, 63–66
  locus of effects, 42–44
  storage mode, 40
Metachromasia, mast cells, 166
Methyl palmoxirate, 51,
Microsmaty fallacy, 357
Midbrain, vocal behavior, 148
Migration and molting, 110, 113
Mitral cells, chemosensory cues, 430
Modulating pheromones, 361
  context dependence, 403
  neuroendocrinology, 389
  new type, 388
Molting and migration, 110, 113
Monogamy
  defining, 326–328
  incidence, 326
  mate choice, 241, 327
  neurobiology, 325
  primates, 328–333
  reproductive behavior, 325
  sexual and genetic, 328
  sociosexual behavior, 325
  species and names, 328
  taxonomic survey, 328–333
  testosterone levels, 106
Mononuclear cells, type and function, 162
Monotremes, 48979
Morphology, sex differences, 257
Mounting behavior, tamarins, 345
Multiple brooding, 91
Musk shrew
  GnRH system, 447, 449
  future research, 463–465
  history and evolution, 457

phylogenetic tree, 458
reproductive behavior, 458
sexual behavior, 459
social interaction, 457
Muskone, perfumes, 396
Myxobacteria, 356

Nasal reflex, 358
Nasopalatine process, 393
Natal clans, male reproduction, 277
Natural habitats, puberty and stability, 20, 21
Natural killer cells, cytotoxicity, 226
Natural selection, 4, 7
Neonates, environment, 488
Nerve growth factor (NGF), 174
Nervous system. See also Central nervous system
  blood-brain barrier, 172
  mast cells, 172
  ringdove studies, 172–175
Nervus terminalis, human pheromones, 394
Nesting phase
  environmental regulation, 101
  integrating information, 102–110
  social cues, 102
  social regulation, 102–110
  supplementary factors, 101
  synchronizing information, 102
  teritorial aggression, 102–105
  testosterone levels, 102
Neural activation, 70, 430, 432–434, 453–455, 484
Neurasthenia, 358
Neurobiology, monogamy, 325
Neuroendocrine system
  behavioral effects, 445
  evolution, 489–492
  photoperiod, 129–131
  seasonal reproduction, 129–131
  sexual behavior, 489–492
  social regulation, 129–131
Neuroendocrinology. See Reproductive endocrinology
Neurohormones, 160
Neuronal system
  chemosensory signals, 423, 427
  GnRH neurons, 134–136, 446–455
  hormonal signals, 423
  male sexual behavior, 423
  mating behavior, 430–432
  networks, 9
  odor-hormone interactions, 427
  pathways, 147
  photoperiod and song, 147
  sex hormones, 6
  steroid effects, 437

Neurons
  alterations, 174
  glandular, 160
  GnRH distribution, 446–448
  magnocellular, 160
  mast cells, 174
  sensitivity, 175
  song-specific, 147
  steroid receptors, 433
Neuropeptides
  brain studies, 161
  energy balance, 19
  infertility, 37
  mast cells, 170
Neuroscience, evolutionary theory, 6
Neurosecretory cells, 160
Neurotransmitters
  chemosensory pathways, 437
  GnRH expression, 451
  hormonal modulation, 437
  infertility, 37
  mast cells, 170, 173
  steroid effects, 437
Neutrophils, function, 162
New World primates, monogamy and taxonomy, 330
Nodose ganglion, 175
Nonneuronal cells, courtship behavior, 164–167
Nonphotoperiod, bird reproduction, 113–116
Norepinephrine, 37, 456
Nonsteroidal estrogens, brain size, 484
Nutritional infertility, 35–39
  animal models, 44–47
  future research, 38
  metabolic signals, 35–37

Odor-hormone interactions, 425
  cell mechanisms, 436, 440
  Fos protein, 432–434
  neural pathways, 427–430
  neuroanatomy, 436
  significance, 439
Odorless compounds, follicular and ovulatory, 371, 379
Odors
  external cues, 424
  gonadal steroids, 425
  ovulatory phase, 377
  preferences, 237
  sexuality, 358
  social interaction, 237
Offspring, annual production, 268, 271
Old World primates, monogamy and taxonomy, 331–333

Olfactory bulbs
  accessory, 393, 429
  Fos activation, 430
  ipsilateral, 436
  main, 434
Olfactory system
  brain schematic, 360
  gender differences, 395
  human pheromones, 395
  sensitivity, 377
  vaginal cues, 295
Opportunistic breeders, 96, 100, 150
Oregon Regional Primate Research Center, 294
Orexin, 37
Organum vasculosum of lamina terminalis, 135
Orgasm, male and female, 311
Ovarian cycle
  estrous cycle, 9, 45, 446
  evolution, 466
  hormone levels, 368
  human and animal, 368
  luteal phase, 372
  menstrual cycle, 304–310
  pheromones, 377–379
Ovary
  function, 362–366
  ovulation, 446
  pheromones, 362, 376
  reproduction, 344–346
  sensitivity, 377–379
  social context, 344–346
  synchrony, 362
Ovulation
  animal behavior, 9
  GnRH activity, 455, 461
  induced, 445, 465–466
  luteinizing hormone, 374
  neuroendocrinology, 446–453
  pheromones, 405
  sexual behavior, 293, 460
  social control, 445
  steroid levels, 460
Ovulatory compounds, 371
Oxidation, reproduction, 66
Oxygen consumption, immune function, 233
Oxytocin, 10

Pair-bonding, callitrichidae, 340–342
Panda's Thumb, The: More Reflections on Natural History (Gould), 5
Parabrachial nucleus, 68
Paraventricular nucleus (PVN)
  blood-brain barrier, 173
  estrogen receptors, 43

extraretinal photoreceptor, 133
GnRH activity, 43
immunoreactivity, 67
luteinizing hormone, 43
Parent-infant bonding, human pheromones, 293–297
Parthogenetic species, 489
Partner preferences, 333–337
Parturition, 10
Peptides. *See* Neuropeptides
Perfumes, sex-specific, 357, 385
Perivascular spaces, brain physiology, 179
Perturbation factors, reproduction, 116
Phenotypes
  adult, 479–483
  environment, 473, 477
  organization, 473, 475
  plasticity, 473–475
  sexuality, 473–475
Pheromones. *See also* Animal pheromones; Human pheromones
  activation, 400–402
  airborne, 402
  aphrodisiacs, 383
  apocrine glands, 399
  behavioral effects, 383–386
  carrier proteins, 401
  chemical structures, 400
  computer simulation, 362–366
  defining, 406
  exudates, 397
  female interactions, 362–366
  follicular and ovulatory, 375
  function, 404, 362–366
  hormone interactions, 405
  individual differences, 379–381
  insect, 356
  learned cues, 404
  male ejaculate, 398
  menstrual cycle, 366–374
  modulating, 361, 388, 403
  neuroendocrinology, 374
  odor concentrations, 395
  odorless compounds, 367–369
  ovarian cycle, 377–379
  ovulation, 405
  phyla, 356
  popular culture, 357
  priming and releasing, 361, 397
  production, 379–381, 400–402
  psychological state, 383–386
  receptor systems, 389
  response detection, 369–371
  science and myth, 357
  sensitivity period, 377–381

sexual impulses, 7
signaling, 361
specialized glands, 398
species-specific, 407
types, 359–361
urine release, 398
using, 406–408
vagility, 402
vaginal secretions, 398
Photic cues, brain secretions, 160
Photoperiod
  bird reproduction, 92, 112, 139
  control, 139–142
  courtship song, 146
  cues and effects, 143
  day length, 201–208
  endocrine changes, 112, 147
  environment, 210
  external coincidence, 134
  food variables, 210
  GnRH system, 134–136, 139–142
  immune function, 227–231
  insensitivity, 208–210
  maternal history, 201–208
  melatonin, 231
  neural response, 132–136
  neuroendocrinology, 129–131
  reproduction timing, 191–195
  seasonal reproduction, 129–131
  species comparison, 139
  steroid hormones, 222
  temperature, 212–214
  winter breeding, 208–210
Photoreceptors, avian brain, 133
Photorefractoriness
  breeding termination, 113
  endocrine basis, 136–138
  physiology, 136
  prolactin, 136
  relative and absolute, 142
  rodents, 196
  seasonal breeding, 132
  thyroid hormones, 137, 140
Photosensitivity, 131
Phyla, pheromones, 356
Phytoestrogens, brain size, 484
Pig pheromones, 407
Pineal gland, photoreception, 132
Pituitary gland
  brain connection, 160
  ovulation, 446
  reproduction, 8
Plasticity
  animal models, 477
  future studies, 143

Plasticity (cont.)
  phenotype, 473–475
  species differences, 142
Plumage, dimorphism index, 109
Plural breeders. *See also* Spotted hyena
  offspring production, 275
  reproductive success, 273–276
Polygyny
  mate choice, 241
  social interactions, 244
  testosterone levels, 106
Population density, immune function, 247
Prairie voles
  mate choice, 241
  monogamy, 325
  pair-bonding, 339
Prazosin, 456
Predictability concept, breeding stage, 95–100
Predictive cues, bird reproduction, 92, 94
Pregnancy
  hormone levels, 487
  immune function, 239
  uterine environment, 487
Preoptic area (POA). *See also* Medial preoptic area
  mRNA expression, 490
  androgen receptors, 490
  estrogen receptors, 43
  GnRH neurons, 454
  sex determination, 481
  whiptail lizards, 484
Primates, behavioral endocrinology, 293–297
Priming pheromones
  context dependence, 403
  follicular and ovulatory, 372
  ovarian-dependent, 403
  physiology, 361
  transduction and responses, 376
Proceptivity, 296, 336
Progesterone
  male sensitivity, 491
  ovarian cycle, 367
  sensitivity period, 379
Prohormones, GnRH production, 453
Prolactin
  day length, 230
  parturition, 10
  refractoriness, 136
Pro-opiomelanocortin, 37, 64
Prosimians, monogamy and taxonomy, 329
Prostaglandins, mast cells, 170
Proteases, types, 171
Pseudosexual behavior, 484, 489
Psychobiology, sexuality, 492
Psychology, behavioral development, 485

Psychoneuroendocrine immunology, 159, 163
Puberty
  adaptationist theory, 15, 28
  animal models, 44–47
  energy reserves, 15–17
  environment, 26
  feeding strategy, 25, 28
  food availability, 16
  functional diagram, 17
  hibernation, 24, 28
  laboratory view, 16
  lipostatic hypothesis, 15
  natural habitats, 20
  nutritionally-delayed, 44
  survival, 17
  thermoregulation, 17
Pulse generator, GnRH system, 66–70
Pyruvate, metabolism, 41

Quail reproduction, photoperiod, 140

Rabbit, GnRH activity, 456
Rainfall, environmental cue, 191
Rape, 292
Rat, GnRH neurons, 447, 449
Realm (fragrance), 384
Red blood cells, function, 162
Red-eared slider, sex determination, 479
Red-sided garter snake, courtship behavior, 485
Reflex ovulators, 9
Relaxation response, human pheromones, 383
Releasing pheromones, 361
  behavioral effects, 383–386
  context dependence, 403
  psychological state, 383–386
Renesting
  reproduction, 113
  hormonal changes, 117
  testosterone levels, 118
Reproduction. *See also* Bird reproduction; Female reproduction
  adipostatic hypothesis, 47, 50
  annual offspring, 268
  biology, 2
  cues and effects, 143
  day length, 195
  energy balance, 35–39
  environment, 13
  etiology, 2
  evolutionary theory, 5–7
  food availability, 13, 45
  function, 116
  GnRH system, 164, 452
  hormone hypothesis, 47, 52

Subject Index

inhibition, 35, 40
interbirth interval, 268
litter size, 268
mast cells, 176
metabolic hypothesis, 49, 62
neural pathways, 147
neuroendocrinology, 8–10
organs, 176
perturbation, 116
pheromonal cues, 454
photoperiod, 191–195
physiology, 143, 454
population density, 247
ringdove studies, 163
seasonal signals, 191–195
sex diferences, 257–259
social cues, 113–116
social interactions, 237–248
spotted hyena, 257, 268
study context, 1–10
success rate, 257, 268, 273
synthesis and integration, 2
termination, 113–116
timing, 191–195
Reproductive behavior
context influences, 340–346
group demography, 342
human pheromones, 355–361
male hyenas, 279
marmosets, 340–346
musk shrew, 458
neuroendocrinology, 489–492
ovarian status, 344–346
plural breeders, 273–276
primates, 325
sex drive, 142
singular breeders, 257
social context, 287, 343
success rate, 273–276
Reproductive neuroendocrinology, 8–10
Retina
hypothalamic tract, 133
photoreception, 132
Rhesus monkeys
behavioral endocrinology, 293
female-bonded society, 299
GnRH neurons, 447, 449
Ringdoves, vocal behavior, 148, 163
Role reversal, spotted hyena, 258

Saccharine, conditioned stimulus, 161
Saki monkeys, monogamy and taxonomy, 331
Salamanders, chemosensory system, 396
Same-sex conspecifics, 339
Seasonal breeding
environment, 219–223
European starling, 131, 141
GnRH activity, 134–136
immune function, 222
log-linear analysis, 96
stress factor, 237–239
Seasonal reproduction
birds, 129–131
environment, 129–131
immune function, 227–231
mammals, 191–195
neuroendocrinology, 129–131
Selective association, monogamy, 328
Self-confidence, human pheromones, 383
Semiochemicals, 386
Sensitivity period, 378
Sertoli cells, 135
Sex. *See also* Sexual intercourse
determination, 478–483
differences, 481
dimorphism, 108–110
drives, 316
evolutionary theory, 5–7
female and male, 316
genetic and gonadal, 473, 492
hormone receptors, 6
metabolic challenges, 46
neural activation, 481
reproduction, 108–110
solicitation, 336
specific perfumes, 385
steroidal control, 435
temperature-dependent, 479–483
testosterone, 108–110
Sexual behavior. *See also* Male sexual behavior; Female sexual behavior
brain secretions, 159
cyclic changes, 302
early views, 291–293
environment, 474
experience and age, 485
fasting, 45
female, 291, 297
fertility, 45, 289
gender-typical, 3
group living, 303
hamster, 45, 424
hormones, 291, 297, 312
human and animal, 291
male, 310–315
musk shrew, 459
neuroendocrinology, 489–492

Sexual behavior. *See also* Male sexual behavior;
  Female sexual behavior (cont.)
 ovulation, 460
 pheromones, 383–386
 primates, 291–293
 reproduction, 278
 ringdove studies, 163
 steroid levels, 460
Sexual desire. *See also* Sexuality
 energy levels, 35
 female, 304–310
 gonadal hormones, 299
 hormones, 299
 ovarian cycle, 304
 primates, 289
 risk, 304–310
 social context, 289
Sexual intercourse
 cyclic variation, 304
 double-mating, 308
 human pheromones, 385
 nonhormonal factors, 309
 risky business, 289
 social context, 289
Sexuality
 adolescent, 310
 autumnal, 132
 biology and psychology, 473
 brain vs. gonad, 489
 evolutionary approach, 5–7
 female, 303–310
 guiding concepts, 475
 hormonal inheritance, 486–489
 human, 473
 individual variation, 476, 486
 male, 310–315
 pair tests, 302
 neuroendocrinology, 489–492
 odors, 358
 phenotype plasticity, 473–475
 pheromones, 405
 transgenerational, 486–489
Seasonal reproduction. *See also* Photoperiod
 environment, 129–131
 European starling, 131
 neuroendocrinology, 129–131
 social regulation, 129
 timing signals, 191–195
Signaling pheromones
 behavioral effects, 386
 context dependence, 403
 marriage partners, 387
 MHC studies, 387
Simulated natural photoperiods, 197

Singleton litters, 268, 270
Singular breeders, reproductive success, 257
Skin cells, pheromone release, 400
Skin tests, hypersensitivity, 226
Social context
 behavioral mechanisms, 421
 male sexual behavior, 423
 reproduction, 287
 sexual desire, 289
Social cues
 breeding stage, 100
 endocrine responses, 143
 nesting phase, 102
 reproduction, 113–116
Social interactions
 breeding stage, 236
 dyadic, 246
 female influences, 265
 genetic differences, 246
 hormones, 297–303
 human pheromones, 402–406
 immune function, 219, 237
 long-term, 242–246
 mating behavior, 240, 299
 nesting phase, 102–110
 nonbreeding stage, 236
 ovulation, 465
 reproduction, 237–248
 seasonal breeding, 129, 219
Social rank
 feeding scores, 266
 food availability, 272
 GnRH challenge, 267
 immune function, 248
 interbirth interval, 268, 270
 mating behavior, 301
 neuroendocrinology, 267
 population density, 248
 reproductive success, 273–276
Socioendocrinology, 346
Sociosexual behavior
 Callitrichidae, 33
 contextual influences, 325, 333
 group demography, 342
 marmosets, 341, 343
 monogamy, 325
 ovarian status, 344–346
 pair-bonding, 340–342
 pheromones, 385
Somatostatin, ovulation, 448
Song
 behavior, 145
 perception, 147
 system, 145

Songbirds
  brain size, 483
  courtship, 146
  photoperiod, 146
  reproduction, 145
  vocal behavior, 145
Spawning behavior, androgens, 492
Specialized glands, pheromones, 398
Species concept, biological, 476
Spotted hyena
  characteristics, 259
  conception rate, 265
  feisty females, 257–259
  meek males, 257–259
  reproduction, 257–259
  role reversal, 258
  study methods, 261–263
  testosterone levels, 488
Steroid hormones. *See also* Estradial; Progesterone; Testosterone
  action sites, 434–436
  chemosensory circuit, 434
  energy metabolism, 230
  female sexual behavior, 459
  Fos activation, 433
  gender differences, 394
  immune function, 243
  implants, 434
  musk shrew, 459
  negative feedback, 9
  olfactory bulbs, 429
  photoperiod interactions, 222
  receptors, 429, 433
  sensitivity, 43
Streptomycetes, 356
Stress
  breeding, 237–239
  immune function, 221
  inherited, 487
  isolation, 242
  types, 238
Substance P, mast cells, 170, 173
Superior cervical ganglion, 133
Supplementary factors
  endocrine response, 143
  predictive cues, 149
  reproduction, 113–116
Suprachiasmatic nucleus, 133
Syrian hamsters, fasting and anestrus, 46

Tachykinins, neuronal sesitivity, 175
Talek hyenas, study methods, 261–263
Tamarins
  monogamy, 331

  pair-bonding, 337
  reproduction, 340–346
  same-sex conspecifics, 339
  sociosexual behavior, 345
  stranger responses, 338
  taxonomy, 331
Taxa, microsomaty fallacy, 357
T-cells
  brain physiology, 178
  function, 162
  immune system, 168
  social interactions, 246
Teenage pregnancy, 310
Telencephalic network, vocal behavior, 148
Temperature
  immune function, 231
  photoperiod, 212–214
  sex determination, 479–483
Terminal nerve. *See* Nervus terminalis
Territorial behavior
  aggression, 102–105
  dimorphism index, 109
  estradiol effects, 113
  nesting phase, 102–105
  reproduction, 248
  testosterone, 102–105
Testis
  function, 313–315
  GnRH levels, 138, 313
  human pheromones, 381
  mast cells, 177
  meadow voles, 205, 213
Testosterone
  biological action, 103
  breeding stage, 105–108
  challenge hypothesis, 105
  dependent traits, 240
  ejaculation, 313–315
  implants, 434
  male aggression, 105, 107
  mating behavior, 105–108
  nesting phase, 102–105
  odor interaction, 427
  secretion levels, 104
  sexual dimorphism, 108–110
  social instability, 108
  species comparison, 108
  temporal patterns, 105–108
  territorial aggression, 102–105
Thermoregulation, puberty, 4, 17
Threatening, female, 300
Thyroid gland, photoperiod, 112
Thyroid hormone, 137
Thyroid-releasing hormone, 161

Thyroxine, 112, 137
Timing signals, reproduction, 191–195
Titi monkeys, monogamy and taxonomy, 330
Tongue displays, tamarins, 336
Transgenerational sexuality, 486–489
Triglycerides, metabolism, 41
Truffles, androstenol, 407
Tyrosine hydroxylase, GnRH activity, 456

Uncoupling proteins, leptin effects, 65
Unisexual species, 489
Urine, pheromone release, 398
Urocortin, 37
Uterine environment, meadow voles, 205, 213

Vagility, pheromones, 402
Vagina
  olfactory cues, 295
  pheromones, 398
  secretions, 398
Vagotomy, 67
Vasoactive intestinal polypeptide (VIP), 136, 170
Ventromedial hypothalamus
  gene expression, 491
  sex determination, 481
  whiptail lizard, 484
Venus of Willendorf (sculpture), 36
Vertebrates, life-history stages, 87
Virchow-Robbins spaces, 179
Visual cues, reproductive physiology, 143
Vocal behavior
  brain areas, 148
  early studies, 144
  endocrine activity, 144
  male, 144–148
  marmosets, 341
  songbirds, 145
Voles. *See* Meadow voles
Vomeromodulin, immunoreactivity, 391
Vomeronasal organ (VNO)
  anatomy, 356, 393
  animal and human, 390
  human pheromones, 384
  location and function, 393
  mating behavior, 427, 429
  septal structures, 391
  testicular function, 382
Vomeropherin, 382
Vomodor, 382

Whiptail lizard
  brain size, 484
  allozyme analysis, 490
  neuroendocrinology, 489

White-crowned sparrow, life-history stages, 86
Wild populations, life spans, 21, 28
Winter breeding, photoperiod, 208–210
Women. *See also* Female reproduction
  body image, 35
  double-mating, 308
  ovarian hormones, 303
  sexual desire, 303–310
  sexuality, 296
Wrens. *See* Marsh wrens

Xenobiotics, brain size, 484

Yerkes Regional Primate Research Center, 296

Zebra finch
  immune function, 233
  testosterone levels, 488